CITIES 2018 International Conference

Spatial Economic Transport Interaction for Sustainable Development

IOP Conference Series: Earth and Environmental Science
Volume 340

Surabaya, Indonesia
24 – 25 October 2018

ISBN: 978-1-5108-9669-7
ISSN: 1755-1307

Printed from e-media with permission by:

Curran Associates, Inc.
57 Morehouse Lane
Red Hook, NY 12571

Some format issues inherent in the e-media version may also appear in this print version.

Institute of Physics
Dirac House, Temple Back
Bristol BS1 6BE UK

Phone: 44 1 17 929 7481
Fax: 44 1 17 920 0979

techtracking@iop.org

Additional copies of this publication are available from:

Curran Associates, Inc.
57 Morehouse Lane
Red Hook, NY 12571 USA
Phone: 845-758-0400
Fax: 845-758-2633
Email: curran@proceedings.com
Web: www.proceedings.com

TABLE OF CONTENTS

IOP Conf. Series: Earth and Environmental Science **340** (2019) 011001 doi:10.1088/1755-1315/340/1/011001

CITIES 2018 International Conference

PREFACE

CITIES conference is an annual conference event held by the Departement of Urban and Regional Planning. CITIES conference has a scale from national to international conference. CITIES conference had been held 7 (seven) times nationally and 4 (four) times Internationally in 2013, 2015, 2017 and 2018. The purpose of this conference is to provide publications of all of the science and technology result based on research and planning practices. The past themes of the CITIES serial are:

- Innovations In Spatial Planning Practices for Development and Decentralization (2005);
- In Search of Integration: Between Planning and Spatial Programming (2008);
- Toward Sustainable, Competitive, and Autonomous Spatial Planning (2009);
- Facing the Future: Innovation In Planning Research and Practices (2010);
- Spatial Planning Research Agenda for Sustainable and Just Urban and Regional Development (2011);
- Facing Global Challenges In The Future Urban Sphere (2012);
- Resilient Cities: Beyond mitigation, preparedness, response, dan recovery (2013);
- Eco City, Utopia or Reality (2014);
- Intelligent Planning Towards Smart Cities (2015);
- Coastal Planning for Sustainable Maritime Development (2016)
- Multi Persepectives on Peri-Urban Dynamics Towards Sustainable Developments (2017);

This **2018** CITIES INTERNATIONAL CONFERENCE is featuring topic in: ***Spatial Economics Transport Interaction for Sustainable Development***. Focusing on the transportation and infrastructure development, spatial – economic development sustainable development, this conference aim to provide the insights in many aspects of the developments towards the future especially in urban transportation development.

Peer review statement

All papers published in this volume of *IOP Conference Series: Earth and Environmental Science* have been peer reviewed through processes administered by the proceedings Editors. Reviews were conducted by expert referees to the professional and scientific standards expected of a proceedings journal published by IOP Publishing.

IOP Conf. Series: Earth and Environmental Science **340** (2019) 012001 doi:10.1088/1755-1315/340/1/012001

Optimization of settlement land use through carbon footprint approach in The North Balikpapan

A Ghozali[1,3], **N Hasanah**[1,3] **and Subchan**[2,3]

[1] Department of Urban and Regional Planning
[2] Department of Mathematics
[3] Institut Teknologi Kalimantan, Jl. Soekarno Hatta Km.15, Balikpapan 76127, Indonesia

Email: ghozali@itk.ac.id

Abstract. Limited land in the downtown area as well as the increasing amount of new activities centre causes residential development leads to North Balikpapan. This area is an urban fringe with vast protected forests as buffer zone and catchment area for the city and surrounding area. Land conversion in this area will increase hazard risk of inundation, water quality decrease and increased CO_2 emissions. Therefore, development should be maintained environmental stability. One of the rights applicated approach is carbon footprint that is capable to measure the balance between production and absorption needs of CO_2 emissions. To find the optimal land allocation, we used carbon footprint calculation from the household activities, identify the factors of settlement growth, and use Linear Programming analysis. Analysis' results show that settlement activities in North Balikpapan produce 108.362,4 tCO_2/year or equivalent with 618,50 Ha green space. Meanwhile, the development of settlement in North Balikpapan is affected by social demographic, developer initiative, environmental condition, public facilities availability, economical structure, and policy factors. According to those factors, optimal allocation of settlement area in North Balikpapan is only about 4,510.01 Ha. With that condition, it still able to absorb CO_2 emissions from inside or outside the area around 2.751 tCO2/year.

Keywords: *Carbon Footprint, CO_2 emissions, Household, Land optimization.*

1. Introduction

The development of the city characterized by population growth and the level of urbanization will trigger rising needs of housing [1]. City spaces availability are limited, which created a natural alternative selection in fulfilling the space needs for living and other functions of activities will always take up spaces in the suburbs [2]. Thus, the limited availability of land in the city center and is supported by the existence of the availability of the infrastructure can spur the development of neighborhoods to the suburbs.

Balikpapan City is a city with a fairly rapid population development. This can be seen from the increasing number of residents of the city of Balikpapan in the last 5 years of 7.58% [3]. Based on Statistic of Balikpapan City, it is known that the composition of the population resides in district of North Balikpapan of as much as 154,908 people in the year 2014 and this number increased by 5.92% in 2015 be 164,081 people. The increase in the number of population will certainly lead to increased needs for housing. It is also aligned with the new growth center around the northern city of Balikpapan which had been contained in the Spatial Plan area Balikpapan City year 2012-2032.

According to the final report of the Development Plan of housing dan Settlement in Balikpapan city 2014 [4], the growth of settlement area also followed by increasing housing units, the number of houses in North Balikpapan are 19,9% out of total houses in 6 districts at Balikpapan and becoming the district with highest housing growth. It is noted that North Balikpapan has 8,83% of housing growth rate increase from the past year, while another districts' increase is only less than 7%. Thus, North Balikpapan becomes a strategic place to be created as a new settlement area.

On the other hand, the North Balikpapan is a suburb which has open green space as well as the vast protected forest which serves to maintain the balance of the environment and the survival of the community and the city of Balikpapan the surroundings. North Balikpapan has two location of the protected forest, those are protected forest of Wein River area 1,967.848 Ha and protected forest of Manggar River area 2,342.945 Ha [5]. Kodoatie & Sjarief [6] state that the development of population and economic growth lead to much going on over the function of land conservation became a waterproof area resulting in a decrease in the quantity and quality of water availability, increased surface flow which led to an increase in potential flood, as well as reduced land sinks of CO2 emissions.

The development of activities in one area turn the land use intensity becomes high which directly or indirectly will affect to the production increasing of greenhouse gas emission [7]. According to Study of an inventory of Greenhouse Gas emissions the energy sector by the Ministry of energy and mineral resources in 2013 [8], the needs of energy in household sector reached for 29% which is the second highest need after industrial sector by the value of 41%. As the impact of citizens number growth, the needs of settlement area in North Balikpapan is also increasing and could be indicated to affects the production increasing of Carbon dioxide (CO2) emission in North Balikpapan.

In order for the development of settlements in the North Balikapapan does not give negative impact to the environment, should consider the environmental conditions so favorable balance and harmony between environment with development of settlements. One of the approaches that can be used to analyze environmental conditions in the North Balikapapan approach is through carbon footprint. Through this approach, knowable conditions environmental balance between the production of CO2 emissions and absorption of CO2 emissions on land. Therefore, this research was conducted to find out the proportion of land suitable for settlement and land use optimization to determine sustainable settlements in North Balikpapan through carbon footprint approach.

2. Methods

The purpose of this study is to determine optimization of settlement land use in North Balikpapan by using analysis method that consists of analyzing land needs to absorb emission from settlement activities through Carbon Footprint approach, analyzing factors that influence settlement development, and analyzing the optimization of settlement area in North Balikpapan.

There are two sub-analysis in land needs as a form of settlement activity consume analysis, those are the calculation of Carbon Footprint analysis in North Balikpapan and analysis in land needs to absorb CO2 emission or the result of Carbon Footprint calculation. The calculation of Carbon Footprint is use to know the value of CO2 emission that created by settlement area. Carbon Footprint calculation divided into 2 type which are, Primary Carbon Footprint and Secondary Carbon Footprint.

Primary Carbon Footprint of settlement activity that calculate in this study is cooking activity carbon footprint by the using of LPG and kerosene also the transportation of total production based on each fuel use. Secondary Carbon Footprint calculated from CO2 household electricity usage total production. The formula of CO2 emission from cooking fuels, transportation fuels, and electricity consumption are shown in equation, (1), (2), and (3).

$$CO_2 \textit{ Emission from Cooking activities} = F_{cy} \, x \, EF_{CO2} \, x \, NCV \quad (1)$$

$$CO_2 \textit{ Emission from Transportation} = EF_{CO2} \times FC \quad (2)$$

$$CO_2 \textit{ Emission from Electrical Consumption} = EF_{CO2} \times EC \quad (3)$$

Explanation:
Fcy : LPG and Kerosene usage (Kg/month)
EFCO2 : CO2 emission factor (gr/MJ)
NCV : Net Calorific Value (MJ/Kg)
FC : Gasoline and Diesel fuel consumes (Liter/month)
EC : Monthly electricity usage (kWh/month)

Emission factor value and Net Calorific Value (NCV) in each kind of fuel is based on standard that set by IPCC Guideline year 2006 [9]. Emission factor and NCV value for cooking fuel are shown in Table 1.

Table 1. Emission factor and Net Calorific Value (NCV) of cooking fuel

No	Fuel	Emission Factor (gr/MJ)	NCV (MJ/Kg)
1	LPG	63,1	47,3
2	Kerosene	71,9	43,8

Source: IPCC (2006)

CO_2 emission factor for transportation fuel according to IPCC Guideline year 2006 [9], Gasoline and Diesel are 69,3 gr/MJ and 74,1 gr/MJ. CO_2 emission factor of electricity usage based on the average of power plant and fuel in Indonesia, the CO2 emission factor achieved is 7,42 x 10-4 tons CO2/kWh.

Land needs analysis is use to know the extents of land that able to absorb CO2 emission as a result of household consume in North Balikpapan. In this study, the type of land cover and absorbance refer to IPCC Guideline year 2006 [9]. From the guideline, the CO2 absorbance in local green open space is close to paddy fields absorbance capacity that reached for 175,20-ton CO2/Ha/year.

Analyzing technique that use to determine influencing factors in North Balikpapan settlement development is Content Analysis technique. Content Analysis technique is an analyze which relies on found codes in data recording text while interview is conducted with the subject on the spot. The step of Content Analysis technique according to Krippendorff [10] are unitizing, stakeholder analysis method sampling, data collecting, coding, data reducing, and concluding.

In formulating the optimization of North Balikpapan settlement area, this study uses Linear Programming analysis technique. Linear Programming is an analysis tools that use to solve optimization problems (maximize and minimize) by using mathematics model in order to solve by considering the existence of limitations [11]. Linear programming analysis stages consist of element determination in Linear Programming (decisions variables, objective functions, and limit functions), formulating objective function model, formulating limit functions, and modelling of Linear Programming through LINGO 11.0 Software.

3. Result and Discussions

3.1. Analyzing land needs as a form of settlement activity consume through carbon footprint approach

There are two sub-analysis in land needs as a form of settlement activity consume analysis, those are the calculation of Carbon Footprint analysis in North Balikpapan and analysis in land needs to absorb CO2 emission or the result of that Carbon Footprint calculation. These are the details of each sub-analysis below.

3.1.1. Carbon footprint analysis in North Balikpapan

According to the results of CO_2 emission calculation through Carbon Footprint approach in North Balikpapan, it's noted that the total of LPG and Kerosene CO_2 emission reached for 1.702,55 tons CO_2/month, with Graha Indah sub-district as the highest with 421,60 tons CO2/month, followed by Muara Rapak sub-district with 344,90 tons CO_2/month, and Batu Ampar sub-district with 307,94 tons CO_2/month. Those three sub-districts are those with the highest number of household in North Balikpapan. Thus, LPG and Kerosene CO_2 emission in North Balikpapan are affected by the number of household in each sub-district.

Whilst the total CO_2 emission in North Balikpapan is 62,07 tons CO_2/month, with the highest emission contributor in Batu Ampar sub-district which is 18,78 tons CO_2/month, followed by Muara Rapak sub-district with 12,04 tons CO_2/month, and Karang Joang sub-district with 11,26-ton CO_2/month. Batu Ampar and Karang Joang are sub-districts that have a quite far distance to the downtown, while Muara Rapak is a sub-district with the highest number of household, therefore it fair to say that CO_2 transportation emission is affected by transportation user mobility and household amount.

Once the CO_2 emission calculation for cooking fuel and transportation fuel are obtained, then the total of primary CO_2 emission in North Balikpapan is obtained by the value of 1.764,62 tons CO_2/month. Next, the total of secondary CO_2 emission in North Balikpapan based on the calculation is 7.265,58 tons CO_2/month, with Muara Rapak as the highest CO_2 emission producer with 2.028,16 tons/CO_2/month.

Therefore, settlement consumption which consist of cooking fuel usage, transportation fuel usage, and electricity usage that produced by 50.766 household in North Balikpapan produce CO_2 emission by the value of 7.502,66 tons/CO_2/ month. CO_2 total emission in North Balikpapan are shown in Table 2.

Table 2. CO_2 Total emission in North Balikpapan Sib District

No	Sub-District	Primary Emission of CO_2 (ton/month)	Secondary Emission of CO_2 (ton/month)	Total Emission of CO2 (ton/month)
1	Muara Rapak	356,94	1.671,22	2.028,16
2	Gunung Samarinda	276,73	1.499,61	1.776,34
3	Batu Ampar	326,72	1.635,77	1.962,49
4	Karang Joang	301,27	729,37	1.030,64
5	Gunung Samarinda Baru	71,20	428,86	500,06
6	Graha Indah	431,76	1.300,75	1.732,51
	Total	**1764,62**	**7.265,58**	**9.030,20**

Source: Analysis Result, 2017

In this example we can see that there are footnotes after each author name and only 5 addresses; the 6th footnote might say, for example, 'Author to whom any correspondence should be addressed.' In addition, acknowledgment of grants or funding, temporary addresses etc might also be indicated by footnotes.

3.1.2. Analysis of land needs to absorb CO_2 emission or the result of carbon footprint calculation

Calculation of land needs is a follow-up analysis from the calculation of CO_2 emission in North Balikpapan. In this analysis, the type of CO_2 absorbance land cover that used is paddy field that well known to have CO_2 absorbance capacity for 175,20 tons/CO_2/ha/year. That choice is based on green space condition in the study area that has less absorption capacity cause the area is not fully- forest, then according to IPCC Guideline (2006) paddy fields has less absorption capacity than forest area. In the previous analysis, it's noted that total emission of CO_2 in North Balikpapan reached for 9.030,20-ton CO_2/month, therefore the total emission for a whole one year is 108.362,4-ton CO_2/year, and the extents of land that needs to absorb CO_2 emission is 618,50 hectares of green spaces in minimum value.

3.2. Analysis of factors that influenced settlement development in North Balikpapan

Determination of factors that influence settlement development in North Balikpapan used as a way to determine the optimization of land needs, especially in formulating the limitation or constraint. Determination of factors that influence settlement development in North Balikpapan in this study conducted by Content Analysis Method with the purpose of confirming and exploring some factors influencing settlement development in North Balikpapan based on opinions of some key stakeholders. Those factors will be used as input data towards equations in Linear Programming analysis in the next step. The results of the content analysis are shown in Table 3.

Table 3. Factors of content analysis results

Indicator	Factor	Factor Discusse (%)*	Number of Stakeholder Confirmed	Node
Demographic	Population	5.7	3	Confirmed
	The number of workforces	4.9	3	Confirmed
Housing Developer Initiative	The number of Housing Developers	3.3	3	Confirmed
	Housing Area	3.3	3	Confirmed
Environmental Condition	Protected Forest Area Availability	11.4	4	Confirmed
	Green Open space Availability	12.2	4	Confirmed
Public Facility Availability	The number of Education Facilities	12.2	4	Confirmed
	The number of Trade and Services	7.3	4	Confirmed
	The number of Health Facilities	2.4	3	Confirmed
	The number of Worship Facilities	2.4	3	Confirmed
	The number of Transportation Support Facilities	3.3	3	Confirmed
	Road Availability	4.1	4	New
Economical Structure	Gross Regional Domestic Product	6.5	5	Confirmed
Urban Planning Policy	The Extents of Allocated Area Based on Spatial Plan	10.6	6	Confirmed
Strategic Location	The Existence of National Project / New Forefront	10.6	6	New

**Calculated from all statement in the interview transcript and from all stakeholders*
Source: Analysis Result, 2017

3.3. Analysis of settlement land use in North Balikpapan

The process of settlement land use optimization in North Balikpapan using Linear Programming analysis through LINGO 11.0 software. This analysis consists of some points such as formulating objective function, formulating constrain function, and running the optimization linier programming based on both functions.

3.3.1. Formulating objective functions
In this study, decisions variable is based on 5 area divisions which are the extent of settlement area (X1), the extent of protected forest (X2), the extent of green area (X3), the extent of industrial area (X4) and other extents (X5). Other extents consist of public facility and trade & services area. Formulating those decisions variables are considering the existing condition and spatial planning in the study area. Since the objective functions are based on CO2 emission then the modelling results minimized the amount of emission towards the environment and each area has a standard CO2 emission coefficient value and CO2 absorption capacity in each hectare of land use. Areas that produce CO2 emission are settlement area (X1), industrial area (X4), and other areas (X5). In settlement and other areas, standard coefficient of emission producer is obtained from the previous carbon footprint calculation, wherein the production of CO2 emission per hectare area is 59,61 tons CO2 per year. It means that in factual condition, the addition of 1 ha of housing area can raise CO2 emission about 60 tons per year.

The standard of CO2 emission measurement produced by industrial activities is used from previous study result. According to Setiawan, et all [12] with their study entitled "Carbon Footprint from Industrial activities in The City of Surabaya", the CO2 emission produced by industrial area is about 98.028,6 tons/year with the extent of industrial area by about 245 Ha, which means that every hectare area of industrial area is provided CO2 emission of 400,12 tons CO2/year.

Furthermore, for the CO2 emission absorbance area consists of protected forest area (X2) and green space area (X3). Absorbance standard in those two kinds of areas are based on IPCC Guideline year 2006, so, the CO2 emission absorbance standard for the protected forest area is 569,40 tons CO2/Ha/year and 175,20-ton CO2/Ha/year for the green space area. Those standard then turned into coefficient in objective function, and the equation of study objective functions is shown in Eq. (4).

$$Min.\ CO_2\ Emission = 59{,}61X_1 - 569{,}40X_2 - 175{,}20X_3 + 400{,}12X_4 + 59{,}61X_5 \quad (4)$$

From the optimization formula it can be seen that every built-up land use will produce emissions and vice versa green land will absorb it. The optimization that will be sought is a condition when the difference in production and absorption of emissions generate a lowest value.

3.3.2. Formulating limit functions or constraint
Limit Functions or Constraint is a linear formula that contains decisions variable and explains the constraint that able to take. In this study, limit functions are obtained from the result of content analysis based on libraries' synthesis and stakeholders/experts' opinions. The results of limit functions or constraint are shown in Table 4.

Table 4. Limit functions or constraint synthesis

No	*Constraint*	**Condition Analysis**
Carbon Footprint Calculation Results		
1	CO2 Emission Absorbance Area Needs X3 ≥ 618,50 0,34X1 ≤ X3	1) The result of green space needs calculation by 618,50 Ha is assumed as the minimum value of area to absorb CO2 emission from settlement activities in North Balikpapan. 2) If 1.817,7 Ha settlement area needs the minimum value of 618,50 Ha green space to absorb CO2 emission, then one-hectare area of settlement requires minimal 0,34 hectare of green space.
Social Demographic		
2	Population 90,27X1 ≥ 243.277	In 2032, the projection number of citizens is assumed to have the same density with current condition which is

No	***Constraint***	**Condition Analysis**
		90,72 people/Ha, so the total area should be able to accommodate the number of projected citizens by 243.277 people in 2032.
3	The number of workforces $446{,}33X4 + 43{,}21X5 \geq 43.775$	Industrial area and the other types of build-up land is assumed to give impacts toward the citizen welfare through job vacancy availability for as the minimum value as the number of current workforces.
Housing Developer Initiative		
4	The number of Housing Developers and The Extent of Housing Area $X1 > 1.710{,}08$	The provision of new settlement area is in the minimum value of housing area that currently planned by 75 developers in North Balikpapan.
Environmental Condition		
5	Protected Forest Area Availability $X2 = 4.310{,}74$	The function of protected forest is extremely fundamental to environment stability and Balikpapan citizens sustainability. In order to take care of this condition, the current extent of protected forest have to be preserved as well.
6	Green Open space Availability $X3 \geq 618{,}50$	From the previous analysis result, the current green open space extent is still able to absorb CO2 emission until the projected year, so they must be nurtured. The minimum extent of green space as the result of carbon footprint calculation in North Balikpapan is 618,50 Ha.
Public Facility Availability		
7	The number of Education Facilities, Trade and Services, Health Facilities, Worship Facilities, Transportation Support Facilities, and Road Availability $0{,}47X1 \leq X5$	The extents of current public facilities are 858,10 Ha with the extents of existing settlement area for 1.817,7 Ha, then one-hectare area of settlement needs 0,47 areas of public facility.
Economical Structure		
8	Gross Domestic Regional Product (GDRP) $658.23X1 + 2.642.767{,}48X4 + 33.003{,}15X5 \geq 69.157.980{,}54$	The use of settlement area, industrial area, and other area must be able to guarantee the increasing of region economy, so in this limitation it assumed that 1.817,7 Ha of settlement area are having Gross Domestic Regional Product by the value of IDR. 1.196.466,16, 15 Ha of industrial area with IDR. 39.641.512,19, and 858,10 Ha of other areas with IDR 28.320.002,19.
Spatial City Planning Policy		
9	The District Area $X1 + X2 + X3 + X4 + X5 = 13.287$	The extent of land provision in North Balikpapan should not violate the provisions of the city planning policy and should be exactly equal with district area.
10	Allocation of Settlement Area $X1 \leq 4.510{,}11$	
11	Allocation of Protected Forest Area $X2 \leq 4.310{,}74$	
12	Allocation of Green Open space	

No	*Constraint*	Condition Analysis
	Area X3 $\leq$ 3.565,6	
13	Allocation of Industrial Area X4 $\leq$ 420,96	
14	Allocation of Office, Trade and Services, Public Services Area X5 $\leq$ 1.047,61	

Source: Analysis Result, 2017

3.3.3. Optimization of land use

Optimization of settlement land use in this study is conducted by using LINGO 11.0 software. All equations which have been formulated in the previous explanation, consists of objective functions and limit function or constraints, are used in linear programming analysis to calculate an optimal and extent allocation. The following table explains the results of the analysis obtained.

Table 5. Solution report from model analysis

Variable	Value	Reduced Cost
X1	4510.01	0
X2	4310.74	0
X3	3565.6	0
X4	15	0
X5	885.55	0
	Infeasibilities	**0**
	Objective Value	**- 2,751,591**

Source: analysis

Based on the results of the solution report in the table above, it can be seen that the reduced cost value (changing in the optimal value of the objective function) for each variable is 0 (zero), so the value of all variables has reached optimal conditions. In addition, Infeasibilities of the model reached zero or the value generated from the analysis does not violate the function of the constraints previously formulated, so that the model analysis is suitable for use.

The allocation of land area in each land use can be known through the values listed in the row value, so that the optimal proportion of land use with the minimum CO2 emissions produced can be achieved when the settlement area (X1) is about 4,510.01 Ha which is followed by a proportion of other types of land use in a row are protected forest area (X2) of 4,310.74 Ha; the area of green land (X3) is 3,565.6 Ha; industrial land area (X4) of 15 Ha; and other land area (X5) of 885.55 Ha. The proportion of land use in this output can be seen more clearly in the following figure 1.

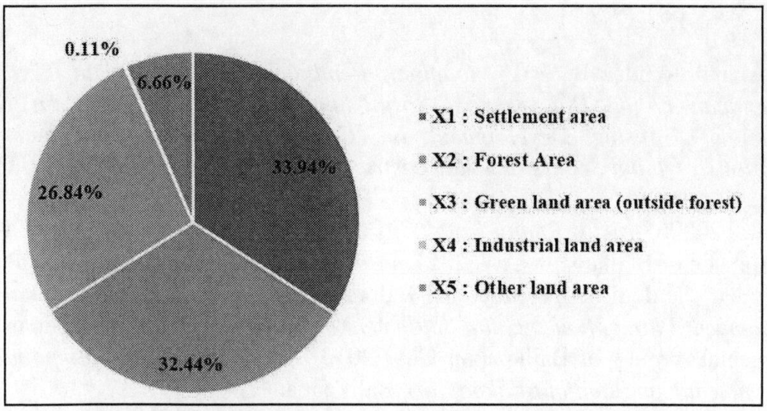

Figure 1. Diagram of the proportion of optimal land use

In this model, settlement has the largest land area of 33%. The land area of the settlement relates to the number of inhabitants and the provision of land to meet the need for housing. North Balikpapan as an area that has experienced rapid development certainly requires a large area of residential land but must still maintain a balance of green land area as a CO2 emitter from residential activities.

With this proportion of land use, North Balikpapan will have a surplus condition as indicated by the objective value in table 1, the substitution value between the land allocation value resulting from the modeling and the objective function. The amount of surplus that occurs is -2.751,591 tons. Thus, the green area in North Balikpapan is strongly able to absorb CO2 emissions produced by built-up land and also able to absorb emissions from outside to around 2.751 ton.

The other consequences from analysis result, it can be obtained that with the proportion of land use North Balikpapan will have population growth about 407.128 people; The citizen density is 3.080,80 people/Km2; the number of workforce is 44.960 people; and get the Gross Domestic Regional Product is IDR 71.836.141,39 Million.

4. Conclusion

Based on the conducted results in this study, it could be concluded that:

- The results of the analysis of the calculation of the carbon footprint in North Balikpapan shows the energy consumption the use of cooking fuel, the use of transportation fuels, and the use of electricity generated from 50,766 households in North Balikpapan contribute 108.362,4 tons of CO2 emissions /year. It minimum requires 18,50 Ha of the green space to absorb the CO2 emissions.
- The factors that influenced the development of settlements in North Balikpapan based on the results of a content analysis are: (a) the social group factors consisting of the factor of demographic factors population and the number of workforces; (b) The initiative housing developers factors consisting of the factor of the number of residential developers and the extent of settlement area; (c) the group of environmental conditions consisting of the factor of protected forest availability and the green space availability; (d) The group of the public facilities availability consisting of the factor of the number of educational facilities, health facilities, trade and services, worship facilities, and supporting transport facilities; (e) the group of local economic structure consisting of the factor of GDRP; and (f) the group of spatial policy plan consisting of the factor of land allocation on Spatial Plan Area documents.
- The Results of the optimization of land through a linear programming analysis showed that the allocation of settlement land area by 4,510.11 Ha; (2) land area protected forest by 4,310.74 Ha; (3) the broad green fields by 3,565.6 Ha; (4) industrial land area by 15 Ha; and (5) the other land area by 885.55 Ha.

5. References

[1] Jauhari A and Render B. 2013. *Dampak Pembangunan Perumahan Terhadap Perubahan Penggunaan Lahan Dan Kondisi Sosial-Ekonomi Penjual Lahan Di Kecamatan Mlati (Impact of Housing Development on Changes in Land Use and Socio-Economic Conditions of Land Sellers in Mlati District)*. Journal of Bumi Indonesia, Volume 2 Number 2.

[2] Yunus, H.S. 2000. *Struktur Ruang Kota (City Space Structure)*. Yogyakarta: Pustaka Pelajar.

[3] Central Statistic of Balikapapn City. 2017. *Balikpapan In Figures Year 2016*. Balikpapan

[4] Housing and Settlement Agency of Balikpapan City. 2014. *The Final Report of the Development Plan of housing dan Settlement in Balikpapan city*. Balikpapan

[5] Environmental Agency of Balikpapan City. 2012 *Buku Hijau Kota Balikpapan* (*The Report of Environment Condition in Balikpapan)*. Balikpapan

[6] Kodoatie, R.J and Syarief, R. 2010. Tata Ruang Air (*Water Management System*). Yogyakarta: ANDI.

[7] Novanda, E. 2015. *Persebaran Spasial Produksi Emisi Karbon Dioksida Dari Penggunaan Lahan Permukiman di Kawasan Perkotaan Gresik Bagian Timur (Spatial Distribution of Production of Carbon Dioxide Emissions from Land Use in Settlements in the Eastern Gresik Urban Area)*. Journal of Teknik ITS Volume 4 Number 1.

[8] Indonesia Ministry of Energy and Mineral Resources. 2013. Review of Inventory of Greenhouse Gas Emissions in Energy Sector. Jakarta: Data and Information Technology Center for Energy and Mineral Resources

[9] IPCC 2006 *Guidelines for National Green House Gas Inventories (*Intergovernmental Panel on Climate Change)

[10] Krippendorff K 2004 *Content Analysis: An Introductions to its Methodology (Second Edition)* (California: Sage Publication)

[11] Heizer, J. and Render, B. 2006. *Operations Management (7th Edition)*. Jakarta: Salemba Empat

[12] Setiawan, R.Y, Boedisantoso R, and Razif M. 2010. *Study of Carbon Footprint from Industrial Activity in Surabaya City*. Proceeding of National Seminar on Environmental Technology VII Surabaya. ISBN: 978-602-95595-2-1. Surabaya, 25 – 26 October 2010.

An insight on Surabaya development: pre colonials, colonial, post colonial and current era

A Pamungkas[1], D Iranata[2], J Yuwono[3] and L M Jaelani[4]

[1] Urban and Regional Planning Department, Institut Teknologi Sepuluh Nopember
[2] Civil Engineering, Institut Teknologi Sepuluh Nopember
[3] Development Initiative for Indonesia (DIFI)
[4] Geomatic Engineering, Institut Teknologi Sepuluh Nopember

Email: adjie.difi@gmail.com

Abstract. Surabaya is the second biggest metropolitan of Indonesia. Its rapid development has established from the northern part to the south. The city has transformed from an emerging industrial city into business and trading city. However, the city is also facing escalating risks and challenges. Therefore, a pressure of development becomes intense which requiring a more careful approach to achieve a desired future sustainable Surabaya. In responding to that, special concerns about making Surabaya more resilience especially to disasters, creating an effective waterfront development management for the city and designing Surabaya as a compact city are keys to a better Surabaya in the future.

Keywords: *Surabaya, urban history, urban development, urban sustainability.*

1. Introduction

With a population of 2.9 million [1], Surabaya, also known as the "City of Heroes" in Indonesia, is the second largest city in the country after Jakarta. It is located on northeastern of Java island, facing the Madura Strait (Figure 1). Surabaya serves as the capital city of East Java Province and administratively, it is divided into 31 districts (Figure 2).

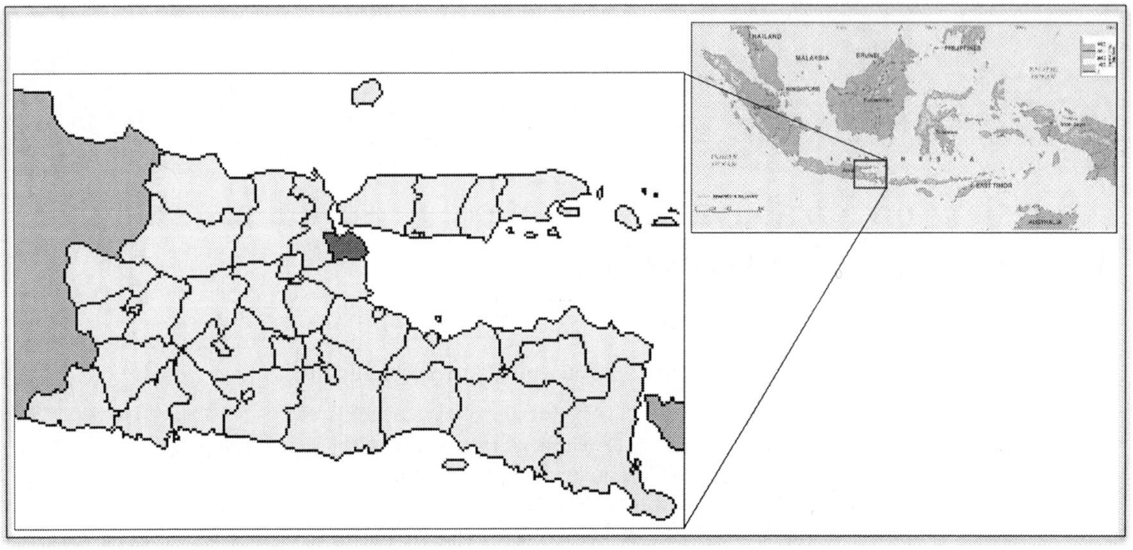

Figure 1. Location of Surabaya City within the within East Java Province and Indonesia
Source: http://www.maps-of-the-world.net and Wikimedia Commons

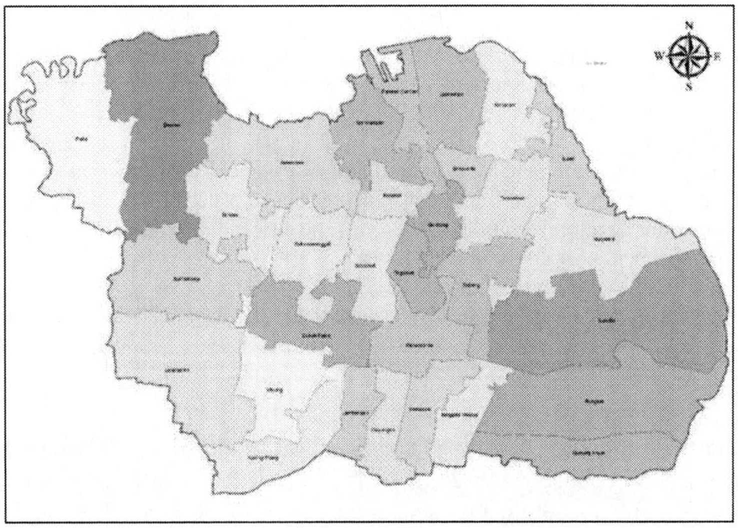

Figure 2. The Districts of Surabaya Municipality
Source: Surabaya Spatial Plan in 2014

As a metropolitan city that keeps growing and improving through years, Surabaya has played a significant role as a social and economic hub for not only the area in its surrounding but also for Indonesia and some other countries in the region. With the increased activities along with the rising number of its population, the city government has seriously attempted to accommodate and facilitate the activities of its residents and at the same time to keep the quality of its ecological system, in order to create a sustainable urban environment. Although the attempts have got some recognition both at national level, for instance by receiving the Indonesian highest environmental award for cities since 2006, known as the Adipura Kencana, and also at international level such as the ASEAN Environmentally Sustainable City Award in 2012, the city is still facing some challenges and striving with many development issues.

Despite its important role and achievements, the study of Surabaya city is still received little attention, especially for a wider international audience. In this article, the discussion about development in Surabaya will be elaborated in a chronological order, from an ancient settlement to a metropolitan city with a high demand for built areas to accommodate its fast economic growth.

2. The Origin of Surabaya

The earliest record of the name *Surabaya* can be found on the Trowulan epigraph from 1358 A.D. in which it was spelt as *Churabhaya* in old Javanese script (Widodo, 2010). However, according to G. H. von Faber—a German-Dutch journalist who was born in Surabaya and the founder of the Mpu Tantular Museum in Surabaya—the city may have already existed a century earlier as a settlement area for the soldiers of Kertanegara, the last king of Singosari Kingdom [2]. With regard to the name of Surabaya, the locals believe that it derives from two words: *sura* and *baya* (respectively means shark and crocodile in Javanese). According to the local myth, those two animals had a fierce battle in the area and they both have been taken as the icons of the city (Figure 3). Based on the official website of the Surabaya Municipality, the myth might be based on the big battle between the army of the Mongolian Kingdom who came from the sea (represented as the white shark) and the Majapahit Kingdom that occupy the land (represented as the giant crocodile). Some historians who were appointed by the Surabaya municipality in 1973 had estimated that on 31 Mei 1293, the Majapahit army succeeded in defeating the Mongols in the battle that took place in the area. This date has since then set by the Surabaya government as the anniversary of the city[1].

Figure 3. Sura (shark) and baya (crocodile) as the icon of Surabaya
Source: allindonesiantourism.com

Although historical evidence about the initial development of Surabaya can rarely be found, there is some evidence to support that in the 16th Century Surabaya was already developed as a major trading port under the government of Kadipaten Surabaya which first was part of the Majapahit Kingdom and later on the Demak Sultanate before it was captured by the Sultanate of Mataram [3], [4]. The city once served as the main gate to enter the Java inner-land in the east from the ocean through the Kalimas River. Due to its strategic location on the busy trading route between Malacca

[1] Official website of Surabaya Municipal Government

IOP Conf. Series: Earth and Environmental Science **340** (2019) 012002 doi:10.1088/1755-1315/340/1/012002

Strait and Maluku islands (known before as the Spices island), Surabaya has emerged as a leading coastal power in the region [4]. As a port city, Surabaya has also been characterized as a multi-cultural city with a mixture of people from various origins since at the early stage of its development [2].

3. The Colonial Era

In 1625, Surabaya was fallen under the control of the Dutch colony through the VOC (the Dutch East India Company) that also dominate most of the area that now parts of Indonesia but back then is called as the Netherlands East-Indies [4]. Later on, the Dutch developed Surabaya not only as the centre of the distribution and trading activities for the products of Java's plantations through the port of Kalimas river (Figure 4) but also as their largest naval base in the colony [5]. During this era, Surabaya has already shown the potential to be a metropolitan city which is dominated by trading and service-based activities. The transformation of Surabaya as a future metropolitan city could also be related to two important things that lead this city to become an industrial city with the establishment of the steel industry and the expansion of rural areas as export commodities producers [2]. The establishment of the steel industry is then followed by the development of other industries, such as construction workshop, wood-cutting industry and traditional wine distillation industry [2]. At this stage, Surabaya has grown not only as a trading city but also an industrial city.

Figure 4. Activities on the port of Kalimas river in the Dutch colonial time
Source: Leiden University Library, Colonial Collection (KITLV)

With those developments, many European have soon afterwards expanded their colony in Surabaya along with the locals and also other ethnics who already came to the area and stayed there to build their business. The VOC grouped the most prominent ethnics in the area into four separate settlements, which included the Dutch and other Europeans, the Arabs, Chinese, and the local Malay or Javanese (see Figure 5). These different settlements have become one of the main features of the city at that time which somehow still can be recognised until now.

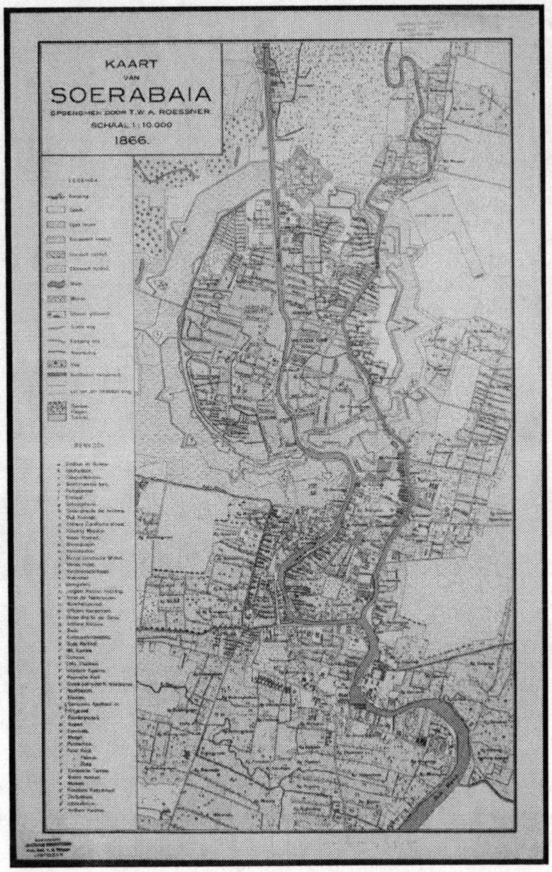

Figure 5. Map of Surabaya in 1866 with the ethnic settlements segregation
Source: Leiden University Library, Colonial Collection (KITLV)

After the dissolution of VOC due to bankruptcy at the end of the 18th century, the control over the Netherlands East-Indies, including Surabaya, was transferred to the Dutch Colonial government. On April 1st, 1906, the Dutch Colonial Government granted Surabaya the status of the municipality [6]. However, only after ten years later the first mayor, A. Meijroos, was appointed to the city. Under the Dutch Colonial, Surabaya was designed as a typical European city (see Figure 6). One of the most influential architects of the city was Cosman Citroen who came to Surabaya in 1915 and later on was appointed as a municipal architect [7].

Figure 6. Remnants of colonial architectural buildings in Surabaya
Source: various under Creative Commons licenses

In order to cope with the growing numbers of populations and activities of the city, many infrastructures were developed which transformed it to become a modern city. One of the important infrastructures was the Tanjung Perak harbour in the estuary of Kalimas river which construction began in 1910 in order to facilitate the increase of trading and cargo traffic (see Figure 7) that could not be accommodated anymore by the port of Kalimas river. The plan to built a bigger port for Surabaya was actually already started in 1875 but the development was postponed due to a shortage of funds [8].

In addition to the new harbour, the city was also equipped with a modern transportation system, for instance, the steam-tram line which was operated by the OJS (Oost Java Stoomtram) Company. In 1924, the electric tramway system in Surabaya was completed (see Figure 8) that connected Tanjung Perak in the north and Wonokromo in the south via Darmo Boulevard and Willemskade, and also the Bagong and Gubeng area in the east side of the city and the Sawahan area in the west[2] [9]. Along with the development of new infrastructures, new and modern facilities such as hospitals, post office, banks, and high schools also developed in the city (see Figure 9) as well as the networks for electricity and piped water supply.

[2] The tram lines in Surabaya were destroyed during the Indonesian National Revolution (see section 4) and the services are discontinued ever since.

Figure 7. The Port of Tanjung Perak in the 1920s
Source: Collection of Koninklijke Marine

Figure 8. Electric tram in Willemskade (now Jembatan Merah street), Surabaya in the 1920s
Source: Leiden University Library, Colonial Collection (KITLV)

Figure 9. Some early modern facilities in Surabaya that were built in the 1920s: (clockwise from left-top) *Simpang* hospital, post office, high school (HBS), and bank.
Source: Leiden University Library, Colonial Collection (KITLV)

4. The Post-colonial Era until the End of 20th Century

During the World War II, Japan dismantled the hegemony of Dutch East-Indies colonial government. Following the surrender of Japan to the allied forces, Soekarno and Mohammad Hatta declared the independence of Indonesia on August 17th,1945. Soon afterwards, the Dutch tried to re-establish their colony with (in the beginning) the help of the British who got a mandate to be the caretaker of the Dutch colony after the surrender of the Japanese. The Indonesian fought fiercely for their independence and start the Indonesian National Revolution. One of the well-known battles of the Indonesian revolution is the Battle of Surabaya which reached its peak on November 10th 1945. The date is later commemorated by the Indonesian as the Heroes Day and Surabaya has since been also called as the City of Heroes by the Indonesians.

After the Dutch eventually recognise the sovereignty of Indonesia at the end of 1949, many cities, including Surabaya, were still facing hard times especially to rise from the devastations caused by the war. Most of the population suffers from poverty and many of them were homeless because their houses were destroyed or taken over by others during the evacuation [10]. At the same time, the number of people who came to the city from the surrounding areas seeking for a better life was also increased since many plantations and the related industries in rural areas have stopped with their operations [11].

The limited available space for settlements and the high demand for new housings due to the rapid growth of population—especially a group of a low income—has led to uncontrolled settlement developments in Surabaya City [11]. As notified by Silas, the settlements of people with low income who come to a city, particularly Surabaya, are mostly concentrated in urban kampung and fringe villages [12]. Silas also added that in Surabaya, kampungs are commonly developed from rural villages that have been included as a part of an urban area due to the urban sprawls; or from empty or abandoned lands in the city that were occupied and claimed by the incoming migrants, mostly in the post-war period. Although in general kampungs are mostly inhabited by low-income communities and they have a minimum quality of urban facilities and services, they are not always the same as the slum

IOP Conf. Series: Earth and Environmental Science **340** (2019) 012002 doi:10.1088/1755-1315/340/1/012002

area [13]. Moreover, since many informal enterprises are based there and their existence has provided many city workers with a place to live, kampungs have become an essential element of the city [12]. Therefore, in order to improve the productivity, sustainability, and the liveability of the city, it is important to improve the condition of urban kampungs.

As a matter of fact, the program to improve the condition of kampung in Indonesia has been introduced in the 1920s by the Dutch colonial government [14]. After the independence, the Indonesian government also recognized the importance of such program and launched the nation-wide Kampung Improvement Program (KIP) in 1969 in Jakarta and Surabaya with the World Bank funding [12]. Although in the beginning KIP only focused on physical upgrading, it arguably has successfully reduced urban poverty (Silas, 1992).

Along with the launch of the KIP, during this era, Surabaya also experienced a new massive wave of real estate developments by private sectors, especially for high- and middle-income communities. This phenomenon is triggered by the increasing demand for new good quality residential areas as a result of the growing economy of the city. The developments of the new real estates in Surabaya mostly extend to the west part of the city because there were still many vacant lands available in this area. One of the most prominent real estate development projects in the area is the CitraLand which construction began in 1993 and occupied an area of 2,000 ha [15].

In addition to the new developments of the residential area as described above, this era also marked by large-scale infrastructure developments in Surabaya. For instance, in 1959, the national government started the development of the new airport, called the Juanda Airport, which was the first airport development project in the country since its independence [16]. The development was finished in 1964 and at first, this airport was mainly used for the naval air base to replace the older one in Morokrembangan. However, over the course of time, the airport also used for civil aviation due to the increased demand for it. The first international flight was opened at the airport in 1987 (see figure 10)

Figure 10. Juanda Airport
Source: Wikimedia Commons/ Fatur Almakawi

Six years after the opening of Juanda Airport, the substantial redevelopment and expansion project was also started for the Seaport of Tanjung Perak (Figure 11). Later in 1991, the port of Tanjung Perak launched a new facility, the international container terminal (Terminal Petikemas Surabaya), with the capacity of two million TEUs/year [2]. With this development, the port has not only contributed significantly to the economic development by accommodating the trade traffics for the city and the province but also the eastern part of Indonesia.

Figure 11. Port of Tanjung Perak
Source: Tempo.co

With regard to the economic development activities, Surabaya was mostly dominated by manufacturing industries during this era. This mainly was supported by the government policy, both at the national and local level, to adopt the idea of industrialization as one of the main strategies in the regional development. In Surabaya, this strategy is implemented through, among others, the development of two industrial areas in Surabaya in the 1970s: Industrial area Ngagel, which actually already exists since the colonial era, in the centre of Surabaya and the Surabaya Industrial Estate Rungkut (SIER) in the south-east part of the city. In addition to those areas, some industrial and warehousing activities also growing in the northern part of the city, for instance in the Tambak Langon - Kalianak – Margamulyo area, which is close to the Port of Tanjung Perak.

5. The Millennial Era

By the end of the 20th Century, Indonesia was hit by economic and political crises that eventually lead to a substantial reform in the country's governance systems. One of the important reforms that has a direct implication to the urban development mechanism is the introduction of the decentralization policies—which came into effect in 2001—that gives the local governments a greater autonomy than before the reform and also the possibility to have direct election for the head of the local government [17]. Although there have been some academic research that explains the negative impact of the decentralization policies to urban and regional development in Indonesia [18], [19], the policies have, evidently, also gave opportunities to some cities to exercise some urban policy innovations that show success stories of urban development [20].

Particularly in Surabaya, one of the success stories of the urban policy innovation is related to the implementation of a waste management program in which the city government develop an international joint-venture cooperation to established several community-based composting centers and successfully encourage community participation in the 3R (reduce, reuse, recycle) campaign (Premakumara et al., 2011). This policy has decreased the total daily municipal waste generation up to almost 25% in only three years [20]. Although this success can be attributed to the exemplary leadership of the current mayor of Surabaya city, it can also be cast as positive changes that stem from electing the right local government leader which made possible by the decentralization policy.

With regard to its economic activities, Surabaya's development in the millennial era was marked by the significant growth of some sectors other than manufacturing industry although the industry is still quite dominant in the city. As shown in Figure 12, since 2003, the commercial sector has been the most dominant sector in Surabaya. In addition, the HORECA and construction sectors also have a significant increase in the last 10 years. This situation can be related to the effort of the municipal government to transform its economy to service and commerce activities. To do so, the municipality, for instance, promoted a slogan of "Sparkling Surabaya" in 2006 as a part of its marketing strategies in attracting tourists and business corporations to come and invest in the city (Purwanti & Genoveva, 2017). As a result, the city has successfully doubled its revenues from tourism activities and in a 2010 national business survey, the city was ranked second among the twenty largest local government economies in Indonesia on a business satisfaction index[3].

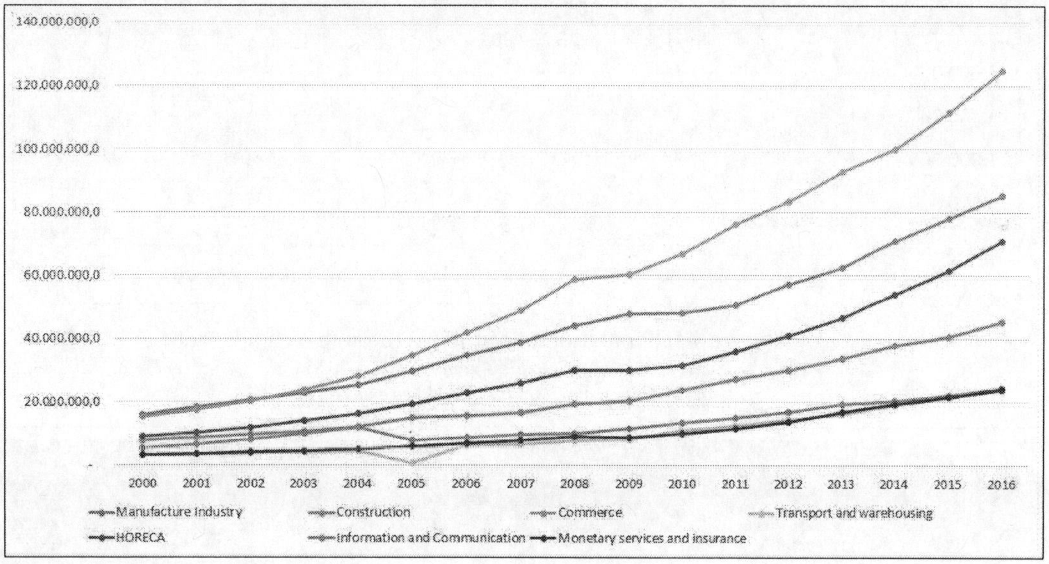

Figure 12. Gross Regional Domestic Products of seven main sectors in Surabaya, 2000 – 2016 (in million Rp.)
Source: Surabaya City Statistical Agency

The millennial era of Surabaya is also marked by its significant population growth (see Figure 13). Since 2000, the populations that migrate to Surabaya reached its highest number, which is 115.594 persons, in 2012. In 2015, the number of citizens that resides in Surabaya has reached more than 3 million.

As an impact of the rapid growth, the land use in Surabaya has also changed, especially to accommodate the demand for settlement area. Based on the land cover analysis, the land coverage for settlement area has reached 16.803,90 Ha in 2015. Compared to the condition in 2001, the demand for settlement area has increased by 4.55,16 Ha, which means that each year 325,44 Ha are converted to settlement area. Meanwhile, the green space has decreased by more than 1.000 Ha for the past 14 years.

The most significant changed areas could be found in the western and eastern part of Surabaya, which is dominantly converted from open space to residential area since 2001. Figure 14 illustrates the comparison of built-up areas in Surabaya between 2001 and 2015. The yellow surface indicates the build-up areas consisting of residential, commercial and industrial areas while the green colour indicates green open space consisting of bare land, parks, paddy fields and farmland. The other dominant land cover is the blue open space that consists of aquaculture, river and pond.

---

3 A profile of Surabaya in the Singapore Straits Times in mid-2010 glossed Surabaya's transformation "From shabby town to sparkling city." The Straits Times, 14 August 2010, D7. Surabaya's strong showing on a "satisfaction index" for business performance appeared in SWA Magazine (12–22 August 2010).

IOP Conf. Series: Earth and Environmental Science **340** (2019) 012002 doi:10.1088/1755-1315/340/1/012002

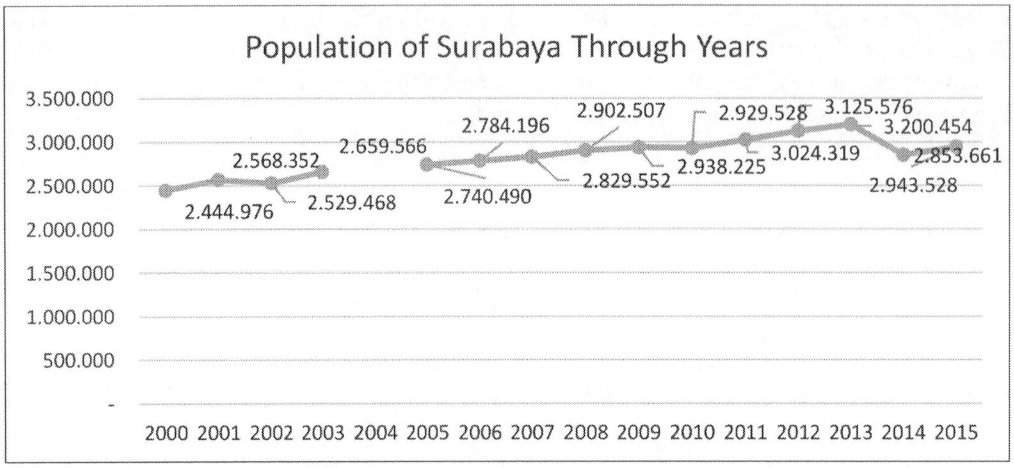

Figure 13. Population of Surabaya through Years
Source: National Statistic Agency, 2002 – 2016
*Notes: *data from 2003 and 2004 were missing*

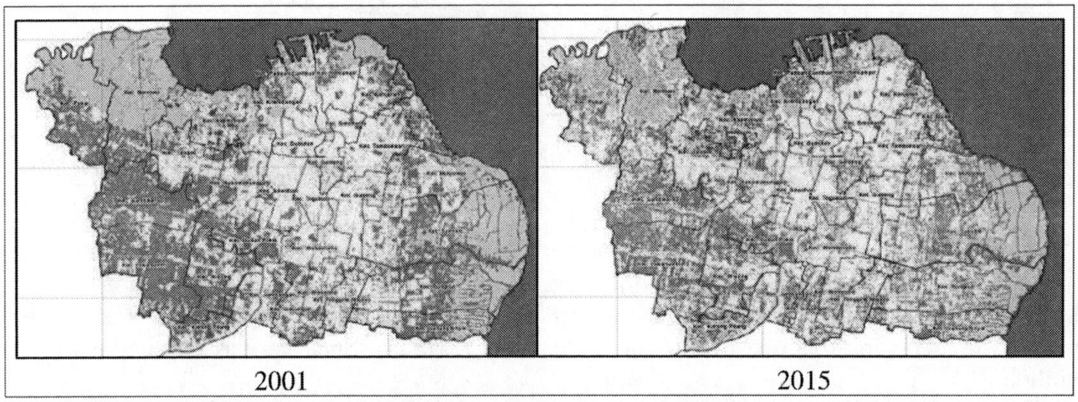

Figure 14. Land coverage of Surabaya
Source: Zulkarnain, 2016

The modern days of Surabaya as a metropolitan city require more spaces to accommodate the complex activities. However, limited space is an issue due to land prices, vacant land availability and land ownership. Thus, the city also grows vertically which can be seen from the increasing number of high-rise buildings. In the near future, the number of a high-rise building is believed to increase significantly. The number application for the building construction certificate has indicated those significant increases. As indicated in Figure 15, there is a high jump number of the legalized building from 2012 to 2015. This increasing pattern is likely to keep occurring in the future. Concerning the distribution of the high-rise buildings location, they are spread out to allover of Surabaya that gives an indication of an unclear pattern of development (Figure 16).

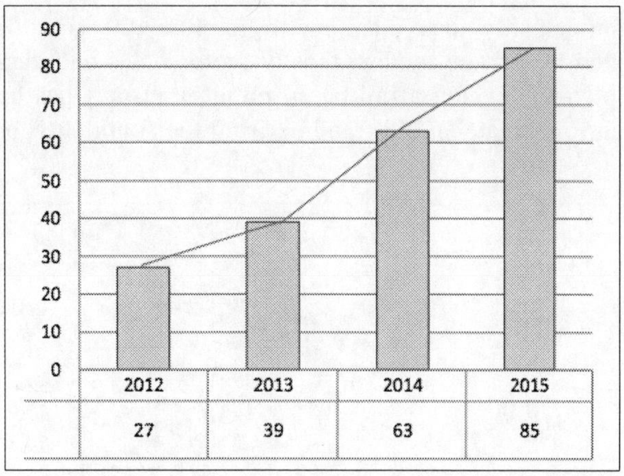

Figure 15. Numbers of legalized high-rise building
Source: Kajian Penyusunan Potensi Pendapatan: Prakiraan Potensi Penerimaan PAD dari Bangunan Tinggi Kota Surabaya, 2016

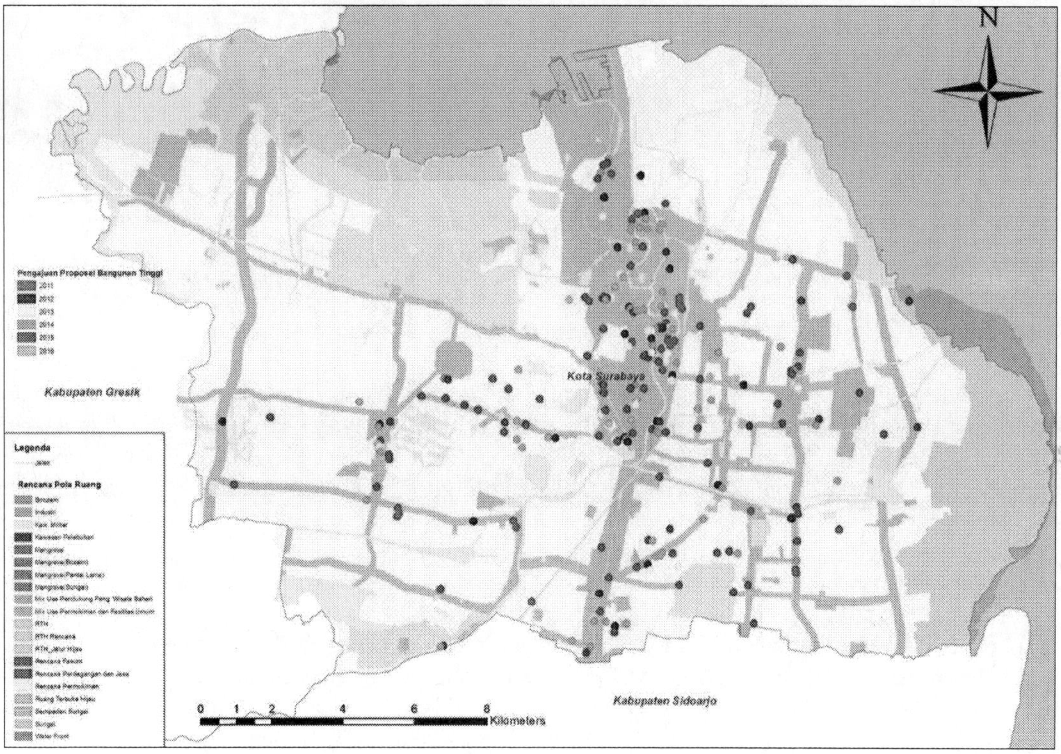

Figure 16. The distribution of legalised high-rise buildings from 2012 to 2015
Source: Kajian Penyusunan Potensi Pendapatan: Prakiraan Potensi Penerimaan PAD dari Bangunan Tinggi Kota Surabaya, 2016

In order to accommodate the rapid growth of population and activities in the city, the municipality has made some plans to develop and improve several main urban infrastructures in the city since many of the existing ones are already inadequate. For instance, to ease the overcapacity of Tanjung Perak port, a new container port, called as the Teluk Lamong terminal, was constructed in 2010 in the western part of Surabaya near the border with the Gresik Municipality. The port is intended mainly to handle domestic cargo. A toll road access is also constructed to connected it with the surrounding area.

IOP Conf. Series: Earth and Environmental Science **340** (2019) 012002 doi:10.1088/1755-1315/340/1/012002

Apart from that, the municipality also attempts to improve the intra-city transportation, which currently is still dominated by private modes, mostly motorcycles and cars, by establishing a mass rapid transportation. A plan to develop a rail-based public transport has been proposed (Figure 17) which also include a plan to recreate an old tramline from the south (Joyoboyo Bus Terminal) to the north (Indrapura Road intersection).

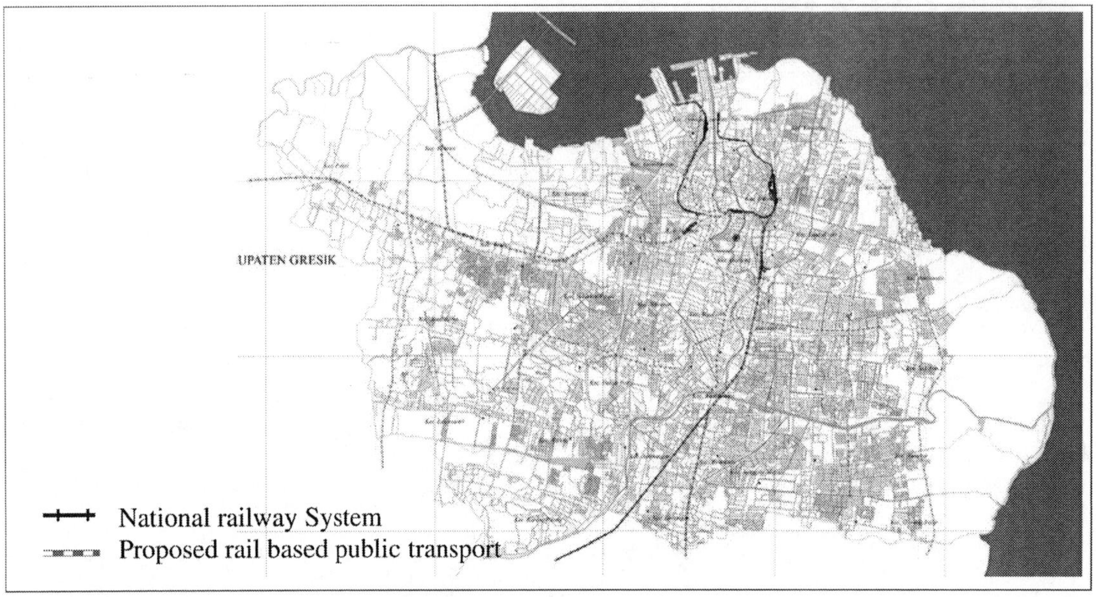

Figure 17. Surabaya plan for public transportation
Source: Regional Regulation No. 12 in 2014 on Surabaya Regional Plan (RTRW)

6. Challenges for Future Surabaya Development

Indisputably, the concept of sustainability has become one of the major transformative drivers of development in many cities and it is also applied to Surabaya. There are at least three urban development approaches or models that can be adopted by Surabaya given the specific characteristics of the city. The approaches are related to the issue of urban resilience, waterfront development, and compact city. In this section, some challenges concerning to the implementation of those approaches are discussed.

6.1. Urban resilience

One of an important aspects in the effort to achieve a desired future sustainable city is the ability of the urban system—which include all of its social, economic, and ecological constituent—to maintain or rapidly return to desired functions in the face of a disturbance or shocks, to adapt to change, and to quickly transform systems that limit current or future adaptive capacity, which also known as the urban resilience [21]. As Surabaya endures to grow and at the same time also faces uncertainties and challenges such as climate change, urban resilience has become an increasingly important concept in its urban sustainability strategy [22]. One of the pressing risks for Surabaya that can be caused by climate change is the rising of the sea level since more than 80% of Surabaya area is lowland with a height between 0 – 8m a.s.l.

Apart from the risk of sea level rise, Surabaya is also prone to the earthquake risk. In September 2017, the National Agency that responsible for recording and reporting earthquake activities in Indonesia issued a report mentioned that besides the Kendeng thrust in the further south of Surabaya which could impact the city, Surabaya is actually also bypassed by two fault lines which they called the Waru fault and the Surabaya fault which can be seen in Figure 18[4]. The faults have the potential to cause an earthquake that can reach up to 6.5 magnitudes. Although based on the Agency's historical earthquake record data, Surabaya has never been hit by a very high magnitude earthquake, there were many incidents of big earthquakes in the neighbouring area of the city in the past as shown in Figure 19. Based on that figure, the closest one was in Sidayu in 1902 and Mojokerto in 1937 with a magnitude between 6 and 7 [23]. Moreover, the latest study on earthquake microzonation in Surabaya has revealed that most of the areas in the city are prone to a moderate to high amplification of earthquake (see Figure 20).

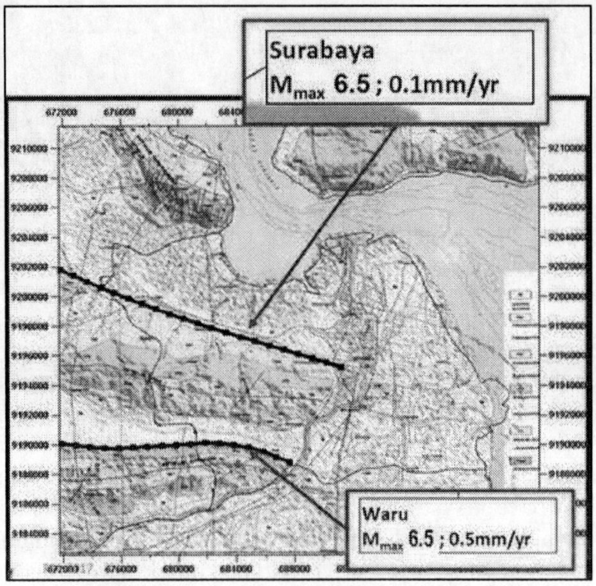

Figure 18. Two fault lines in Surabaya area
Source: Indonesian Agency for Meteorology, Climatology and Geophysics (BMKG), 2017

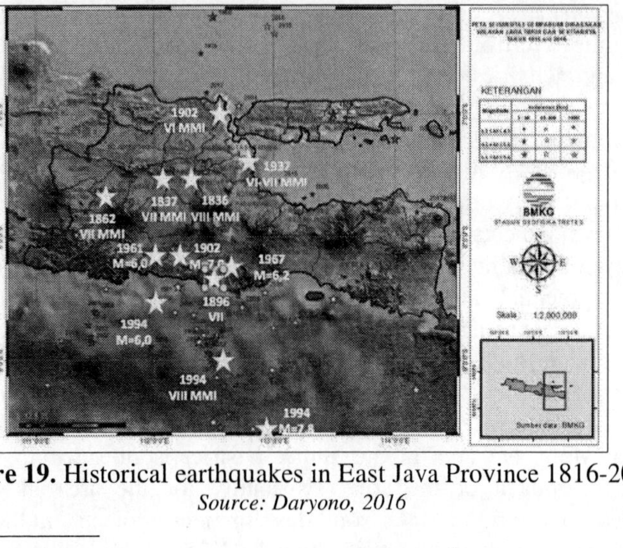

Figure 19. Historical earthquakes in East Java Province 1816-2016
Source: Daryono, 2016

[4] Press release of the Head of BMKG Geophysics Station Class II: Tretes-Pasuruan on November 9, 2017

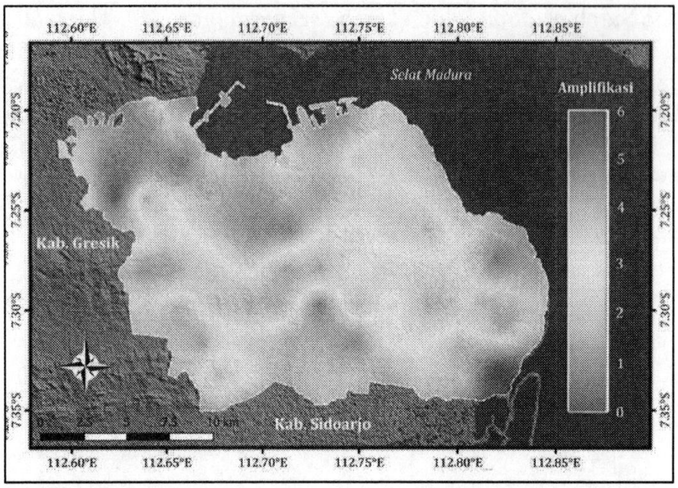

Figure 20. Amplification level in Surabaya based on earthquake micro zonation study
Source: Solikhin, 2016

In spite of this alarming situation, there is no significant effort at the moment to prepare Surabaya from the earthquake risk for instance in the form of regulation development that acknowledges this risk. Although there are some regulations at the national level related to the design standard for building development to reduce the potential threat of earthquake, its specific implementation rules to Surabaya is still missing. Therefore, it is highly important for the Surabaya municipal government to adopt a building code and have a system in place for building regulation at the local level that will result in adequate earthquake-resistant design and construction, more importantly for the high-rise buildings which number is growing in the city. Without this kind of regulation, many buildings in Surabaya might continue to be constructed without adequate protection, leaving people in these communities at considerable risk.

6.2. Waterfront development

As described earlier, the riverfront and waterfront areas in Surabaya have been intensely used and thriving with people and activities since the early development of the city. With the industrialisation especially during the post-colonial era, the relationship with the water was interrupted and cities begin to expand into the mainland. The development of port activities also blocked access to the water and many places next to the port activities areas are becoming less attractive and obsolete.

In many cities around the world, the (re)development of urban waterfronts has been one of the important economic developments [24]–[26]. Waterfront sites evidently offer many opportunities not only for economic development but also for public enjoyment. The municipal government of Surabaya is also aware of those potentials and tries to capitalise them by, as stated in the city's latest spatial plan, setting up three strategic areas for the city waterfront developments, which located in the districts of Benowo, Asemrowo, and Bulak (see Figure 21). It is also mentioned in the plan that the areas in the Benowo and Asemrowo Districts will be developed to support the activities of the Port of Teluk Lamong Terminal and the area will be equipped with facilities and infrastructure that are necessary for those activities. Meanwhile, the area in the Bulak District is intended to support marine tourism activities and the development of the area. Especially for the area in the Bulak District, the municipality has tried to offer the waterfront development projects in the area to some private investors and also prepared an investment of about Rp 175 billion (or about USD 12.5 million) to build road infrastructures and parks [27]. Apart from the development of recreation facilities, the private investors might also be interested to develop some high-rise apartments and condominiums that optimizing the utilization of view's orientation to the water in the area, especially for the mid and high-income group.

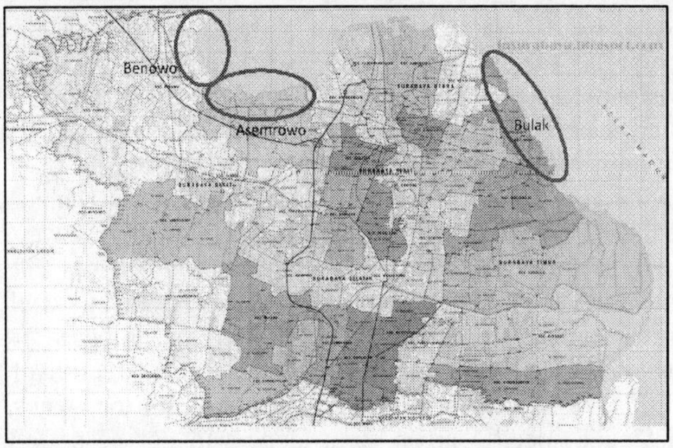

Figure 21. The proposed waterfront development areas in Surabaya
Source: Surabaya Spatial Plan 2014-2034

Although the development of waterfront area has always been hailed as a promising strategy, it should definitely be taken with cautions due to some complexities that involve different actors and considerations to the human and non-human environment in its processes. For instance, as argued by Bunce & Desfor [28], the connectivity of the process and contents of the development with the local people as well as the balance between the environmental benefits and the human needs are among the key factors in the waterfront development. Moreover, concerning the high-rise building that is going to be built in the area, some site-considerations should be taken into account, for instance they should not cast significant shadows on public parks, plazas, waterways, beaches or playgrounds and each building must not only be evaluated on its shape, height and orientation, but they should also focus on environmental factors, by creating better relationships and harmony between buildings and their urban and environmental contexts [29]. At the moment, the Surabaya municipal government already has some regulations that provide a guideline for the maximum building height and there is also a consideration to make a specific high-rise buildings zonation in the city[5]. However, specific principles to improve the connectivity of the buildings with the local people and to preserve the ecosystem should also be considered as a part of the building regulations in the future.

6.3. Compact city

The significant growth of populations and their activities in Surabaya has evidently lead to an increasing demand for land. With a limited available land especially in the city centre, urban sprawl has become unavoidable. This has resulted, among others, to an increase of green/blue spaces conversion to build up areas which can cause a higher risk of future floods due to an increased runoff also costs for the provision of public infrastructure, utilities, and services, as well as the social segregation that in end would decrease the livability and sustainability of the city. In another side, the increasing demand for development land, for instance, to build high-rise buildings as happens in Surabaya, indicates that the city is still economically attractive for new investments. Regulating this development attraction could sustain the development process in Surabaya.

Creating Surabaya as a compact city could be a promising response to this situation through the reduction of land conversion and replacement of the horizontal development with vertical

[5] Kajian Penyusunan Potensi Pendapatan: Prakiraan Potensi Penerimaan PAD dari Bangunan Tinggi Kota Surabaya, 2016

development. Compact development can be rooted from several concepts such as transit-oriented development (TOD) and pedestrian-oriented development (POD). Both development concepts give a priority to the non-motorized and public transportation mode together with a mixed-use land use allocation in order to create a balance interaction among varieties of land uses with no/limited distance. The interaction between different land use types and a walking distance can also be considered as a solution particularly in making the city sustainable [30], [31].

As mentioned earlier, there is already a tendency that Surabaya will also grow vertically with the increasing number of high-rise building developments. However, those developments are spread throughout the city without a clear pattern of development which could also lead to sprawling activities. The consideration to make a zone for high-rise buildings that was mentioned earlier could initiate a way to prevent the sprawling and support the compact development in the city. The zone—which can be done in a form of node, corridor or block development—could allocate the buildings based on its height in the certain precinct of the city. With the proposed zone, the development of high-rise building can have a specific pattern that forms the spatial structure of the city.

It should be noted that, as argued by many scholars—e.g. Burton, Jenks, & Williams, [32], and Neuman [33], establishing a specific urban form alone will not be sufficient as a strategy for a sustainable urban development. A serious consideration towards the process in which how the city is functioning which include the processes of living, consuming and producing in cities as well as the understanding of the idea that a city is a process of co-creation and co-evolution would have a more imperative role to attain a sustainable city [34], [35].

7. Conclusion

Surabaya has been one of the major cities since at the beginning of the Indonesian history. The early development of the city cannot be separated from the role of port activities, first along the Kalimas river and then at the Tanjung Perak port. Although the role of the port in Surabaya development is still high, the current economic profile of the city has changed, mainly from an industrial city in the olden days to a commerce and service city in the current era.

Surabaya also has varieties of challenges from time to time with some continuing issues and also escalating to a certain level as well new issues that have not been considered before. The fast-growing population in the previous era has generated development pressure—particularly for settlement availability and infrastructures support and created a phenomenon like urban kampung in the city. This issue now is still continued and modified with the new challenges in the modern day. High demand on a high-rise building, fast rate of land conversion and a new potential of fault zone make the settlement and infrastructure issues become more complex in an urban scale.

The approach to deal with the complex issues in urban scale should be responded in a comprehensive way. Such urban development concepts and strategies like urban resilience, waterfront development, and the compact city could be considered to offers some solutions for the challenges in achieving a desired future sustainable Surabaya. The urban resilience strategy is important to improve the capacity of the city to survive, adapt, and grow given the future stresses and acute shocks from the devastating events like the rise of sea level and earthquake. The waterfront strategy that ensures the connectivity of the process and contents of the development with the local people as well as the balance between the environmental benefits and the human needs, could create many economic and social advantages for the city. While a compact development strategy could accommodate development pressure in certain areas and limited development in specially designated land such as the risky area of the fault zone, green/blue open space and infrastructures allocation land. A compact development concept with a consideration towards the process in which the city is functioning would also manage the 'big scale' of development including high-rise building that then shapes the urban form in a good manner and also distributes the needs of infrastructures in a more efficient way.

8. Reference

[1] B. P. Statistik, *Kota Surabaya dalam Angka 2017*. Surabaya: BPS, 2017.

[2] Handinoto, *Perkembangan Kota dan Arsitektur Kolonial Belanda di Surabaya 1870-1940*, 1st ed. Surabaya: LPPM Univ. Kristen Petra Surabaya dan Penerbit Andi, 1996.
[3] T. G. T. Pigeaud and H. J. de Graaf, *Islamic states in Java 1500-1700, A Summary, Bibliography and Index: Eight Dutch Books and Articles by H.J. de Graaf (Verhandelingen van het Koninklijk Instituut voor Taal-, Landen Volkenkunde, 70)*. Martinus Nijhoff, 1976.
[4] M. C. Ricklefs, *A History of Modern Indonesia since c . 1200*. Houndmills: Palgrave Macmillan, 2008.
[5] P. J. Rimmer and H. Dick, *The City in Southeast Asia: Patterns, Processes and Policy*. NUS Press, 2009.
[6] Gemeenteraad Soerabaja, *Verslag der Gemeente Soerabaja over 1917: Met beknopte verslagen over 1915 en 1916*. Surabaya: E. Fuhri & Co., 1918.
[7] J. T. Santoso, "Gubeng bridge in Surabaya, a Dutch Architect C. Citroen's work in the 1925 Paris Exhibition," in *MATEC Web of Conferences*, 2017, vol. 101, p. 5028.
[8] Berlian, "Dari Kalimas hingga Tanjung Perak," *Majalan Dermaga*. [Online]. Available: http://www.majalahdermaga.co.id/post/14/dari_kalimas_hingga_tanjung_perak. [Accessed: 10-Sep-2017].
[9] H. Dick, *Surabaya City of Work: A Socioeconomic History, 1900-2000*. Ohio: Ohio University Press, 2002.
[10] P. Basundoro, "Antara baju loreng dan baju rombeng: kontrol tentara terhadap rakyat miskin di kota surabaya tahun 1950-an," *Masyarakat, Kebud. dan Polit.*, vol. 24, no. 4, pp. 309–317, 2011.
[11] P. Basundoro, *Dua Kota Tiga Zaman: Surabaya dan Malang Sejak Zaman Kolonial Sampai Kemerdekaan*. Yogyakarta: Ombak, 2009.
[12] J. Silas, "Marginal Settlements in Surabaya, Indonesia : Problem or Potential?," *Environ. Urban.*, vol. 1, no. 2, pp. 60–70, 1989.
[13] R. Damayanti, "'Kampung Kota' as Third Space in an Urban Setting : The Case Study of Surabaya, Indonesia," in *Transdisciplinary Urbanism and Culture : From Pedagogy to Praxis*, 1st ed., Q. Mahtab Zaman and T. Igea, Eds. Springer International Publishing, 2018, pp. 127–139.
[14] B. Setiawan, "Kampung kota dan kota kampung: tantangan perencanaan kota di Indonesia," 2010, p. 24.
[15] S. Fikriyah and Wisnu, "Perkembangan Kawasan Real Estate di Surabaya Barat Tahun 1970-2000 (Kontribusi Citraland dalam Perkembangan Kawasan Real Estate di Surabaya Barat)," *AVATARA, e-Journal Pendidik. Sej.*, vol. 1, no. 3, pp. 503–518, 2013.
[16] Info BANDARA, "Juanda International Airport," *Berjaya Blog*. [Online]. Available: http://info-bandara.blogspot.com/2010/08/juanda-international-airport.html on 6 February 2018. [Accessed: 06-Feb-2018].
[17] T. Firman, "Potential impacts of Indonesia's fiscal decentralisation reform on urban and regional development : Towards a new pattern of spatial disparity," *J. Sp. Polity*, vol. 7, no. 3, pp. 247–271, 2003.
[18] T. Firman, "Decentralization Reform and Local-Government Proliferation in Indonesia : Towards a Fragmentation of Regional Development," *Rev. Urban Reg. Dev. Stud.*, vol. 21, no. 2–3, pp. 143–157, 2010.
[19] S. M. Murshed, M. Z. Tadjoeddin, and A. Chowdhury, "Is Fiscal Decentralization Conflict Abating ? Routine Violence and District Level Government in Java , Indonesia," *J. Oxford Dev. Stud.*, vol. 37, no. 4, pp. 397–420, 2009.
[20] T. Bunnell, M. A. Miller, N. A. Phelps, and J. Taylor, "Urban Development in a Decentralized Indonesia: Two Success Stories?," *Pac. Aff.*, vol. 86, no. 4, pp. 1–24, 2013.
[21] S. Meerow, J. P. Newell, and M. Stults, "Defining Urban Resilience : A Review," *Landsc. Urban Plan.*, vol. 147, pp. 38–49, 2016.
[22] J. Carmin, N. Nadkarni, and C. Rhie, "Progress and Challenges in Urban Climate Adaptation

Planning : Results of a Global Survey," Cambridge, MA, 2012.
[23] Daryono, "Aktivitas Gempa Bumi di Jawa Timur," Surabaya, 2016.
[24] B. Hoyle, "Global and Local Change on The Port-City Waterfront," *Geogr. Rev.*, vol. 90, no. 3, pp. 395–417, 2000.
[25] U. Lehrer and J. Laidley, "Old Mega - Projects Newly Packaged? Waterfront Redevelopment in Toronto," *Int. J. Urban Reg. Res.*, no. December 2008, pp. 786–802, 2009.
[26] T. C. Chang and S. Huang, "Reclaiming the City : Waterfront Development in Singapore," *Urban Stud.*, vol. 48, no. 10, pp. 2085–2100, 2011.
[27] P. Widarti, "Surabaya Bangun Water Front Rp 175 Miliar, Berikut Masukan Pengusaha," *Bisnis.com.* [Online]. Available: http://surabaya.bisnis.com/read/20140331/17/69873/surabaya-bangun-water-front-rp175-miliar-berikut-masukan-pengusaha. [Accessed: 24-Feb-2017].
[28] S. Bunce and G. Desfor, "Introduction to 'Political ecologies of urban waterfront transformations,'" *Cities*, vol. 24, no. 4, pp. 251–258, 2007.
[29] D. L. A. Gordon, "Planning , Design and Managing Change in Urban Waterfront Redevelopment," *TPR*, vol. 67, no. 3, pp. 261–290, 1996.
[30] Y. R. Jabareen, "Sustainable Urban Forms : Their Typologies, Models, and Concepts," *J. Plan. Educ. Res.*, vol. 26, no. 3, pp. 38–52, 2006.
[31] A. M. Hassan and H. Lee, "Land Use Policy Toward the Sustainable Development of Urban Areas : An Overview of Global Trends in Trials and Policies," *Land use policy*, vol. 48, pp. 199–212, 2015.
[32] E. Burton, M. Jenks, and K. Williams, Eds., *The Compact City : A Sustainable Urban Form?*, 1st editio. Routledge, 2003.
[33] J. M. M. Neuvel and A. van den Brink, "Flood risk management in dutch local spatial planning practices," *J. Environ. Plan. Manag.*, vol. 52, no. 7, pp. 865–880, 2009.
[34] K. McCormick, L. Neij, O. Mont, C. Ryan, H. Rodhe, and R. Orsato, "Advancing Sustainable Solutions : an Interdisciplinary and Collaborative Research Agenda," *J. Clean. Prod.*, pp. 1–4, 2016.
[35] J. M. Wittmayer, F. Van Steenbergen, A. Rok, and C. Roorda, "Governing sustainability : a dialogue between Local Agenda 21 and transition management," *Local Environ.*, pp. 37–41, 2015.

Acknowledgements

Authors wishing to acknowledge ITS as the main funder of the research under the scheme of Penelitian Unggulan – Dana Lokal ITS – Centre for Earth, Disaster and Climate Change. Furthermore, authors also thank Urban and Regional Planning Department, ITS and Radboud University for research collaboration.

Constructing Input-Output (I-O) models from less than perfect datasets: an economic impact analysis of garage construction project in Gresik , East Java Province, Indonesia

A Irawan[1] and P R Satiawan[1]

[1] Urban and Regional Planning Department, Institut Teknologi Sepuluh Nopember

Email: hicakandi@gmail.com

Abstract. This paper demonstrates techniques to construct input-output (I-O) models to perform impact analysis under a situation of data limitation. Of policy makers' main interests regarding project proposals is to have some idea on the economic impact of the projects. I-O based model is an useful analytical tool to do impact analysis. However, the I-O table from which the analysis is based on is often not available for the region of interest, in this case, Gresik, a municipal in East Java. This paper shows how to derive the I-O table for Gresik out of the East Java's I-O table with four approaches, namely: 1) location quotient (LQ); 2) cross-industry location quotient (CIQ); 3) Flegg location quotient (FLQ); and 4) augmented Flegg location quotient (AFLQ). We suggest that these approaches are suitable for a situation facing time and cost constraints, which is a typical one in city/municipal-level policy domain.

Keywords: *input-output, impact analysis, transportation garage*

1. Introduction

In real public policy setting, a situation where impact analysis is urgently needed under imperfect information is not uncommon. Neither an action of doing nothing nor an attitude unrealistically expecting all information would be perfectly available is wise. There are indeed methods to deal with imperfect information can be utilized [1]. In the case of impact analysis based on I-O modeling framework, this paper aims at showing that we could still conduct impact analysis in the absence of the study region's I-O table by deriving the I-O table for the study region from the I-O table of the supra region where the study region is part of.

This paper uses a real case study where the municipal government of Gresik needs to assess the economy-wide impact of a construction of large garage for road transportation fleet. However, the Indonesian statistical agency (Badan Pusat Statistik, BPS) only makes the I-O table available at national and provincial level. It is not made available at city or municipal level. Therefore, there is no such thing as the I-O table of Gresik. This situation of imperfect data poses a challenge to the policy need. The main contribution of this paper is to provide a methodological solution to overcome the challenge.

The main approach for the impact analysis in this paper is based on the I-O framework. The main information in the I-O modeling is the I-O table which is a comprehensive measure to give an aggregate (macro) economic picture of a region in a given year in aggregate. Based on the data contained in the I-O table, we can carry out various economic analyzes that are useful as a basis in the development planning process.

The analysis produced by I-O modeling is practical and quantitative. Descriptive analysis of table I-O provides a general overview of the economy of a region that is the focus of the study. Descriptive analysis can show the role of goods and services of domestic and imported production in fulfilling domestic and foreign demand, describing the role of each sector in the economy and the linkages between sectors (backward-forward linkages), and identifying leading sectors. In addition to descriptive analysis, another type of analysis that can be derived from table I-O is impact analysis. Impact analysis calculates the effect of a final change in demand on the output of the economy. Changes in final demand occur because of a shock in the economy. Surprises can be positive (for example: bridge construction projects, factories, and various other infrastructure) or negative (e.g. natural disasters, factory closures, or divestments). Impact analysis is used in this study and is the main focus in this section.

Impact analysis firstly requires the calculation of multiplier coefficient. The transportation and warehousing sector (which the fleet garage project belongs to) has a positive multiplier coefficient and is greater than one (1). This means that every Rp. 1 planted will have an impact greater than Rp. 1 for the economy as a whole. Based on the various methods used, the range of the output and multiplier of the transportation and warehousing sector is 1.11 - 1.33. Investment injection for the construction of a truck garage valued at around Rp. 23.55 billion will produce an additional output of Rp. 31.98 to Rp. 45.51 billion for the economy of Gresik Regency.

In succession, after the introduction section, this section will display the basic concept sections of the I-O model, the construction of the Gresik I-O table, the results section, and the conclusion section.

2. Basic concepts of the I-O model

The structure of the I-O table is briefly explained in figure 1 below. This figure is taken from Blair and Miller [2].

	Interindustry Sales (Z)	Sales to Final Demand (f)	Total Sales (x)
	z_{11} z_{12} ... z_{1n}	c_1 i_1 g_1 e_1	x_1
	z_{21} z_{22} ... z_{2n}	c_2 i_2 g_2 e_2	x_2
	⋮	⋮	⋮
	z_{n1} z_{n2} ... z_{nn}	c_n i_n g_n e_n	x_n
Value Added	l_1 l_2 ... l_n ov_1 ov_2 ... ov_n	Transactions between the payments sectors (including imports) and final demand	L OV
Imports	m_1 m_2 ... m_n		M
Total Outlays (x)	x_1 x_2 ... x_n	C I G E	

Figure 1. I-O table structure.

Basically, I-O tables are a combination of several matrices. The main matrix in the I-O table is the Z matrix (top left corner) which records transactions between sectors. From the Z matrix, we can produce an Matrix A or commonly referred to as the technology coefficient matrix. Matrix A is also commonly referred to as the 'production recipe' matrix because it shows the composition of the use of inputs in the production of a sector. Each element in matrix A, namely a_{ij}, shows how much (in monetary value, for example in rupiah) the sector buys input from sector i for each monetary unit of output produced by sector *j*. So, the matrix A element gives a kind of recipe, how much input from sector *i* is needed for each monetary unit of sector output *j*.

This Matrix A is needed to calculate matrix multipliers or also commonly referred to as Leontief Inverse or B matrix. Mathematically, the Matrix B is formulated as follows: $(I - A)^{-1}$, where matrix I is the identity matrix, that is a symmetrical matrix where the diagonal element is one (1) and the other element is zero (0). This Matrix B is very important in impact analysis, which is to see changes in output in the economy, ΔX, if there is a change in the final demand, Δf, due to an economic shock. The impact calculation formula is as follows:

$$\Delta X = (I - A)^{-1}.\Delta f \tag{1}$$

The change in the final request is Rp 23.55 billion, which is worth the investment in the construction of a truck garage, in the transportation and warehousing sector, assuming other sectors have not experienced a change in final demand. So changes in output in the economy will depend on the matrix A.

As will be explained in the next section, I-O tables are derived or constructed from East Java I-O tables through four different ways and two parameters of the location quotient (LQ). So we will get several types of matrix A. The difference in this method, besides functioning as a robustness check model, will also provide a range of estimated impacts, not just one estimated impact. This is useful for decision making because we minimize the situation of over/under estimation.

3. Construction of table I-O Gresik

Table I-O of Gresik Regency is not available. Due to time and cost constraints, construction of survey-based I-O tables for Gresik is not possible. Following Flegg, Webber, and Elliot [3,4,5], therefore the construction of the Gresik I-O table must be done with a non-survey approach, which is derived from the East Java I-O table through four methods, namely: 1) location quotient (LQ); 2) cross-industry location quotient (CIQ); 3) Flegg location quotient (FLQ); and 4) augmented Flegg location quotient (AFLQ). Following are the stages of construction of the Gresik I-O table:

1. Reclassification of sectors in the East Java I-O table and Gresik Gross Regional Domestic Product (GRDP).
2. Calculating LQ) output of Gresik's economic sectors relative to the same sectors in East Java's economy.
3. Calculating matrix A for Gresik based on the four methods above.

3.1 Sector Reclassification

The East Java I-O Table in 2015 consisted of 110 x 110 sectors, while the Gresik GRDP in 2105 consisted of 46 sectors which could be integrated into 17 sectors. Therefore, both data must be reclassified so that they have the same number and sector nomenclature. This process is called crosswalk. Figure 2 below describes the process briefly. Table 1 shows the crosswalk sector results.

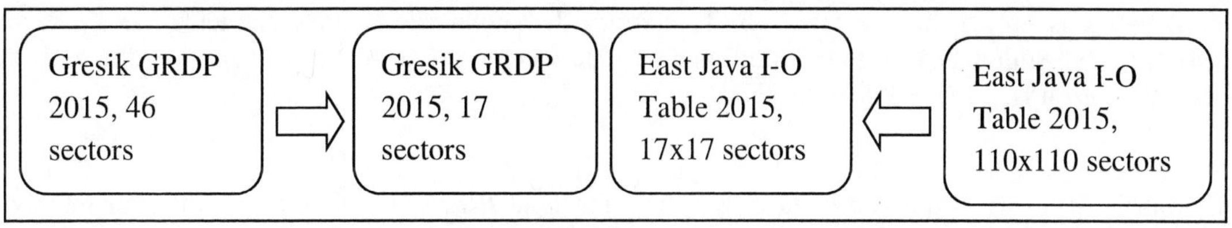

Figure 2. Crosswalk between two data

Table 1. 17 sectors *crosswalk result*

Code	Sectors
1	Agriculture, Forestry and Fisheries
2	Mining and Excavation
3	Processing industry
4	Procurement of Electricity and Gas
5	Procurement of Water, Waste Management, Waste and Recycling
6	Construction
7	Large and Retail Trade; Car and Motorcycle Repair
8	Transportation and Warehousing
9	Provision of Accommodation, Food and Beverage
10	Information and Communication
11	Financial and Insurance Services
12	Real Estate
13	Company Services
14	Government Administration, Defense and Mandatory Social Security
15	Education Service
16	Health Service and Social Activity
17	Other Services

3.2. LQ calculation

LQ is basically the ratio between the proportion of a sector in the total economy of a region and the proportion of a sector that is the same in the total economy of a larger region where the first region is part of it. LQ is a measure to see the similarity of the economic structure of a region with a supra-region that surrounds it. In this case study, the area of concern is Gresik and the supra-region, East Java. LQ is calculated based on the following formula:

$$LQ_{i,Gresik} = \frac{s_{i,Gresik} / \sum_{i=1}^{n=17} s_{i,Gresik}}{s_{i,East\ Java} / \sum_{i=1}^{n=17} s_{i,East\ Java}} \quad (2)$$

The sector output for Gresik is taken from Gresik GRDP based on the 2015 sector, 17 sectors. While the output of the East Java sector is taken from the East Java 2015 I-O table, 17 x 17 sectors.

There are two values that will be used as denominators in the use of formula (2) above. First, the sector value is based on the total vector matrix as shown in figure 1 of the structure of the I-O table above. Second, sector value is based on intermediate demand. In figure 1 above, the temporary request vector matrix is not explicitly displayed. Temporary demand vector matrix elements can be easily calculated, i.e. summing of cross column lines per line (sector per sector) in matrix Z. In the end, we will get two groups of LQ parameter values.

3.3 Calculating matrix A Gresik

The Matrix A Gresik is derived from matrix A in the East Java I-O table through the four methods above (LQ, CIQ, FLQ, AFLQ) and two groups of LQ parameter values. This process is illustrated in figure 3 below.

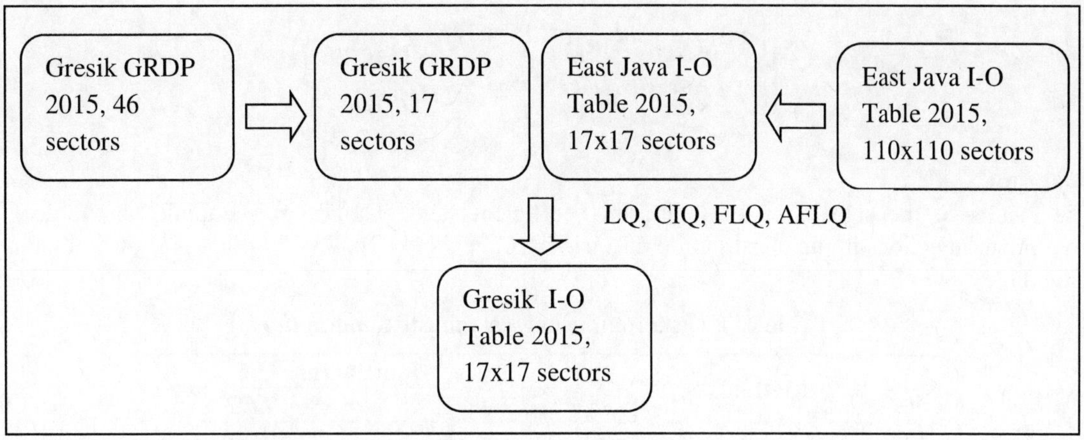

Figure 3. Construction of 2015 Gresik I-O, 17 x 17 sectors

The derivation through the LQ method is to use the value of $LQ_{i,Gresik}$ obtained from formula (2) and further adjustments are made as follows:

$$a_{i,j}^{Gresik} = \begin{cases} a_{i,j}^{East\ Java} \ if\ LQ_{i,Gresik} \geq 1\ (for\ j = 1,2,\dots,n) \\ (LQ_{i,Gresik})a_{i,j}^{East\ Java} \ if\ LQ_{i,Gresik} < 1\ (for\ j = 1,2,\dots,n) \end{cases} \tag{3}$$

The derivation through the CIQ method requires first cross-industry LQ calculations, where:

$$CIQ_{ij}^{Gresik} = {LQ_{i,Gresik}}\Big/{LQ_{j,Gresik}}, i \neq j \tag{4}$$

and adjusting matrix A elements is done in the following way:

$$a_{i,j}^{Gresik} = \begin{cases} a_{i,j}^{East\ Java} \ if\ CIQ_{ij}^{Gresik} \geq 1 \\ (CIQ_{ij}^{Gresik})a_{i,j}^{East\ Java} \ if\ CIQ_{ij}^{Gresik} < 1 \end{cases} \tag{5}$$

The derivation through the FFQ method based on the following criteria:

$$a_{i,j}^{Gresik} = \begin{cases} a_{i,j}^{East\ Java} \ if\ FLQ_{ij}^{Gresik} \geq 1 \\ (FLQ_{ij}^{Gresik})a_{i,j}^{East\ Java} \ if\ FLQ_{ij}^{Gresik} < 1 \end{cases} \tag{6}$$

Where

$$FLQ_{ij}^{Gresik} = (\lambda)CIQ_{ij}^{Gresik};\ \lambda = \{log_2[1 + \left(\frac{worker^{Gresik}}{worker^{East\ Java}}\right)]\}^{\delta} \tag{7}$$

and $0 < \delta < 1$ in empirical studies, commonly used $\delta = 0.3$

Finally, the augmentation in AFLQ is as follows

$$AFLQ_{i,j}^{Gresik} = \begin{cases} [log_2(1 + LQ_{i,Gresik})]FLQ_{ij}^{Gresik} \ if\ LQ_{i,Gresik} > 1 \\ FLQ_{ij}^{Gresik} \ if\ LQ_{i,Gresik} \leq 1 \end{cases} \tag{8}$$

So that,

$$a_{i,j}^{Gresik} = \begin{cases} (AFLQ_{i,j}^{Gresik})a_{i,j}^{East\ Java} & if\ LQ_{i,Gresik} > 1 \\ (FLQ_{ij}^{Gresik})a_{i,j}^{East\ Java} & if\ LQ_{i,Gresik} \leq 1 \end{cases} \quad (9)$$

4. Result

The results of the calculation of multiplier coefficients and changes in economic output due to investment injections in the construction of a truck garage worth Rp. 23.55 billion is shown in table 2 and table 3.

Table 2. Transportation and warehousing, *multipliers*

Methods	Parameter	
	LQ1	*LQ2*
LQ	1.31	1.33
CIQ	1.23	1.34
FLQ	1.11	1.11
AFLQ	1.16	1.12

Table 3. Transportation and warehousing, *impact* (Rp. Billion)

Methods	Parameter	
	LQ1	*LQ2*
LQ	39.28	40.15
CIQ	43.33	45.51
FLQ	33.33	31.98
AFLQ	34.64	32.95

Table 2 shows that the range of the output and multiplier of the transportation and warehousing sector is 1.11 - 1.33. The implication is that the investment injection for the construction of a truck garage worth around Rp. 23.55 billion will produce an additional output of Rp. 31.98 to Rp. 45.51 billion for the economy of Gresik Regency (see table 3).

As a percentage, the additional output is less than 1%, in the range (0.032% to 0.045%). The construction of this truck garage should also increase the regional revenue (PAD) of Gresik. If the change in regional income (PAD) of Gresik changes with the same growth rate as the output growth rate above, then PAD Gresik will increase in the range of Rp. 0.25 billion - Rp. 0.36 billion. Table 4 below lists the PES Gresik (delta) scenarios with the 2015 PAD Gresik baseline realization of around Rp. 799.78 billion.

Table 4: Transportation and warehousing, *Additional Regional Original Income* (Rp. Billion)

Methods	Parameter	
	LQ1	*LQ2*
LQ	0.312	0.319
CIQ	0.344	0.361
FLQ	0.265	0.254
AFLQ	0.275	0.262

Conceptually, the impact of changes in output and PAD will occur in the same year as the project implementation or one (1) year ahead, assuming the project can be completed in a short time. This modeling is based on 2015 data. If the economic structure of Gresik and East Java in 2018 does not change drastically, this modeling can still be used to estimate the impact of changes in output and PAD due to the construction of a garage truck.

5. Conclusion

Impact analysis based on the I-O model is practical and quantitative, because it is able to calculate the effect of a change in final demand on economic output. I-O modeling provides various methods for analyzing in less than perfect data. These methods allow us to construct an I-O table for a region as a derivative of a larger region of I-O tables that shelter it. This procedure is very useful in situations of limited resources, especially time and financial resources.

This concise study applies the above procedure to analyze the impact of the construction of a truck garage on the overall economy of Gresik. Impact analysis through various methods gives its own advantages, namely we will get a range of estimated values, not just one estimated value. As shown in the previous section, the construction of truck garages has a positive and greater multiplier effect on the transportation and warehousing sector. In the end, the construction of a truck garage worth Rp.23.55 billion will increase Gresik's economic output of Rp. 31.98 to Rp. 45.51 billion; and adding PAD of Rp. 0.25 billion - Rp. 0.36 billion.

6. References

[1] Hulu, Edison and Geoffrey J. D. Hewings. 1993. “The Development and Use of Interregional Input-Output Models for Indonesia Under Conditions of Limited Information,” Review of Urban and Regional Development Studies, 5, 135–153.

[2] Miller, Ronald and Peter Blair. 2009. “Input-Output Analysis: Foundations and Extensions”. Cambridge University Press.

[3] Flegg, A. T., C. D. Webber and M. V. Elliott. 1995. “On the Appropriate Use of Location Quotientsin Generating Regional Input-Output Tables,” Regional Studies, 29, 547–561.

[4] Flegg, A. T. and C. D. Webber. 1997. “On the Appropriate Use of Location Quotients in Generating Regional Input-Output Tables: Reply,” Regional Studies, 31, 795–805.

[5] Flegg, A. T. and C. D. Webber. 2000. “Regional Size, Regional Specialization and the FLQ Formula,” Regional Studies, 34,563–569.

IOP Conf. Series: Earth and Environmental Science **340** (2019) 012004 doi:10.1088/1755-1315/340/1/012004

Analytical tools beyond Gini index to study inequality: a case of City of Blitar, East Java, Indonesia

A Irawan[1] and E B Santoso[1]

[1] Urban and Regional Planning Department, Institut Teknologi Sepuluh Nopember Surabaya, Indonesia

Email: hicakandi@gmail.com

Abstract. This paper demonstrates the use of some analytical tools beyond usual Gini index to study inequality in the City of Blitar (henceforth, Blitar), East Java, Indonesia. The Blitar's mid-term development plan has set social and macroeconomics indicators as basis to evaluate how local development is carried out by the city government. Of the government's primary concerns is how the city's economic pie is distributed between its citizens. As in many other places or context, Gini index has long been a to-go indicator to gauge inequality at first glance. Gini is in fact a never obsolete since it has many desirable properties to be a good indicator for inequality. However, it is necessary to go deeper so the policy makers would better understand the inequality decomposition and dynamics. Attempts to understand inequality should also include an assessment whether the observed growth is pro-poor. Better equality and pro-poor growth make quality growth, which is a primary dimension of sustainable development goals. This paper offers rather comprehensive yet practical guidance for policy analysts or planning practitioners to study inequality in a more meaningful way.

Keywords: *inequality, pro-poor growth, Gini, inequality indices.*

1. Introduction

Based on Minister of Home Affairs Regulation Number 54 of 2010 about Implementation of Government Regulation Number 8 of 2008 with reference to Stages, Procedures for Organization, Control and Evaluation of Implementation of Regional Development Plans, in the preparation of regional development planning documents, regional economic analysis is required in order to assess to what extent the regional development realization can influence the economic performance and how far the economic indicators realization are in accordance with what is assumed in medium-term development planning. The analysis is also used as one of the main inputs in making regional financial analysis. Furthermore, the results of the analysis forms the basis for determining the formulation of priorities and development goals, including being one of the references in formulating development issues.

In order to determine government policies which are on target as described above, the preparation of the inequality analysis needs to be done as a reference study in the preparation of a quality development planning. In this regard, the inequality analysis was carried out to help the Regional Development Planning Board of Blitar City preparing the Gini Analysis of the City of Blitar in 2018. The preparation of the Blitar Gini index analysis in 2018 was intended to provide the distribution of welfare indicators analysis, mainly with approaches, but not limited to, Gini index, so that it can be one of the basic considerations for policy makers in carrying out regional development planning, especially regarding equitable development policies.

The scope of the analysis includes: 1). Calculation of the Gini index and entropy inequality index; 2). Gini index projection for the next 5 (five) years; and 3). Policy recommendations to reduce the value of the Blitar Gini index. The purpose of the preparation of the analysis are: 1). Compile analysis of the Blitar Gini index; 2). Based on the results of the analysis, identify problems surrounding inequality; and 3). Develop policy recommendations to encourage increased equity.

2. Basic concept of inequality measures

According to Atkinson [1,2], inequality is a broader concept than the concept of poverty because its measurement is based on the entire population, not only for the part (subset) below a certain poverty line. Most indicators of inequality do not depend on the average distribution of welfare indicators which is used (e.g., per capita expenditure). The mean feature of independence is a desirable feature of the size of inequality. In addition to per capita expenditure, measurement of inequality can also use other indicators, for example: income, land, assets, tax payments, and many other economic demographic variables.

The simplest measure of inequality is to rank the population based on selected indicators (eg expense), from the poorest to the richest; and then the percentage of expense (or income) that corresponds to each group (quintile, quintile) or tithe (decile, decile) of the population. The poorest quintile usually accounts for 6-10 percent of all expenses, the top quintile is 35-50 percent. The World Bank (World Bank) standard in measuring inequality is by seeing the contribution of the poorest 40% in the distribution of expenditure as shown in Table 1 [3].

Table 1. Inequality level standard from the World Bank

Income Distribution	Inequality Level
The 40% poorest group have < 12% expense of all expenses	High
The 40% poorest group have 12%–17% expense of all expenses	Average
The 40% poorest group have > 17% expense of all expenses	Low

The popular measure of inequality is the Gini coefficient or Gini index, whose values range from 0 (perfect equality) to 1 (perfect inequality), but usually in the range of 0.3 to 0.5 for per capita expenditure indicators. The Gini coefficient comes from the Lorenz curve, which groups the population from the poorest to the richest, and shows the cumulative proportion of the population on the horizontal axis and the proportion of cumulative expense (or income) on the vertical axis. The Gini coefficient meets many indicators of good inequality indicators, including:

1. Mean independence. If the expense (income) of all members of the population is doubled, the Gini coefficient value will not change.
2. Population size independence. If the population size changes, but other variables remain (ceteris paribus), the Gini index value does not change.
3. Symmetry. If two people in the population exchange expenses (income), the Gini index value does not change.
4. Pigou-Dalton Transfer sensitivity. Based on these criteria, if there is a transfer from the rich group to the poor, the Gini coefficient value will decrease, meaning that there will be a decrease in inequality.

Table 2 below provides a benchmark level of inequality based on the Gini coefficient value.

IOP Conf. Series: Earth and Environmental Science **340** (2019) 012004 doi:10.1088/1755-1315/340/1/012004

Table 2. The inequality level based on Gini coefficient score

Gini coefficient score	Inequality Level
Gini < 0,4	Low
0,4 < Gini < 0,5	Average
Gini > 0,5	High

In addition, there are also various other measures of inequality. The most common measure of entropy for assessing inequality is Theil's T and Theil's L, both of which allow someone to decompose inequalities that occur within regions (for example, urban and rural) and inequality caused by differences between regions (for example, village income inequality cities), as well as sources of changes in inequality over time. Usually, at least a third of inequality in a country caused by inequality in the group, and the rest is differences between groups. The size of Theil's inequality has a feature that the Gini coefficient does not have, named decomposability. The size of class inequality by Atkinson is also used sometimes. The decile dispersion ratio is also a measure of inequality which is easy and popular, but is a rough measure. The decile dispersion ratio is defined as the expense (or income) of the richest deciles divided by the poorest deciles, is a popular and easy method but is a very rough measure. For visualization of the inequality dynamics, Pen's Parade Charts can be useful in showing how distribution of expense (income) changes over time.

3. Methodology

The most widely used measure of inequality is the Gini coefficient. The Gini coefficient is derived from the Lorenz curve (see Figure 1), named the cumulative frequency curve that compares the distribution of certain variables (for example, expense) with a uniform distribution that represents equality. To construct the Gini coefficient, graph the cumulative percentage of households (from poor to rich) on the horizontal axis and the cumulative percentage of expense (or income) on the vertical axis. Diagonal lines represent perfect equality. The Gini coefficient is defined as the ratio of area A / (area A + area B), where A and B are the areas shown in the image. If A = 0, the Gini coefficient becomes 0, which means absolute equality, whereas if B = 0, the Gini coefficient becomes 1, which means absolute inequality [4]. In this example, the Gini coefficient is around 0.35. Some users, including the World Bank, multiply this number by 100, in this case it will be reported as 35.

In mathematical formulas [5], Gini coefficients can be calculated using the following formula:

$$Gini = 1 - \sum_{i=1}^{N} (x_i - x_{i-1})(y_i + y_{i-1}). \tag{1}$$

Where x_i is a value in the x axis, and y is a value in y axis.

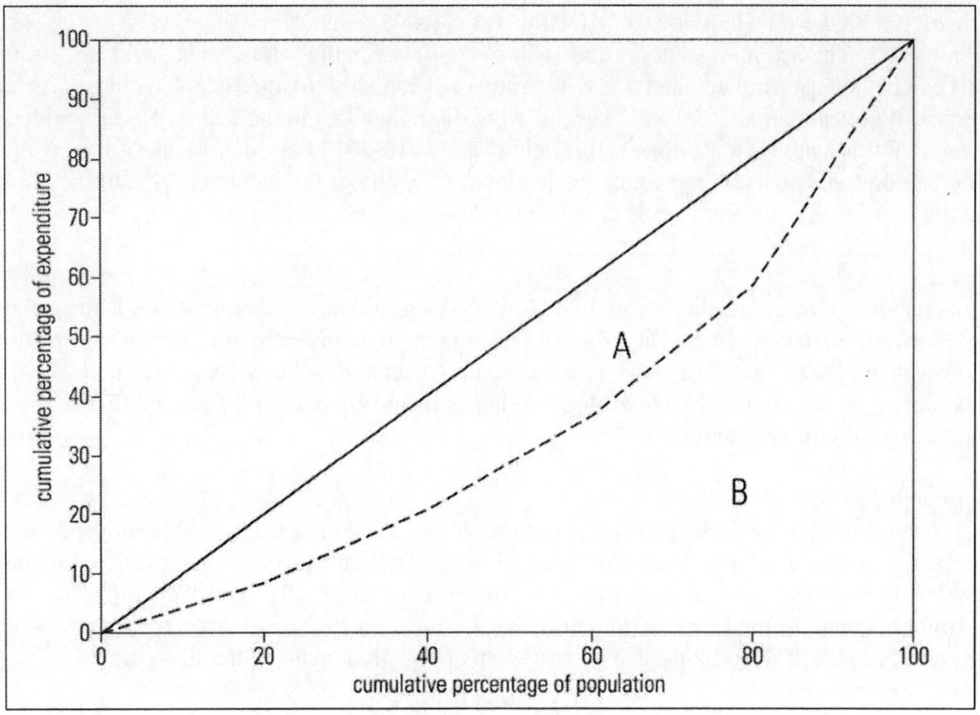

Figure 1. Lorenz Curve and Gini coefficient

These following is a mathematical formula to calculate the entropy inequality index.
Formula (2): General Entropy

$$GE(\alpha) = \frac{1}{\alpha^2 - \alpha}\left\{\frac{1}{N}\sum_{i=1}^{N}\left(\frac{y_i}{\bar{y}}\right)^{\alpha} - 1\right\} \quad (2)$$

N= population size;
y_i= individual expense (income);
α= weighting factor, given to the distance between expenses (income) between various different positions in the distribution

Formula (3): GE(1) or Theil's T index

$$GE(1) = \frac{1}{N}\sum_{i=1}^{N}\frac{y_i}{\bar{y}}\ln\left(\frac{y_i}{\bar{y}}\right) \quad (3)$$

Formula (4): GE(1) or Theil's L index or Mean-log deviation

$$GE(0) = \frac{1}{N}\sum_{i=1}^{N}\ln\left(\frac{\bar{y}}{y_i}\right) \quad (4)$$

The main data for the preparation of the Gini analysis is the micro data of the National Socio-Economic Survey (Susenas). Without this Susenas data, this Gini analysis work cannot be done. Susenas data is usually stored in SPSS sav and dBASE dbf formats. Susenas data consists of three groups of files, namely: KORIND (individual), KORRT (household), and three blocks: 4.1, 4.2, and most importantly 4.3. The scope of data needed is Susenas data for the whole East Java. So, comparative analysis can be done, to find out the relative position of Blitar City to other selected cities /districts, both the whole city/regency in East Java or especially the districts/cities in the former Kediri residency. The data needs for this analysis to be done is throughout the 2010-2017 period or longer if it is available. So that you can do trend analysis and future projections. Analysis of the inequality dynamics can be done by making two graphs: namely Parade Graphs and Pro-poor Growth Charts [6].

4. Results

This chapter shows the calculation result and Gini analysis, and others inequality indicators. Especially the time series Gini data of Blitar City, the data used is the data given by the statistics bureau of Blitar City. Other than that, all calculation and output used the Susenas data based on the method which already been explained in the section 3 methodology. What is meant by per capita expenditure is the average per capita expenditure per month.

4.1. Gini and Entropy

Table 3 shows the Gini coefficient and GE (α) entropy index of Blitar City throughout the 2008 to 2017 period. Based on the categorization of inequality levels based on the magnitude of the Gini coefficient (see Table 2 in Section 2), the inequality level in Blitar City in 2008-2017 is still classified as low. There was a slight decrease in the Gini coefficient in 2017 compared to 2016. However, if we look at it in a longer period (2012-2017 or 2008-2017), there seems to be an increasing trend.

Table 3. Gini and GE(α) Blitar City, 2008-2017

Year	Gini	GE(0)	GE(1)
2008	0,32	-	-
2009	0,32	-	-
2010	0,33	0,13	0,14
2011	0,34	0,19	0,21
2012	0,38	0,33	0,54
2013	0,40	-	-
2014	0,35	-	-
2015	0,37	0,27	0,30
2016	0,41	0,29	0,34
2017	0,39	0,26	0,28

Note: "-", the calculation can not be done
Because of the data limitations.

As explained in Section 2, the GE (α) entropy index is additively decomposable, while the Gini coefficient is not. Because it has additively decomposable features, the decomposition analysis of the entropy index can be used to understand deeper the character and process of visible inequality. The trend of increasing inequality seems to occur because there is an increase in inequality between expenditure groups (40% down, 40% medium, and 20% above), as indicated in Table 4.

Table 4 shows the total magnitude and decomposition of the GE entropy index (α), which consists of within groups inequality and between groups inequality. The ratio of inequality between groups to total inequality is also shown (numbers in dark blue). In the 2012-2017 period, the ratio of inequality between groups to total inequality increased. Contribution of inequality between expense groups to overall inequality is more dominant than the contribution of inequality in groups.

IOP Conf. Series: Earth and Environmental Science **340** (2019) 012004 doi:10.1088/1755-1315/340/1/012004

Table 4. Entropy Index Decomposition, GE(α)

GE(α): total, within, between, share	**2012**	**2017**
GE(0) total	0,33	0,26
GE(0) within group	0,07	0,04
GE(0) betwen group, bw	0,26	0,21
Share GE(0) bw to total	**0,78**	**0,83**
GE(1) total	0,54	0,28
GE(1) within group	0,26	0,06
GE(1) betwen group, bw	0,28	0,22
Share GE(1) bw to total	**0,52**	**0,78**

Figure 2 shows the trend of inequality of Blitar City in the period 2010-2017 based on the Gini index and GE entropy (α) visually. In a short period, for example from a year compared to one year before, the Gini index looks smoother (smooth) compared to the two types of GE entropy index (α). Nonetheless, the three seem to move in tandem, where: i) there is a trend of increasing inequality; ii) the level of inequality at the end of the period is greater than at the beginning of the period.

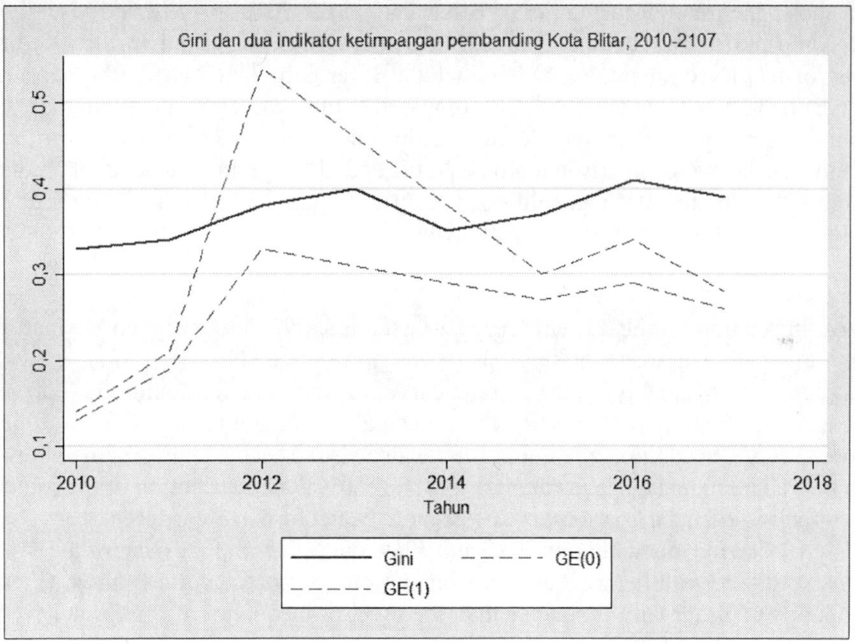

Figure 2. Blitar City inequality trend, 2010-2017

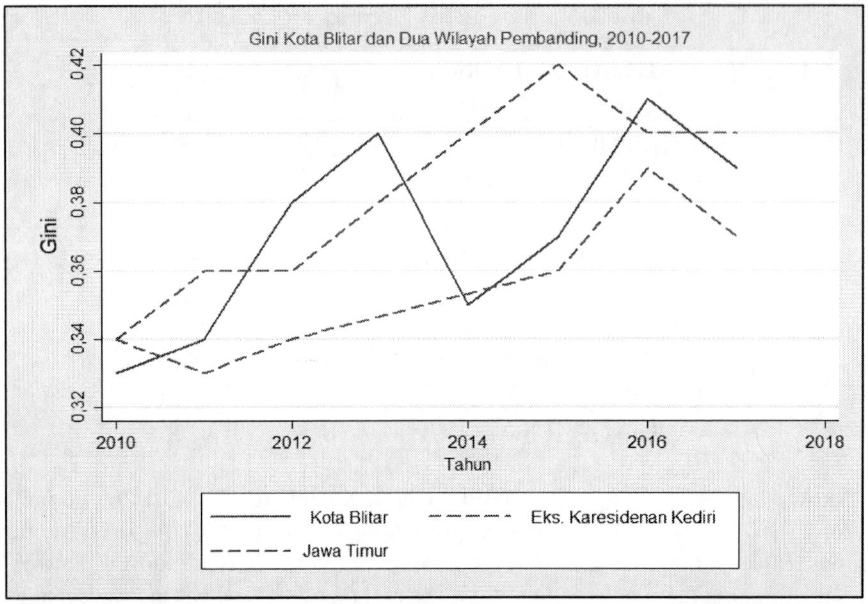

Figure 3. Blitar City inequality trend vs. the comparison regions, 2010-2017

Figure 3 shows the trend of inequality of Blitar City in the 2010-2017 period visually based on the Gini index compared to two comparison regions, named East Java and former Kediri Residency. Calculations for two fixed comparison areas include Blitar City data. During the period, the trend of inequality in Blitar City seemed more volatile compared to the trend of inequality in the two comparison regions. At the beginning of the period, the level of inequality in Blitar City was lower than the level of inequality between the two comparison regions. At the end of the period, the level of inequality in Blitar City was higher than the level of inequality of the former region. The Kediri Residency, although still lower than the level of East Java inequality as a whole.

4.2. Lorenz Curve

As explained in Section 2, the Lorenz curve is a cumulative frequency curve that compares the distribution of certain variables, in this case per capita expenditure per month, with a uniform distribution representing equality. In the Lorenz curve, the uniform distribution is represented by a 45 degree diagonal straight line. Mathematically, the actual magnitude of the Gini coefficient is derived from the Lorenz curve. In addition, the Lorenz curve also visualizes inequality, named how far the actual distribution of welfare indicators is used relative to the equivalent distribution of the indicator.

In this section is shown the Lorenz curve of Blitar City and former Kediri Residency as a comparison area for 2010-2017 period, not including 2013 and 2014, due to incomplete data for the district/city level in the two years. In line with Figure 3 above, from a group of Lorenz curves shown (Figure 4 a-f), the level of inequality of Blitar City is smaller than the level of inequality of the former Kediri residency only in 2010 (Figure 4 f)

The gray area in the Lorenz curve is an area of 95% confidence range. That is, if a measurement is made 100 times, 95 times that of 100 measurements, the actual Gini coefficient value will be located within that confidence level. Generally, the greater the variation in data distribution, the wider the range of trust or the greater the gray area shown in the figure.

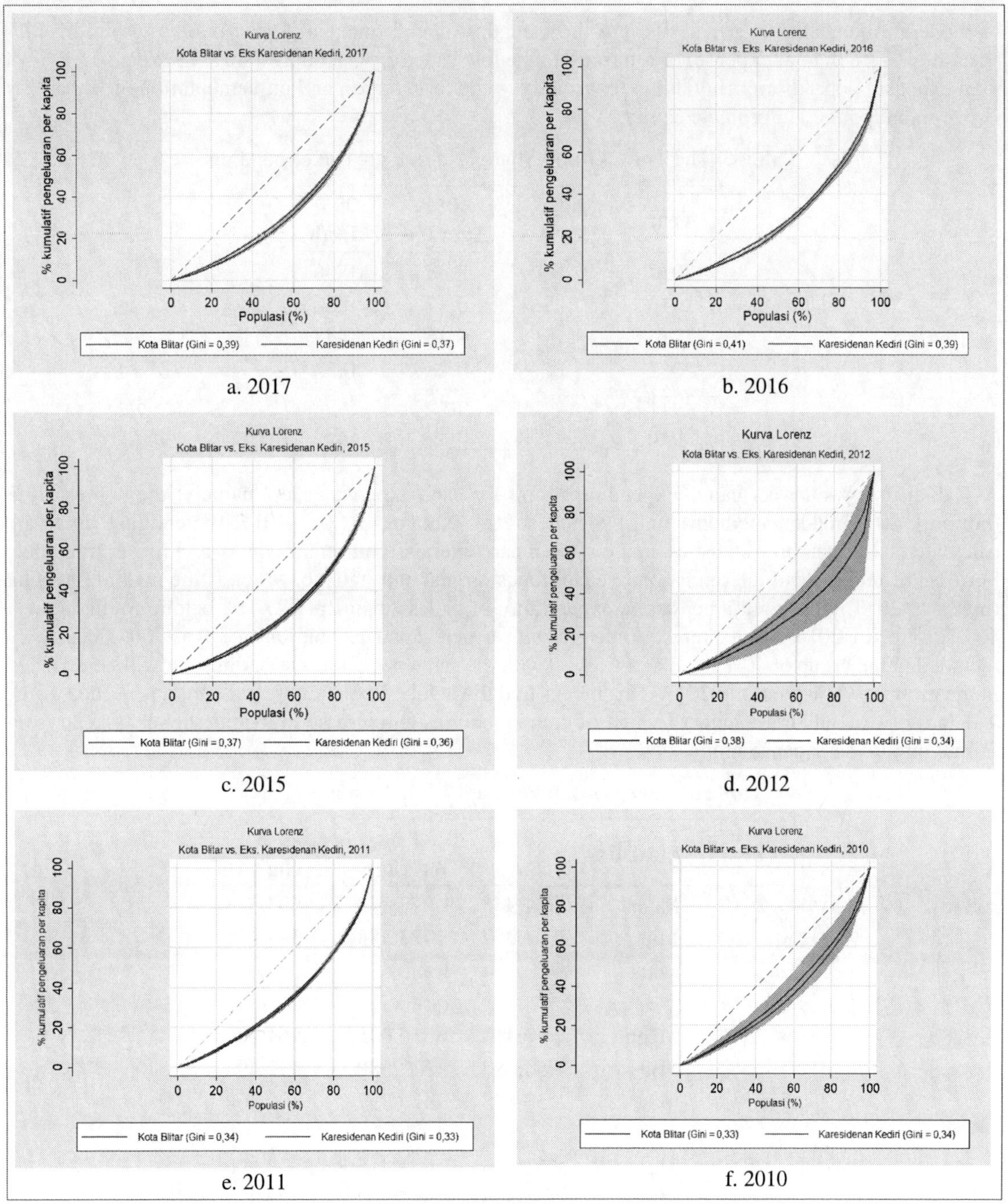

Figure 4. The Lorenz Curve of Blitar City vs. Former Kediri Residency

4.3. The inequality standard of World Bank

In line with the trend of inequality in Table 3 and Figure 2, the trend of inequality in Blitar City according to World Bank standards also shows movement from the low category at the beginning of the period to the medium category (see Table 1 in Section 2) at the end of the period. The group expenditure ratio of 40% down to total expenditure in 2017 is 0.16 (16%). This ratio is close to the upper limit of 17% for the medium category, given that the World Bank categorizes moderate inequality if the poorest group of 40% of expenditure is 12% -17% of total expenditure. The transition to the lower limit of the category

IOP Conf. Series: Earth and Environmental Science **340** (2019) 012004 doi:10.1088/1755-1315/340/1/012004

of moderate inequality or even falls into the category of high inequality (the poorest group of 40% spending <12% of total expense) or returning to the low category (the poorest 40% of expense >17% of total expense) depends on many things, especially in the formulation and implementation of Blitar City Government policies to promote equity.

Table 5. The Three Group Share Expense on Total Spending

Year	40% Low	40% Average	20% High
2017	0,16	0,38	0,46
2016	0,15	0,37	0,48
2015	0,15	0,36	0,48
2012	0,14	0,28	0,58
2011	0,20	0,36	0,44
2010	0,22	0,39	0,39

Table 6 shows the per capita expenditure limit for each expense group, namely: the poorest 40% (bottom), 40% middle (medium), and the richest 20% (above), in 2015-2017. Expenditure limits are calculated based on three statistics, namely the mean (average), minimum value, and maximum value. For ease of reading and interpretation, we can focus on just one statistic, which is the maximum value (blue). In 2017, those with a per capita expenditure of or less than Rp. 723,164 belong to the poorest 40% group. In 2016, for example, someone with a per capita expenditure of Rp. 700,000 already included in the group of 40% middle; however, with the same per capita expenditure, it will be included in the group of 40% poorer in 2017. This means that the level of expenditure per capita is needed or the standard of expenditure is higher (which of course requires an increase in income level) to keep from falling into lower expense group.

Table 6. Three Group Per Capita Expenditures (IDR)

Year	Statistics	40% Low	40% Average	20% High
2017	Mean	484.945	1.072.143	2.693.513
	Min	197.057	723.485	1.588.667
	Max	723.164	1.583.622	12.000.000
2016	Mean	440.418	1.016.045	2.764.930
	Min	253.828	633.580	1.431.498
	Max	630.688	1.429.957	12.600.000
2015	Mean	461.925	1.021.015	2.639.310
	Min	231.666	697.702	1.527.377
	Max	694.127	1.514.036	20.200.000

4.4. Inequality dynamics: pro-poor growth?

The trickle down effect thesis argues that growth will be distributed downward. But this thesis does not explain how the growth that occurs will trickle down. In other words, this thesis does not offer a complete explanation of how the drop down works or how existing growth will be distributed among various groups of people. The group here is percentile or quintile of the welfare indicator, for example per capita expenditure as used in preparing this Blitar City Gini analysis analysis report.

Quality growth has many dimensions. These dimensions include improvements in the quality of Human Resources (HR) and employment, environmental quality, and reduced inequality / improvement in equity, especially for the poorest groups. This report focuses on the dimensions of equity. One of the available analysis tools to check whether the growth that occurs is also enjoyed proportionally by the poorest groups is a diagram called the growth incidence curve or growth curve for the income group.

The Blitar City growth incidence curve between 2012 and 2017 is stated in Figure 6. Growth is calculated based on the real value of per capita expenditure in 2012 and 2017. This means that inflation in per capita expenditure in 2017 has been neutralized based on the 2012 base year, where the 2012 consumer price index (CPI) is expressed as 100, or commonly stated as follows , 2012 = 100.

From the data, the absolute negative growth actually only occurs in the first percentile (the poorest 1%) and at the 99th percentile (the richest 1%). The other percentiles still experience positive growth (> 0%). However, if the standard values are used (mean; 50.5%) and the median value (median; 51%) growth of all percentiles, around 50% of the population experiences growth below that standard. The group that experienced growth far above the standard was the group between the 60-80 percentile and between the 90-100 percentile. In essence, the growth performance of the poorest 50% group is still below the 50% growth performance of the richest community.

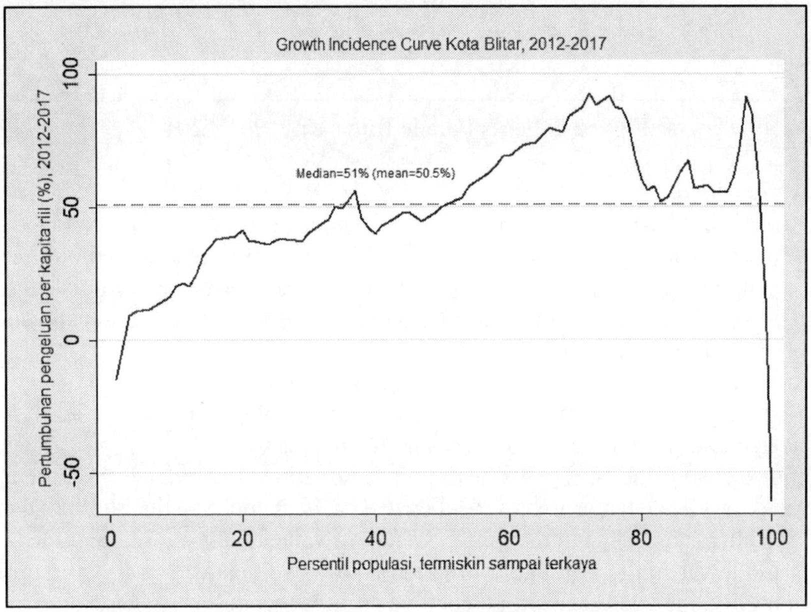

Figure 5. Growth incidence curve of Blitar City, 2012-2017

The pattern of growth distribution as depicted in the growth incidence curve is in line with Figure 6 below or commonly referred to as the Parade diagram. The Parade as in Figure 4.5 also uses the real value for the 2012 base year. The level of expenditure per capita in all 2017 percentiles (in blue) is compared with the level of expenditure per capita in all of the 2012 percentiles (red). As expected and because there is indeed a positive absolute growth between the level of real per capita expenditure between 2012-2017 as illustrated in Figure 4.4 before, the level of expenditure per capita 2017 is always above the 2012 per capita expenditure level.

However, the gap between the two is increasing for the expensegroups above the poorest 20% group. Gap enlargement is increasingly progressive for the upper middle class (above the poorest 40% group). The pattern of gap enlargement such as open scissor indicates a divergence process. In simple terms, divergence occurs when a group of wealthier individuals experiences a higher level of expense growth compared to the growth rate of the poor. The process of divergence will maintain or even increase inequality.

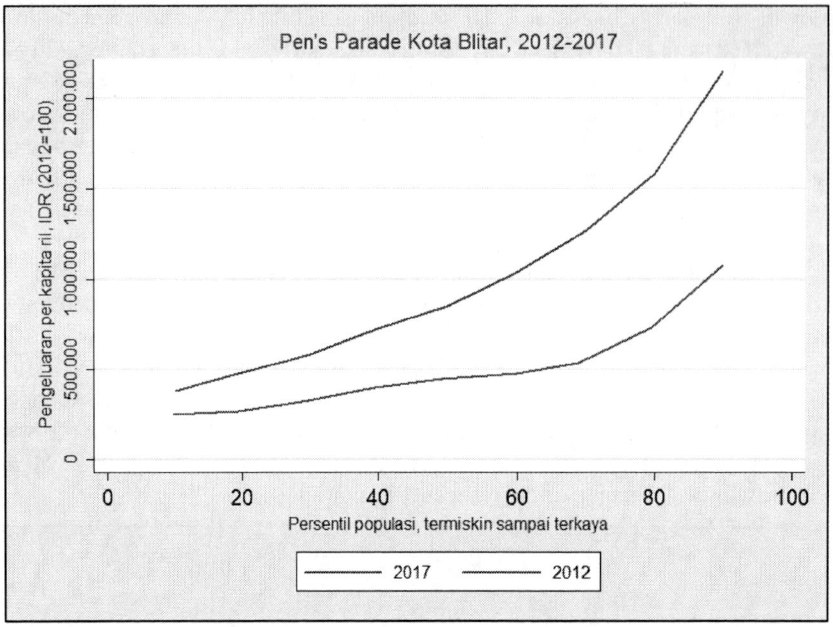

Figure 6. Pen's Parade Blitar City, 2012-2017

4.5. Inequality projection

Because the existing time series data range is not too long, the projections that can be done are only up to the next five years, namely 2018-2022 (see Table 7). The least squares method is used to produce estimates for all three models, namely: 1) linear model; 2) non-linear models with 10 log-based transformations, Log; 3) non-linear models with natural number-based log transformations, Ln. The three models produce almost the same estimates every year.

The projection results indicate the movement of the level of inequality moves from the low category to the medium category. This indication is in line with the trend of shifting categories of inequality based on World Bank standards as outlined in sub-section 4.3 above.

The estimation shown is the average value or point estimation. The range estimation upper and lower limits with a 95% confidence range are not displayed. That is, it could be that the actual Gini coefficient in the future is smaller (greater) than the point estimation value displayed in the table. More important than the upper and lower limits of range estimation is how we behave towards the projection results. This projection needs to be addressed with a positive attitude and interpreted as a sign of early warning. The projection results in Table 7 should be used as a motivation to formulate and implement policies to encourage equity to be more effective, so that the level of inequality in Blitar City does not shift into the medium category.

Table 7. Gini Coefficient Projection, 2018-2021

Year	Model 1: linier	Model 2: non-linier (Log 10)	Model 3: non-linier (Ln)
2018	0,41	0,41	0,41
2019	0,42	0,42	0,42
2020	0,43	0,44	0,43
2021	0,44	0,45	0,44
2022	0,45	0,46	0,45

5. Conclusion

Some conclusion which can be addressed are :

1. Various inequality indicators whether it is Gini or entropy index show that there is an increase in inequality trend.
2. Contribution of inequality between expense groups to overall inequality is more dominant than the contribution of inequality within groups.
3. Throughout the study period, the level of inequality in Blitar City is more varied compared to the level of inequality in East Java as a whole or the former region of Kediri Residency.
4. The inequality level in Blitar City facing an shifting risk from the low category to the average category.
5. Middle upper class groups experience higher rates of growth per capita expenditure compared to the poor.

These are the policy recommendations which need to be considered:

1. Carry out fiscal expansion, such as the Family Hope Program (PKH) at the national level, to increase the purchasing power of the poorest groups of society. Increasing purchasing power of the poor in turn will also increase the economic growth of Blitar City, because consumption is still the main growth engine of Blitar City.
2. In order for the program and activities in the first point to be more targeted, it is necessary to improve the database of the poorest groups of people. This database should not only feature by name, by address, but also by Figure. Improving and verifying this database must be an annual program.

6. References

[1] Atkinson, A. B. 1970. "On the Measurement of Inequality." *Journal of Economic Theory* 2 (3):244–63.

[2] ———. 1983. *The Economics of Inequality*, 2nd edition. Oxford: Clarendon Press.

[3] Bourguignon, François, Francisco Ferreira, and Nora Lustig, eds. 2005. *The Microeconomics of Income Distribution Dynamics in East Asia and Latin America*. Washington, DC: World Bank and Oxford University Press.

[4] Gottschalk, P., and T. Smeeding. 2000. "Empirical Evidence on Income Inequality in Industrial Countries." In *Handbook of Income Distribution. Volume 1. Handbooks in Economics, vol. 16,* ed. A. B. Atkinson and F. Bourguignon. Amsterdam, New York, and Oxford: Elsevier Science/North-Holland.

[5] Duclos, Jean-Yves, and Abdelkrim Araar. 2006. *Poverty and Equity: Measurement, Policy and Estimation with DAD*. New York: Springer, and Ottawa: International Development Research Centre.

[6] Ravallion, Martin, and Shaohua Chen. 2003. "Measuring Pro-Poor Growth." *Economics Letters* 78 (1): 93–9.

IOP Conf. Series: Earth and Environmental Science **340** (2019) 012005 doi:10.1088/1755-1315/340/1/012005

Comparative analysis on access and egress distances for Semi BRT Trans '*Mamminasata*' and Commuter Train '*SUSI*'

A Susanti[1], R A A Soemitro[2] and H Suprayitno[2]

[1] Civil Engineering Departement
[2] Institut Teknologi Sepuluh Nopember Surabaya, 60111, Indonesia

Email: anitasusanti.pasmar@gmail.com

Abstract. The existence of the Semi Bus Rapid Transit (BRT) Network and the Commuter Train Network, as part of Urban Mass Transportation is developed with the aim of providing better service on commuter trips. To estimate passenger demands on new mass transit services, especially in areas without prior service, Access and Egress distances data are needed to design an appropriate mass transit. This underlies the reason why a comparative analysis of Access and Egress distance of BRT Trans '*Mamminasata*' in Makassar and Commuter Train Surabaya – Sidoarjo was necessarily done. Which of the two had further Access or Egress distance; was it the BRT Trans '*Mamminasata*' Makassar or the Commuter train Surabaya-Sidoarjo? Data were obtained through interviews and data management from related department. The results found that both Access and Egress of the BRT Trans '*Mamminasata*' in Makassar was shorter than the Commuter Train Surabaya-Sidoarjo.

Keywords: *Comparison, Access, Egress, BRT Trans Mamminasata, Commuter Train*

1. Introduction

Increasing numbers of motorcyclists have caused traffic congestion in several major cities in Asia, such as Taiwan, Indonesia, Malaysia and Vietnam [1]. This problem, of course, requires the city mass transport services, such as BRT (Bus Rapid Transit) and commuter (urban train). In several major cities in Asia, such as in Bangkok Thailand, BRT services have been developed and equipped with park and ride facilities [2]. However, in Makassar Indonesia, the BRT Trans '*Mamminasata*' services are still in the developmental stages [3]. While in Surabaya Indonesia, commuter service (Surabaya-Sidoarjo) is also in the developmental stage similar to the BRT Trans '*Mamminasata*' in Makasar. As part of Urban Railway in Surabaya, this mode is far from being evenly extended throughout the region [4].

Developing mass transit services, such as BRT Trans '*Mamminasata*' in Makassar and commuter train Surabaya-Sidoarjo in underserviced areas need to be executed in order to minimize the possible issues, such as high-cost construction, human resources, as well as city layout.

Underserviced area is an area that has not been served by any mass transportation and located in the skirt of influenced area. While, the influenced area is an area where travellers (passengers) are comfortably to head and leave the stations/shelter on foot, by motorbikes, cars, or other public transportation [5]. The influenced area is measured from the distance between the designated location and stations/shelter or the contrary. The distance from stations/shelters to designated area is called as the Access distance, while the distance from designated area to stations/shelter is called the Egress distance (i.e [2][6][7]).

On the BRT Trans '*Mamminasata*' in Makassar, Access and Egress data were only limited to Access distance from residential areas to BRT stops (shelter) approximately 400 meters and the Egress distance from residential areas to BRT stops (shelter) approximately 150 meters [2]. Several factors affecting Access distance in the BRT Trans '*Mamminasata*' were parking capacity, parking facilities, and shelter location [8]. The one of the importance of Access distance data is to actualize the City Park Accessibility Concept [9] and to estimate the Catchment Area from mass-transit stops [5].

Following the prior studies, another study on mass transportation in Aceh found that the Trans Koetaradja has longer Access distance (0 - 1.5 km) than its Egress distance (0 - 1 km) [7]. This results were supported the Access and Egress of BRT Trans '*Mamminasata*' in Makassar where the Access distance was provably longer than the Egress distance. The fact in which the Access distance was longer than the Egress distance became an interesting issue that can be developed and examined further. This study was to compare the Access and Egress distances of two different mass-transit modes, BRT (Bus Rapid Transport) Trans '*Mamminasata*' in Makassar) and Commuter Train Surabaya-Sidoarjo in Surabaya. As for the further purpose, it is to provide data for both academic purposes in transportation and developmental purposes to help government or related parties to develop mass-transport and urban layout design.

2. Methods

Data were taken from prior studies of the BRT Trans '*Mamminasata*' in Makassar (i.e. [3] [7]) and from interviews of 50 passengers of the commuter train Surabaya-Sidoarjo on March 29, 2017. These interviews obtained the passengers experiences from-to shelters/stations (commuter stops). Data obtained were tested using Chi Square statistical test to determine whether there was difference among the observed frequency [10] and Curve Pattern Acceptance (CPA) Test to determine the acceptance of curve patterns. Both were used to compare standard deviation values of the data analyzed (i.e. [11] [12]). The purposes of both tests were to figure out whether or not there was differences on both Access and Egress on the BRT Trans '*Mamminasata*' in Makassar and on the Commuter Train Surabaya-Sidoarjo; to find the length of the difference; to find out the factors that may influence in this condition; and to synthesize the result into percentage and cumulative point to record the Access and Egress of two modes.

3. Result and discussion

3.1. Access distance and egress distance on BRT Trans 'Mamminasata' Makassar at morning departure hours

Figure 1 shows the trip network model heading to the original station was called the H trip to the Main Mode (PHK), while the trip leaving the destinated station called the Connecting Trip from Main Mode (PHD) (i.e [6] [13]). The Access distance was longer than the Egress distance. The Access distance in corridor 1 and corridor 2 was about 1-2 km with total percentage of 48% (see Table 1). The Egress distance was shorter than the Access distance. The Egress distance in corridor 1 and corridor 2 was about 0.0-0.5 km with total percentage of 71% (see Table 2) [3].

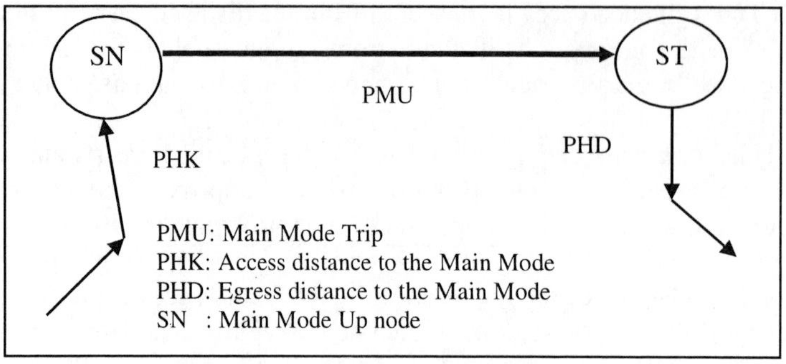

Figure 1. A Travel Network Model
Source:[6,13]

Table 1. Access distance of BRT Trans '*Mamminasata*' Makassar

No.	Distance	Corridor 1	Corridor 2	Total	Percentage	Cumulative (%)
1.	0.0 – 1.0 Km	12	17	29	36%	36%
2.	1.0 – 2.0 Km	20	18	38	48%	84%
3.	2.0 – 3.0 Km	8	5	13	16%	100%
	Total	**40**	**40**	**80**	**1**	

Source:[3]

Table 2. Egress distance of BRT Trans '*Mamminasata*' Makassar

No.	Distance	Corridor 1	Corridor 2	Total	Percentage	Cumulative (%)
1.	0.0 – 0.5 Km	29	28	57	71%	71%
2.	0.5 – 1.0 Km	11	6	17	21%	92%
3.	1.0 – 1.5 Km	0	6	6	8%	100%
	Total	**40**	**40**	**80**	**100%**	

Source: [3]

3.2. Access and egress distance on Commuter Train Surabaya-Sidoarjo

The interview conducted on 50 passengers of Commuter Train Surabaya-Sidoarjo found that the Access distance was about 4.0-50 km with a cumulative percentage of 78% longer than the Egress distance which was about 3.0-4.0 km with a cumulative percentage of 86% [4] (see Table 3, Table 4 and Figure 2).

Table 3. Access distance of Commuter Train Surabaya - Sidoarjo

No.	Distance (km)	Commuter Train Surabaya-Sidoarjo (org)	Percentage %	Cumulative %
1	0.0 - 1.0	6	12%	12%
2	1.0 - 2.0	5	10%	22%
3	2.0 - 3.0	9	18%	40%
4	3.0 - 4.0	8	16%	56%
5	4.0 - 5.0	11	22%	78%
6	5.0 - 6.0	0	0%	78%
7	6.0 - 7.0	2	4%	82%
8	7.0 - 8.0	5	10%	92%
9	8.0 – 9.0	4	8%	100%
Total		**50**	**100%**	

Source: [4]

Table 4. Egress Distance of Commuter Train Surabaya-Sidoarjo

No.	Distance (Km)	Commuter Train Surabaya-Sidoarjo (org)	Percentage %	Cumulative %
1	0.0 - 1.0	20	40%	40%
2	1.0 - 2.0	8	16%	56%
3	2.0 - 3.0	6	12%	68%
4	3.0 - 4.0	9	18%	86%
5	4.0 - 5.0	3	6%	92%
6	5.0 - 6.0	0	0%	92%
7	6.0 - 7.0	2	4%	96%
8	7.0 - 8.0	2	4%	100%
Total		**50**	**100%**	

Source: [4]

3.3. Comparison between access and egress distance of BRT Trans 'Mamminasata' Makassar and Commuter Train Surabaya-Sidoarjo

Study to compare the Access and Egress distances between BRT Trans '*Mamminasata*' in Makassar and Commuter Train Surabaya-Sidoarjo was carried out in order to find out the longer distance for both types of mass transportation (see Figure 2).

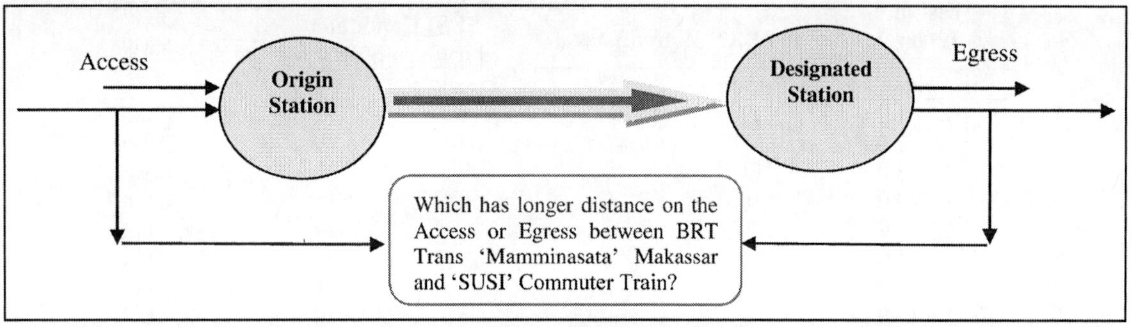

Figure 2. The Access and Egress of BRT Trans '*Mamminasata*' Makassar and Commuter Train Surabaya-Sidoarjo

The results showed that there was different point of Access and Egress of the BRT Trans '*Mamminasata*' Makassar and the Commuter train Surabaya-Sidoarjo. The Access distance of BRT Trans '*Mamminasata*' was 0-2 km (cumulative 84%) and the Egress distance of BRT Trans '*Mamminasata*' was 0-0.5 km (cumulative 75%). While the Access distance of Commuter Train Surabaya-Sidoarjo was 4.0-5.0 km (cumulative 78%) and the Egress distance of Commuter Train Surabaya-Sidoarjo was 3.0-4.0 km (communicative 86%). The difference of Access and Egress distances were analyzed using Chi Square statistical test. The difference was synthesized if H0: $\chi2 \leq \chi2$ table and the difference was not synthesized if H1: $\chi2 \geq \chi2$ table (i.e. [11] [12]).

Table 5 shows the comparison of the Access distance. The results of the Access distance of BRT Trans '*Mamminasata*' and Commuter Train Surabaya-Sidoarjo were H1: $\chi2$ (46.016) $\geq \chi2$table (14.1), meaning that the Access distance of both mass transportation could not be synthesized (see Table 6). Table 7 shows the comparison of the Egress distance. The results of the Egress distance of BRT Trans '*Mamminasata*' and Commuter Train Surabaya-Sidoarjo were the H1: $\chi2$ (43.675) $\geq \chi2$ table (11.1) meaning that the Egress distance could not be synthesized. Table 8 shows the calculation of the Egress distance on BRT Trans '*Mamminasata*' and Commuter Train Surabaya-Sidoarjo.

Table 5. Comparison access distances between BRT Trans '*Mamminasata*' and Commuter Train Surabaya-Sidoarjo

No.	Distances (Km)	BRT Trans Mamminasata (org)	Commuter SUSI Surabaya-Sidoarjo (org)
1	0.0 - 1.0	29	6
2	1.0 - 2.0	38	5
3	2.0 - 3.0	13	9
4	3.0 - 4.0	0	8
5	4.0 - 5.0	0	11
6	5.0 - 6.0	0	0
7	6.0 - 7.0	0	2
8	7.0 - 8.0	0	5
9	8.0 – 9.0	0	4
	Total	**80**	**50**

Source: [3,4]

Table 6. Calculation of access distance of BRT Trans '*Mamminasata*' and Commuter Train Surabaya-Sidoarjo

No.	Distance Interval	% Distribution				Calculation χ2
		Ref BRT Mamminasata		SUSI Commuter Surabaya-Sidoarjo		χ^2
	Km	Person	%	Person	%	
1	0.0 - 1.0	29	36.25	6	12	16.222
2	1.0 - 2.0	38	47.5	5	10	29.605
3	2.0 - 3.0	13	16.25	9	18	0.188
4	3.0 - 4.0	0	0	8	16	0.000
5	4.0 - 5.0	0	0	11	22	0.000
6	5.0 - 6.0	0	0	0	0	0.000
7	6.0 - 7.0	0	0	2	4	0.000
8	7.0 - 8.0	0	0	5	10	0.000
9	8.0 – 9.0	0	0	4	8	0.000
	Total	**80**	**100**	**50**	**100**	
CPA		$\upsilon = 7$				**46.016**
		$\chi_0{}^2{}_{(7,95\%)} = 14.1$				**Rejected H0**

Source: Data Analysis, 2018

Table 7. Comparison egress distances between BRT Trans '*Mamminasata*'and Commuter Train Surabaya-Sidoarjo

No.	Distance (Km)	BRT Trans Mamminasata (org)	SUSI Commuter Surabaya-Sidoarjo (org)
1	0.0 - 1.0	74	20
2	1.0 - 2.0	8	8
3	2.0 - 3.0	0	6
4	3.0 - 4.0	0	9
5	4.0 - 5.0	0	3
6	5.0 - 6.0	0	0
7	6.0 - 7.0	0	2
8	7.0 - 8.0	0	2
	Total	**80**	**50**

Source: [3,4]

Table 8. Calculation of Egress distance of BRT Trans '*Mamminasata*' and Commuter train Surabaya-Sidoarjo

No.	Distance Interval	% Distribution				Calculation χ2
		Ref BRT Mamminasata		SUSI Commuter Surabaya-Sidoarjo		χ^2
	Km	Person	%	Person	%	
1	0.0 - 1.0	74	92.5	20	40	29.797
2	1.0 - 2.0	6	7.5	8	16	9.633
3	2.0 - 3.0	0	0	6	12	0.000
4	3.0 - 4.0	0	0	9	18	0.000
5	4.0 - 5.0	0	0	3	6	0.000
6	5.0 - 6.0	0	0	0	0	0.000
7	6.0 - 7.0	0	0	2	4	0.000
8	7.0 - 8.0	0	0	2	4	0.000
Total		**80**	**100**	**50**	**100**	
CPA		$\upsilon = 6$				**39.431**
		$\chi^2_{\text{tabel }(6,95\%)} = 12.6$				**Rejected H_0**

Source: Data Analysis, 2018

3.4. Analysis result of access and egress distances on BRT Trans 'Mamminasata' in Makassar and Commuter Train Surabaya-Sidoarjo

The Chi Square statistical tests showed that there was a difference in the distance between the BRT Trans '*Mamminasata*' in Makassar and the Commuter Train Surabaya-Sidoarjo. The Access and Egress distances on BRT Trans '*Mamminasata*' in Makassar were shorter than on Commuter Trains Surabaya-Sidoarjo. The Access distance of BRT Trans '*Mamminasata*' in Makassar was 0-2 km with a cumulative of 84% and was shorter than Access distance of the Commuter Train Surabaya-Sidoarjo by 0.0-5.0 km with a cumulative of 78%. The Egress distance of the BRT Trans '*Mamminasata*' in Makassar was 0.0 - 0.5 km with a cumulative of 75% and was shorter than the Egress distance of the Commuter Train Surabaya-Sidoarjo by 0.0 - 4.0 km with a cumulative of 86% (see Table 9 and Figure 3). The cumulative percentage value is used as reference to determine the influence area. Due to less than 100% cumulative percentage of influence area of mass transportation, the influence area with 100% cumulative percentage was ignored. Figure 4 shows an overview of the boundaries between the influence area and underserved areas, while Figure 5 shows the benefits of this study.

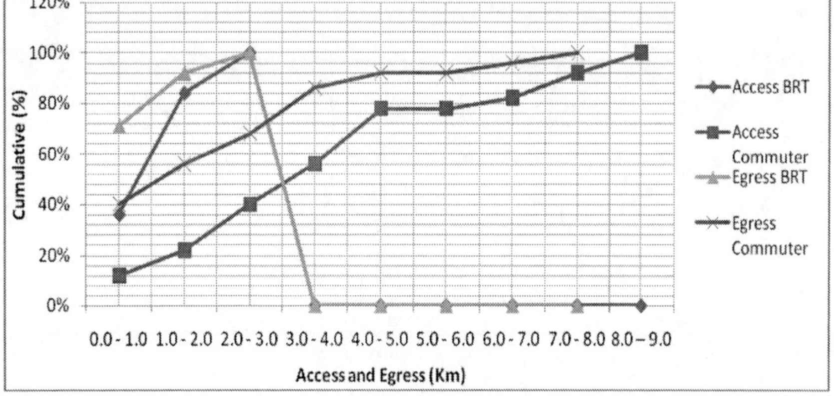

Figure 3. Cumulative value of access and egress BRT Trans '*Mamminasata*' and Commuter train Surabaya-Sidoarjo

IOP Conf. Series: Earth and Environmental Science **340** (2019) 012005 doi:10.1088/1755-1315/340/1/012005

Table 9. Comparison of the access and egress distances on BRT Trans '*Mamminasata*' Makassar and 'SUSI' Commuter Train

BRT Trans Mamminasata Makassar			SUSI Commuter Train		
Access (Km)	**Egress (Km)**	**Benefits**	**Access (Km)**	**Egress (Km)**	**Benefits**
0.0 – 2.0 (84%)	0.0 – 0.5 (78%)	1. Knowing the influence area from the original location to the original bus station 2. Knowing the influence area from the designated station to the designated location	0.0 – 5.0 (78%)	0.0 – 4.0 (86%)	1. Knowing the influence area from the original location to the original bus station 2. Knowing the influence area from the designated station to the designated location
Access and Egress on BRT Trans Mamminasata Makassar and SUSI Commuter Train on 100% cumulative					
0.0 – 3.0 (100%)	0.0 – 2.0 (100%)	The Access distance was longer than the Egress distance	0.0 – 9.0 (100%)	0.0 – 8.0 (100%)	The Access distance was longer than the Egress distance
Result on Statistical Test obtained the H1: $\chi^2 \geq \chi^2_{tabel}$, meaning that it cannot be synthesized due to the high different of Access and Egress distance between BRT Trans '*Mamminasata*' Makassar and '*SUSI*' Commuter Train					

Source: [3], Primary Survey, 2017; and Data Analysis, 2018

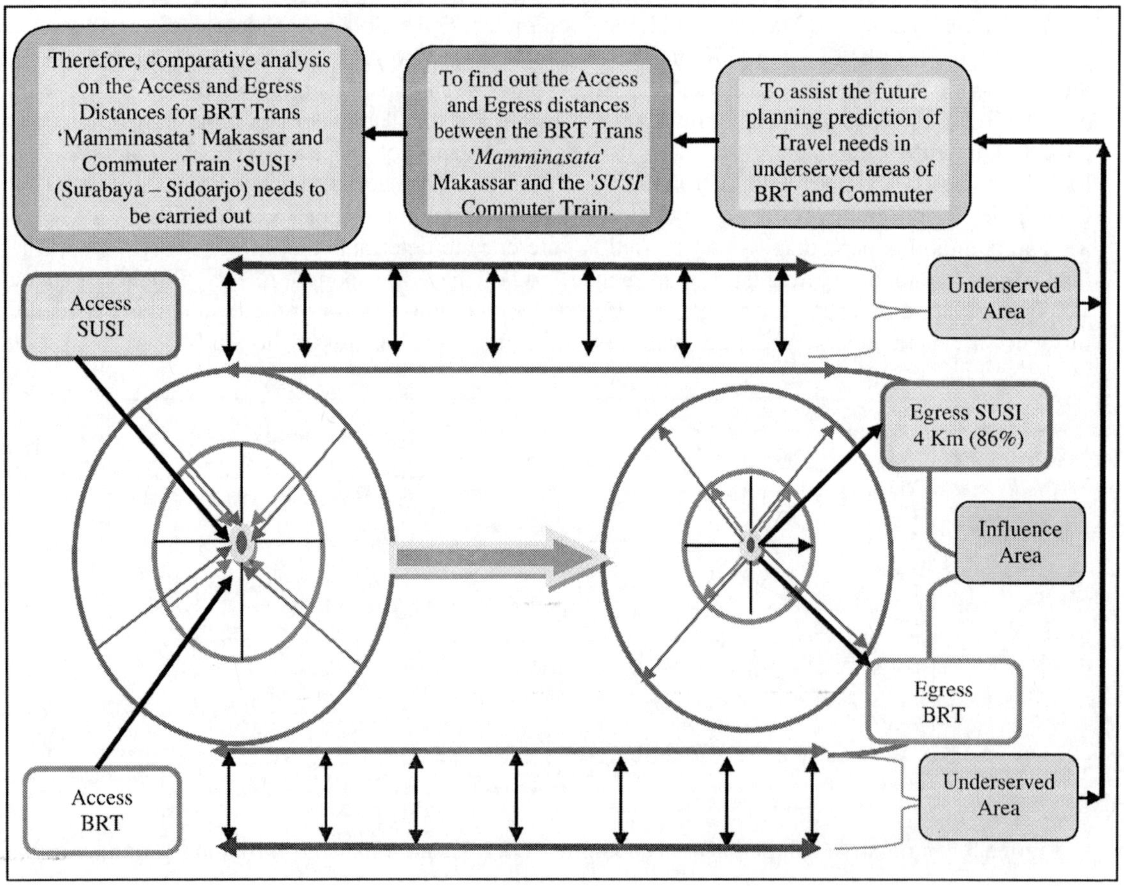

Figure 4. Influence area and underserved area

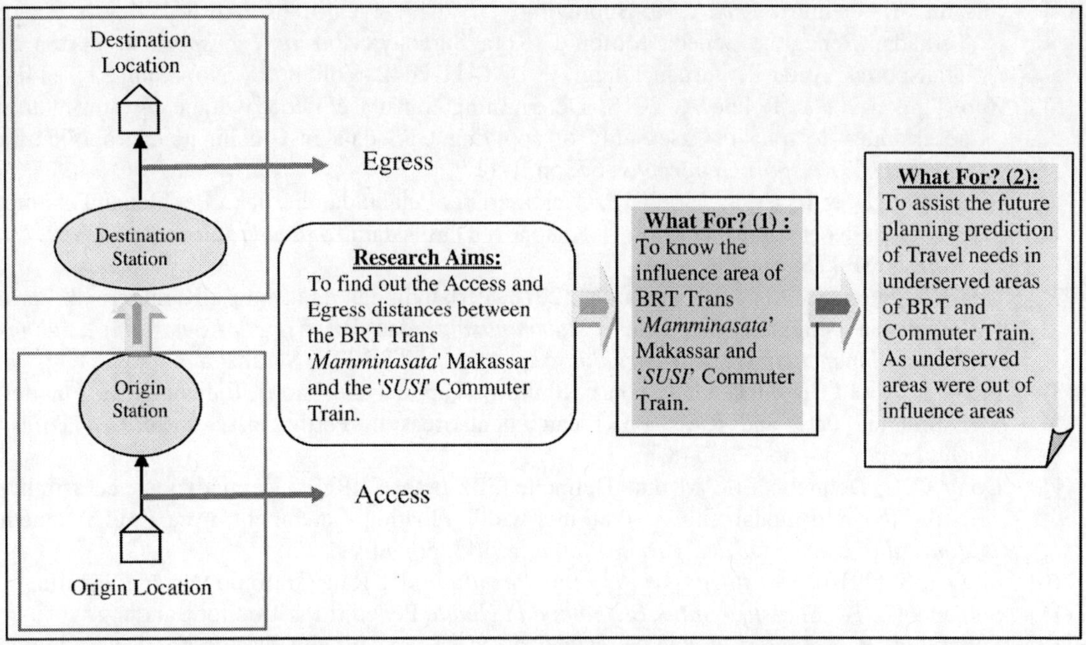

Figure 5. Benefits on knowing the access and egress distances

4. Conclusion

Comparison of the Access and Egress Distance between BRT Trans '*Mamminasata*' in Makassar and commuter train Surabaya-Sidoarjo found:

- Access distance on BRT Trans '*Mamminasata*' in Makassar was 0–2 km (75%) shorter than Access distance on Commuter train Surabaya-Sidoarjo as 0–5 Km (78%).
- Egress distance on BRT Trans '*Mamminasata*' in Makassar was 0 – 0.5 Km (71%) shorter than Egress distance on Commuter train Surabaya-Sidoarjo as 0 – 4 Km (86%).
- Generally viewed from 100% cumulative, there was different on Access and Egress distances between BRT Trans '*Mamminasata*' in Makassar and Commuter train Surabaya-Sidoarjo as follow:
  1. Access distance of BRT Trans '*Mamminasata*' in Makassar from 100% cumulative was about 0–3 km, while on Commuter train Surabaya-Sidoarjo was about 0–9 km.
  2. Egress distance of BRT Trans '*Mamminasata*' in Makassar from 100% cumulative was about 0–2 km, while on Commuter train Surabaya-Sidoarjo was about 0–8 km.

This study found that the Access and Egress distances between BRT Trans '*Mamminasata*' in Makassar was shorter than Commuter train Surabaya-Sidoarjo. Therefore, further research is needed to find out the factors causing the Access and Egress distances from each mode.

5. References

[1] Le T.Q. & Nurhidayanti Z.A. 2016. A Study of Motorcycle Lane Design in Some Asian Countries, *Procedia Engineering*, 142, pp. 292-298.

[2] Satiennam T., Jaensisrisak S., Satiennam W. & Detdamrong S. 2016. Potential for Modal Shift by Passenger Car & Motorcycle Users Towards Bus Rapid Transit (BRT) in an Asian Developing City, *IATSS Research*, 39, pp. 121-129.

[3] Suprayitno H., & Upa V.A. 2016. Mamminasata BRT User Trip Characteristics for the Design of Demand Modelling Method for a New BRT Line, *The Journal for Technology & Science,* Vol. 27, No. 3, December, FTSP-ITS, Surabaya.

[4] Susanti A, Soemitro R.A.A. & Suprayitno H. 2017. Identifikasi Awal Wilayah Pengaruh Terhadap Pengguna Sepeda Motor di Kota Surabaya, *Jurnal Transportasi.* Forum Studi Transportasi Antar Perguruan Tinggi, ISSN 1411-2442, Volume 17, No. 3, pp. 235-244.

[5] Zuo T, Wei H., & Rohne A. 2018. Determining transit service coverage by non-motorized accessibility to transit: Case study of applying GPS data in Cincinnati metropolitan area, *Journal of Transport Geography*, 67, pp. 1-11.

[6] Suprayitno, H. & Ryansah, M. (2018). Karakteristik Pelaku dan Perilaku Perjalanan Penumpang Bis Trans Koetaradja, Koridor 1: Keudah - Darulsalam, *Submit to Jurnal Aplikasi Teknik Sipil.* FTSP-ITS, Surabaya.

[7] Upa VA, Suprayitno H., & Ryansah M. 2018. Perbandingan dan Sintesa Karakteristik Perilaku Perjalanan Pengguna Bis Trans '*Mamminasata*' dan Bis Trans '*Koetaradja*', *Submit to Jurnal Manajemen Aset Infrastruktur*. September, FTSP-ITS, Surabaya.

[8] Lin T G, Xia J C, Robinson T P, Olaru, Smith B, Taplin J and Cao B. Enhanced Huff model for Estimating Park and Ride (PnR) catchment areas in Perth, WA, *Journal of Transport Geography*, 54, pp. 336-348.

[9] Dony C C, Delmelle Eric M and Delmelle Elizabeth C. Re-conceptualizing accessibility to parks in multi-modal cities: Avariable-width Floating Catchment Area (VFCA) method, *Journal of Landscape and Urban Planning*, 143, pp. 90-99.

[10] Siregar, S. (2016), *Statistik deskriptif untuk Penelitian,* PT Raja Grafindo Persada, Jakarta.

[11] Sudjana. (2004), *Statistika untuk Ekonomi dan Niaga*, Penerbit PT Tarsito, Bandung.

[12] Suprayitno H, Ratnasari V and Saraswati N. 2017. Experiment Design for Determining the Minimum Sample Size for Developing Sample Based Trip Length Distribution, *IOP Conf. Series: Materials Science and Engineering 267, doi:10.1088/1757-899X/267/1/012029*, pp. 1-13.

[13] Buchari E. 2018. Hierarchical Service for Integrating Multimodal Public Transport System in Palembang, Indonesia, *Journal of Technology and Social Science (JTSS)*, Vol. 2, No. 1, pp. 17-23.

Acknowledgment

This article was supported by previous researches and studies conducted by Research Team from Institute Technology of *Sepuluh* November. Best regard is delivered to all parties who motivate and support until this study was concluded.

The study of Larantuka urban infrastructure service level to accommodate the connectivity of surrounding islands

A M Gai[1], M C Endarwati[1] and A R T O Lamapaha[1]

[1] Urban and Regional Planning Department, Institut Teknologi Nasional (ITN Malang), Address: Jl.Sigura-gura No.2, Malang

Email: ardiyanto_maksimilianus@lecturer.itn.ac.id

Abstract. Infrastructure is important and as a solution to the problem of high price disparity and the inequality in the percentage of GDP product value nationally, in particular, in the western and eastern regions of Indonesia which have been a challenge that needs to be overcome. East Flores regency is one of the islands in the Eastern Indonesia, precisely in East Nusa Tenggara with Larantuka as its capital. This area is facing problems related to low quality strategic infrastructure and inadequate regional transportation infrastructure. This study aims to assess urban infrastructure service level in Larantuka to accommodate the connectivity of surrounding islands. This study used Linkage, Integrate Rural Accessibility Planning, and Importance Performance Analysis as the methods. The results showed that there is a connection between road infrastructure and the port as well as airport. In addition, there is a connection between marine infrastructure (collection ports) and local feeder ports in the surrounding islands and a connection between spoke airport infrastructure (air) and the hub airport. The condition of urban infrastructure in Larantuka is considered "good," with service level of; "less satisfactory" for land infrastructure and "somewhat satisfactory" for air and sea infrastructure.

Keywords: *infrastructure, Larantuka, planning*

1. Introduction

Transportation is one of the sectors which significantly affects socio-economic development and growth of standard living [1]. In recent years, the availability and affordability of goods in the global market has become important; the production is no longer just a metric. Logistics is one of the most important components in every supply chain. With the development of manufacturing industry and the rapidly growing demand for specialized supply chain management services, the logistics industry is growing with the national economy [2]. The growth of logistics industry demands logistics infrastructure improvement which means as investment in certain nodes of transport network such as ports, airports, and linear infrastructure (road and rail link) [3].

Since East Flores regency is an archipelago, an infrastructure that can connect the islands is needed to support economic activities and reduce disparities in East Flores development. At the beginning of East Flores regency establishment, it consisted of Flores Island (mainland), Adonara Island, Solor Island, and Lembata Island, however, in 1999 with the enactment of Law Number 52 of 1999 concerning the establishment of Lembata regency which was inaugurated by the Governor of East Nusa Tenggara, then

East Flores regency only consists of Flores Island (the eastern part of the mainland), Adonara Island, and Solor Island.

Larantuka City, besides being the capital of Larantuka sub-district, is also the capital of East Flores regency. Judging from the direction and development trends of the city, Larantuka is a transit city, especially from the mainland (Flores Island) to the surrounding islands, and vice versa. In addition, Larantuka is also a center of functional activities oriented to regional service scale in which its development is along the coast including port area, trading, and airport. While the area oriented to the regional road covers part of the settlement area (Bappeda NTT).

The problems existing in East Florest regency, especially in Larantuka city related to regional development, in terms of the Regional Government Work Plan (*Rencana Kerja Pemerintah Daerah--RKPD)* 2016 are as follows: (1) The quality of strategic infrastructure is still low, (2) Regional transportation infrastructure is inadequate. Based on those problems, a research entitled "The Study of Larantuka Urban Infrastructure Service Level to Accommodate the Connectivity of Surrounding Islands" is needed.

2. Literature Review

2.1. Regional development

In this globalization era, a correlation between infrastructure development and regional development acceleration is needed. However, there are some issues related to the development, one of them is regional disparity. This disparity can be seen from the GDP number, per capita income, and the number of poor people per island which shows regional economic growth and infrastructure gap between regions. Infrastructure disparity in Indonesia is shown by infrastructure development concentrated in Java and Bali, while Kalimantan, Sumatera, Nusa Tenggara, Maluku, and Papua are still facing infrastructure limitation. Related facts also show that high national economic growth in 2010, around 6.1%, was not followed by equitable regional economic growth. This disparity occurs mainly between Java and outside Java, between urban and rural areas, as well as between fast-growing regions and underdeveloped areas including border areas. Due to the unfulfilled infrastructure in a region, it may cause expensive transportation fares resulting in inefficient production and distribution. The availability of infrastructure is divided into competitiveness (economy) and basic needs (equity). Furthermore, that infrastructure availability must be developed continuously along with regional development based on sovereign local wisdom that is rich in traditions [4].

2.2. Infrastructure and regional connectivity

Transportation infrastructure is an integral part of the city or country transportation system. In relation to community development and international relation intensification due to globalization process, the importance of transportation as a factor for economic and social development has been increased [5]. Transportation infrastructure is often used as a means of achieving political goal which is to promote strong and sustainable regions. Implicit mechanism between infrastructure and population growth is that infrastructure investment enlarges the area as to improve all aspects [6].

The transportation system framework is very complicated, it consists of various infrastructure such as terminal facilities, travel methods, transportation vehicles, and information system. Transportation has become one of critical infrastructure sectors in most countries in the world. Transportation is considered as one of the most important infrastructure sectors at the national and international level.

Transportation must be open and accessible (high level of user access), broad and ubiquitous (large number of physical infrastructure and assets; most infrastructure are protected and unattended; millions of vehicles and containers are distributed through the network), effective and adaptable. Critical transportation infrastructure is important for the national economy, security, and state function. This is important for:

- national priorities,
- providing basic services, reliability and availability,

- ensuring trade flows,
- supporting economic growth.

Transportation infrastructure has many critical points including elements, objects and nets, as well as vastly geographically distributed assets [1]. While connectivity infrastructure is an infrastructure connecting and facilitating people and goods from one place to another which includes road, bridge, port, and airport [7].

3. Methodology

This research used descriptive analysis method with a quantitative approach. Descriptive research is a research that attempts to describe a phenomenon or event occurred at the present time. Descriptive research obtains data based on the factors supporting the object of research, then these factors are analyzed to determine their roles [8]. Quantitative approach is an approach used in research by measuring indicators of research variables to obtain an overview between these variables in the form of research results in numbers with certain meanings.

3.1. Linkage

Linkage is a false line connecting one element to another, one node to another, or one district to another [9]. This line is usually in the form of road network, pedestrian path, and open space in various forms. Linkage theory involves organizing lines connecting parts of the city and "spatial datum" design from the building line to space. Spatial datum can be the site line, movement direction, axis, and building edge which together form a linkage system in a spatial environment. An urban linkage can be observed in different ways and approaches, there are three urban linkage approaches:

- Visual linkage,
- Structural linkage,
- Collective linkage.

3.2. IRAP (Integrated Rural Accessibility Planning)

Accessibility is the ease or difficulty in reaching social, economic, transportation, or other types of facilities and services. IRAP is a multi-sector integrated method including human, transportation system, and existing travel pattern. This is used in identifying priority design process in developing rurality due to villager capacity in achieving the easiest access to basic demands and other economic service facilities [10]. In addition, IRAP can support the development of local infrastructure planning process to improve the accessibility in rural area in Indonesia. Parameters believed to influence community needs in rural area demanding accessibility include: mobility, water, energy/electricity, education, health care, agriculture, basic need production, fishery production, small industry, market, and transportation [11].

There are three steps in IRAP analysis, namely determining accessibility indicators, determining indicator quality, and calculating indicator value.

- Determine accessibility indicators
  The first step in determining priority area and sector is determining accessibility indicators that will be reviewed. Then, assessment and category of those accessibility indicators are made. Giving score to each indicator based on the accessibility condition. The lowest score shows that the indicator is in the best condition, while the highest score means that the indicator condition is very bad. Indicator score is determined based on survey results and existing condition.
- Determine indicator value.
- Indicator value is obtained from respondents' assessment result based on the importance of each indicator.
- Calculate accessibility value which is obtained by multiplying indicator score and average indicator value.

3.3. IPA (Importance Performance Analysis)

Originally, IPA was developed to support a synchronous analysis of two components with different customer satisfaction service importance, and provider performance toward them. Instead, it was proposed to have IPA feature of "importance" axis displaying customer analysis result, and "performance" axis displaying expected value for the provider [12].

IPA Steps are:

- Sum the score of each item of all respondents
- Calculate the average of each item by dividing the score result by the number of respondents
- Sum the average of each item then divide them by the number of items
- Make Importance Performance Analysis (IPA) diagram to determine each item category

Before making importance performance analysis cartesian diagram, first find the intersection for the X axis (interest) and Y axis (satisfaction) using the following equations:

$$\overline{\overline{X}} = \frac{\sum_{i=1}^{n} Xi}{k} \quad (1)$$

$$\overline{\overline{Y}} = \frac{\sum_{i=1}^{n} Yi}{k} \quad (2)$$

Then, make IPA (importance performance analysis) diagram:

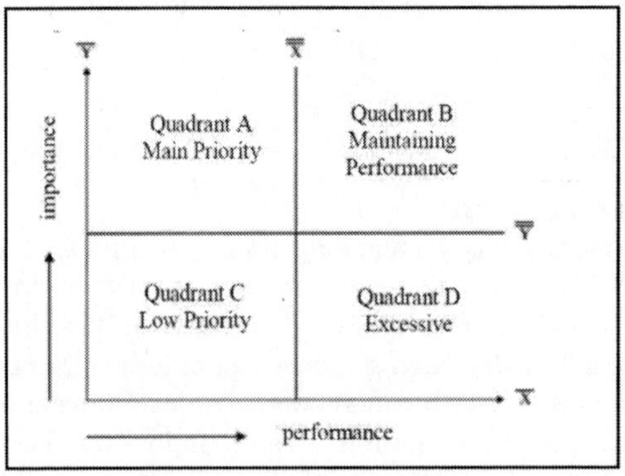

Figure 1. IPA diagram

4. Result and discussion

Result and discussion consist of the analysis results of connecting infrastructure, infrastructure condition, and urban infrastructure service level in Larantuka city to accommodate surrounding islands. The analysis of connecting infrastructure in Larantuka city was conducted to determine the connectivity between land infrastructure and other area around Larantuka city, as well as connecting Waibalun ferry port, the sea port, JPT pallo beach, and airport/ Style and Spacing airport.

4.1. The analysis of connecting infrastructure in Larantuka City

There are three connecting infrastructure in Larantuka city, namely land, marine, and air infrastructure. The following is the analysis of the connecting infrastructure in Larantuka city.

Table 1. Connecting infrastructure analysis

No	Linkage	Description	Maps
1	**Land infrastructure linkage**	The basis of land connecting infrastructure in Larantuka city is seen from transportation movement system connecting Waibalun ferry port to Larantuka sea port, JTP pallo beach, and Gewayantana airport Larantuka.	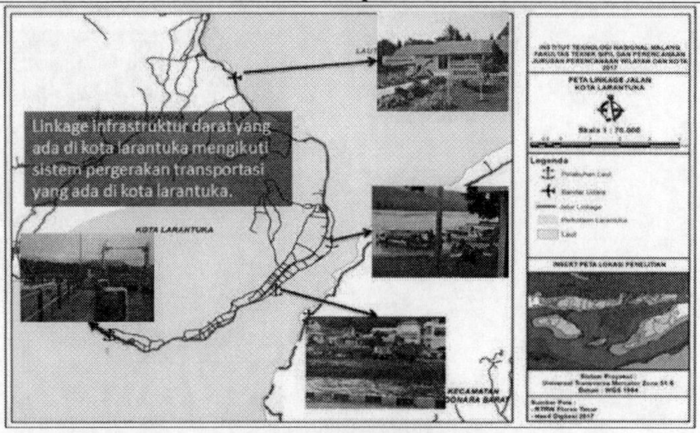
2	**Marine infrastructure linkage**	**Local Service Marine Infrastructure Linkage**	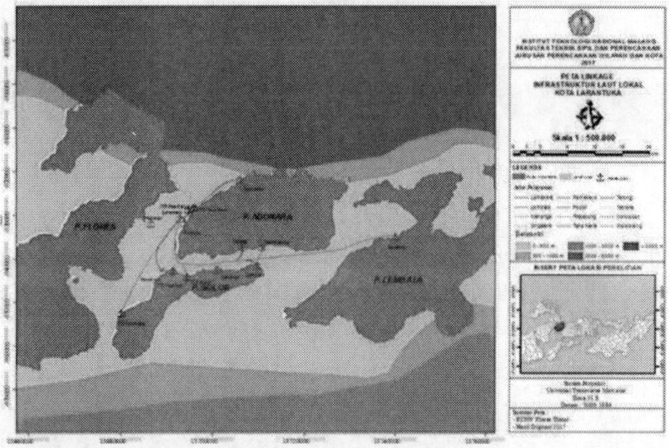 In Larantuka city, there are two types of port, namely Larantuka sea port and Waibalun ferry port. In Larantuka sea port, the linkage used is seen from the hierarchy and port location, in which the hierarchy and Larantuka sea port status is as collection port, while the ports in surrounding islands (Adonara island and Solor island) are local feeder ports.
		National Service Marine Infrastructure Linkage	For national service infrastructure linkage, seen from national transportation 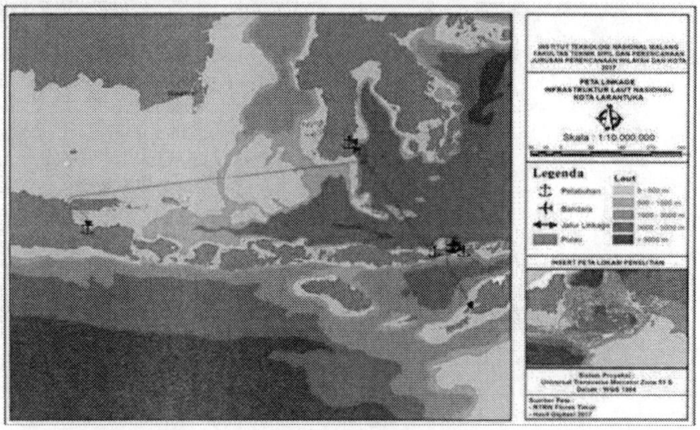

No	Linkage	Description	Maps
			system, Larantuka sea port is in national shipping system with service route of Surabaya-Makasar- Maumere-Larantuka-Lembata-Kupang (round trip).
		Regional Service Marine Infrastructure Linkage	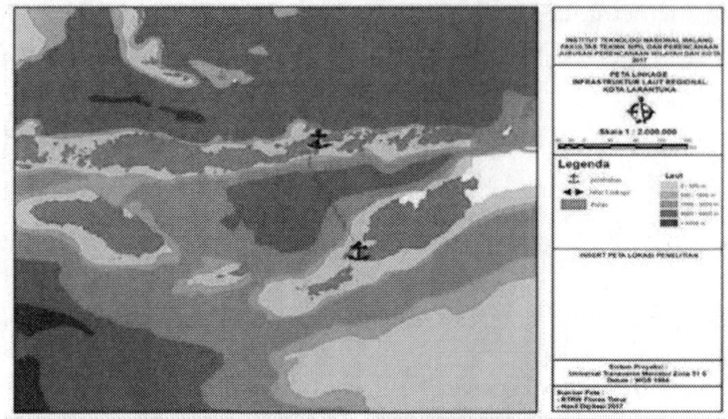 Waibalun ferry port is class III ferry port of which functions are: 1. Port that functions as moving bridge (crossing) 2. Serves as a bridge connecting primer 2 collector road, primer 3 collector road, primer 4 collector road, and roads are not in port criteria for class I and class II crossing transport. 3. The location is not at the conception of the national crossing belt. 4. Port that has not been commercially operated. The linkage of this ferry port connects between Waibalun ferry port Larantuka (class III port) with Bolok port (class I port) in Kupang district as an access to Kupang city.
3	**Air infrastructure linkage**		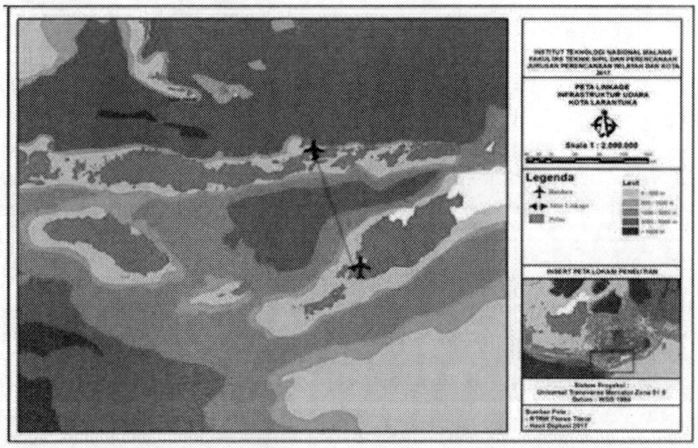 Air infrastructure in Larantuka, in this case Gewayanatana airport, only serves Kupang – Larantuka – Larantuka – Kupang flight route. It was caused by the hierarchy of Gewayantana airport Larantuka as spoke airport to serve hub airport that is El Tari airport kupang (Transportation Minister Regulation Number: PM 69 of 2013 concerning Airport). Spoke airports are: a. Airports with service coverage and affect local economy. b. Destination airport or supporting airport from the collecting airport. c. Airport as one of the supporting infrastructure service for local activities. Gewayantana airport larantuka has a runway of 1.400 m length and 30 m width that is only able to serve aircraft with short/medium distance flight. Operators that manage the flights from Kupang to Larantuka are Trans Nusa and Wings Air operators. The aircraft serving Gewayantana airport Larantuka is Fokker 50 for Trans Nusa operators, while Atr 72 for Wings Air.

Source : Analysis Result, 2017

4.2. Analysis of infrastructure condition in Larantuka City

The analysis of infrastructure condition in Larantuka city used IRAP (Integrated Rural Analysis Planning) analysis method. The results can be seen as follows:

a. Land Infrastructure

Accessibility value was obtained by multiplying each indicator score by each value score then the result was divided by the number of questionnaire to get the average score. Each result then summed and the total average score can be obtained. Last, the total average score was divided by the number of respondents to obtain the accessibility value.

Table 2. Land Infrastructure Accessibility

Infrastructure Type	Total Average I x V	Accessibility Value
Land Infrastructure	1390,75	21,40

Source: Analysis, 2017

Table 3. Land Infrastructure Condition in Larantuka City

Land Infrastructure	Condition			Remarks
	Good	M	B	
Road	√			

Source: Analysis, 2017

With accessibility value of 21,40, it can be said that land infrastructure in Larantuka is in good condition.

b. Marine Infrastructure

Accessibility value was obtained by multiplying each indicator score by each value score then the result was divided by the number of questionnaire to get the average score. Each result then summed and the total average score can be obtained. Last, the total average score was divided by the number of respondents to obtain the accessibility value.

Table 4. Marine Infrastructure Accessibility

Infrastructure Type	Total Average I x V	Accessibility Value
Marine Infrastructure	1539,00	23,68

Source: Analysis, 2017

Table 5. Marine Infrastructure Condition in Larantuka City

Marine Infrastructure	Condition			Remarks
	Good	M	B	
Ports	√			

Source: Analysis, 2017

With accessibility value of 23,68, it can be said that marine infrastructure in Larantuka is in good condition.

c. Air Infrastructure

Accessibility value was obtained by multiplying each indicator score by each value score then the result was divided by the number of questionnaire to get the average score. Each result then summed and the total average score can be obtained. Last, the total average score was divided by the number of respondents to obtain the accessibility value.

Table 6. Air Infrastructure Accessibility

Infrastructure Type	Total Average I x V	Accessibility Value
Air Infrastructure	726,75	11,18

Source: Analysis, 2017

Table 7. Air Infrastructure Condition in Larantuka City

Marine Infrastructure	Condition			Remarks
	Good	M	B	
Airport	√			

Source: Analysis, 2017

With accessibility value of 11.18, it can be said that air infrastructure in Larantuka is in good condition.

4.3. Analysis of infrastructure service level in Larantuka City

This analysis used IPA (Importance Performance Analysis) analysis method. This analysis aimed to determine infrastructure (land, marine, and air infrastructure) user satisfaction level in Larantuka city.

a. Land Infrastructure Service Level

Service level in this analysis was user satisfaction level in using land infrastructure. Cartesian diagram was made by inputting average score of each item of interest level (X) and satisfaction level (Y). The detail can be seen in the following table and diagram:

Table 8. Average Score of Interest Level (X) and Satisfaction Level (Y)

Item	Interest Level	Satisfaction Level
1	1.68	3.32
2	1.48	3.52
3	1.40	3.58
4	1.57	3.43
5	1.40	3.60
6	1.40	3.60
7	1.40	3.60
8	1.57	3.43
9	1.42	3.58
10	1.40	3.60
11	1.62	3.38
12	1.60	3.40
Total Average	**1.49**	**3.51**

Source: Analysis Result, 2017

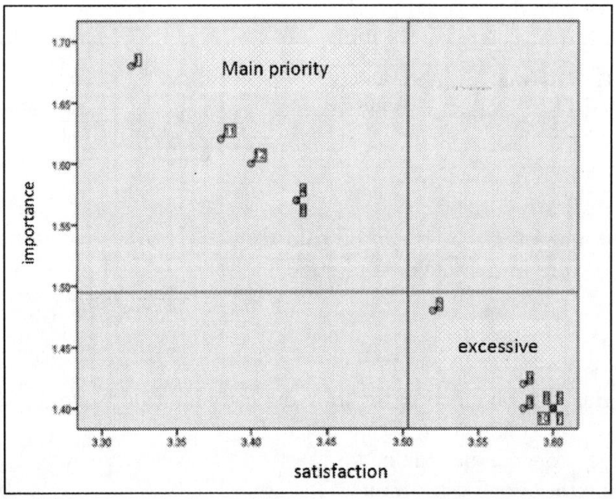

Figure 2. Importance performance analysis of land infrastructure
Source: Analysis Result, 2017

Based on the the result of cartesian diagram, two quadrants were obtained, namely main priority quadrant and excessive quadrant. Main priority quadrant is the quadrant containing items considered important by the community but still lacking in user satisfaction, while excessive quadrant contains items with relatively low interest but relatively high satisfaction. Items in main priority quadrant include: item 1, item 4, item 6, item 8, item 11, and item 12. While items in excessive quadrant include: item 2, item 3, item 5, item 6, item 7, item 9, and item 10. In general, land infrastructure users in Larantuka felt **"less satisfied"** with the existed service level. It can be seen from the total value which reached **42.61%.**

b. Marine Infrastructure Service Level

Service level in this analysis is marine infrastructure user satisfaction. Cartesian diagram was made by inputting average score of each item of interest level (X) and satisfaction level (Y). The detail can be seen in the following table and diagram:

Table 9. Average Score of interest level (X) and satisfaction level (Y)

Item	Interest Level	Satisfaction Level
1	1.80	3.20
2	1.82	3.18
3	2.08	2.86
4	2.18	2.82
5	1.66	3.34
6	1.83	3.17
7	1.91	3.09
8	1.83	3.17
9	2.00	2.98
10	1.49	3.51
11	1.69	3.29
12	1.92	3.06
Total Average	**1.85**	**3.14**

Source: Analysis Result, 2017

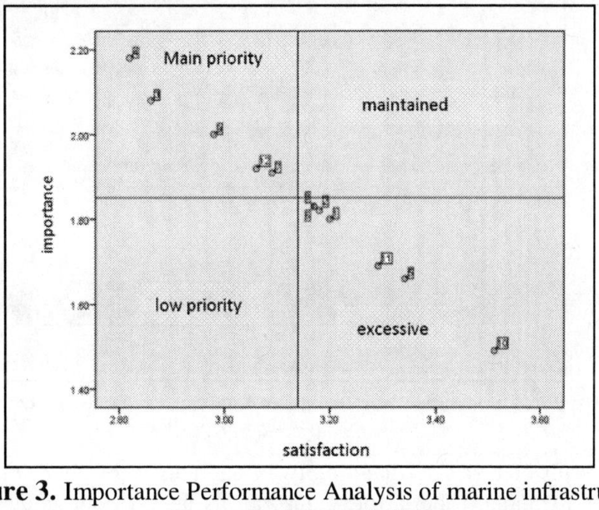

Figure 3. Importance Performance Analysis of marine infrastructure
Source: Analysis Result, 2017

Based on the the result of cartesian diagram, two quadrants were obtained, namely main priority quadrant and excessive quadrant. Main priority quadrant is the quadrant containing items considered important by the community but still lacking in user satisfaction, while excessive quadrant contains items with relatively low interest but relatively high satisfaction. Items in main priority quadrant include: item 3, item 4, item 7, item 9 and item 12. While items in excessive quadrant include: item 1, item 2, item 5, item 6, item 8, item 10, and item 11. In general, marine infrastructure users in Larantuka felt "**quite satisfied**" with the existed service level. It can be seen from the total value which reached **58.96%**.

c. Air Infrastructure Service Level

Service level in this analysis is air infrastructure user satisfaction. Cartesian diagram was made by inputting average score of each item of interest level (X) and satisfaction level (Y). The detail can be seen in the following table and diagram:

Table 10. Average score of interest level (X) and satisfaction level (Y)

Item	Interest Level	Satisfaction Level
1	1.80	3.20
2	1.80	3.20
3	1.40	3.60
4	2.00	3.00
5	1.80	3.20
6	1.83	3.20
7	1.63	3.40
8	2.40	2.60
9	1.60	3.40
10	1.40	3.60
11	1.40	3.60
12	1.80	3.20
Total Average	**1.74**	**3.27**

Source: Analysis Result, 2017

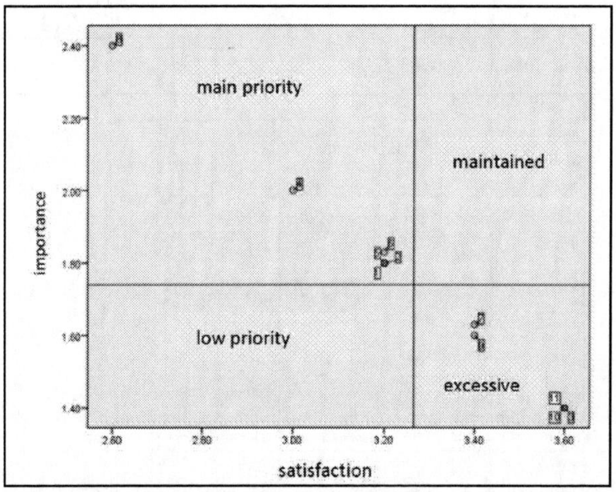

Figure 4. Importance Performance Analysis of Air Infrastructure
Source: Analysis Result, 2017

Based on the the result of cartesian diagram, two quadrants were obtained, namely main priority quadrant and excessive quadrant. Main priority quadrant is the quadrant containing items considered important by the community but still lacking in user satisfaction, while excessive quadrant contains items with relatively low interest but relatively high satisfaction. Items in main priority quadrant include: item 1, item2, item 4, item 5, item 6, and item 8. While items in excessive quadrant include: item 3, item 7, item 9, item 10, item 11. In general, marine infrastructure users in Larantuka felt "**quite satisfied**" with the existed service level. It can be seen from the total value which reached **52.23 %.**

5. Conclusion

Based on the analysis that has been done, it was known that:

a) Connecting infrastructure in Larantuka city consisted of land, marine, and air infrastructure.
   - For land infrastructure (road), existing linkage seen from transportation movement system and the connectivity between Waibalun ferry port, Larantuka sea port, JTP pallo beach, and Gewayantana airport Larantuka. While marine infrastructure linkage was

divided into two namely Larantuka sea port linkage for local and national service, and Waibalun ferry port for regional service. Local service of marine infrastructure linkage is seen from the hierarchy and port location in which Larantuka sea port is the collection port while the ports in surrounding islands (Adonara island and Solor island) are local feeder ports. National service of marine linkage infrastructure is seen from national transportation system in which Larantuka sea port is in national transportation system with service of Surabaya-Makasar-Maumere-Larantuka- Lembata-Kupang.

- Waibalun ferry port is seen from the port status in which this port is class III port that is not in national service belt and has not been commercialized.

b) Infrastructure condition in Larantuka city
- Based on the analysis results, it was known that infrastructure in Larantuka city (land, marine, and air infrastructure) are in "good" condition.

c) Infrastructure service level in Larantuka city
- Based on the analysis result, it was known that the users of land infrastructure in Larantuka city are "less satisfied" while for marine and air infrastructure they are "quite satisfied."

6. References

[1] Janušová, Lucia, and Silvia Čičmancová. 2016. Improving Safety of Transportation by Using Intelligent Transport Systems. 9th International Scientific Conference Transbaltica 2015. University of Zilina, 1. mája 32, 01001 Zilina, Slovakia.

[2] Acar, A. Z., and Yurdakul, H. (2015). *Evaluation of competitive power of logistics industry in Turkey*. Journal of Caspian Affairs , 1 (1), 1-22.

[3] Acar, A. Z., Bentyn, Z., and Kocaoglu, B. (2015). *Turkey as a Regional Logistics Hub in Promotion of Revivaling Ancient Silk Route Between Europe and Asia*. Journal of Management, Marketing and Logistics , 2 (2), 94-109.

[4] Putri,Andina Oktavia Sulistya. 2012. *Peran Infrastruktur dalam Percepatan Pembangunan Wilayah Sesi II*. Seminar Nasional Kota Untuk Rakyat.

[5] Skorobogatova, Oksana and Irina Kuzmina-Merlino*. 2017. *Transport Infrastructure Development Performance*. 16th Conference on Reliability and Statistics in Transportation and Communication,RelStat'2016, 19-22 October, 2016, Riga, Latvia. Procedia Engineering 178 (2017) 319 – 329

[6] Aarhaug, Jørgen and Frants Gundersen. 2016. *Infrastructure investments to promote sustainable regions*. 44th European Transport Conference 2016, ETC 2016, 5-7 October 2016, Barcelona, Spain. Transportation Research Procedia 26 (2017) 187–195

[7] Anonim. 2014. *Kementerian PPN Bappenas - Konektivitas Infrastruktur Wilayah dan Antar Wilayah*. Tarakan

[8] Arikunto, Suharsimi. 2010. *Prosedur Penelitian: Suatu Pendekatan Praktik.* Jakarta : Rineka Cipta

[9] Markus Zahnd. 2006. *Perancangan Kota Secara Terpadu.* Yogyakarta: Kanisius

[10] Oktaviana, Maria Goretti.2011. *Development Strategy For Rural Transportation Case Study Province Of West Papua. J. Basic. Appl. Sci. Res.*, 2(1)179-184, 2011. ISSN 2090-4304 Journal of Basic and Applied Scientific Research

[11] Bakri ,M. Djaya, Harnen Sulistio, Achmad Wicaksono, Ludfi Djakfar. 2015. *Infrastructure Development Program in Rural Areas using the Integrated Rural Accessibility Planning (IRAP.)* Australian Journal of Basic and Applied Sciences, 9(23) July 2015, Pages: 456-463.

[12] Rondinia. Alice, Marco Bertonib, Giuditta Pezzottaa. 2017. *An IPA based method for PSS design concept assessment.* The 9th CIRP IPSS Conference: Circular Perspectives on Product/Service-Systems. Procedia CIRP 64 (2017) 277 – 282

Sustainable pedestrian ways in Central Business District of Tunjungan Surabaya: Can principles of new urbanism be applied?

A M Navastara[1,3] **and V Mandasari**[2,3]

[1] Lecturer and Researcher, Department of Urban and Regional Planning
[2] Research Assistant, Department of Urban and Regional Planning
[3] Institut Teknologi Sepuluh Nopember

E-mail: ardy.navastara@gmail.com

Abstract. Pedestrian ways of Tunjungan which is known as Central Business District is an important accessibility element in Surabaya. According to Guidelines for the Provision and Utilization of Pedestrian Infrastructure and Facilities in Urban Area (2014), pedestrian ways is important component that should be provided for increasing effectiveness of citizen mobility in urban area. In general fact, pedestrian ways of Surabaya is in good condition. In 2007 Surabaya has received Wahana Tata Nugraha Award from The Ministry of Transportation. It has been avowed that pedestrian ways of Surabaya is first ranked of 20 cities in Indonesia (Detiknews, 2007). However pedestrian ways in Surabaya especially in Tunjungan district frequently misused (Kompas, 2010). That condition shows unmaximize use of pedestrian ways in this district. Objectives: This study will focus on LOS (Level of Service) to determine walkability in Tunjungan. Hereafter, value ranking of this LOS and what solution should be done will be confirmed to an expert. Finally, the concept of walkable city will be determined based on solution of LOS characteristic. Expected output in this study is alternative concept of sustainable pedestrian ways. It will be influenced by factors which are decided by some stakeholders. Finally, the concept of walkable city will be determined based on solution of LOS characteristic.

Keywords: *New urbanism, sustainable pedestrian ways, walkable city, CBD of tunjungan*

1. Introduction

All of the human activities need space. Based on Hamid Shirvani elements of open space are landscape, hardscape (street, pedestrian ways, etc), park and recreation space of city [1]. Spreiregen (1965) said that walking is the best transportation system though it has limitation on speed (about 3-4 km/hour) and physic affordability.

In reducing traffic jam, pedestrian ways in Surabaya was created to support Mass Rapid Transportation [2]. RTRW Surabaya 2009-2029 says about masterplan of *Light Rapid Transit* (LRT) on Joyoboyo Terminal – Darmo Street – Urip Sumoharjo Street –Basuki Rachmad Street – Embong Malang Street – Bubutan Street – Tugu Pahlawan Street. Trem network will service North-South corridor because there are more passenger than the others. Whereas, West-East corridor will be serviced by monorail [3]. Tunjungan Area will be the object of this research because it is an important accessibility for CBD area in Surabaya. In contrast, the utilization's not maximized yet, and frequently misused as vehicle trajectory and informal activities (Kompas, 2010). In result, it has negative effect on pedestrians safety.

This research use New Urbanism concept as a basic theory. New urbanism is an alternative for urban sprawl phenomenon starting from neighbourhood development [4]. In 1990, urban and environment designers discussed about suburban sprawl, including financing disparity, environment degradation, and excessive vehicle using. This concept began in the 1970s and 1980s in America. Based on this concept, urban sprawl phenomenon need integrated transportation system, mixed-landuse, and environmetn preservation. One of New Urbanism principle is walkability. Walkability can be measured by one of the analysis tool called pedestrian LOS (Level of Service). LOS may be used, for example, to evaluate the performance of a sidewalk and determine the need to redesign it (change its width, relocate, replace or remove street furniture, etc.), but it really depends on environment characteristic of the location (Bloomberg, 2006). In other words, the range number of pedestrian LOS (A-F) is debatable for every different location. The universal use of pedestrian LOS can't be applied carelessly.

2. Methodology

This research use secondary and primary data. Below Table 1 which shows the parameters, indicators, variables (sub-variables).

Table 1. Parameters, indicators, variables and sub-variables

Parameters	Indicators	Variables	Sub-variables	Unit
Safety	Availabilty of fly over bridge facility	Amount of fly over bridge		Unit
		Fly over bridge location		Location
		The travel time for crossing pedestrians		Minutes
	Availabilty of signages and street lighting	Amount of signages		Unit and Location
		Signages location		
		Amount of street lighting		Unit
		Street lighting location		
Comfort	Physical condition	Path dimension		Meter
		Amount and location of street furniture	Amount of trees	Unit and Location
			Shade trees location	
			Amount of seats	
			Seats location	
			Amount of trash	Unit and Location
			Trash location	Unit
	Level of service Pedestrian paths	Average flows of pedestrian	Amount of Most pedestrians at certain time intervals	Pedestrians Meter
			The effective width of pedestrian space	Meter/minute
		Pedestrian space	Average speed of space	Meter/minute

Parameters	Indicators	Variables	Sub-variables	Unit
			Pedestrian flows	Pedestrians/meter/minute
	Social Interaction	Utilization of pedestrian paths by a particular social group		Persons
		The existence of pedestrian paths that encourage social interaction	The existence of the pavement as a place to interact	Meter
			The existence of the corridors of the building the shopping center as a place to interact	Meter
			The existence of open spaces as a place to interact office	Meter per square
			The existence of the edge as a green line to interact	Meter
Attraction	Attractive Activities	Frequency of visits to the trade center by a pedestrian		Pedestrians/day
		Frequency of Visits to a service center by pedestrian		Pedestrians/day
	Visual	Pavement Design		
		Vegetation		
		Murals		Unit/location
Connectivity	Availability of public transport	Type of transport moda		Unit
		Amount of transport moda		
		Ease of pedestrians to reach the location of public transport stops		Minutes
	Continuity and accessibility of pedestrian paths	Pedestrian facilities to reach the destination		Minutes
		The type and length of the special pedestrian paths difable		Meter
Land use	Trade and Services	Types of land use	Macro land use	Hectare
			Micro land use	Unit
		Intensity of Space	Building	Percentage (%)

Parameters	Indicators	Variables	Sub-variables	Unit
		Utilization	Coverage Ratio (BCR)	
			Floor Average Ratio (FAR)	Percentage (%)
			Building's perimeter line	Meter

Source: Mandasari and Navastara, 2014

This study is identificating the characteristics of pedestrian ways and furthermore LOS analysis which analyze service's level of Tunjungan Area especially at Basuki Rachmad corridor as a sample for pedestrian traffic counting. Traffic counting area divided into 6 segments (A1, A2, A3, B1, B2, B3). Meanwhile, the recording time divided into 3 times per day (sampling 1 day for weekday and 1 day for weekend). Researcher recorded each segment with digital camera then counted its pedestrian flow. Pedestrian flow influenced by peak hour and surrounding landuse. That's why each segment showed different characteristic. The pedestrian flow then formulated with other elements such as widht of pedestrian ways, pedestrian modul space, and pedestrian average speed. After knowing the LOS of each segment then it could be confirmed to the urban landscape expert about the solution for each segment with different LOS characteristic.

3. Results and discussion

3.1. Characteristics pedestrian ways in CBD of Tunjungan

Characteristics of pedestrian ways in CBD of Tunjungan can be described below:

1. Walkability and Connectivity

Pedestrian ways in the corridor Basuki Rahmat are included Side Street types (Sidewalk) , the system of pedestrian paths from the edge of the highway up to the outermost edge of land owned buildings. Conditions pedestrian path on the east side of the road can be seen below.

Figure 1. *Sidewalk* of Tunjungan
Source: Mandasari and Navastara, 2014

Based on field observations, it has a green line/shade consistent that located along the pedestrian path. Furthermore, there is also a kind of pedestrian space arcade on the east and west side of the road as in the image below.

Figure 2. *Arcade*
Source: Mandasari and Navastara, 2014

Pedestrian ways in term of Arcade type is a pedestrian space that is adjacent to the buildings on either or both sides. In Tunjungan, this type is discontinuous and has inconsistent of green line/shade. The length for each type can be seen below:

Table 2. Type of pedestrian ways

No.	Type	Route Length
1.	Sidewalk	2.084,43 meters (0,2 km)
2.	Arcade	197 meters (0,01 km)

Source: ArcGIS Calculation, 2014

Walking path has length about 2,230 meters and it width varying from 3-4 meters.

2. Safety and Comfort

Safety and comfort of pedestrian ways of Tunjungan can be seen from the availability of facilities consisting of signage, street lighting and street furniture. Here's an explanation of each of these facilities:

a. Signage in the pedestrian area of research is already quite a lot and informative. Signage as prohibited parking, bike lanes, and are prohibited from stopping greatly affect the comfort and safety of pedestrians. In addition to traffic signs, signage can also be a billboard. Billboards in the observation area located at several locations with a number of 4-5 units billboard / point location.
b. Facilities available at the track crossing pedestrian crosswalk and observed as the pedestrian bridge. There is a pedestrian bridge at the 4 point locations as in Figure 4. Map of Facilities Distribution . Meanwhile, there is a zebra crossing in 3-point location. One measure of comfort and security of a facility is the length of time pedestrians to cross. Based on field observations, a long time to cross in this area <10 minutes with a pedestrian bridge, and 10 seconds by using a zebra crossing traffic lights. These facts indicate that the pedestrian was safe enough to cross from one point to another. Meanwhile, the travel time to the facility ranged crossings 3-4 minutes.
c. The lighting in the area of observation in the form of LPJU (street lighting lamps) with a distance of 20-30 meters between the lamp. Model lights in the observation area is divided into two such as: (1) model single pole lights located along both sides of the corridor Basuki Rahmat and (2) model forked light pole located in the median of the road
d. Furnishings pedestrian paths available along the corridor Basuki Rahmat is trash, seating and shelter. Garbage bins in Basuki Rahmat Corridor is located on the east side and west along the sidewalk with their respective bins distance is 50 meters. Garbage bins are divided into two (2) parts, namely garbage bins for dry waste, with yellow and bins for wet garbage, blue. Along the

corridor Basuki Rahmat, there are three (3) stop which all have to be on the side of the shelter. Whereas, for seating only in two point locations.

e. Model green line in the observation area is divided into two, namely the plants in pots permanent and single shade trees that lined pedestrian corridor based on RTBL of Basuki Rahmat corridor, it is making permanent pot in the middle of the pedestrian pathway aims to anticipate that sidewalks are not used to *"Pedagang Kaki Lima"* (street informal vendors). Types of plants in the region varied observations, namely angsana, king palm, tanjung, and bamboo (Department of Public Works, 2006)

Figure 3. Facilities of pedestrian ways
Source: Mandasari and Navastara, 2014

Table 3. Facilities of pedestrian ways

No.	Facilities	Amount
1.	Signages	
	- No parking/stop	21
	- Bike pathway	4
2.	Zebra Cross	4
	Fly over bridge	4
3.	Lighting	
	Street lighting	57
	Traffic lights	1
4.	Open space furniture	
	Trash bins	19
	Seats	2
	Halte	4

Source: Mandasari and Navastara, 2014

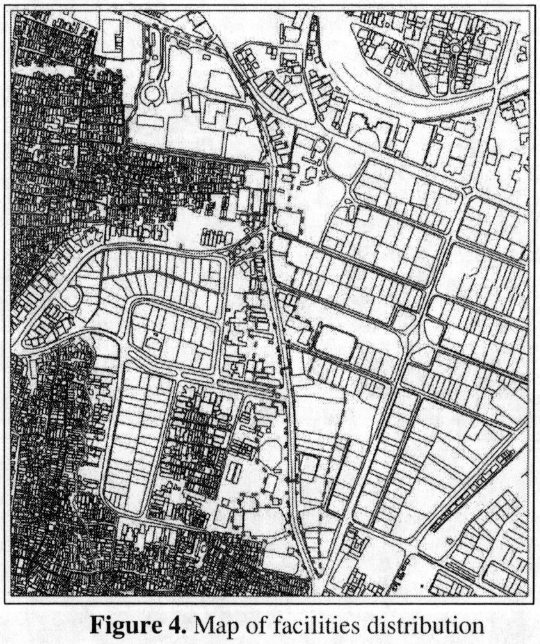

Figure 4. Map of facilities distribution
Source: Mandasari and Navastara, 2014

3. Mixed Land Use

Land uses are dominated by trade and services. The corridor Basuki Rahmat at Tunjungan which is the golden triangle area of Surabaya grow into the area of trade, services and offices that provide a major contribution in the field of economy of the city. Physically , the centers of both administrative and business activities are scattered along this corridor so that the facilities for mobility is more prevalent in the corridor Basuki Rahmat which is a north - south corridor . Land use along the corridor Basuki Rahmat includes the designation of trade and services, green open space and the provision of municipal facilities, among others, general government facilities, offices and public facilities and services trade. The percentage of proportion of land use in the area of research is shown as below:

Table 4. Land use proportion

No.	Land use	Area (Ha)	Percentage (%)
1.	Settlement	0,34	1,65
2.	Trade and Service	19,05	92,4
3.	Facilities	1,04	5,04
4.	Green Open Space	0,178	0,86
	Total	**20,60**	**100**

Source: ArcGIS Calculation, 2014

4. Attraction

Social interactions that occur in the study area occurs because the public space to communicate. Facilities such as restaurant / cafe became strong pull to the public interact. In addition, *"Pedagang Kali Lima"* (street informal vendors) that sell at some point locations of pedestrian paths a special attraction for walkers / non - walkers interact with each other . Based on field observations, other social interaction occurs at night, especially Saturday sidewalks are used for the gathering of motorcycle groups which is quite a lot and occupies almost the entire sidewalk along the corridor Basuki Rahmat:

a. Activities Interest Walking Along Path
   - Frequency of visits to Shopping Center
     Tunjungan Plaza I - IV as one of the commercial centers in the city of Surabaya, the main attraction for the public . Shopping center and an area of 125,000 m² is able to accommodate 78 125 people with the number of people by the standards of human space of 1.6 m² / person.
   - Frequency of Visits to Parks

b. Visual scenery Along Walking Path
   Based on field observations visual scenery in the research area is vegetation, murals, lighting , pavement design, and billboards. Lights, vegetation and billboards in addition to the visual aspects of the city as well as serves as a pedestrian facilities that have been discussed above. Explanations related murals and pavement design can be seen in the following table and figure:

Table 5. Murals and pavements

No.	Pedestrian ways Facilities	Information
1.	Murals	Mural as one aspect of the city visualization can be a main attraction for hikers . In the existing condition of the murals is only found on several sides of the building alone
2.	Pavement design	Pavement design looks dynamic with combination natural stone and ceramics material as well as the provision of motifs and colors in design

Source: Mandasari and Navastara, 2014

Figure 5. Pavements and murals
Source: Mandasari and Navastara, 2014

Public transportation modes that pass through this corridor consists of the MPU, bus, and taxi .

Table 6. Transportation modes

No.	Transport Moda	Information
1.	Bus	Damri : Perak-Bungurasih
2.	Car (MPU)	• Lyn V coklat : Krampung-Joyoboyo • Lyn RT ijo : Rungkut-Bratang-Pasar Turi • Lyn E ijo : Balungsari-Tidar-Karang Menjangan • Lyn DA kuning : JMP-Pasar Atum • Lyn W ijo toska : Karang Menjangan
3.	Taxi	By order

Source: Mandasari and Navastara, 2014

MPU which has quite a lot of modes of transportation routes is the most widely used by the community. Below it shows condition of public transport.

Figure 6. Public transport
Source: Mandasari and Navastara, 2014

3.2. Level of Services (LOS) Analysis

The result of pedestrain traffic counting for each segment (weekday) is :

a. The highest pedestrian flow for segment A1 happened on 12-1 a.m and 5-6 p.m
b. The highest pedestrian flow for segment B1 happened on 7-8 a.m and 12-1 a.m
c. The highest pedestrian flow for segment A2 happened on 12-1 a.m and 5-6 p.m
d. The highest pedestrian flow for segment B2 happened on 7-8 a.m
e. The highest pedestrian flow for segment A3 happened on 7-8 a.m and 5-6 p.m
f. The highest pedestrian flow for segment B3 happened on 7-8 a.m and 5-6 p.m

The result of pedestrain traffic counting for each segment (weekend) is :

a. The highest pedestrian flow for segment A1 happened on 7-8 p.m
b. The highest pedestrian flow for segment B1 happened on 7-8 p.m
c. The highest pedestrian flow for segment A2 happened on 7-8 p.m
d. The highest pedestrian flow for segment B2 happened on 10-11 a.m
e. The highest pedestrian flow for segment A3 happened on 7-8 p.m
f. The highest pedestrian flow for segment B3 happened on 7-8 p.m

All the pedestrian flow above influenced by suroundling landuse especially shopping centre Tunjungan Plaza, restaurant, showroom, bookstore and hotel. After knowing the pedestrian flow we can identify LOS of each segment. LOS of segment A1 is F, means that pedestrians walk very slowly with limited speed. However, based on interview (2015) most pedestrians feel comfort with the capacity. This phenomenon due to the pedestrian movement has not reached the limit with pedestrian speed of 27 m/minute. But, the problem is damaged pedestrian way and median trees along the pedestrian way. The other reason is the interval of each pedestrian is far away due to difference bussines hours of each activities.

LOS of segment B1 is E, means that pedestrians have same speed because of numerous pedestrians. Pedestrians began to feel uncomfortable with this condition. LOS of segment A2, B2, and B3 is B, means pedestrians walk comfortably and fast. LOS of segment A3 is A, means that pedestrians can walk free. It's shown below for more details:

Table 7. LOS per segment

No.	Segment	LOS	Pedestrian Average Speed (m/minute)	Pedestrian Flow (pedes/minute/meter)	Pedestrian Modul Area (m^2/pedes)
1.	A1	F	0,25	2,35	0,11
2.	B1	E	0,71	0,84	0,85
3.	A2	B	1,80	0,31	5,81
4.	B2	B	1,80	0,33	5,45
5.	A3	A	3,00	0,20	15,00
6.	B3	B	1,93	0,31	6,22

After knowing LOS ranks it should be confirmed to an expert how the best solution is. Therefor, researcher did an interview with one of the urban landscape in Surabaya about the walkability concept. The concepts are:

Table 8. Concept according LOS

No.	LOS	Concept
1.	F	Pedestrians space widening is needed; without disruption of vehicles
2.	E	Prioritize pedestrians existence than vehicles
3.	B	Enhancement of experience quality is needed such as increasing of street furniture
4.	A	Equation of pedrestrian ways model on the corridor

4. Conclusion

Pedestrian ways in Tunjungan still need development and improvement with emphasis on walkability and connectivity; safety and comfort; increasing mixed use and attractions. Various level of problems on design, technical and policy requiring treatment simultaneously and holistically. Beside that also the most important is how to push the awareness of all parties such as user to increase *"walking motivation"* and *"green actitude"*, as the local government to increase the commitment to run pedestrian ways programs which are innovative and sustainable.

5. References

[1] Shirvani, Hamid. 1985. *The Urban Design Process*. Michigan: Van Nostrand Reinhold

[2] Surabaya.go.id. 2014. *World Bank Apresiasi Persiapan Pembangunan AMC*. http://www.surabaya.go.id/berita/index.php?Page_page=9&&tot_Hasil=2452&selSortBy=&txtsearch=&act=. Diakses 19 Maret 2014

[3] Suarasurabaya. 2012. *Trem dan Monorel Mulai Pembangunan Fisik 2014*. http://kelanakota.suarasurabaya.net/news/2012/108022-Trem-dan-Monorel-Mulai Pembangunan-Fisik-2014. Diakses 19 Maret 2014

[4] Leccese, Michael, Kathleen, McCormick. 2000. *Charter of The New Urbanism*. McGraw-Hill, Inc. Pennysylvania

[5] Carr, Stephen. 1992. *Public Space: Environment and Behavior*. Cambridge: Cambridge University Press

[6] Fruin, J. (1971). *Pedestrian Planning and Design*. New York: Metropolitan Association of Urban Designers and Environmental Planners, Inc.

[7] Hendrayan et al,. 2013. *Analisis Pelayanan Fasilitas Pejalan Kaki*. Jurnal Ilmiah Elektronik Infrastruktur Teknik Sipil. Volume 2

[8] Kerridge, J., Hine, J., & Wigan, M. (2001). *Agent-based modelling of pedestrian movements: The questions that need to be asked and answered*. Environment and Planning B: Planning and Design, 28(3), 327-341.

[9] Lenakoly, Steven. 2007. *Penilaian 20 Kota, Pedestrian Surabaya Rangking Satu*. http://news.detik.com/read/2007/09/03/100707/824728/466/penilaian-20-kota-pedestrian-surabaya-rangking-satu?nd771104bcj. Diakses 3 April 2014

[10] Lynch, Kevin. 1960. *The Image of The City*. MIT Press. Cambridge

[11] Nicholson, Geoff. 2009. *Lost Art of Walking*. Riverhade Trade. USA

[12] *Pedoman Perencanaan, Penyediaan, dan Pemanfaatan Prasarana dan Sarana Jaringan Pejalan Kaki di Kawasan Perkotaan*. 2014. Direktorat Jenderal Penataan Ruang Kementerian Pekerjaan Umum

[13] Puskarev & Zupan. 1975. *Urban Space for Pedestrian*. The MIT Press. Cambridge

[14] Southworth, Michael. 2005. *Designing the Walkable City*. Journal of Urban Planning and Development:4(246)

[15] Speck, Jeff. 2012. *Walkable City: How Downtown Can Save America, One Step at a Time*. Farrar,

[16] Transportation Research Board. 1985. *Highway Capacity Manual*

[17] *UU No 26 Tahun 2007 tentang Penataan Ruang*. 2007. Sekretariat Badan Koordinasi Tata Ruang

IOP Conf. Series: Earth and Environmental Science **340** (2019) 012008 doi:10.1088/1755-1315/340/1/012008

Analyzing the occupant's awareness, behavior and obstacle in achieving energy efficiency in a campus building

A G Prafitasiwi[1] and M A Rohman[1]

[1] Department of Civil Engineering, Institut Teknologi Sepuluh Nopember, Surabaya, Indonesia

Email: avishagita@gmail.com

Abstract. Campus building is essential to be the pioneer of the energy saving-system in the society. Occupant's behavior should be considered to better formulate an appropriate energy-saving strategy in the campus building. This paper identifies the social parameters in achieving energy efficiency in campus building using a quantitative approach. Data will be collected through survey involving the campus building's occupants. Series of statistical tools consists mean ranking, Analysis of Variance (Anova) and factor analysis will be applied in this research. It is expected that this finding can be used to formulate an appropriate energy-saving system based on the occupant's behavior evaluation.

Keywords: *sustainable construction, awareness, behavior, building occupant, energy efficiency, green building*

1. Introduction

Buildings contribute to energy consumption and gas emissions which have a significant impact to the environment. According to EPA (Environment Protection Agency), the building sector contributes to 39% of total energy use, 12% of total water consumption, 68% of total electricity consumption and 38% of total carbon emissions in all of America. In Indonesia, this sector is responsible for 50% of total energy use, and more than 70% of overall electricity consumption (EECCHI, 2012). From the amount of total energy use, the building sector contributes 30% of gas emissions in Indonesia.

Various programs have been established to anticipate the environmental problems mentioned above. The Ministry of Public Works and Public Housing has set the guidelines for implementing the sustainable construction principle in infrastructure (Permen PUPR No. 5, 2015). Sustainable construction is the implementation of the principle of sustainable development in the construction industry with one of aspects considered is environmental aspect. In building sector, the concept of green building is one of the solutions to environmental aspects.

Green Building is a building concept that is based on the principles and methodology of sustainable construction [1]. Green Building Council Indonesia (GBCI) states that a building that follows the Green Building standard that refers to Greenship is able to make energy savings between 26% to 40% every month. However, although green building principle come with solution to environmental aspects, more attention should be given to social parameters such as awareness and behavior of building occupant as part of a comprehensive energy-saving effort [2]. The energy-saving systems would not be enough to reduce energy consumption if the building occupant were not involved in the process [3].

This research will investigate the social parameters of Institut Teknologi Sepuluh Nopember (ITS) building occupant. As academics, the future thinking about environmental issues is most awaited by the community, because good environmental quality will sustain a good life [4]. Universities have a unique characteristic which require the development of specific research agendas for promoting pro-environmental behavior to achieve sustainability goals [5].

ITS has an initiative in creating a sustainable development with the Smart Eco Campus strategy that has been developed since 2011. Smart Eco-Campus, defined as reflection of the participation of the entire occupant to give more attention in creating safety environment around them, have a program to evaluate and revitalize ITS with a green building standard. However, as highlighted before, the energy-saving systems would not be enough to reduce energy consumption if the building occupant were not involved in the process. Rhodes University recorded more energy reductions of up to 9 percent when student change into greener behavior [6]. It shows that occupant behavior influence energy-saving achievement. As it is important to know the social parameters of the building occupant, this study includes measuring awareness and behavior of ITS building occupant which is conducting of lecturer, staff and student [7] and comparing based on the socio-demographic factor such as gender, age and education which also influence the green living behavior [8].

2. Literature review

Three theories underpinning this research are the principle of green building, energy saving as sustainable goal and energy-saving awareness and behavior as part of the social parameters of energy-saving. To get similar understanding about the concepts, these three theories are reviewed next.

2.1. The principle of green building

Green building is a philosophy and associated project and construction management practices that seek to: (1) minimize or eliminate impacts on the environment, natural resources, and nonrenewable energy sources to promote, the sustainability of the built environment; (2) enhance the health, wellbeing and productivity of occupants and whole communities; (3) cultivate economic development and financial returns for developers and whole communities; and (4) apply life cycle approaches to community planning and development [9]. The reference for green building application in Indonesia refers to GREENSHIP which was formed by Green Building Council Indonesia (GBCI). Green building assessment aspects that refer to GREENSHIP are: 1) appropriate site development (ASD); 2) energy efficiency and conservation (EEC); 3) water conservation (WAC); 4) material resources & cycle (MRC); 5) indoor air health & comfort (IHC); 6) building and environment management (BEM). Therefore, the terms of green building sometimes also referred to as sustainable design, sustainable construction, and other terms previously listed.

As it became part of sustainable construction, a building which referred to green building concept built with an emphasis on seven core principles across the building's life cycle such as reducing resource consumption, reusing resources, using recyclable resources, protecting nature, eliminating toxics, applying life cycle costing, and focusing on quality. It is showed in Figure 1.

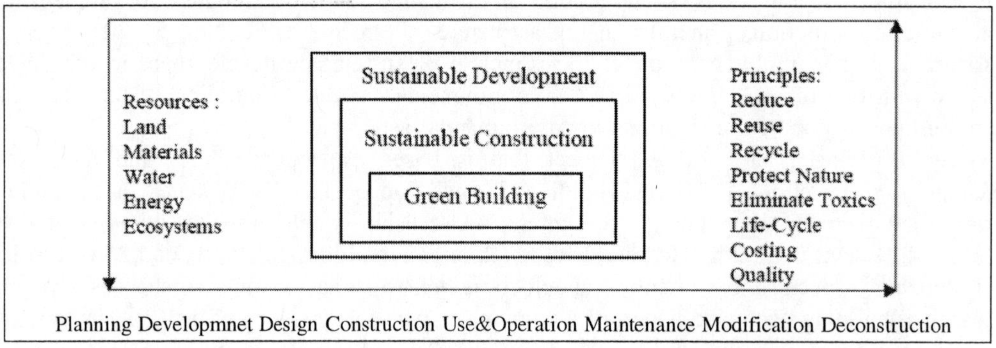

Figure 1. Framework for green building as sustainable construction

In the context of ITS, Smart Eco Campus Program has been initiated to promote energy efficiency across the campus building. Furthermore, several studies had also been conducted to evaluate energy consumption based on the Greenship Standard such as Civil Engineering Department building [10] and Rectorate building [11].

2.2. Energy-saving as sustainable goal

Although sustainable construction generally referred to sustainable development which is considered to environment aspect, social aspect and economic aspect, in 1994, Conseil International du Batiment (CIB) explained that the ultimate goal of sustainable construction is to create and operate a development environment based on resource efficiency and ecological design. Therefore, green building as part of sustainable construction also has a responsibility in achieving resource efficiency such as energy efficiency.

The Green Building Council Indonesia explains that energy efficiency and conservation in the application of green building concepts refers to the aim of encouraging energy savings and encouraging the use of new and renewable energy sources sourced from within the building site. As defined by IEC (Indonesia Environment and Energy Center) in 2018, efficiency is related to the use of less energy to get the same or even more benefits, or use the same energy but produce more benefits.

2.3. Social parameters of energy-saving

This can be possible to gain successful in sustainable construction if all stakeholders cooperate in its implementation and be supportive to changes. As showed in figure 2, not only client and contractor but also government, consultant, and end-user have a responsibility to creating sustainable in construction. The first factor that will start sustainable movement are awareness and knowledge, then behavior changes will only come when stakeholder commitment to it. Two main problems associated with the lack of a user's knowledge about sustainability design and operation with regards to performance are: (1) lack of occupant knowledge about environmental control and operation of the building systems affecting the energy efficiency of the building as originally intended by architects and (2) occupant poor understanding about why the building was designed in a particular manner and how to operate the appliances of the buildings which impact on comfort and satisfaction levels in a sustainable building [12]

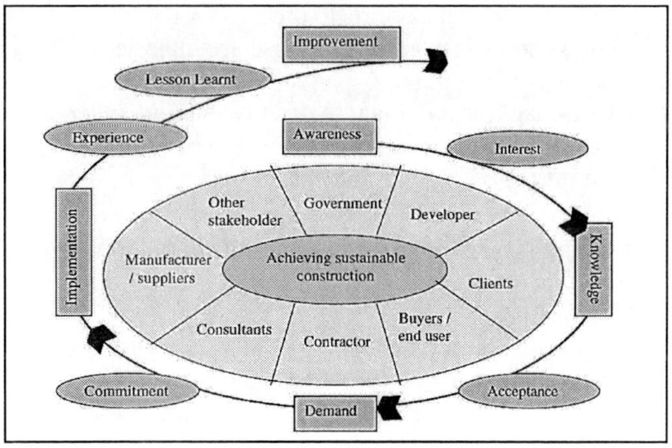

Figure 2. The path for achieving sustainable construction [13]

In social psychology, because there are several things that influence human behavior such as attitudes towards behavior, perceived behavior control, subjective norm, and intention, sometimes it's caused complexity [14]. Several literatures explain that social parameters have become a fundamental barriers in achieving sustainable in construction because its complexity [15][16].

However, in relation to the success in achieving energy saving, the behavior of building occupants has a very important role. In the building sector, social parameters, such as behavior, lifestyles and culture, are one of the factors that most influence energy consumption [2]. Energy-saving behavior refers to consumer decision-making processes related to energy consumption practiced to reduce energy consumption [17]. Energy-saving behavior changes of building occupant can save electricity consumption by up to 10% [18].

Within Indonesia context, there is still limited research has been conducted to investigate energy-saving behavior from the building occupant's perspective. One of the study, Rohi etc in 2013, is measuring the green living behavior in Surabaya residential [19]. The finding of the research shows that there is a not too large gap between the knowledge and the behavior of electricity use. But, the research does not discuss why there are behavior that are contrary to energy saving efforts. Therefore, this research is needed because universities have an opportunity for promoting pro-environmental behavior to achieve sustainability goals, especially in Indonesia. As academics, the future thinking about environmental issues is most awaited by the community.

3. Research methodology

The research consists several steps such as variable identification, preliminary survey, main survey and data analysis.

3.1. Variable identification

Variables identified on this study are awareness and behavior variables that reflect energy efficiency efforts. In this research, will also be investigate the obstacle in green behaving which might be faced by the occupant. These variables are the results of literature review at the initial identification stage, it is showed in table 1, 2 and 3. Table 1 shows the awareness variables, table 2 shows the behavior variables and table 3 shows the obstacles variables from the literature review. These variables will be used in preliminary survey to determine the variables relevance was then performed by interviewing experts.

Table 1. Awareness variables from the literature review

No	Awareness Variable	Source
1	I know that energy saving influences the economic development of a nation	[20]
2	I should recognize the seriousness of resource depletion and make efforts to prepare for it	[20][21]
3	I know most of the energies we use are imported from overseas	[20]
4	I know that it is important to practice energy saving to solve the issue of energy-resource depletion	[20][21]
5	I am interested in environmental issues	[20]
6	I have information and knowledge on energy saving	[20][21]
7	I want practice energy saving cause it will help reduce energy imports, even though it is only I who practices this	[20][21]
8	If I practice energy saving, it will help solve the weather problems resulting from global warming	[20][21]

Table 2. Behavior variables from the literature review

No	Behavior Variable	Source
1	Turning off the tap after use	[22]
2	Alerting staff of a damaged tap	[22]
3	Take shorter shower	[23]
4	Turning off the lights when leaving the WC	[22]
5	Turning off the lights when leaving the classroom	[22][23]

No	Behavior Variable	Source
6	Alerting staff of a blown lamp	[22]
7	Checking if the computer is off when leaving the classroom	[22]
8	Checking if the projector is off when leaving the classroom	[22]
9	Turning off the computer when it is no longer necessary in the classroom	[22]
10	Turning off the projector when it is no longer necessary in the classroom	[22]
11	Turning off the public computer after using it	[22]
12	Unplug chargers and devices when not in use	[6][19]
13	Full use of daylight	[6]
14	Install low energy bulb(s)	[19][21]
15	Use a proper reading lamp	[21]
16	Turn off laptop when not in use	[6][19]
17	Apply electricity-saving mode in laptop/computer if possible.	[18]
18	Set temperature of AC as high as possible	[18]
19	Turn off AC when leave the room	[21]
20	Close the windows and the doors when the AC in use	[21]

Table 3. Obstacles variables from the literature review

No	Obstacle Variable	Source
1	Lack of environmental concern	[24]
2	Lack of awareness of energy saving	[15][16]
3	Lack of policy and legislative measures	[15][24]
4	Lack of communication from stakeholders	[24]
5	Objections from the vested interests groups	[15]
6	Inadequate data and information	[15][16][24]
7	Lack of incentive support and motivation	[15][16][24]
8	Lack of social norms and lifestyles	[24]

3.2. Preliminary survey

As we highlighted before, the variables from literature review will be used in preliminary survey to determine the variables relevance by interviewing experts. Expert is individual who were knowledgeable about energy issues including the role of energy in the design and operations of buildings at least in five years of experience. The number of expert who will be used to preliminary survey is minimum of three people [7]. By using valid and reliable variables, it is hoped that the results of the research will also be valid and reliable [25]. The finding of this stages is used to develop the main survey questionnaire which will distribute to the respondent.

3.3. Main survey

Primary data is the perception of respondents, where data collection carry out at the main survey by distributing questionnaires to the respondent. The respondent of this study is ITS building occupant which is referred to lecturer, staff and student. Secondary data is electric and water consumption bills of ITS for around 1-2 years. The secondary data is used to do energy consumption audit. In this study, the number of sample is taking from certain population, so the sampling method refers to the model developed by Isaac and Michael [25].

In this main survey will be carried out an assessment of the behavior of ITS building occupants in energy efficiency efforts. The questionnaire had closed-ended questions using a 1-5 Likert Scale with 1 representing as "strongly disagree", 2 as "disagree", 3 as "neutral", 4 as "agree" and 5 as "strongly agree".

3.4. Data analysis

Based on the background and the purpose, this study was categorized in descriptive-quantitative. Some explanation techniques of observed data can be explained by using table or diagram. Furthermore, it can be explained by using mean, modus and median [25]. But previously, data analysis would begin with a factor analysis process. Generally, factor analysis is used to reduce and summarize the data. [25] Factor analysis can be done with statistical software such as Statistical Product and Service Solutions (SPSS) using the KMO and Bartlett's test and Anti-image test. [26]

4. Conclusion and recommendation

From the literature review, the research gap has been identified that need to measure the social parameter such as awareness and behavior of the occupant building, especially in campus building. It is also identified that the measurement of energy-saving social parameter in Indonesia has not been widely explored, especially in campus building sector. It is expected that this research finding explains the ITS building occupant's awareness and behavior in supporting energy-saving system. This finding can be used as the improvement tool as well as help the decision-maker to formulate appropriate energy-saving strategy based on the obstacles found to achieve energy–saving target.

5. References

[1] Kibert C J 2005 *Sustainable Construction Green Building Design and Delivery* (New Jersey: John Wiley & Sons, Inc.)

[2] Allouhi A, El-Fouih Y, Kousksou T, Jamil A, Zeraouli Y and Mourad Y Energy consumption and efficiency in buildings: current status and future trends 2015 *Journal of Cleaner Production* **109** 118-130

[3] Siero S, Bakker A, Dekker G and Van den Burg M Changing organizational energy consumption behavior through comparative feedback 1996 *Journal of Environmental Psychology* **16** 235- 246.

[4] Lakswendra A 2015 *Eco campus: studi deskriptif tentang perilaku mahasiswa ITS terhadap program eco campus.* Surabaya : Universitas Airlangga

[5] Adomssent M, Godemann J and Michelsen G Transferability of approaches to sustainable development at universities as a challenge 2017 *International Journal of Sustainability in Higher Education* **8** 385-402.

[6] Bulungaa A A L and Thondhlana G Action for increasing energy saving behaviour in studentnresidences at Rhodes University, South Africa 2018 *International Journal of Sustainability in Higher Education* **19**

[7] Marans R W and Edelstein J Y The human dimension of energy conservation and sustainability A case study of the University of Michigan's energy conservation program 2009 **11**

[8] Mehmetoglu M Factors influencing the willingness to behave environmentally friendly at home and holiday settings 2010 *Scandinavian Journal of Hospitality and Tourism* **10** 430-447

[9] Robichaud L B and Anantatmula V S Greening project management practices for sustainable construction 2011 *Journal of Management in Engineering* **27** 48-57

[10] Putri A A, Rohman M A and Utomo C Penilaian kriteria green building pada gedung teknik sipil ITS 2012 Jurnal Teknik ITS **1**

[11] Darmanto D and Wiguna I P A Penilaian kriteria green building pada gedung rektorat ITS 2013 Jurnal Teknik Pomits **2**

[12] Brown Z and Cole R J Influence of occupants' knowledge on comfort expectations and behaviour'

in Building 2009 Research & Information **37** 227-245.

[13] Abidin N Z Investigating the awareness and application of sustainable construction concept by Malaysian developers 2010 Habitat International **34** 421-426

[14] Kassin S, Fein S and Markus H R 2011 *Social Psychology* (USA : Wadsworth Cengage Learning)

[15] Wang G H, Wang Y X and Zhao T Analysis of interactions among the barriers to energy saving in China 2008 *Energy Policy* **36** 1879-1889

[16] Crossley D J Identifying barriers to the success of consumer energy conservation policies 1983 *Energy* **8** 533-546

[17] Brandon G and Lewis A Reducing household energy consumption: a qualitative and quantitative field study 1999 *Journal of Environmental Psychology* 1999 **19** 75-85

[18] Ouyang J and Hokao K Energy-saving potential by improving occupants' behavior in urban residential sector in Hangzhou City, China 2009 *Energy and Buildings* **41** 711–720

[19] D Rohi and J E Luik 2013 Kesadaran masyarakat surabaya untuk memiliki gaya hidup ramah lingkungan "green living" melalui menghemat penggunaan energi listrik (Surabaya: Universitas Kristen Petra)

[20] Kang N N, Cho S H and Kim J T The energy-saving effects of apartment residents' awareness and behaviour 2012 *Energy and Buildings* **46** 112–122

[21] Jiang P, Chen Y, Xu B, Dong W and Kennedy E Building low carbon communities in China: The role of individual's behaviour change and engagement 2013 *Energy Policy* **60** 611–620

[22] Soares N, Pereira L D, Ferreira J, Conceição P and da Silva P P Energy efficiency of higher education buildings: a case study 2015 International Journal of Sustainability in Higher Education **16** 669-691

[23] Zou P X W and Yang R J Improving sustainability of residential homes: occupants motivation and behavior 2014 *Energy Sector Management* **8** 477-491

[24] Sony M and Mekoth N A qualitative study on electricity energy-saving behaviour 2018 *Quality Management of Environmental Quality* 1477-7835

[25] Sugiyono 2007 *Statistika untuk Penelitian* (Jawa Barat: Alfabeta)

[26] Santoso S 2010 *Statistik Multivariat Konsep dan Aplikasi dengan SPSS* (Jakarta: Elex Komputindo)

IOP Conf. Series: Earth and Environmental Science **340** (2019) 012009 doi:10.1088/1755-1315/340/1/012009

Multi Automated Guided Vehicle (AGV) cardboard carrier using wireless communication

B E Putra[1], A Z Arfianto[1], L Subiyanto[1], Z M A Putra[1], M K Hasin[1], D A Utari[1], M B Rahmat[1], D Hidayat[2], M Taali[3], A Y Lusiandri[4] and V Y P Ardhana[5]

[1] Politeknik Perkapalan Negeri Surabaya
[2] Universitas Harapan Medan
[3] Politeknik Negeri Madiun
[4] Politeknik Pelayaran Surabaya
[5] Universitas Qamarul Huda Badaruddin

Email: afif@afifzuhri.com

Abstract. This paper presents Automated Guided Vehicle (AGV) to move cardboard from one place to another using modul communication method as communication medium between AGV. In previous agv usage agv operation is done automatically with human monitoring to allow agv to operate within its working area and no accident occurs between agv. In this study, the input signal value of the received signal strength indicator (RSSI) method is used to control and regulate the alternate workspace. The test results show that the use of the received signal strength indicator (RSSI) method can optimize AGV work with work orders that are run interchangeably when some agv go to the same point. Furthermore, Line Follower system can accelerate the operation of Automated Guided Vehicle (AGV) on the determination of the operation path.

Keywords: *Automated Guided Vehicle, Received Signal Strength Indicator, NRF24l01*

1. Introduction

In the modern era, technological developments are getting increasingly advanced in various fields of industry. In the industrial world robotics technology is used to help solve existing problems. Today's technology, automatic vehicle to vehicle communication and vehicle to infrastructure communication using nrf24l01 module [1], has illustrated how to measure the distance between Automated Guided Vehicle (AGV) using NRF24l01 as a module. In this research, it is about how to find the distance using value received signal strength indicator (RSSI) to communicate between agv. Currently the application of control theory becomes more frequent to solve real problems.

Rusdinar et al [2] reduced the disruption of vehicle that was vertically measured from horizontal view using extended Kalman filter and artificial landmark on ceiling. This research discusses the movement control algorithm of weighted vehicle by trolley using Fuzzy Inference System [3]. Sensing and tracking of a moving object/human by a robot is an important topic of research in the field of robotics and automation for enabling collaborative work environments [4]. Jaiganesh et al entitled Automated Guided Vehicle with Robotic Logistics System[5] measured the distance between two nodes (we used NodeMCU as sensor node). The NodeMCU is an Arduino type board which runs on ESP8266 module

[6] indoor positioning is an interesting topic and an enabler in several fields of application, and consequently subject to a significant number of research activities [7]. Vehicle intelligent position systems based on Received Signal Strength Indicator (RSSI) in Wireless Sensor Networks (WSNs) are efficiently utilized [8]. Received Signal Strength Indicator (RSSI) location algorithm is a range based localization algorithm. It uses the connection between communication distances and the received signal strength to calculate the coordinates of unknown nodes[9].

Various techniques have been proposed in recent years, Broadbent et al [10] proposed the shortest time path planning problem without collision; MinLv et al [11] proposed a multi-AGV path planning algorithm based on directed graph. The biggest advantage of this algorithm is its efficiency. However, the shortcomings are obvious. It is only applicable to the directed map, making map utilization low. QiSun et al [12] used the improved Dijkstra algorithm to achieve global path planning. There are many guidance systems available and the selections are based on need, application, and environmental constraints. It is better to have guidance system which can be modified at low cost than high cost of modifying fixed-path equipment. Many researchers, tried to implement different methods in guiding the AGV on the production area such as magnetic sensor [13], On-board Camera [14][15]. Changing the values of RSSI due to the transmitters and receivers non idealities [16] and the influence the variability and accuracy of RSSI measurements. The device used as a wireless transmission medium that is on the wireless sensor network can be regarded as a transceiver module. Transceiver module utilizes radio frequency in digital data transmission. Examples of transceiver modules include nRF24L01, Xbee and Wifi Module ESP2866.[17].

In this research, we proposed an RSSI modification with the NRFl2401 module to control communication between AGVs. This study was conducted to overcome accidents and goods in the form of cardboard. We changed the speed by determining the direction and velocity caused by the creation of instability motion. Movement instability directly affects the related items. Motor speed is moving to make the vehicle move steadily. Input from the system results from ultrasonic motion. The selection of the sensors is placed on the location of the vehicle, which is outside the room. Because it is very susceptible to disturbances such as light, nrfl240 communication modem is used. The module becomes the best choice because it can be connected with AGV of this result and then into system input. Data input is processed in Primary Control System. The output is motor speed and parameter for PID controller.

In this paper, an innovative concept about a robot chain for cardboard carrier based on wireless communication has been presented. The multiple robots which coordinate and cooperate as a team is called robot swarm. There are a lot of related studied about robot swarm, such as aggregation collective behavior, pattern formation collective behavior, chain formation collective behavior, and self-assembly collective behavior [18], [19].

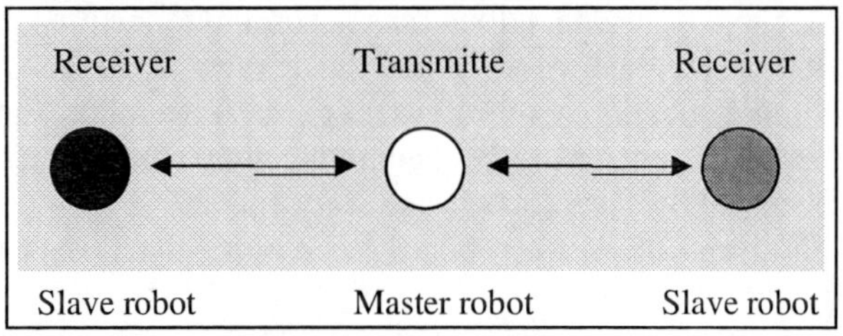

Figure 1. Master-slave-based robot chain wireless communication

IOP Conf. Series: Earth and Environmental Science **340** (2019) 012009 doi:10.1088/1755-1315/340/1/012009

2. Methodolgy

2.1. Hardware design (Automated Guided Vehicle)

The purpose of this research is to identify In hardware parts which have arduino as control system to run automated guided vehicle, to carry the cardboard using gripper in motion using servo motor and at the bottom there is a color sensor to detect the track path with the propulsion using 12v dc motor. In this system, communication wireless is used to find signal so communication between automated guided vehicle can occur.

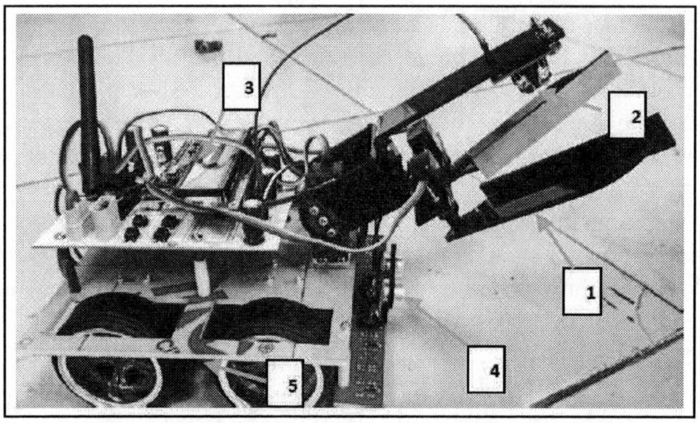

Figure 2. Hardware design system
Caption :
1. Gripper
2. Color sensor TCS3200
3. Main Control
4. Ultrasonic sensor
5. 12v DC Motor

2.2. Software – hardware integration

- Prepare materials and equipment to be created
- Prepare the required software install
- Construct the photodiode and other required hardware
- Open the arduino software and program the hardware to be created
- Incorporating the program on the master Robot and on the Slave Robot 1 and Slave Robot 2
- Construct the required track path
- Running Master Robot, Slave Robot 1 and Slave Robot 2.
- Data received by the photodiode which then runs according to the predetermined route
- The master robot meets the slave robot 1 and slave robot 2. Then the signal data from the master robot using communication wireless is sent to the slave robot 1 and slave robot 2.
- In master robot, the data is processed by communication wireless that produces signal with output of distance accepted by slave robot 1 and slave robot 2.
- After the slave robot 1 and the slave robot 2 receive the signal data from the master robot, the slave robot 1 and the slave robot 2 walk out the track and stop
- The master robot moves the item according to the color to the destination

IOP Conf. Series: Earth and Environmental Science **340** (2019) 012009 doi:10.1088/1755-1315/340/1/012009

2.3. System design control

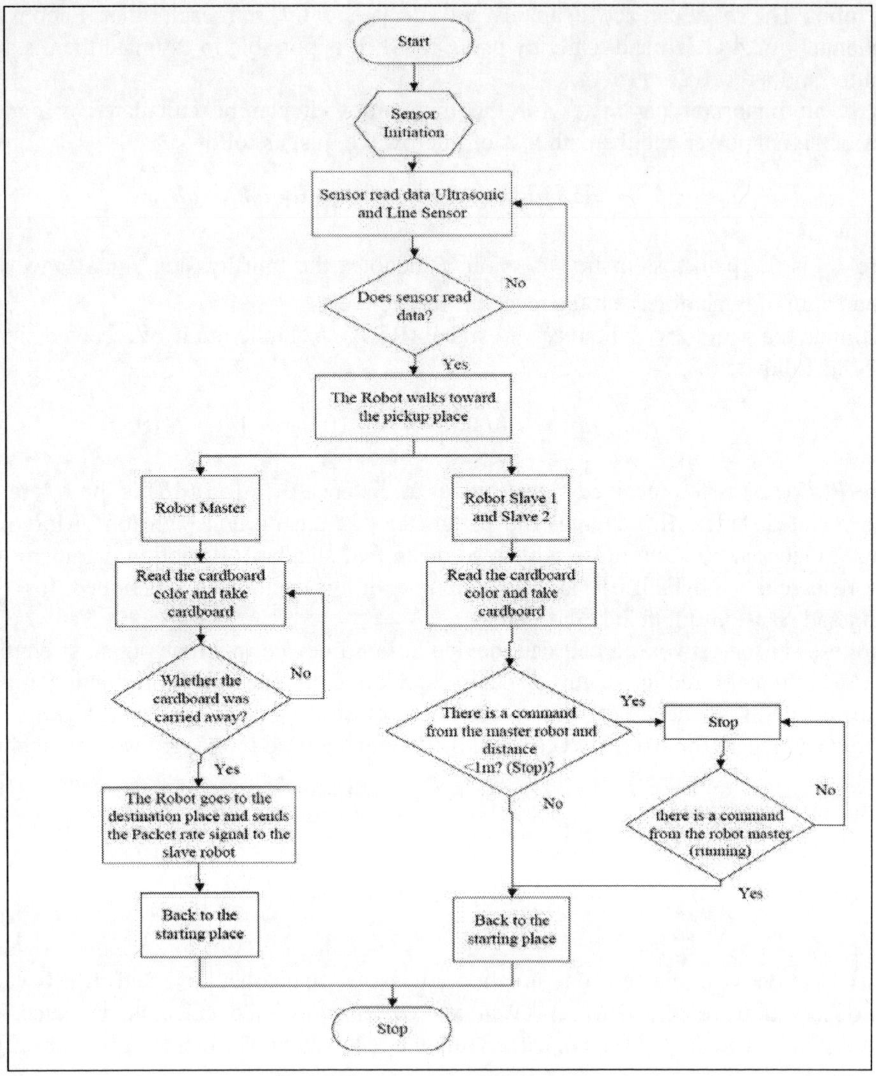

Figure 3. System Design Control

The picture above shows the system design of robot movement Automated Guided Vehicle (AGV) Master and Slave from the beginning to the end position. The system is continuous so when the robot has reached the destination, it is re-positioned again. Here the sensor is used to help the robot to be able to run well on its track and deliver the goods in accordance with the specified color, and the line sensor is used as a marker of the destination point of the robot and as a counter. Here the robot will start running when the goods have been received by the sensor, once it is read, the robot will walk to the location addressed by bringing goods according to color and send the RSSI signal to the slave robot, and then the slave robot will stop. The master robot and the slave robot will return to the original position when the task is fulfilled.

2.4. RSSI-based reliable communication

RSSI is a parameter indicating the power assigned to the channel frequency shown. This research is used to search for the strongest value and to communicate with the robots. In our proposed method, each agv robot is equipped with only a 2.4-2.5 GHz NRF24L01 module for communication between robots, which is also used for estimating the short distance between the robot and other adjacent robot. The robot can act separately and independently from each other. Each robot owns its unique channel of 2.4 GHz band. Only by using RSSI, it is possible to estimate the distance between a transmitter and a receiver of robot.

RSSI is an important parameter for the transmitter. It can be calculated using the measured transmission signal power and the path loss of microwave, just as follows:

$$RSSI(dB) = P_i(db) + G_i + G_r - L_{bf}(dB) \tag{1}$$

Where L_{bf} is the path loss of transmission, P_i denotes the transmission signal power, G_i and G_r is the antenna gain of transmitter and the receiver, respectively.

To estimate the path loss of the received signal (RSSI), a classic mode of received signal power can be defined as follows:

$$P(d_i)(db) = P(d_o)(db) + 10\gamma 10 \log_{10} \left(\frac{d_i}{d_o}\right) + N(0, \sigma) \tag{2}$$

Where $P(di)(dB)$ is the received signal power at distance di, $P(do)(dB)$ is the reference received signal power at distance di. γdenotes the parameter of transmission path loss, Moreover, $N(0, \sigma)$ represents a Gaussian random noise with zero mean and standard deviation parameter of σ. In our previous research, a NRF24L01 based communication system was developed for continuously generating and transmitting in 2.4 GHz band.

In wireless sensor network, a cell data device is radio device in which signal strength and signal quality are both measured in dBm(i.e. decibels relative to one milliwatt) and connected to the cellular tower. RSSI is a negative dBm value, values closer to 0 dBm are strong signals. Here we get an idea how and where the NodeMCU (NRF24L01) may be used in modern wireless technology.

$$P_{out} = 10^{\frac{dBm}{10}} \tag{3}$$

$$dBm = 10 \times log_{10}[\frac{P_{out}}{1mW}] \tag{4}$$

In this section, we discuss our methodologies to map the observed RSSI vector, **rv**[n], into the observation vector, **O**[n]. *A. Distance Observations* To calculate distance, we employ one of the standard equations used in calculating the received power in a RF channel [20]:

$$\begin{aligned} P_{r,dBm} &= P_{t,dBm} + G_{dB} - \mathcal{L}_{ref} - 10\eta \log_{10} \frac{d^m[n]}{d_{ref}} + \psi \\ P^{ref}_{r,dBm} &= P_{t,dBm} + G_{dB} - \mathcal{L}_{ref} + \psi \\ \Longrightarrow \frac{d^m[n]}{d_{ref}} &\approx 10^{\frac{(P^{ref}_{r,dBm} - P_{r,dBm})}{10\cdot\eta}} \end{aligned} \tag{5}$$

where Pr,dBm = avg {**rv**[n]} is the average received power in our system, Pt,dBm is the transmitter power in dBm, GdB is the average antenna gain in dB, $\mathcal{L}ref$ is the path loss at the known reference distance $dref$ in dB, η is the path loss exponent, $dm[n]$ is the distance between the transmitter and receiver, $\psi$is the random shadowing and multipath fading noise in dB, and $Pref$ r,dBm is the pre- estimated received power at the reference distance ($dref$). More comprehensive details can be found in [21].

3. Testing and data analysis

The testing mechanisms include Collision Simulation, Tracking of the AGV cardboard carrier and simulating the distance between robots AGV by using RSSI to know the desired data analysis.

3.1. Collision simulation

Testing the whole system is done by using a sample of three prototypes, the scenario carried out in this experiment is simulated collision agv with colliding direction. The success parameter of this test is the occurrence of change in the position of the track by observing the distance and time of the experimental results. Then for parallel experiments it takes three prototypes that already have agv control circuit. The three prototypes will later be summed up as agv with different direction of movement as illustrated below:

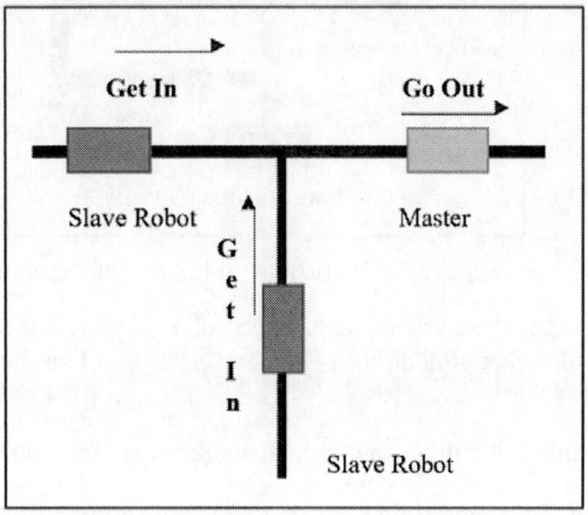

Figure 4. Testing of prototype collision simulation

The tests require a large enough space of up to 10 square meters, with slippery floor. The reference point used during the test is a reference point for movement in the Get In and Go Out areas. Further testing of the work response tool can be done by following the scenario according to the table below:

Table 1. Training data to find simulated collision testing scenarios

No	Prototype			Work Response
	Slave Robot 1	Slave Robot 2	Master Robot	
1	Get In	Get In	Get In	Work
2	Go Out	Go Out	Go Out	Not Work
3	Stop	Get In	Get In	Work
4	Stop	Get In	Go Out	Work
5	Stop	Go Out	Get In	Work
6	Stop	Go Out	Go Out	Work
7	Get In	Stop	Get In	Work
8	Get In	Stop	Go Out	Work
9	Go Out	Stop	Get In	Work
10	Go Out	Stop	Go Out	Work

3.2. Track of the AGV cardboard carrier

Overall testing by means of testing hardware that has been connected with another test is done by using the path and goods that have been determined. This test is intended to find the right program to be used on the robot in order to run as desired.

IOP Conf. Series: Earth and Environmental Science **340** (2019) 012009 doi:10.1088/1755-1315/340/1/012009

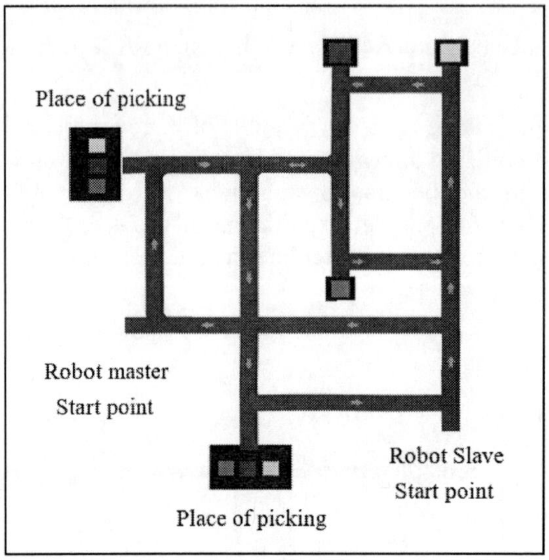

Figure 5. The path the robot passes through

In this experiment, the color sensor can capture the color. If the color sensor has not been able to capture the desired color, a darker similar color will be selected so that the goods carried is based on the color specified.

Table 2. Training data to find simulated appropriate color

No	Traveling time	Appropriate Color	%Error
1	21.47	Corresponding	20 %
2	20.62	Corresponding	
3	19.55	Corresponding	
4	21.06	Corresponding	
5	20.59	Corresponding	
6	Fail	Fail	
7	Fail	Fail	
8	20.66	Corresponding	
9	20.04	Corresponding	
10	20.22	Corresponding	

3.3. Simulating the distance packet rate between robots AGV using wireless communication

Testing simulation of distance between agv using rssi method is then used for communication between agv. Below is the graph obtained from the three agv and the result is a good or small data value.

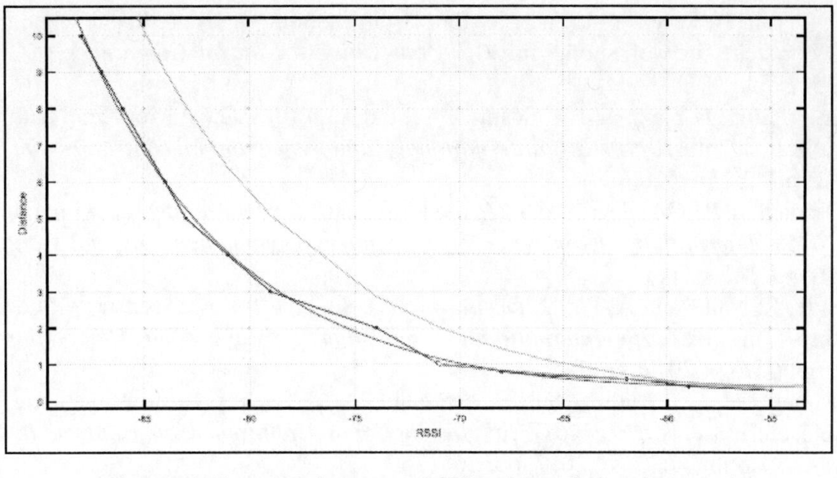

Figure 6. Compare between AGV estimated signal strength

4. Conclusion

We propose an AGV system of cardboard carriers and communication between wireless-based AGV, to autonomously communicate system between agv transmitter and receiver NRF24L01, and show our system performance and error statistics. However, there are many research questions that need to be addressed in our work in the future. First, we want to develop a carrier strategy with the right trade- off between Optimism and Pragmatism, which has the potential to improve performance. Second, we want to conduct a thorough evaluation of our system through a series of large-scale experiments. Lastly, we want to explore the domain of game theory to see if better or stronger predictions about the Leader's movement can improve performance.

5. References

[1] N. Pothirasan and M. P. Rajasekaran, "Automatic vehicle to vehicle communication and vehicle to infrastructure communication using NRF24L01 module," *2016 Int. Conf. Control Instrum. Commun. Comput. Technol. ICCICCT 2016*, pp. 400–405, 2017.

[2] S. Silvirianti, A. S. R. Krisna, A. Rusdinar, S. Yuwono, and R. Nugraha, "Speed control system design using fuzzy-pid for load variation of automated guided vehicle (agv)," *Proc. 2017 2nd Int. Conf. Front. Sensors Technol. ICFST 2017*, vol. 2017–Janua, pp. 426–430, 2017.

[3] R. K. A. Sakir, A. Rusdinar, S. Yuwono, A. S. Wibowo, Silvirianti, and N. T. Jayanti, "Movement control algorithm of weighted automated guided vehicle using fuzzy inference system," *2017 2nd International Conference on Control and Robotics Engineering (ICCRE)*. IEEE, pp. 135–139, Apr-2017.

[4] P. Ghosh, J. A. Tran, and B. Krishnamachari, "ARREST: A RSSI Based Approach for Mobile Sensing and Tracking of a Moving Object," 2017.

[5] V. Jaiganesh, J. Dhileep Kumar, and J. Girijadevi, "Automated guided vehicle with robotic logistics system," *Procedia Eng.*, vol. 97, no. December, pp. 2011–2021, 2014.

[6] S. Barai, D. Biswas, and B. Sau, *Estimate distance measurement using NodeMCU ESP8266 based on RSSI technique*, vol. 2018–Janua. 2018.

[7] A. De Angelis *et al.*, "Design and Characterization of a Portable Ultrasonic Indoor 3-D Positioning System," *IEEE Trans. Instrum. Meas.*, vol. 64, no. 10, pp. 2616–2625, 2015.

[8] Z. Cao, Z. Yuan, and S. Zhang, "Experimental exploration of rssi model for the vehicle intelligent position system," *J. Ind. Eng. Manag.*, vol. 8, no. 1, pp. 51–71, 2015.

[9] S. Jia and C. Yang, "Received Signal Strength Indicator Node Localization Algorithm Based on Constraint Particle Swarm Optimization," *TELKOMNIKA (Telecommunication Comput. Electron. Control.*, vol. 13, no. 1, p. 221, 2015.

[10] C. Liu, J. Tan, H. Zhao, Y. Li, and X. Bai, "Path planning and intelligent scheduling of multi-AGV systems in workshop," in *2017 36th Chinese Control Conference (CCC)*, 2017, pp. 2735–2739.

[11] Q. Sun, H. Liu, Q. Yang, and W. Yan, "On the design for AGVs: Modeling, path planning and localization," in *2011 IEEE International Conference on Mechatronics and Automation*, 2011, pp. 1515–1520.

[12] I. S. Derpich and J. M. Sepulveda, "A model for storage facility design with energy costs," in *2016 6th International Conference on Computers Communications and Control (ICCCC)*, 2016, pp. 147–150.

[13] K. Khade, R. Naik, and A. Patil, "Design of all color line follower sensor with auto calibration ability," in *2017 7th International Symposium on Embedded Computing and System Design (ISED)*, 2017, pp. 1–5.

[14] P. Karimaghaee, A. Amidi, and E. Roshandel, "A high precision magnetic based linear displacement sensor," in *2017 5th International Conference on Control, Instrumentation, and Automation (ICCIA)*, 2017, pp. 265–269.

[15] J. Sankari and R. Imtiaz, "Automated guided vehicle (AGV) for industrial sector," in *2016 10th International Conference on Intelligent Systems and Control (ISCO)*, 2016, pp. 1–5.

[16] D. Capriglione, L. Ferrigno, E. D'Orazio, V. Paciello, and A. Pietrosanto, "Reliability analysis of RSSI for localization in small scale WSNs," in *2012 IEEE International Instrumentation and Measurement Technology Conference Proceedings*, 2012, pp. 935–940.

[17] U. J. Shobrina, R. Primananda, and R. Maulana, "Analisis Kinerja Pengiriman Data Modul Transceiver NRF24l01, Xbee dan Wifi ESP8266 Pada Wireless Sensor Network," *J. Pengemb. Teknol. Inf. dan Ilmu Komput.*, vol. 2, no. 4, pp. 1510–1517, 2018.

[18] M. P. Vicmudo and E. P. Dadios, "Artificial neural network controller for maintaining underwater swarm robots' wireless connections," in *2015 International Conference on Humanoid, Nanotechnology, Information Technology, Communication and Control, Environment and Management (HNICEM)*, 2015, pp. 1–6.

[19] A. Marjovi, J. Nunes, P. Sousa, R. Faria, and L. Marques, "An olfactory-based robot swarm navigation method," in *2010 IEEE International Conference on Robotics and Automation*, 2010, pp. 4958–4963.

[20] T. S. Rappaport and Theodore, *Wireless communications: principles and practice*. Prentice Hall PTR, 2002.

[21] P. Ghosh, J. A. Tran, and B. Krishnamachari, "ARREST: A RSSI Based Approach for Mobile Sensing and Tracking of a Moving Object," Jul. 2017.

IOP Conf. Series: Earth and Environmental Science **340** (2019) 012010 doi:10.1088/1755-1315/340/1/012010

Analysis of air quality influenced by traffic density using web based geographic information system (Study case: Central Surabaya)

B M Sukojo[1] and M Zulfa[1]

[1] Geomatics Engineering, Institut Teknologi Sepuluh Nopember

Email: bangunms@gmail.com

Abstract. Central Surabaya is a city center where there are a lot of shopping centre and office buildings which populate the street. High traffic rate increases air pollution caused by motor vehicle, thus resulting in bad air quality for health. Carbon emission that has the biggest effect is CO and CO2, which can amplify global warming rate. The process of air quality monitoring itself, however, lacks a lot of things, especially in keeping data log and information delivery, causing inequal information spread for citizen about air pollution in Surabaya. Geographic Information System (GIS) technology, which is served in web platform, is one way to assist information delivery to Surabaya citizen about the air quality. The result of this study shows that the air quality is still good and doesn't go beyond the threshold of national ambient air quality. Primary arterial road category, Jalan Diponegoro, has highest emission rate of 1635.64 ton/year CO and 35516.32 ton/year CO2 caused by high volume of vehicle, which is 285388 smp/hour. For secondary arterial road category, Jalan Raya Darmo, has highest emission rate of 1491.72 ton/year CO and 31155.44 ton/year CO2 caused by vehicle 732395 smp/hour vehicle. Result of this study is served in web platform.

Keywords: *Air Quality, Web GIS, Central Surabaya*

1. Introduction

Surabaya is the biggest city in Indonesia, only second to Jakarta, with the number of citizen reaching 3.105.679 millions people in a 33306,30 Ha wide area which has to cater to the needs of people with various social and economic condition [1]. The development of technology and lifestyle of people creates dynamics to the people's life pattern in the area, especially true in Surabaya. In social sector urbanization increases people's growth. This causes the degradation of living environment quality and water as well as air pollution that becomes a problem every year. Air quality suffers great damage in particular, as Central Surabaya, being the centre of the city economics shown by numerous office building, malls, and markets requires streams of vehicles to commute in mass.

The air quality degradation is caused by air pollution (emission) from motor vehicle populating the traffic. the highest influence comes from carbon emission. This is from the usage of fossil fuel in transportation activity. Controlling effort is continuously being done by Surabaya City Government to ensure the pure air quality and to mitigate negative impact towards the health and welfare of human, animals, plants, and materials. Negative impacts as in thinning of ozone layers, decrease in atmosphere oxydation, and global warming [2].

Currently the air quality monitoring in Surabaya City is done through installation of ambient air quality monitor network. This network will do the measurement directly to various pollutant area. The measurement result will be processed into average value of one day and will be informed to the citizens as value of the day without prediction of the next day's value. In that process of environment development monitoring, the aspect of data keeping and information delivering about air pollution in Surabaya as hardcopy and analog data still have a lot to improve, as the current method doesn't deliver the information evenly to the citizens.

One of applicable technology that can be a solution to that shortcoming is Geographic Information System (GIS) served in web form so that the information given will be spread evenly. WebGIS in this study gives information about air quality based on traffic density and pollution rate especially in arterial road. Traffic density is the division result of traffic number to area with unit/area of a certain area [3]. Arterial road is public road to serve main transportation which travels far and can be passed with vehicle with high average velocity and efficiently limited by distribution service for native citizen. Arterial road is divided into primary arterial road and secondary arterial road. Map using is also important in WebGIS for this study to get clear bearing of traffic density and air pollution rate location.

2. Method

2.1. Study location

The location of this study is Central Surabaya, precisely in 7°12' - 7°21' LS and 112°36'- 112°54' BT.

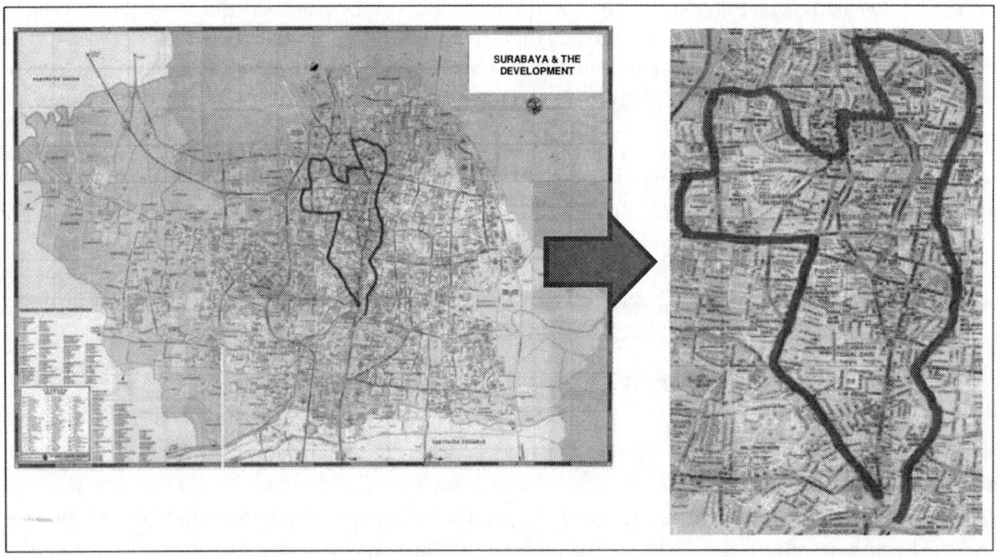

Figure 1. Study Location

Surabaya Central Border

North side	: North Surabaya
East side	: East Surabaya
South side	: South Surabaya
West side	: West Surabaya

2.2. Tools

In this Final Project study, tools to process data are required, as such:

a. Hardware
   1. Laptop
   2. Printer

3. GPS Handheld

b. Software

1. Operation System Windows 7.
2. Word processor for report documenting.
3. Number processor for calculation.
4. Flowchart making programme.
5. Conversion programme to convert format ".shp" to ".jpeg".
6. Map processing Software for spatial data processing, which is digital map editing.

2.3. *Materials*

Materials or data used in this study is spatial data. Spatial data is data with geographical orientation, with base references as certain coordinate system. Spatial data used are as follow:

a. Digital Map of Central Surabaya 1:25.000 from Dinas Cipta Karya Kota Surabaya.

b. Attribute data of CO and CO_2 emission load from Lembaga Penelitian dan Pengabdian Masyarakat (LPPM) ITS.

2.4. *Data Processing Step*

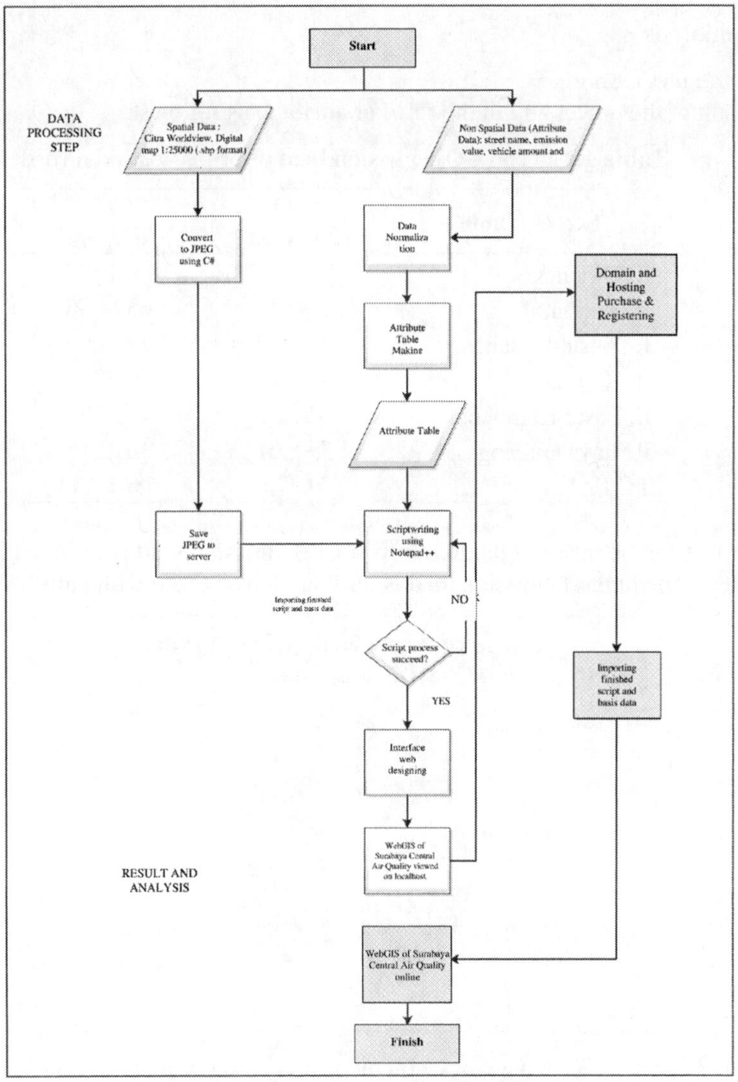

Figure 2. Data Processing Flowchart

1. Basis data making and normalizing data retrieved from field survey using Microsoft Access to make the emerged data more structurized and appropriate according to GIS standard.
2. Map with .shp format will be formatted as JPEG because it doesn't use Mapserver. The convert process from shp to jpeg uses C# programming language with software Microsoft Visual Studio. The result of those converts are saved into folder images in server.
3. After the images are converted and saved in folder images in server, the images will be accessed by php programme which will be made at script writing.
4. The script writing process uses Notepad++. If error or failure happen at the application test then the written script needs to be observed to fix the web interface thus ease the user in doing web functions.
5. Interface web making is done to design the desired website appearance.
6. After all the scripts are done and interface web is designed, webGIS can be viewed on localhost (XAMPP).
7. Hosting and domain are needed to publish the web. Hosting is a place to store files that have been made. Domain is a web address. This Final Project uses the domain name www.emisisurabaya.com.

3. Result and analysis

3.1. Attribute data processing

Below is the result of the processing in the form of attribute as followst:

Table 1. CO and CO_2 emission load of primary arterial road

Street Name	CO (ton/year)	CO_2 (ton/year)
Jl. Diponegoro	1635,64	35516,32
Jl. Kalibutuh	296,47	6874,86
Jl. Kusuma Bangsa	329,32	7744,06
Jl. Kapasari	185,14	4179,5
Jl. Pasar Kembang	467,32	9690,90
Jl. Raya Gubeng	215,61	4612,17
TOTAL	**3129,5**	**68617,82**

From Table 1, it can be incurred that the lowest CO_2 emission load is on Jalan Kapasari with value of 4179,5 ton/year. The highest emission load is on Jalan Diponegoro with value of 35516,32 ton/year.

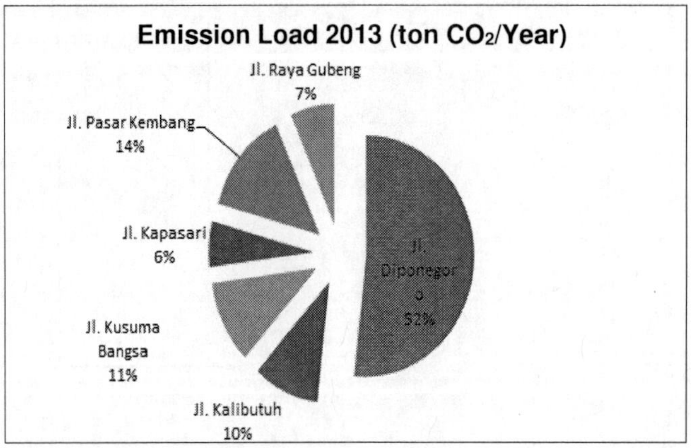

Figure 3. Percentage of CO_2 emission load on primary arterial road

From Figure 3 it can be incurred that the biggest percentage is on Jalan Diponegoro lane, 52%, and the smallest is on Jalan Kapasari, 6%.

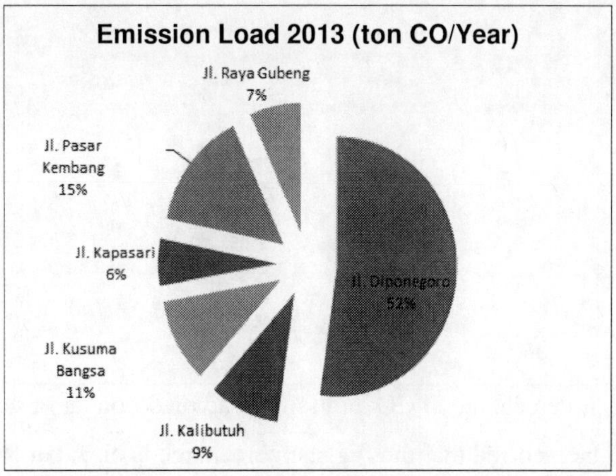

Figure 4. Percentage of CO emission load on primary arterial road

From Figure 4 it can be incurred that the biggest percentage is on Jalan Diponegoro lane, 52%, and the smallest is on Jalan Kapasari, 6%. Both CO and CO_2 emission load in 2013 shows the same value, with the highest percentage on Jalan Diponegoro and the lowest percentage on Jalan Kapasari.

Table 2. Carbon CO and CO_2 Emission Load on Secondary Arterial Road

Street Name	CO (ton/year)	CO_2 (ton/year)
Jl. Blauran	89	1747,13
Jl. Bubutan	989,67	28862,64
Jl. Embong Malang	86,91	2123,16
Jl. Gemblongan	76,87	1365,99
Jl. Gubernur Suryo	190,31	4570,86
Jl. Jagalan	34,19	809,51
Jl. Basuki Rahmat	428,5	10936,21
Jl. Kalianyar	45,91	884,23
Jl. Kapasan	98,42	2390,19
Jl. Kramat Gantung	67,83	1498,69
Jl. Ngaglik	96,76	1720,33
Jl. Pahlawan	126,26	2688,33
Jl. Pandegiling	113,18	1631,09
Jl. Panglima Sudirman	1099,14	28091,98
Jl. Pasar Besar	71,03	1658,58
Jl. Raya Darmo	1491,72	31155,44
Jl. Tunjungan	192,38	4463,07
Jl. Urip Sumoharjo	547,4	15611,56
TOTAL	**5845,48**	**142209,00**

From Table 2, it can be incurred that the lowest CO_2 emission road is on Jalan jagalan with 809,51 ton/year value. The highest value is on Jalan Raya Darmo with 31155,44 ton/year.

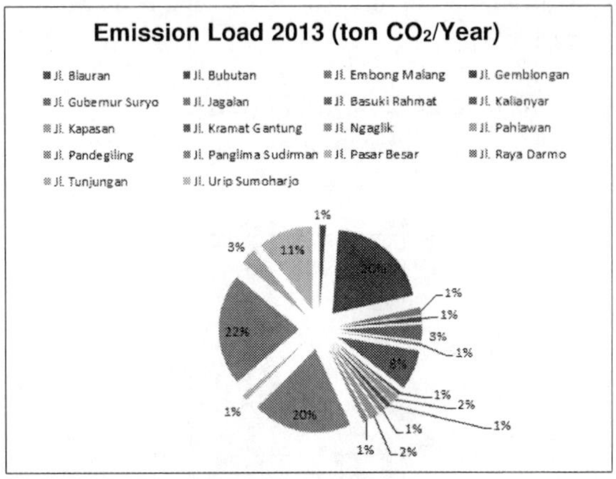

Figure 5. Percentage of CO_2 emission load on secondary arterial road

From Figure 5 it can be incurred that the biggest percentage is on Jalan Raya Darmo, 22%, and the lowest is on Jalan Pasar Besar, Blauran, Embong Malang, Kramat Gantung, Jagalan, Kalianyar, Gemblongan, Ngaglik, and Pandegiling with 1%.

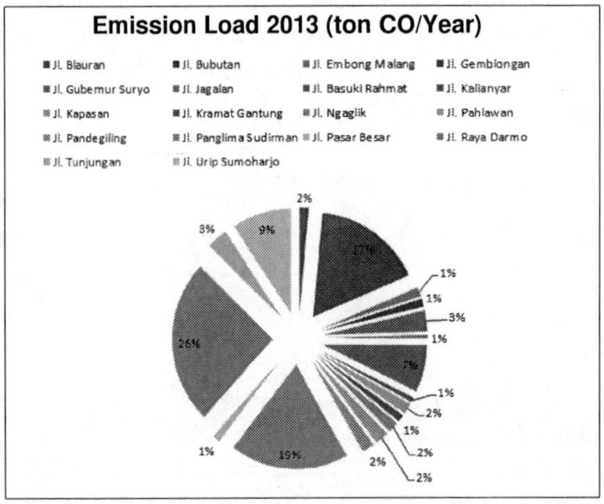

Figure 6. Percentage of CO emission load on secondary arterial road

From Figure 6 it can be incurred that the biggest percentage is on Jalan Raya Darmo, 26% and the smallest is on Jalan Pasar Besar, Embong Malang, Kramat Gantung, Jagalan, Kalianyar, and Gemblongan with 1%.

Table 3. Vehicle volume of primary arterial road

Street Name	CO (ton/year)	CO_2 (ton/year)	Vehicle (smp/hour)
Diponegoro	1635,64	35516,32	285388
Kalibutuh	296,47	6874,86	144480
Kusuma Bangsa	329,32	7744,06	100407
Kapasari	185,14	4179,5	98928
Pasar Kembang	467,32	9690,90	284887
Raya Gubeng	215,61	4612,17	43956
Total	3129,5	68617,82	958046

Table 4. Vehicle volume of secondary arterial road

Street Name	CO (ton/year)	CO_2 (ton/year)	Vehicle (smp/hour)
Blauran	89	1747,13	128400
Bubutan	989,67	28862,64	202563
Embong Malang	86,91	2123,16	59815
Gemblongan	76,87	1365,99	84291
Gubernur Suryo	190,31	4570,86	162700
Jagalan	34,19	809,51	31120
Basuki Rahmat	428,5	10936,21	172443
Kalianyar	45,91	884,23	56840
Kapasan	98,42	2390,19	50080
Kramat Gantung	67,83	1498,69	51162
Ngaglik	96,76	1720,33	54920
Pahlawan	126,26	2688,33	51160
Pandegiling	113,18	1631,09	29920
Panglima Sudirman	1099,14	28091,98	265757
Pasar Besar	71,03	1658,58	51920
Raya Darmo	1491,72	31155,44	732395
Tunjungan	192,38	4463,07	107800
Urip Sumoharjo	547,4	15611,56	282404
Total	5845,48	142209,00	2575690

On the primary arterial road table it shows that the street with most traffic is Diponegoro street with vehicle amount 285388 (smp/hour) that it causes it having highest emission value with CO value 1635,64 (ton/year) and CO_2 35516,32 (ton/year). On the secondary arterial road table it shows that the street with the most traffic is Jalan Raya Darmo with vehicle amount 732395 (smp/hour) that it causes it having highest emission value with CO value 1491,72 (ton/year) and CO_2 31155,44 (ton/year)

3.2. Spatial data processing

In this study, the spatial data processing uses programme C# because of its usage of HTML, PHP, and Javascript language. This programme can't read .shp data, however, thus map conversion is done from .shp to JPEG so the map can be viewed on web.

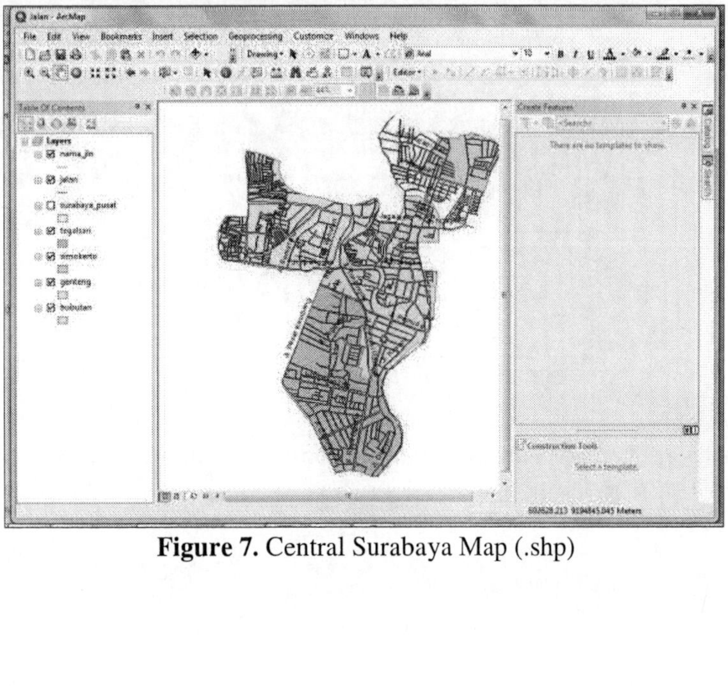

Figure 7. Central Surabaya Map (.shp)

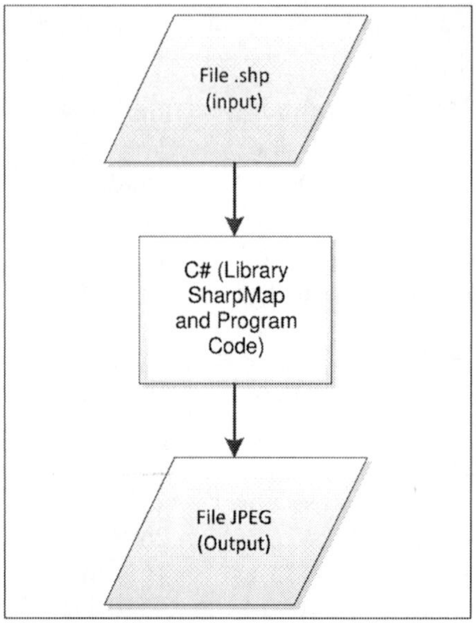

Figure 8. Conversion Diagram

Steps to convert (.shp) map into JPEG form are as follows:

1. Open .shp file from C# programme (Visual studio)
2. Determine the zoom on the picture
3. Save the .shp file in the C# programme into JPEG file type

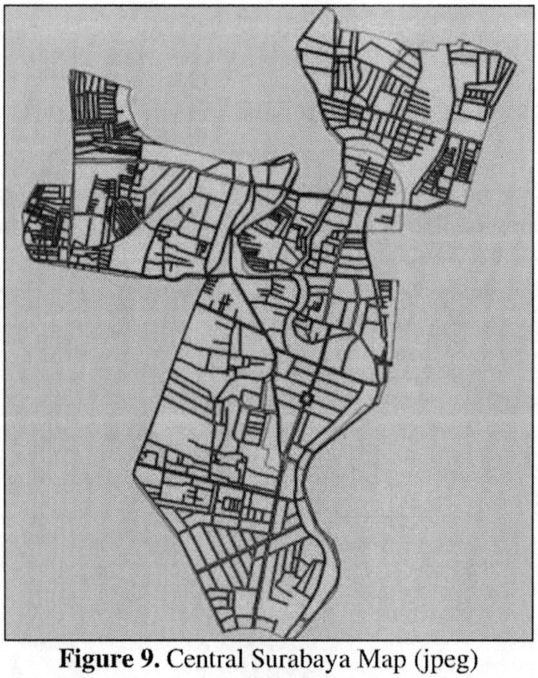

Figure 9. Central Surabaya Map (jpeg)

3.3. Basis Data Design

In this study 6 tables are used, which will be explained in emission information and air quality part to ease the explanation of basis data design.

a. Conceptual Basis Data Design

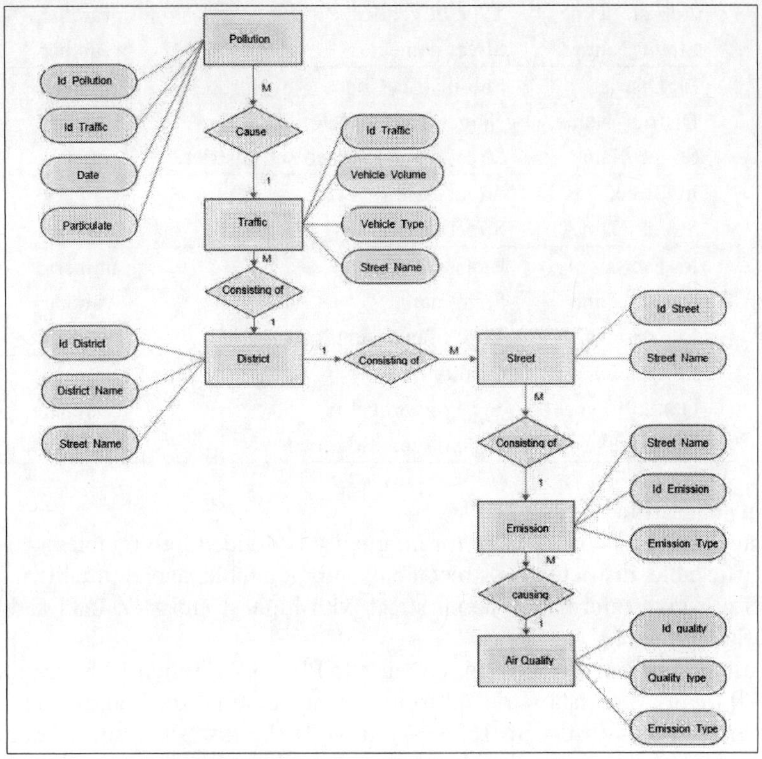

Figure 10. Conceptual Basis Data Design

The conceptual basis data design has table structure as follows:

- Pollution (Id_Pencemaran, Id_Lalulintas, Tanggal, Partikulat) [Id_Pollution, Id_Traffic, Date, Particulate]
- Traffic (Id_Lalulintas, Volume_Kendaraan, Jenis_Kendaraan, Nama_Jalan) [Id_Traffic, Vehicle_Volume, Vehicle_Type, Street_Name]
- District (Id_Kecamatan, Nama_Kec, Nama_Jalan) [Id_District, District_Name, Street_Name]
- Street (Id_Jalan, Nama_Jalan) [Id_Street, Street_Name]
- Emission (Id_Emisi, Nama_Jalan, Jenis_Emisi) [Id_Emission, Street_Name, Emission_Type]
- Air Quality (Id_Kualitas, Jenis_Kualitas, Jenis_Emisi) [Id_Quality, Quality_Type, Emission_Type]
-

b. Fiscal Basis Data Design

On fiscal basis data design, model basis data is pictured in tabel form as follows:

Table 5. Fiscal Model Table

Entity	Attribute	Description	Type Data	Key
Pollution	Id_Pollution	Pollution code	numeric	PK
	Id_Traffic	Traffic code	numeric	
	Date	Record date	varchar	
	Perticulate	Pollution parameter	varchar	
Traffic	Id_Traffic	Traffic code	numeric	PK
	Vehicle_Volume	The number of vehicles per road	varchar	

Entity	Attribute	Description	Type Data	Key
	Vehicle_Type	Type of vehilce	varchar	
	Street_Name	Street name	varchar	
District	Id_District	Sub-district Code	numeric	PK
	District_Name	Name of sub-district	varchar	
	Street_Name	Street name for each sub-district	varchar	
Street	Id_Street	Street code	numeric	PK
	Street_Name	Street name	varchar	
Emission	Id_Emission	Emission code	numeric	PK
	Street_Name	Street name	varchar	
	Emission_Type	Type of emission load	varchar	
Air Polution	Id_Quality	Quality code	numeric	PK
	Quality_Type	Type of air quality	varchar	
	Emission_Type	Type of emission load	varchar	

Conclusion and suggestion

a. In this study the basis data design for air quality is divided into 6 tables which are pollution table, traffic table, district table, street table, emission table, and air quality table.
b. WebGIS can give information about street with highest emission load and the impact in Central Surabaya area.
c. As in national ambient threshold in reference to PP No. 41 Tahun 1999, the threshold for CO is: 30.000 $\mu g/m^3$. The data retrieved from Primary Arterial road holds Jalan Diponegoro as the highest emission load with 1635,64 ton while the lowest being in Jalan Kapasari with 185,14 ton. Thus the air quality of primary arterial road in Surabaya Central is below the threshold, meaning good. For the secondary arterial road, the highest CO emission is on Jalan Raya Darmo with 1491,72 ton CO/year and the lowest is on Jalan Jagalan with 34,18 ton CO/year. Thus, the air quality of secondary arterial road in Surabaya Central is below the threshold, meaning good.
d. The flaw of this web is the inability to answer if the user means to find road with highest emission in Surabaya Central therefore software with multiple query trait is needed.
e. Usage of public transportation can decrease traffic in Surabaya, especially the centre area because this area is the center of Surabaya citizen activity, where there are many shopping malls, hotels, and office building.

4. References

[1] Dinas Kependudukan dan Catatan Sipil Kota Surabaya. 2013. *Grafik Penduduk Surabaya.* <URL:http://www.dispendukcapil.surabaya.go.id>. Visited on 30 October 2014, 11:00 am.

[2] Pemerintah Kota Surabaya. 2010. *RPJMD (Rencana Pembangunan Jangka Menengah Daerah) Kota Surabaya Tahun 2010-2015.* <URL:http://www.surabaya.go.id/dinamis/?id=3721>. Visited on 28 October 2014, 09:00 am.

[3] Suwardjoko. 2002. "*Pengelolaan Lalu Lintas dan Angkutan Jalan*", Jakarta.

Acknowledgement

The author appreciates greatly Lembaga Penelitian dan Pengabdian Masyarakat ITS which has given guidance in doing this study.

Appendix

Figure 11. WebGIS of air quality environment of Central Surabaya

IOP Conf. Series: Earth and Environmental Science **340** (2019) 012011 doi:10.1088/1755-1315/340/1/012011

Evaluation of slum potential settlement data collection system toward Surabaya smart government (case study: Keputih Sub District)

B U Aulia[1,3], A Muklason[2,3], E B Santoso[1,3] and N Farikha[1,3]

[1] Department of Urban and Regional Planning
[2] Department of Information System
[3] Institut Teknologi Sepuluh Nopember, Surabaya 60111, Indonesia

Email: b3ltown@gmail.com

Abstract. The existence of slum areas is one of the main problems that occur in the city of Surabaya. The Surabaya City Government continues to make efforts to renew the slum area data collection system in order to provide appropriate and targeted treatment. This research was conducted to evaluate the existence of the data collection system, the process and quality, also the sustainability of the slum settlement data collection system through a Likert scale assessment. The results of this study show that the existence of data collection and involvement aspects (score: 4.44) has a better performance in Keputih Sub District compared to other aspects, which are process and quality aspects (score: 4.11) and sustainability aspects (score: 4.33). Based on the evaluation results on the criteria of each aspect shows the need for improvement in the following matters: 1) potential slum settlement data collection program (score: 3.33) in the aspect of data collection and involvement; 2) the data collection process (score: 2.67) in the process and quality aspects in the data collection system; 3) improving the quality of human resources involved in data collection (score: 2.67) in the aspects of process and quality in the data collection system; 4) data recapitulation process (score: 2.67) in the sustainability aspect of data collection.

Keyword: *evaluation, slum, data collection system, smart government*

1. Introduction

Slum settlements are defined as uninhabitable settlements due to high levels of building density, building irregularities, high population density, low living standards of the population, and the lack of availability of housing facilities and infrastructure [1, 2]. Slum settlements develop due to the urbanization flow into the city but are not balanced by the existence of affordable residential land, so many people live on non-legal lands or under-habitable housing environments [3]. The existence of slums itself will lead to a bad paradigm of governance, especially in terms of regulating the lives of its citizens, besides that the existence of slums will lead to social inequality and a decline in the quality of life of the community [4].

The city of Surabaya itself has been committed in handling slums since 1968 through the existence of environmental improvement programs in low-income areas in Indonesia or better known as the Kampung Improvement Project (KIP) [5]. KIP itself is the first program that is successful and globally recognized, and is the largest improvement program in several countries for more than three decades [6]

[7]. The main component of KIP itself is by integrating the role of government and society in improving the village environment and reducing poverty in urban areas [5].

As for data collection on slums, the City Government of Surabaya through the Social Agency is conducting a data collection on Persons with Social Welfare Issues (PMKS) and the Potential of Social Welfare Resources (PSKS) with one of the data collection observations that are uninhabitable houses. The observations aspects in data collection of uninhabitable houses include the building ownership, space density, availability of health facilities (defecation facilities, fecal disposal sites, sources of drinking water, main lighting sources, availability of kitchens, and types of fuel for cooking a day), as well as the condition of the house in the form of floor walls and roof.

This study was conducted to evaluate the extent of the slum data collection system, the process and quality of the data produced, and how the sustainability of the data collection program that was seen from the point of view of stakeholders at the District and Sub District levels with a case study on Keputih Sub District. In Keputih Sub District, there were still several slum locations scattered among others in RW 01, RW 02, RW 03, and RW 08 [8]. By knowing the result of the evaluation, it is expected to be an input for the Surabaya City Government for implementing a more effective and efficient data collection program in the future.

2. Research methods

The research method was used in this study is a qualitative method with a descriptive approach. This method was used with the aim to be able to produce a final formula more systematically, factually, and accurately about the properties of the research object to be used [9]. This study focuses on slum areas in Kelurahan Sub District. The data collection method used is structured interviews with the Keputih Sub District and Sukolilo District. Then the data analysis method used is to assess the answers of the respondents with a Likert scale.

The evaluation of potentially slum settlement data collection systems is based on 3 aspects, namely: 1) the existence of data collection and involvement; 2) aspects of data processing and quality; 3) aspects of sustainability. Each of the three aspects has their own criteria and indicators. The scoring system of the questionnaire asked the respondents is shown in Table 1.

Table 1. List of questions and assessment system on the questionnaire

No	Questions	Score		Information
1.	Is there a slum area data collection program in your area?	Good	5	There is a special slum data collection program
		Moderate	3	There is a data collection program where there is one point about slums
		Bad	1	There is no data collection program
2.	How the involvement of the Government in the slum data collection process?	Good	5	There is an active involvement of the Government
		Moderate	3	The government is passively involved
		Bad	1	The government is not involved
3.	How is the involvement of the community in the data collection process?	Good	5	The community is actively involved
		Moderate	3	The community is passively involved
		Bad	1	The community is not involved
4.	How is the entire survey process carried out in your area?	Good	5	Integrated with a system
		Moderate	3	Semi manual
		Bad	1	Manual
5.	Are there a guidelines in carrying out the survey?	Good	5	There are technical guidelines with measurable criteria
		Moderate	3	There are guidelines
		Bad	1	There is no guidelines
6.	How the human resource needs in the data collection process?	Good	5	<10 HR involvement in data collection process
		Moderate	3	Involvement of 10-20 HR in data collection process
		Bad	1	>20 HR involvement in data collection process

No	Questions	Score		Information
7.	How is the quality of human resources in the data collection process?	Good	5	There is special training for selected HR
		Moderate	3	There is special training
		Bad	1	There is no training in advance
8.	How long is the data collection period?	Good	5	<3 months
		Moderate	3	3-6 months
		Bad	1	>6 months
9.	What is the quality of the data that produced?	Good	5	All data is collected and valid
		Moderate	3	Data is collected in part and valid
		Bad	1	All data is collected but invalid
10.	How about the data recapitulation process?	Good	5	Integrated with a system
		Moderate	3	Semi manual
		Bad	1	Manual
11.	Who is the person in charge in the data recapitulation?	Good	5	Kecamatan/Kota
		Moderate	3	Kelurahan
		Bad	1	Society
12.	Are the data updated properly?	Good	5	Data is updated every year
		Moderate	3	Data is updated every two years
		Bad	1	Data not updated
13.	How the follow up on the data collection process?	Good	5	There are a further activities in order to assistance the slums to the residents
		Moderate	3	There are a follow-up activities
		Bad	1	There are no follow-up activities

3. Result

Evaluation of the potential slum settlement data collection system in Keputih Sub District was conducted through questionnaire interviews on the parties involved. Respondents from Keputih Sub District was 3 people and from Sukolilo District was 3 people. Respondents in the evaluation of the data collection system for potentially slum settlements in Keputih Sub District are shown in Table 2.

Table 2. List of respondents

Instantion	Name	Position
Kecamatan Sukolilo	Gingin Ginanjar	Head of Development Section
	Yayuk Hestimingrum	Staff of People's Welfare Section
	Machfud Hidayat	General Staff
Kelurahan Keputih	Mudita Dhira Widaksa	Head of Keputih Sub District
	Adi Priyanto	Head of Development Section
	Yulianita Setiawati	Staff of People's Welfare Section

The results of the evaluation of a potentially slum settlement data collection system based on 3 aspects of data collection showed in Figure 1. Out of the three aspects of the data collection evaluation, the existence of data collection and involvement (score: 4.44) had better performance in Keputih Sub District compared to other aspects, namely process and quality aspects (score: 4.11) and aspects of sustainability (score: 4.33).

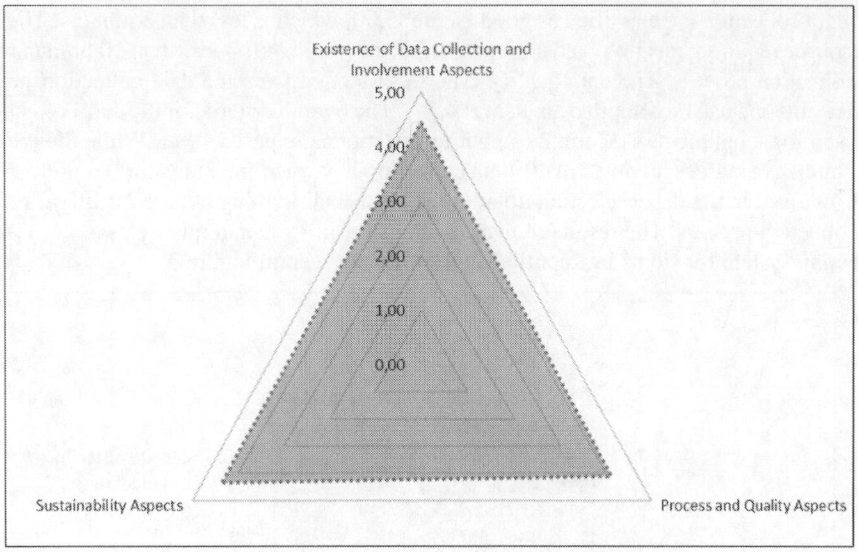

Figure 1. Evaluation score of data collection aspects

In the aspect of the existence of data collection and involvement in the potential data collection system of slums in Keputih Sub District produces a dominant score on the criteria of community involvement and criteria for government involvement. Evaluations on the criteria for community involvement and criteria for government involvement in data collection on slum potential (score: 5.00) indicate that the community and the government are fully and actively involved. While the criteria for the availability of slum data collection (score: 3.33) have not specifically led to slum data collection. The existing data collection is integrated in other data collection with one of the points about slums and uninhabitable settlements. The results of the evaluation of the existence of data collection and involvement in the potential data collection system for slums in Keputih Sub District are seen in Figure 2.

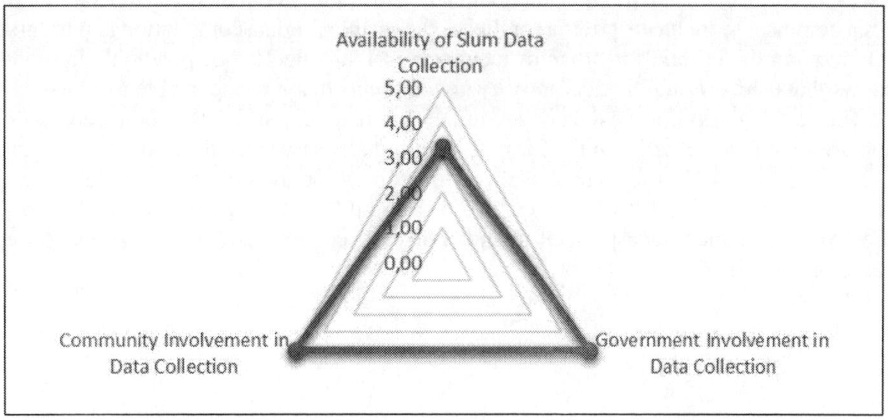

Figure 2. Evaluation score of existance of data collection and involvement aspects

In the aspect of process and quality in the potential data collection system the slums in Keputih Sub District produce a dominant score on the criteria for the availability of technical guidelines and implementation period. Evaluation on the criteria for the availability of technical guidelines in the collection of potential slum areas (score: 5.00) indicates the availability of technical guidelines in data collection with measurable criteria from other parties, namely the Social Service. The criteria for the period of data collection are classified as good (score: 5.00) which is less than 3 months. The criteria for human resource requirements involved are more effective (score: 4.67) less than 10 human resources in the data collection process. The data quality criteria generated from the data collection process show valid data results and can be compiled all (score: 4.67). The overall criteria for the data collection process showed a semi-manual process (score: 2.67) with a questionnaire based sheet. While the criteria for the quality of human resources involved in the data collection have medium indicators (score: 2.67). Human resources involved in the data collection do not receive special training but are directly directed during the data collection process. The results of evaluation of the process and quality aspects in the potential data collection system for slums in Keputih Sub District are seen in Figure 3.

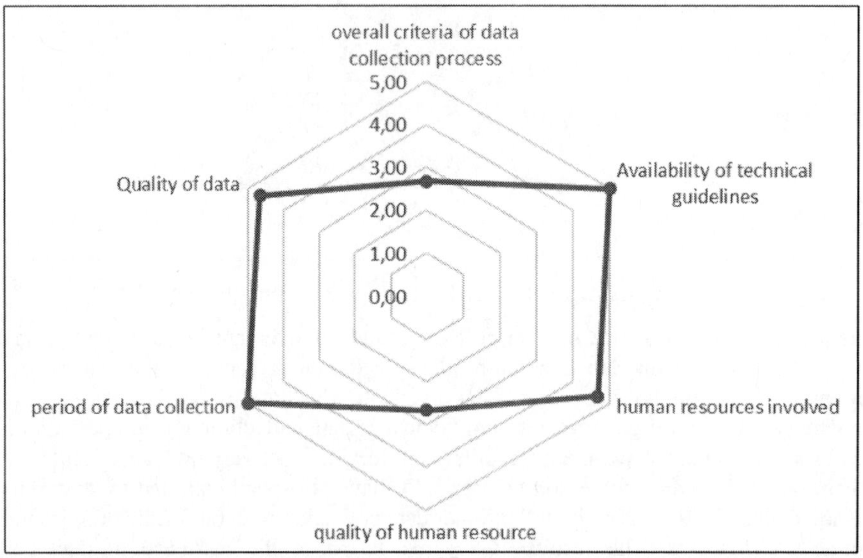

Figure 3. Evaluation score of process and data quality aspects

In the aspect of sustainability in the potential data collection system of slums in Keputih Sub District produces a dominant score on the criteria for those responsible for data compilation and program follow-up. Evaluation on the criteria for program managers in the collection of potential slum areas (score: 5.00) shows that there are sub-districts and city governments that are responsible for the data collection process. The sustainability for the data collection program are classified as good (score: 5.00), namely that there are follow-up activities in the form of handling and assistance to residents in the slums. Data update criteria (score: 4.67) shows updates every 1 year. While the criteria for the data recapitulation process show a semi-manual process (score: 2.67) compiled in an excel sheet. The results of the evaluation of the sustainability aspects of the potential data collection system for slums in Keputih Sub District are seen in Figure 4.

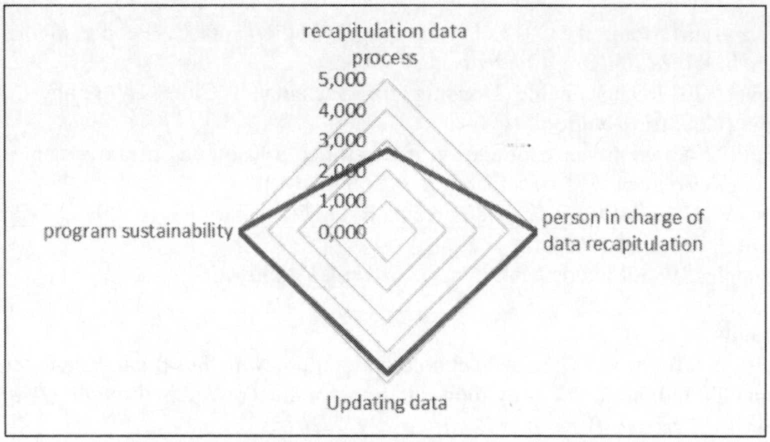

Figure 4. Evaluation score of data sustainability aspects

Based on the three aspects of the data collection evaluation above, it is found that it is necessary to improve the quality of the criteria: 1) potential slum settlement data collection program; 2) the data collection process; 3) improving the quality of human resources involved in data collection; 4) data recapitulation process. These criteria need to be improved in accordance with their respective indicators to realize an effective and efficient data collection system to support the Surabaya Smart Government program.

4. Conclusion

Based on the results of the data collection from the respondents showed an evaluation of the data collection process of potential slum areas in Keputih Sub District. The aspect of data collection and involvement (score: 4.44) has better performance in Keputih Sub District compared to other aspects, namely process and quality aspects (score: 4.11) and aspects of sustainability (score: 4.33). The criteria in each aspect indicate the need for improvement, namely on the criteria: 1) potential slum settlement data collection program (score: 3.33) in the aspect of data collection and involvement; 2) the data collection process (score: 2.67) in the process and quality aspects in the data collection system; 3) improving the quality of human resources involved in data collection (score: 2.67) in the aspects of process and quality in the data collection system; 4) data recapitulation process (score: 2.67) in the sustainability aspect of data collection.

Following up on the results of the evaluation of the data on potential slum areas in Keputih Sub District, it needs to be integrated with the Smart City framework. In the smart city framework in terms of monitoring the condition of urban settlements which in real time requires an application to report the condition of slums. The resulting application should be used by both the community and the related SKPD so that it can facilitate the data collection and monitoring of slum areas. This implementation is realized by integrating the Smart Government concept in data collection on the potential of slums.

5. References

[1] Kuswartojo, Tjuk. 2005. Perumahan dan Permukiman di Indonesia "Upaya membuat Perkembangan Kehidupan yang Berkelanjutan". Penerbit ITB. Bandung

[2] What are slums and why do they exist? UN-Habitat, Kenya (April 2007)

[3] Singh, P.O., Dhote, K.K. and Soni, N., 2013. Development of typologies of slum settlements: the case of a million plus city of India. *The Sustainable City VIII (2 Volume Set): Urban Regeneration and Sustainability*, *179*, p.1153.

[4] Turley, R., Saith, R., Bhan, N., Rehfuess, E. and Carter, B., 2013. Slum upgrading strategies involving physical environment and infrastructure interventions and their effects on health and socio-economic outcomes. *status and date: New, published in*, (1).

[5] Silas, J. and Ernawati, R., 2013. Liveability of Settlements by People in the Kampung of Surabaya. In *World Building Congress.*
[6] UN-Habitat. 2012. Sustainable Housing for Sustainable Cities: A policy Framework for Developing Countries. Nairobi.
[7] Silas, J., 1992. Government-community partnerships in Kampung improvement programmes in Surabaya. *Environment and Urbanization*, 4(2), pp.33-41.
[8] Koswara, A Y., Ariastita, P G., Handayeni, K D M E., Farikha, N. 2018. Typology of Slum Settlement in Kelurahan Keputih
[9] Suryabrata, S., 1983. Metode penelitian. *Jakarta: CV Rajawali.*

Acknowledgments
The authors are grateful to the Institut Teknologi Sepuluh Nopember and Kementerian Riset dan Teknologi Republik Indonesia for providing funding for this research through *Hibah Pengabdian Kepada Masyarakat Berbasis Penelitian Dana Lokal ITS 2018.*

IOP Conf. Series: Earth and Environmental Science **340** (2019) 012012 doi:10.1088/1755-1315/340/1/012012

Pedestrian exposure to Nitrogen Dioxide (NO_2) and Carbon Monoxide (CO): A case study of Surabaya, Indonesia

B N Dewi[1]**, A D Syafei**[1] **and T N Ciptaningayu**[1]

[1] Department of Environmental Engineering, Institut Teknologi Sepuluh Nopember

Email: dipareza@enviro.its.ac.id

Abstract. The increase in motor vehicles used every day also increases the rate of pollution. Motor vehicles are the main source of the nitrogen dioxide (NO_2) and carbon monoxide (CO) pollutants that can be dangerous for human health in excessive amounts. This study investigates the patterns of NO_2 and CO concentrations in the ambient air, specifically on the sidewalks of two big streets in Surabaya. Measurements were taken over a period of 15 hours using a portable monitoring device. The results indicated that the NO_2 concentration would decrease by 0.008 mg/m^3 and 0.012 mg/m^3, respectively, at peak times during the day and night. Meanwhile, the CO concentration would increase by 1.486 mg/m^3 during the peak time at night. The factors that significantly influenced NO_2 were day variation (weekday or weekend), morning peak hour, noon peak hour, night peak hour, and temperature. Regression with these factors had an R^2 value of 30.7%. For CO, the significant factors were night peak time, temperature, wind direction, and the number of vehicles, with an R^2 value of 16.9%.

Keywords: *NO_2, CO, plants, vehicles*

1. Introduction

Surabaya is the biggest city in Indonesia, second only to Jakarta. The city has over than 2 million people based on the population census in 2010, which showed at the time that the population had increased significantly (by 0.63%) since the previous census [1]. Consequently, there has been an increase in the use of various transport modes to support daily activities, e.g., motor vehicles. Based on data from the Transportation Agency of Surabaya City, there were 4.5 million vehicles in Surabaya in 2014, and that number was increasing by 17,000 every month.

There are six main pollutants in Surabaya, which are Carbon monoxide (CO), Lead (Pb), Nitrogen dioxide (NO_2), Ozone (O_3), Particulate matter (PM), and Sulfur dioxide (SO_2) [2]. Motor vehicles, especially the old ones, are the main cause of NO_2 and CO pollution [2]. Therefore, any increase in vehicles will increase the NO_2 and CO gasses in the ambient air. Pedestrians, particularly those who walk along roads with heavy traffic, have been exposed to these gases, especially when they exceed threshold levels.

In Lagos City, Nigeria, pedestrians are exposed to CO at levels of 8.84 ± 1.40 ppm during rush hour and 6.80 ± 1.48 ppm during normal traffic periods [3]. Meanwhile, pedestrian exposure to NO_2 on busy streets is relatively high at an average of 38 mg/m^3. These high amounts of pollutants are due to poorly tuned engines and street canyon conditions. The lowest exposure to NO_2 was 23 mg/m^3, which occurred in places where pedestrians walked alongside low-intensity traffic and pollutants spread well, such as residential neighborhoods [4].

Another study was conducted in Brisbane, Australia. There, CO and NO_2 concentrations peaked from 08.00 AM to 09.00 AM on workdays. On the weekend (Sunday), the concentration increased starting at 06.00 PM and kept increasing until the peak night hour in the period 11.00 PM-01.00 AM. These peak concentrations were influenced by the increase in vehicles during these periods. On workdays, most people commute to work in the morning; therefore, the number of vehicle was greatest in the morning, increasing the concentrations of pollutants as well. Meanwhile, during the weekend (Sunday), the usual activity was leisure activities, thus the number of vehicles increased at night [5].

CO can poison humans in the form of COHb (carboxihemoglobin) in the blood [6]. The affinity for CO is greater than that for oxygen, causing Hb to bond with CO instead of oxygen. This can disrupt the nervous system and slow down reflexes. CO can also cause continuous seizures, leading to unconsciousness and death. Inhaling air with a high concentration of NO_2 disrupts the respiratory channel in the human respiratory system. This kind of exposure, even in the short term, can cause respiratory diseases, especially asthma, which, in turn, can lead to other respiratory symptoms (such as coughing and difficulty breathing) [7]. Considering the above facts and also the massive increase in vehicle ownership in Surabaya, which has led to heavy traffic on some roads, the present study determines the concentration patterns of NO_2 and CO and their correlations with traffic volume.

2. Methods

2.1 Location selection

The study was done on Gemblongan St. and Embong Malang St., Surabaya, where we measured concentrations on a workday and weekend (Sunday). In each location, we used two difference scenarios, one with vegetation separating traffic and pedestrians, the other without vegetation.

Gemblongan St.: Based on Mayor Decree Number 46, year 2000, concerning the class of roads in Surabaya (Keputusan Walikota Surabaya Nomor 46 Tahun 2000 tentang Kelas Jalan di Kota Surabaya), Gemblongan is a street with a Class 3B designation, a collector street where motor vehicles, including cargo vehicles up to 8 tons, are allowed to pass through. The street goes in one direction, every lane has a width of 2.5 m, and it is divided into five lanes. On both sides of the street, there are sidewalks with widths ± 3 m. This street goes on for 0.74 km. Plants are placed on one side of the road with a distance between plants of ± 6 m.

Embong Malang St.: Embong Malang is a street with a Class 3C designation, meaning that it is a local street where motor vehicles with a cargo width of not more than 2,100 mm can pass through, and the heaviest cargo allowed is 8 tons. The street goes in one direction only, every lane is 2.5 m wide, and it is divided into five lanes. On the right and left sides of the street there were sidewalks with widths of ± 6 m and ± 2 m, respectively. This street goes on for 1.63 km. Plants were placed along the side of the road with the distance between plants was ± 4 m.

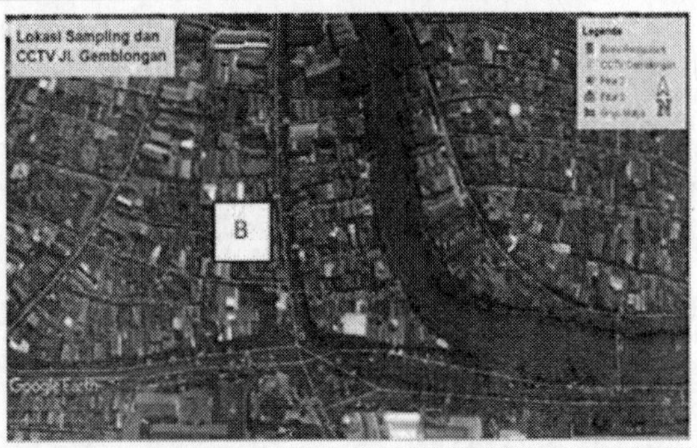

Figure 1. Gemblongan St.

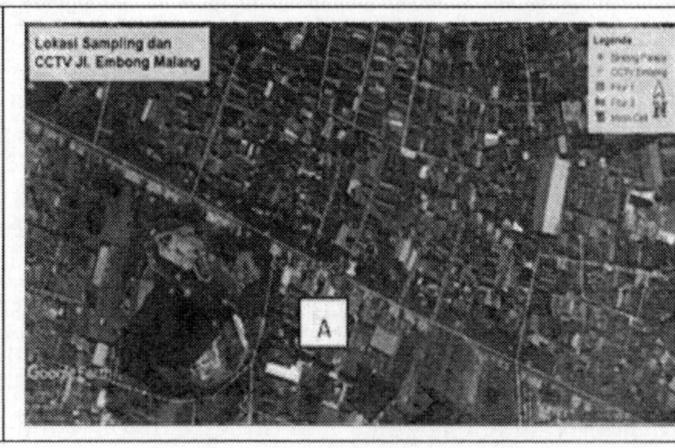

Figure 2. Embong Malang St.

2.2 Sample measurements

When sampling without plants, the tool was placed on a tripod with a ± 1.5 m height (as high as the average human respiratory system). The distance between the tool and the road was 1 m. Meanwhile, in an area with plants, the tool was placed 1 m behind the plants at a ± 1.5 m height. The distance between plants was ± 30 cm. This distance was measured from the outmost plant stem to the next one. The plants were planted so that there would be no gaps in the canopy. The plants used were plants with heights around 1.5 m – 2.0 m, which had a good ability to absorb pollutants, and which were aesthetically pleasing. This study used croton plants (*Codiaeum variegatum*). Each sampling period was 15 hours long (06.00 AM-09.00 PM West Indonesian Time, WIB) on both Sunday (weekend) and during weekdays (Monday, Tuesday, or Wednesday). In total, measurements were taken four (4) times. To obtain concentrations of CO and NO_2, a factory-calibrated portable Aeroqual S500 was used. Aside from sampling, vehicle counting was done using CCTV footage from the Transportation Agency of Surabaya. CCTV cameras were placed in crossroads above the traffic lights to ensure that they captured the traffic volume in all lanes. Wind direction and wind speed were measured using a Kestrel 5500. Temperature and relative humidity were also measured.

2.3 Data analysis

We identified patterns in NO_2 and CO concentrations, and, with the analysis from the CCTV footage, possible sources were also identified. The possible sources of pollution were divided into groups, which were motorbikes, gas-fueled cars, solar-fuel cars, busses, trucks, and others. The volume in each group was counted. The CCTV footage was also used to identify other pollutant sources; these may include smokers, food sellers, trash burning, and so forth.

To obtain variables that may affect the concentrations, multiple linear regression was employed. This analysis was used to see the significance of influence from meteorological condition (wind speed and wind direction), vehicles, day variations, and plant existence towards NO_2 and CO concentrations. Nine independent variables were tested, and six of them were dummy variables. These were the day variable (weekend = 0; weekday = 1), peak time in the morning (1 = yes), peak time at noon (1 = yes), night peak time (1 = yes), vegetation (with plant = 0, without plant = 1), and wind direction. The other three variables were continuous, i.e., temperature, wind velocity, and traffic volume (total number of vehicles).

3. Results

3.1 NO_2 and CO concentration patterns

Figure 3 displays the NO_2 concentrations on Embong Malang St., showing that the weekend and weekdays do not differ much in terms of there being plants or not. It is noted that concentrations on weekdays were higher than they were during the weekend in both cases.

Whenever vegetation was available, the highest concentration on a weekday on this road was 0.288 mg/m^3 at 11.45 AM. During peak concentration, the total number of vehicles was 457 in a span of 5 minutes, of which 268 had motorbikes, 139 were gas-fueled cars, 45 were diesel cars, 4 were trucks, there was 1 other vehicle, and no buses went by. The average concentration was 0.136 mg/m^3 and the lowest concentration was 0.085 mg/m^3.

Whenever there was no vegetation, the highest concentration was slightly lower than with vegetation, at 0.24 mg/m^3. The total number of vehicles was 452, with 289 motorbikes, 110 gasoline cars, 50 diesel-cars, 2 buses, 1 truck, and no other vehicles. The peak occurred at 16.20 WIB. However, the lowest and average concentrations were the same or higher than they were with vegetation. The concentrations were 0.136 mg/m^3 and 0.178 mg/m^3, respectively. This suggests the possibility that there was a surge in pollutants when no plants were present, probably due to rapid traffic increases within a short time or other possible causes, such as smoking activities on the part of the pedestrians.

During the weekend, on the same street, the highest concentrations of NO_2 without plants was 0.195 mg/m^3, higher than when vegetation was present, which was 0.167 mg/m^3, possibly due to a

IOP Conf. Series: Earth and Environmental Science **340** (2019) 012012 doi:10.1088/1755-1315/340/1/012012

significant traffic volume difference. The highest concentrations occurred at 03.35 PM (without vegetation) and 10.00 AM (with vegetation). Traffic during no-plant measurements was at 533 units, significantly higher than during the plant scenario. On average, the concentration was higher whenever vegetation was not present. The average and lowest concentrations were 0.154 mg/m^3 and 0.097 mg/m^3 (without plants), respectively, compared with 0.119 mg/m^3 and 0.076 mg/m^3 (with plants), respectively.

Based on Figure 4, it can be inferred that the NO_2 concentrations on Gemblongan St. showed results similar to those for Embong Malang St., where the weekend and weekday concentrations did not differ by much, even though, generally, the NO_2 concentration on weekdays was higher than it was on the weekend.

The highest concentration with plants on the weekend on this road was 0.244 mg/m^3 at 09.40 AM. At this peak concentration, there were 412 vehicles on the road. The average concentration was 0.163 mg/m^3, and the lowest concentration was 0.141 mg/m^3. On the weekdays, the highest concentration was 0.244 mg/m^3 at 05.50 PM. At this peak concentration, the number of vehicles was 423. The lowest concentration was 0.142 mg/m^3, and the average concentration was 0.175 mg/m^3.

In comparison when vegetation was not available, during the weekend, the highest concentration was 0.168 mg/m^3, the lowest concentration was 0.077 mg/m^3, and the average concentration was 0.118 mg/m^3. Unsurprisingly, these values where lower than found on weekdays, where the highest concentration was 0.199 mg/m^3 at noon when there were 489 vehicles on the road. The average concentration was 0.129 mg/m^3, and the lowest concentration was 0.092 mg/m^3.

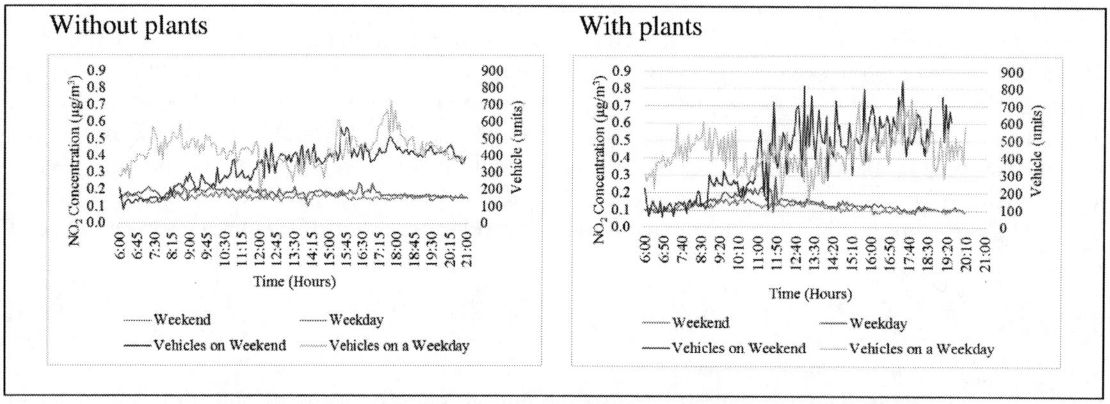

Figure 3. Concentration patterns for NO_2 on Embong Malang St.

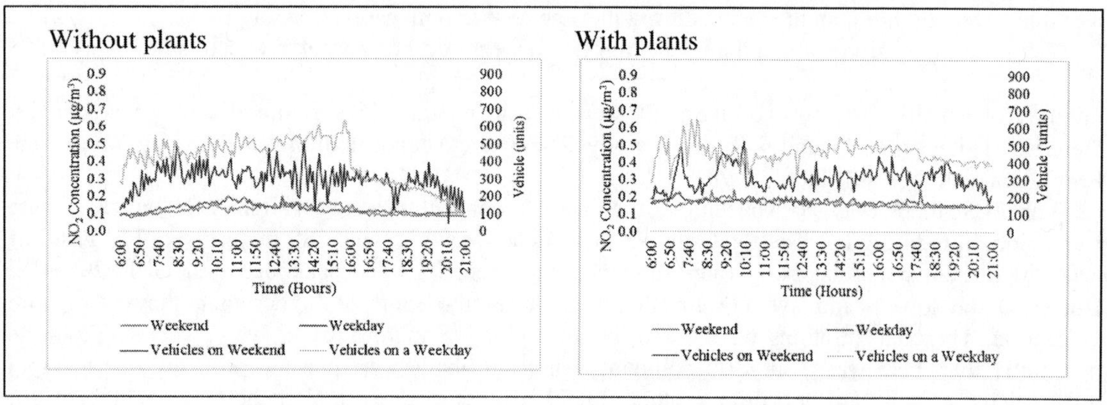

Figure 4. Concentration patterns for NO_2 on Gemblongan St.

Figure 5 displays that CO concentration on weekdays was higher than on the weekday. The highest concentration on the weekend happened in a location without plants and was 13.11 mg/m^3 at 10.40 WIB. At this peak concentration, there were 273 vehicles present. The average concentration was 1.73 mg/m^3. For an area with plants, on the weekend, the highest concentration was 4.09 mg/m^3. The number of vehicles passing by was 395.

During weekdays, the highest concentration was 14.18 mg/m^3 in the area with plants. This happened at 11.30 WIB when 373 vehicles were passing by. The average concentration was 1.54 mg/m^3, and the lowest concentration was 0.00 mg/m^3. The highest concentration in the location without plants was 14.05 mg/m^3, occurring in the afternoon with 583 vehicles passing by. The average concentration was 3.03 mg/m^3, and the lowest concentration was 0.00 mg/m^3.

Based on Figure 6, it can be inferred that the highest concentration on a weekday in Gemblongan St. where there were plants was 21.71 mg/m^3 at 17.15 WIB and with 462 vehicles present. The vehicles included 307 motorbikes, 135 gas-fueled car, 15 solar-fueled cars, 1 bus, 3 trucks, and 1 other vehicle. The highest concentration without plants was 6.96 mg/m^3 at 09.40 WIB. The number of vehicles then was 459 with 315 motorbikes, 117 gas-fueled cars, 15 solar-fueled cars, 1 bus, 8 trucks, and 3 other vehicles. The average concentration with plants was 1.73 mg/m^3, and without plants it was 1.64 mg/m^3. The lowest concentration was 0.00 mg/m^3.

For a weekend on Gemblongan St., the CO concentration showed 0.00 mg/m^3 for all the highest, average, and lowest concentrations. For the location with plants, the highest concentration on the weekend was 6.71 mg/m^3 at 09.35 AM when there were 380 vehicles passing by which consisted of 332 motorbikes, 29 gas-fueled cars, 1 solar car, 8 trucks and no busses or other vehicle. The average concentration was 0.54 mg/m^3, and the lowest concentration was 0.00 mg/m^3.

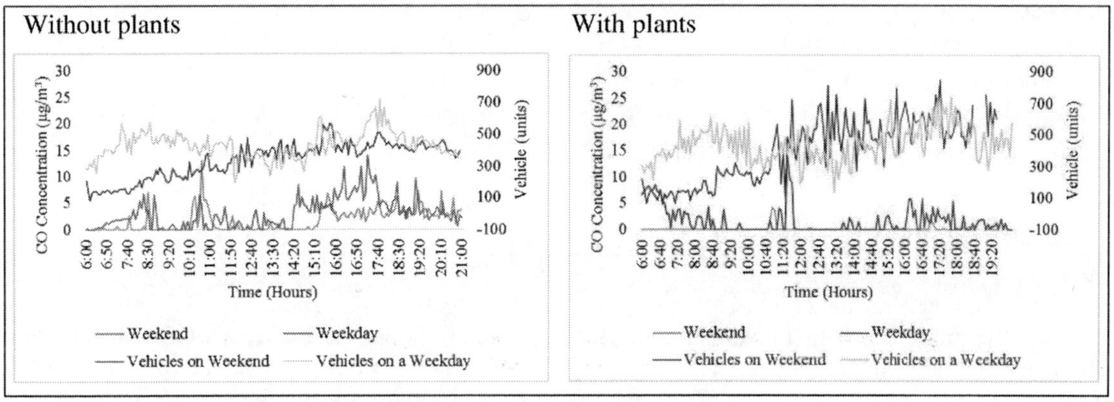

Figure 5. CO Concentration patterns on Embong Malang Street

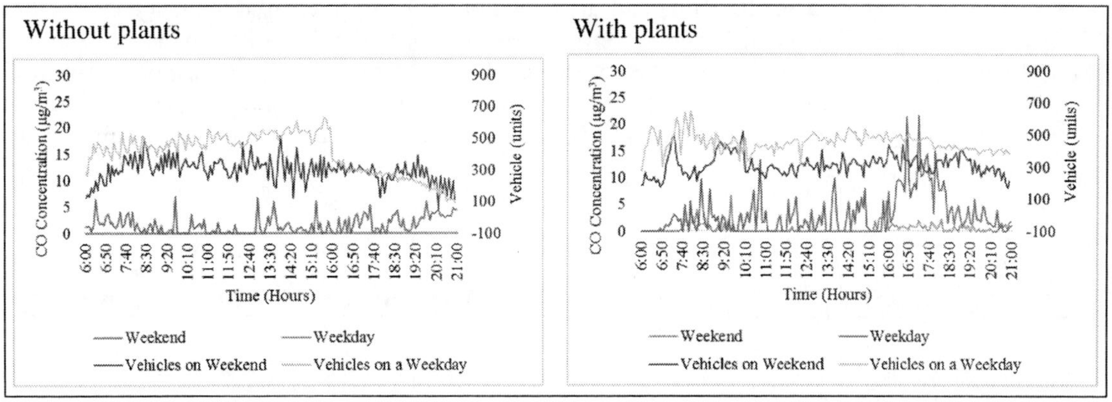

Figure 6. CO Concentration patterns on Gemblongan Street

3.2 Variables affecting concentrations of the pollutants

Table 1 shows the result of multiple linear regression for NO_2. Based on the results, it can be inferred that the existence of plants, wind direction, wind velocity, and the number of vehicles did not influence the NO_2 concentration significantly. The NO_2 concentration was affected significantly by the day variable, morning peak hour, noon peak hour, night peak hour, and temperature. The morning peak hour could increase the NO_2 concentration by 0.013 mg/m^3, while the noon and night peak hours could decrease the NO_2 concentration by 0.008 mg/m^3 and 0.012 mg/m^3, respectively. It can also be inferred that the NO_2 concentration was bigger on weekdays compared to the weekend, indicated by the day variable coefficient of 0.016. Everyone degree increase in temperature increased the NO_2 concentration by 0.005 mg/m^3. As shown in the R^2 value of 0.307, the influence of the existing variables (day variable, morning peak hour, noon peak hour, night peak hour, and temperature) can explain 30.7% of the variation in NO_2 concentration.

Table 1. Coefficients of variables affecting the concentration of NO_2

Variable	Coeff.	Std. Error Coeff	p-value	Model R^2
Constant	-0.030	0.008	0.000	0.307
Day variable (dummy variable, weekend = 0, weekday = 1)	0.016	0.002	0.000	
Morning peak (dummy variable, peak time = 1)	0.013	0.003	0.000	
Noon peak (dummy variable, peak time = 1)	-0.008	0.003	0.011	
Night peak time (peak time = 1)	-0.012	0.003	0.000	
Plant (without plant = 1)	0.002	0.001	0.145	
Temperature	0.005	0.000	0.000	
Wind direction	0.003	0.002	0.161	
Wind velocity	-0.002	0.001	0.258	
Volume of Vehicles	4,28E-6	0.000	0.513	

Table 2 shows the linear regression results for CO, which differ from those of NO_2. Based on Table 2, it can be inferred that there were five variables with a significant influence on CO concentration; these are: the day variable, night peak hour, temperature, wind direction, and number of vehicles. The night peak hour saw an increase in CO concentration of 1.486 mg/m^3. From the coefficient for the day variable, it can be inferred that the CO concentration was higher on a weekday compared to the weekend by 1.65 mg/m^3. Wind direction and the number of vehicles increased the CO concentration by 0.397 mg/m^3 and 0.001 mg/m^3, respectively. Everyone degree increase in temperature decreased the CO concentration by 0.067 mg/m^3. The other variables, i.e., the morning peak hour, noon peak hour, existence of plants, and wind velocity, did not have a significant influence on CO concentration. Table 2 also shows an R^2 value of 0.169. This means that the day variable, night peak hour, temperature, wind direction, and amount of vehicle can explain 16.9% of the variability in CO concentration.

Table 2. Coefficients of the variables affecting the concentration of CO

Variable	Coeff.	Std. Error Coeff.	p-value	Model R^2
Constant	2,063	0.705	0.004	0.169
Day variable (dummy variable, weekend = 0, weekday = 1)	1,650	0.132	0.000	
Morning peak (dummy variable,	0.175	0.241	0.466	

Variable	Coeff.	Std. Error Coeff.	p-value	Model R^2
peak time = 1)				
Noon peak (dummy variable, peak time = 1)	0.226	0.247	0.360	
Night peak time (peak time = 1)	1,486	0.245	0.000	
Plant (without plant = 1)	0.136	0.125	0.277	
Temperature	-0.067	0.019	0.001	
Wind direction	0.397	0.169	0.019	
Wind velocity	-0.116	0.115	0.314	
Volume of Vehicles	0.001	0.001	0.035	

3.3 Discussion

The NO_2 concentration shows a bimodal pattern, as it increases in the morning, decreases in the day, and increases again in the evening. Apart from the traffic contribution, the decrease of NO_2 in the day is caused by the NO_2 photolytic cycle [8]. The photolytic cycle phase with a direct interaction between sunlight and NO_2 causes NO_2 to absorb energy in the form of ultraviolet rays and sunlight. The absorbed energy breaks NO_2 molecules down into NO molecules and oxygen (O) atoms. The oxygen atoms formed are very reactive and react with atmospheric oxygen (O_2), forming ozone (O_3) which is another pollutant [9]. The CO concentration in an open location increases when an atmosphere inversion occurs. This phenomenon occurs when colder air is trapped below a warmer layer of air. An inversion acts like a cover, preventing pollution from mixing in the atmosphere and trapping it on the surface. High CO concentration also occur during the morning and evening rush hours [10].

The day variable has a positive influence on both NO_2 and CO concentrations. The influence is relatively similar, i.e., a higher concentration on a weekday compared to the weekend, and the p-value indicates that the increase during a weekday is significant. This result is comparable to the research showing that the NO_2 concentration was higher by 16% on workday compared to the weekend (Sunday) in Brisbane, Australia [5]. This finding also corresponds to the number of vehicles on the road, as the number was 48% higher on weekdays compared to weekends (Sunday). For CO, the research result was similar to the study that found that the average CO decreased every hour from 06.00 AM to 03.00 PM on Sunday compared to Wednesday on Gemblongan St. [11]. After taking measurements for 9 hours (06.00 AM-03.00 PM), the results of this study are in agreement with other studies, i.e., the CO decreases by 28% from weekday to weekend. The CO concentration on a workday was higher by 21% compared to the weekend (Sunday), which is proportional to the increase in vehicles on workdays [5]. From the research of Blanchard and Tanenbaum, In Southern California, the average CO was more or less 0-17% lower on Saturday and about 12-32% lower on Sunday compared to workdays [12]. Overall, the results show that vegetation reduces both NO_2 and CO concentrations.

The results for the NO_2 concentration at noon fit the above statement where the NO_2 concentration would decrease when the noon peak hour occurred. This is caused by the photolytic cycle. The increasing CO concentration at night was potentially caused by other pollutant sources aside from vehicles. Since the sample measurement was taken from a pedestrian point of view, during the measurement session, most of the time during the afternoon and night, often people, some smoking, would pass the location, get close to the sampling device and even linger. This could have led to a surge in concentration.

Differently, during the night peak hour, the NO_2 decreased because the number of vehicles continued to decrease the later the hour became. Based on Figures 1-4, it can be inferred that the trend line for the number of vehicles in the night generally decreased. Therefore, the NO_2 concentration also decreased, unlike the CO concentration, which increased due to other factors, such as smokers.

The R^2 values obtained for the regression models are relatively small, i.e., 30.7% for NO_2 and 16.9% for CO. The small R^2 values are caused by other variables influencing the result, such as traffic jams, surrounding activities, and other factors not accounted for in this research. Therefore, to increase the R^2 value, other variables could be added that could probably affect the NO_2 and CO concentrations.

4. Conclusion

Pedestrians are susceptible to exposure to pollutants, especially that coming from traffic activities. Emissions from large and small vehicles have a potentially dangerous impact on health. Despite high concentrations of pollutants during the morning and evening, exposure will be less whenever the sidewalks are lined with vegetation, which acts as natural barrier for pollutants. The wind direction and temperature both influence concentrations of pollutants. It is also noted that traffic volume plays a great role in the ambient concentration.

Considering these facts, regulations should be enacted to protect pedestrians from long exposures to pollutants. The least that policy makers could do is to plant ample vegetation along the sidewalks. This will reduce pedestrian exposure and reduce the health impact, especially during the morning peak times.

5. References

[1] Badan Pusat Statistik 2017 Surabaya dalam Angka. BPS Kota Surabaya, Surabaya

[2] Tugaswati, A.T. 2007. Emisi Gas Buang Kendaraan Bermotor dan Dampaknya terhadap Kesehatan http://www.kpbb.org/makalah_ind/Emisi%20Gas%20Buang%20Bermotor%20&%20Dampaknya%20Terhadap%20Kesehatan.pdf Accessed 27 Oktober 2017

[3] Odekanle E L, Fakinle B S, Jimoda L A, Okedere O B, Akeredolu F A, Sonibare, J A 2017 In-Vehicle and Pedestrian Exposure to Carbon Monoxide and Volatile Organic Compounds in A Mega City *Urban Climate* **21**: 173-182

[4] Zak M, Wolny E M, Widziewicz K 2016 The Exposure of Pedestrian, Drivers and Road Transport Passengers to Nitrogen Dioxide *Atmospheric Pollution Research* **8(4)**: 781-790

[5] Morawska L, Jayaratne E R, Mengersen K, Jamriska M, Thomas S 2002 Differences in Airbone Particle and Gaseous Concentration in Urban Air between Weekdays and Weekends *Atmospheric Environment* **36**: 4375-4383

[6] Soedomo M 1999 Kumpulan Karya Ilmiah Mengenai Pencemaran Udara ITB, Bandung

[7] US EPA 2014 Integrated Review Plan for the Primary National Ambient Air Quality Standards for Nitrogen Dioxide. Research Triangle Park, North Carolina

[8] New Jersey Department of Environmental Protection 2017 2016 Nitrogen Dioxide Summary

[9] Damayasa I G O 2013 Dampak NOx terhadap Lingkungan *Jurnal Ilmiah Kurva Teknik*

[10] New Jersey Department of Environmental Protection 2017 2016 Carbon Monoxide Summary

[11] Blanchard C L, Tanenbaum, S J, Lawson D R 2008 Differences between Weekday and Weekend Air Pollutant Levels in Atlanta, Baltimore, Chicago, Dallas-Fort Worth, Denver, Houston, New York, Phoenix, Washington, DC, and Surrounding Areas *Journal of the Air and Waste Management Association* **58 (12)**: 1598-1615

[12] Blanchard C L dan Tanenbaum, S J 2003 Differences between Weekday and Weekend Air Pollutant Levels in Southern California *Journal of the Air and Waste Management Association* **53 (7)**: 816-828

IOP Conf. Series: Earth and Environmental Science **340** (2019) 012013 doi:10.1088/1755-1315/340/1/012013

Transformation of urban flood modelling from hydrodynamic to system dynamics approach

C Susetyo[1], H Idajati[1] and A M Navastara[1]

[1] Urban and Regional Planning Department, Institut Teknologi Sepuluh Nopember

Email: cahyono_s@urplan.its.ac.id

Abstract. Flood problem is one of the key issues in the urban water management practice. Currently, there is a gap between flood prediction techniques, which mostly by Hydrologic Modelling, and the water management modelling, which mostly use the System Dynamic approach. This paper evaluates the possibility of the integration of hydrodynamic modelling to System Dynamics, which is one of the most common method to develop an urban water modelling. By integrating hydrodynamic modelling within a larger system dynamics-based urban water modelling, the hazard aspect of urban water management can thoroughly be examined. In this paper, the integration of Hydrodynamic Modelling into system dynamics-based urban water modelling is achieved by implementing the flow inputs of hydrodynamic modelling as a variable in the system dynamics.

1. Introduction

One of disturbances that may occur in an urban area is when there is an unexpected body of water on the surface, due to the amount of water entering the area is higher than its capacity to discharge it. Flood problem in urban areas is induced by various factors, such as the network system of drainage within the urban area, higher rate of rainfall than anticipated, and changes of volume of water carried by rivers passing through the city. In general, causes and types of urban flooding can be categorized to [1] Lack of drainage infrastructure, Blockage of the drainage system, Flooding in low-lying areas, Backup due to elevated downstream water levels, and Inundation caused by high river water levels. Mitigation measures for those problems are related to other aspects of urban management, such as provision of freshwater, drainage system, and water storage. For example, during rainy season, most water that enter an urbanized area should be discharged to its outer part, while in a dry season, incoming water should be stored to maintain the supply of freshwater. Therefore, we can argue that flood management is an integral part of the urban water management.

Efforts to manage urban flooding problems mostly started with predicting the flood itself, by using flood modelling to produce predictions about potential flood that may occurs. The most common type of flood modelling is the Hydrodynamic Flood Modelling, which calculates the characteristics of the water flow, either in channels or on surface. Despite this approach is proven to be useful for municipalities to manage flood problems, hydrodynamic approach is not suitable to be developed as a basis for urban water management. The main drawback of hydrologic modelling is this approach considered water as a continuous flow, from inlet to outlet. It simplifies the situation by assuming that a problem occurs when the volume of water at inlet is bigger than at outlet.

Urban water management is actually beyond that, because water is not only passing, but also may utilized for various purposes, such as household activities, water-based services, and even for industries. Therefore, there is a need to explore how hydrologic modelling can be integrated to urban water management modelling. This paper evaluates the possibility of the integration of hydrodynamic modelling to System Dynamics, which is one of the most common method to develop an urban water modelling. By integrating hydrodynamic modelling within a larger system dynamics-based urban water modelling, the hazard aspect of urban water management can thoroughly be examined

2. Hydrodynamic approach for flood modelling

This paper implements a Meta-Synthesis approach, which analyses and synthesise the key elements in researches related to the concept of a resilient city and the implementations of measures to achieve it. The aim of the meta-synthesis in this paper is to transform findings in each of the research and implementation to construct a new concept and generalization [10], [11]. In this paper, available methods and practices to improve or achieve the adaptive aspects of a resilient city were examined to make a categorization of those methods and practices. While previous works related to implementations of hazard mitigation measures in urban areas are abundant, a specific research that summarizes those measures and categorizes them based on their characteristic is not available.

To establish an overview of methods and techniques related to adaptability measures in urban areas, this paper first formulates the underlying characteristic of those measures. This step will produce general categories of adaptability measures. The main hypothesis of this paper is that by comparing the main characteristics of adaptability measures toward a resilient city, a pattern will emerge, that will make a categorical grouping of those measures possible. After the general categories of adaptability measures are established, the next step is to perform an in-depth analysis of each category, to explore the similarity and variance of adaptability measures that fall within the same category. The result of this phase will give a detailed explanation about types of adaptability measures, how they are implemented, and examples of their implementations.

2.1. 1D flood modelling

In the 1D approach, water flow is assumed to occur in one dominant spatial dimension aligned with the centre line of the main river channel. The geometry of the problem is represented in the model by channel and floodplain cross-sections perpendicular to the channel centreline. Measured distances between these cross-sections are also required by the computer model [2]. According to [3], 1D Flood model is capable of capturing the downstream propagation of a flood wave and the response of flow to free surface slope, which can be described in terms of continuity and momentum equations as:

$$\frac{\partial Q}{\partial x}+\frac{\partial A}{\partial t}=q \quad (1)$$

$$S_0=\frac{n^2P^{4/3}Q^2}{A^{10/3}}-\frac{\partial h}{\partial x}=0 \quad (2)$$

Where Q is the volumetric flow rate in the channel, A the cross-sectional area of the flow, q the flow into the channel from other sources (i.e. from the floodplain or possibly tributary channels), S_0 the down-slope of the bed, n Manning's coefficient of friction, P the wetted perimeter of the flow, and h the flow depth. One source of the technical requirements in an urban flood model is from [4], which stated that data requirements for 1D flood modelling are:

- Dynamic flow description: when urban flooding occurs, surface water can flow in both street and pipe systems with flow exchange between these two systems through manholes. This means that simulation of backwater effects is needed in modelling of urban flooding. By using a dynamic wave model, the model includes backwater effects and surcharge from manhole including rapid change of water level.
- Parallel flow routing: while surface flooding takes place, water from the pipe system flows through

manholes or catch pits to street system. Flow along the street (e.g. right above the pipes) can be in either direction along the streets, i.e. it can flow following a slope of the street or against it. It is not necessary that the flow direction in the street has to be the as the flow direction in the pipe system.

- GIS interface: GIS is important in simulation urban flooding. It is used as a tool to provide data and display simulation results. Surface storage for simulating surface flooding can be calculated by the application of GIS together with the DEM the study area, i.e. find area–elevation relation from DEM. In addition, results of the simulation can be easily understood in form of flood inundation maps. Model output in term of water along the streets are transferred to GIS and with interpolation routine, water surface is able developed. Flood inundation maps can be generated by overlaying of water surface and introducing flood depth map which is a method to visualise flood situations.

The main drawback of 1D flood modelling approach is an inaccuracy in the treatment of street channels and in the case of water flow on a surface. Both cases still considered as one dimensional. When the channels are overtopped, the flow of water is not only no longer 1D, because water move around 2D plane.

2.2. 2D flood modelling

As an addition to 1D flood modelling, 2D flood modelling can be used to analyse flow behaviour on a surface, both for water depth and velocity. 2D-modelling of out-channel floods clearly has the potential to revolutionise understanding of high-magnitude spatial and temporal hydraulics and high-magnitude flow phenomena, geomorphological and sedimentological processes, and hence rapid fluvial landscape change. This potential for new understanding is because of the now wide availability of high-resolution DEM data for large and often inaccessible areas, and the availability of remotely-sensed data [5]. Models to predict flood inundation based on the 2D shallow water equations are classed here as 2D approaches and solve for water level and depth-averaged velocities in two spatial dimensions. Unlike 1D models there is no need to prescribe a particular direction for the flow in 2D models. This brings advantages in terms of providing more detail on the nature of floodplain inundation but also introduces a number of limitations, e.g., difficulties in including hydraulic structures such as weirs and bridges in the models, longer computer run times and greater data requirements [2]. The system of 2D flood model consists of three equations: one equation for continuity and two equations for the conservation of momentum in the two orthogonal directions [6].

$$\frac{\partial h}{\partial t}+\frac{\partial (hu)}{\partial x}+\frac{\partial (hv)}{\partial y}=0 \tag{3}$$

$$\frac{\partial (hu)}{\partial t}+\frac{(hu^2)}{\partial x}+\frac{\partial (huv)}{\partial y}+gh\frac{\partial h}{\partial x} = -gh\frac{\partial z_b}{\partial x}-gn^2\frac{u\sqrt{u^2+v^2}}{h^{1/3}}+v_{eff}\frac{\partial}{\partial x}\left(h\frac{\partial u}{\partial x}\right)+v_{eff}\frac{\partial}{\partial y}\left(h\frac{\partial u}{\partial y}\right) \tag{4}$$

$$\frac{\partial (hv)}{\partial t}+\frac{(huv)}{\partial x}+\frac{\partial (hv^2)}{\partial y}+gh\frac{\partial h}{\partial y} = -gh\frac{\partial z_b}{\partial y}-gn^2\frac{v\sqrt{u^2+v^2}}{h^{1/3}}+v_{eff}\frac{\partial}{\partial x}\left(h\frac{\partial v}{\partial x}\right)+v_{eff}\frac{\partial}{\partial y}\left(h\frac{\partial v}{\partial y}\right) \tag{5}$$

In equation above, h is water depth, u and v are velocities along horizontal x and y axes, z_b is bottom level, n is Manning's roughness coefficient, g is gravity acceleration and V_{eff} is effective cinematic viscosity. Usually, it is assumed that the viscosity is constant throughout the flow field. Although 2D flood modelling can accurately describe the behaviour of flow on a surface, this model also has some drawbacks. First, calculation method is considerably more complicated than 1D flood model, and second, river and drainage network sometimes has a complicated morphology that cannot incorporated sufficient enough in 2D flood model.

2.3. 1D2D flood modelling

Because both 1D and 2D flood modelling has its own advantage and disadvantages, there are extensive approaches to integrate 1D and 2D models resulting in hydrodynamic model of floodplains and integrated 1D (channel flow) and 2D (overland flow), or widely known as 1D2D flood modelling. The main idea of 1D2D flood modelling is integrating 1D hydrodynamic modelling technologies, Digital Elevation Models and GIS systems is to take advantage of the best combination of 1D hydrodynamic data for rivers together with 2D terrain data, and presenting them in the GIS as maps [3].

The integrated one-dimensional and two-dimensional (1D-2D) model development focuses on the extension of model capabilities in order to simulate flooding situations more accurately. This includes improving flood wave propagation over initially dry land, improving the presentation of hydraulic control (levees and embankment) in the floodplain and integration of one-dimensional hydraulic elements (pumps, bridges and regulator gates). The combined 1D-2D modelling opens up the possibilities for studying flood control measures, flood forecasting, and development of flood evacuation plans. Main advantages of 1D2D flood modelling model are [7]:

- Flow computation on initially dry land, without using any special drying or wetting procedures.
- Accurate and stable flow computation on very steep slopes, such as dike walls and other manmade structures.
- Especially suitable for short event predictions (hours and days).
- Realistic flood predictions of dike break due to heavy rainfall or other natural hazards.
- Pre- and post-processing within a GIS environment.

For 1D-2D flood modelling, there are four parameters needed for the input which are [8];

- Digital Elevation Model to represent topography and river cross sections.
- Roughness data to represent the resistance to water flow on.
- Embankment and other artificial structure to make the model work as close as to the real-world condition.
- Boundary data in which time series of flow or water levels or Q-H relation can be defined.

3. System dynamics

System dynamics is method of thinking that can help to understand a complicated relationship in a system that consists of multiple key factors, constructing and visualizing the dynamic relationship between those factors within the system. [9]. After the system is constructed and visualized, users can modify the values of the factor to evaluate the behaviour of the system, and even find the optimum setting for each factor. System dynamics approach is unique from other method to construct a complex system in a way how its simulates and predicts a complex nonlinear system, and has been widely implemented in scientific research and engineering application. System dynamics is also already applied in solving the difficulties in researches, modelling and simulation is implemented during the analysis [10]. System dynamics was developed based on systems theory, which is a scientific field that tries to understands a system by focusing on its key components, relationship between components, and its internal mechanism. Implementation of System dynamics to represent real-world systems gain a world-wide attention when it is proven effective to develop an efficient management strategies for private companies.

In a computer-based construction of System dynamics, key factors or components in a complex system are represented as Stocks, Flows, Converters, and relationship among those components. The relationship among components are represented graphically, and modelled by a mathematical finite difference equations. A specific value of each component is calculated for each time step that pre-defined during the construction of the model. Computer-based simulation for system dynamics methodology mostly constructed using software modelling that are object-oriented in nature, such as Powersim, Vensim, Stella, and AnyLogic [11].

In visualizing a complex system of urban water management, system dynamic can be used to integrate a various types of sub model both from quantitative and qualitative data. System dynamic analysis respects the complexity of relationships between research objects by accommodating them through an arrow pointed from one subsystem as represented by a variable to others. The arrows reflect the relationships between variables. The positive marks signal the reinforcing value of relationships while the negative marks are utilized for balancing the relationships. The relationships resemble the reality of a multi-system within the community. Figure 1 below shows a simplified version of relationships in system dynamic analysis.

In order for a system dynamic analysis in urban water infrastructure to be conducted, first a modelling process should be completed. (Sterman 2014) suggests that there are five major steps of modelling, as follows:

- Problem articulation
- Formulation of dynamic hypothesis
- Formulation of a simulation model
- Testing
- Policy design and evaluation.

Models in system dynamics are continous, with a constant time steps. The time advancement is regular, the values of the components changes during the time advance, and the system records those values at any time the simulation run [12]. Therefore, System dynamics simulation models can be considered as discrete-time models. At any time-step *t*, the state of the model and the values of its components is an exact value, and must be unambiguously can defined. This model's state at time-step t will determines the state of the model at time-step (*t+dt*), where *dt* is the later time-steps.

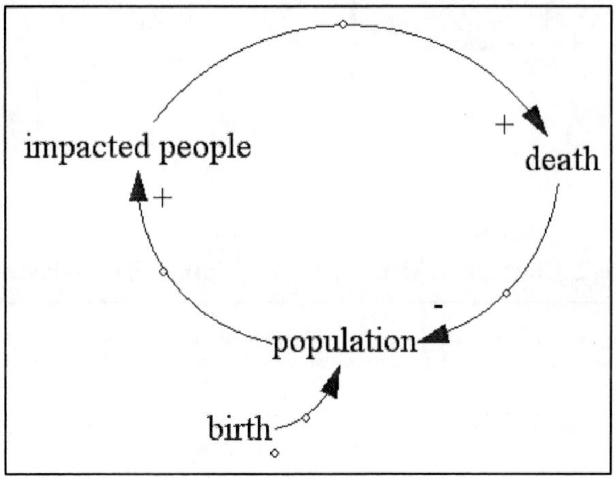

Figure 1. An Example of Relationships Between Factors in System Dynamics.

4. Results

From the discussion in the previous section, we can assume that system dynamics approach may be applied to develop a dynamic hydrodynamic systems such as urban flood modelling. Characteristics of elements of system dynamics, one of them is Flow, is directly compatible with the concept of hydrodynamic modelling. The key aspect of System dynamics, which is the constant time-step, is also linear with the concept of hydrodynamic modelling, because the time-step is associated with periodic nature of observations on urban water infrastructures. The principle of interconnection between components also can be used for realistic modelling and simulation of urban flooding. Example of System dynamics in urban flood and water management can be seen in [13], which developed a simulation of the operation of reservoirs for managing flood hazard. Another example is shown in [14], where a System dynamics-based hydrodynamics model is constructed in order to evaluate the system performance, especially its loss, due to flooding that caused by extreme rainfall, as well as how they can

IOP Conf. Series: Earth and Environmental Science **340** (2019) 012013 doi:10.1088/1755-1315/340/1/012013

potentially recoverd due to a city's adaptive measures.

Key components of System dynamics can be used to construct a complete hydrodynamic model. Flow component of System dynamics can be used to model simulate a wide array of water infrastructures that carry water into the city or discharge it to outer part of the city. Stocks are implemented to calculate the volume of the water that collected in the urban area. Connections can be implemented to model the water and drainage network within the area. The conceptual method of how hydrodynamic model is constructed in a System dynamics approach can be seen in figure 2 below.

Figure 2 shows the key components of hyrdologic modelling that represented as a flow diagram, a method that widely used to sketch the model of complex system. The inflow of water into the urban area is primarily from local rain that may collected in the drainage system or became a run-off water on the surface of the terrain. The inflow water can also from rivers that entering the city that connected to the drainage system. Even when there is no local rain, an excessive amount of water from incoming rivers can also caused flood in the urban areas. After the flow diagram is constructed, the next step is to develop a computer-based model of the flow, as seen in figure 3.

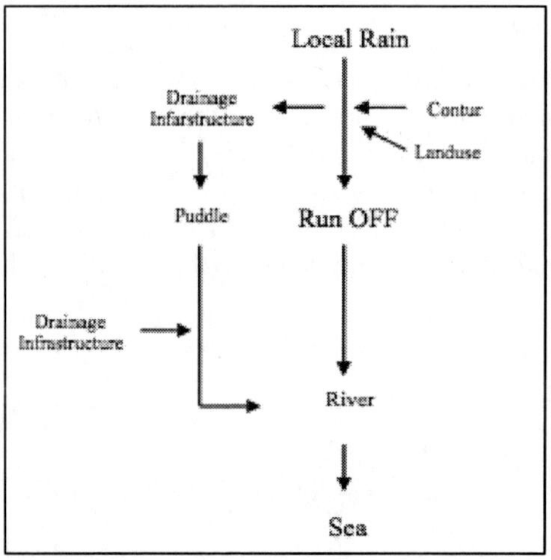

Figure 2. Conceptual Method of Flood Modelling in System Dynamics

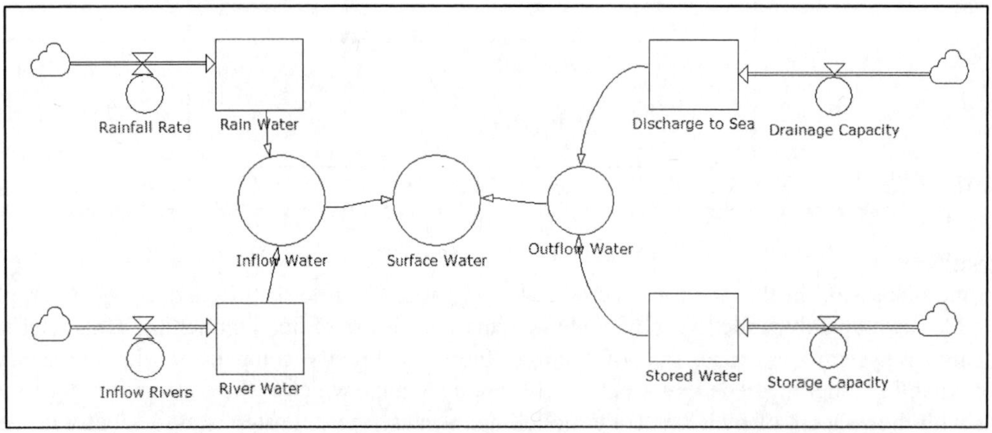

Figure 3. Example of Flood Modelling in System Dynamic Software (Powersim)

In Figure 3 above, The main components of the model are:

(a) Levels or Stocks, which consiscts of Inflow Water, Outflow Water, and Surface Water
(b) Flows, that consists of Inflow Rivers, Rainfall, Drainage Network, and Water Storage
(c) Connectors that simulating the flow of water from one components to another.

The values of those three components can be calculated by a 1D Hydrologic model. The volume of water can be entered directly from data source to the system dynamic model, while the capacity of water infrastructures must be calculated first by using equation (2) before implemented as an input for the System dynamic model. The functionality of System dynamic approach for flood modelling cannot extend to evaluate the water behaviour above the terrain surface. Hence, 2D or 1D2D hydrologic modelling cannot transformed into a System dynamics model. However, this drawback should be compensated by the increased accuracy and the visualization of the water system within the System dynamics model.

5. Conclusions

Adaptive aspects were seldom given priority in promoting a resilience city, due to those aspects' complexity compared to the vulnerability aspects. Because a city's resilience is measured in both adaptive and vulnerability values, the lack of adaptive efforts was not emerged as a concern, as long as the vulnerability-reducing efforts were adequate. By focusing not only on vulnerability but also towards adaptive aspects, the quest to make a city resilient will be improved. Recently emerging technologies provide an opportunity to further increase and promote the adaptive measures in facing disturbances and disasters. With these technologies, various applications, either desktop or mobile, can give suggestions to users to avoid disturbances, such as traffic jam, flood, etc. It is up to the user whether to use these suggestions or not.

The main challenges of utilization of those technologies are the security aspect of the technologies itself because the implementation of applications to increase a city's adaptability will also increase in its vulnerability, because technologies, especially in form of computer systems, are prone to malicious attacks, i.e. from hackers and computer viruses. Therefore, implementation of new technologies in promoting a resilient city must be taken very cautiously.

This paper shows the possibility constructing a System dynamics-based flood modelling based on input both from raw data and 1D flood modelling. The visualization funcionality of System dynamics should give an added value to the flood modelling practice, because planners and decision makers can evaluate the existing condition of flood hazard and generate scenarios to various possible conditions. Even though the 2D and 1D2D hydrologic modelling is not possible to be transformed into a System dynamics, the visualization of the water network can give a lot of advantages.

For our future work, we planned to construct a complete model of System dynamic-based flood modelling, based on the existing 1D flood models for a particular city. By contructing a multiple System dynamics-based flood modelling from various 1D flood, we hope that we can evaluate the accuracy of the flood modelling if it is constructed using a System dynamic approach. An integrated model of System dynamics and 1D flood modelling is also possible, that would bring a complete new approach in flood modelling for urban environment.

6. References

[1] Parkinson, "Urban Stormwater Management in Developing Countries, IWA Publishing, London, UK," 2005.

[2] G. Pender and S. Neelz, "Use of computer models of flood inundation to facilitate communication in flood risk management," *Environ. Hazards*, vol. 7, no. 2, pp. 106–114, 2007.

[3] M. S. Horritt and P. D. Bates, "Evaluation of 1D and 2D numerical models for predicting river flood inundation," *J. Hydrol.*, vol. 268, no. 1–4, pp. 87–99, 2002.

[4] O. Mark, S. Weesakul, C. Apirumanekul, S. B. Aroonnet, and S. Djordjevic, "Potential and limitations of 1D modelling of urban flooding," *J. Hydrol.*, vol. 299, no. 3–4, pp. 284–299, 2004.

[5] J. L. Carrivick, "Application of 2D hydrodynamic modelling to high-magnitude outburst floods: An example from Kverkfjoll, Iceland," *J. Hydrol.*, vol. 321, no. 1–4, pp. 187–199, 2006.

[6] E. Mignot, A. Paquier, and S. Haider, "Modeling floods in a dense urban area using 2D shallow

water equations," *J. Hydrol.*, vol. 327, no. 1–2, pp. 186–199, 2006.
[7] T. Shaviraachin, "Flood simulation : a case study in the lower Limpopo valley, Mozambique using the SOBEK flood model," ITC, Enschede, 2005.
[8] M. Bin Usamah, "Modelling flood events in the lower Bicol floodplain," ITC, Enschede, 2005.
[9] M. Faezipour and S. Ferreira, "Assessing Water Sustainability Related to Hospitals Using System Dynamics Modeling," *Procedia Comput. Sci.*, vol. 36, pp. 27–32, 2014.
[10] Z. Chen and S. Wei, "Application of System Dynamics to Water Security Research," *Water Resour. Manag.*, vol. 28, no. 2, pp. 287–300, 2014.
[11] S. Park, B.-J. Kim, and S.-Y. Jung, "Simulation Methods of a System Dynamics Model for Efficient Operations and Planning of Capacity Expansion of Activated-sludge Wastewater Treatment Plants," *Procedia Eng.*, vol. 70, pp. 1289–1295, 2014.
[12] J. Pecherska, "A System Dynamics Model for Basin Level Forecasting on the Basis of a Digital Elevation Model," *Information Technology and Management Science*, vol. 16. p. 101, 2013.
[13] S. Ahmad and S. Simonovic, "System Dynamics Modeling of Reservoir Operations for Flood Management," *J. Comput. Civ. Eng.*, vol. 14, no. 3, pp. 190–198, Jul. 2000.
[14] C. K. Gotangco *et al.*, "Quantifying resilience to flooding among households and local government units using system dynamics: a case study in Metro Manila," *J. Flood Risk Manag.*, vol. 9, no. 3, pp. 196–207, Sep. 2016

IOP Conf. Series: Earth and Environmental Science **340** (2019) 012014 doi:10.1088/1755-1315/340/1/012014

Priority study of infrastructure development at Suburban Pekanbaru (Case study: Tambang District)

D Ismiyanti[1], F Asteriani[2], and P Astuti[2]

[1] Master of Regional and Urban Development, Diponegoro University, Semarang, Indonesia
[2] Urban and Regional Planning Departement, Universitas Islam Riau, Pekanbaru, Indonesia

Email: ismiyantidesy@gmail.com; pujiastutiafrinal@eng.uir.ac.id

Abstract. The problems in the field of infrastructure are the main problems in the suburban of Pekanbaru, making it difficult to determine the priority of infrastructure development most needed by the community. The Analytical Hierarchy Process method and weighting of variables in priority setting are used in infrastructure development. The results of this study indicate that the average development of infrastructure is 20.75%. The priority of infrastructure development is road network (0.165), housing (0.095), educational facilities (0.089), economic facilities (0.086), health facilities (0.075), clean water (0.074), bridges (0.067), electricity grids (0.066) (0.059), telecommunication network (0.051), government facilities (0.049), garbage and drainage facilities (0.042), and sports facilities (0.041).

1. Introduction

Peri-urban area is an area that is located between two regions that are very different in their environmental conditions, its between regions that have urban appearance on one side and areas that have a rural appearance on the other. Because urban and rural areas have such complex dimensions of life that generally show different attributes, then in this area, special attributes emerge which are hybrids of both [1].

Pekanbaru City, which is known as the center of trade and services, the center of government, and other centers of activity have always evolved over time. This continuous development has resulted in areas that are directly bordering the city of Pekanbaru, which have been heavily influenced by the city (urban sprawl). There are many suburbs of Pekanbaru City, its bordered by three regencies; Kampar, Siak and Pelalawan. Tambang District is part of Kampar Regency, directly adjacent to Pekanbaru, is one of the suburbs of Pekanbaru. That has been affected by the development of Pekanbaru. The selection of this area to become a research area based on the scope of the area that was predominantly affected by the development of Pekanbaru. It was based on the limitations of researchers to conduct an overall study of the area included in the suburbs.

Tambang District is a suburban area in many different developments with the center of Pekanbaru. The distance between these two regions is very close, limited by the administrative boundary gates of the two regions. Problems have arisen, ranging from issues of settlements, infrastructure, facilities and infrastructure, human resources and so on, apart from the location of the sub-district which is directly adjacent to the City of Pekanbaru, this area is also located on the national crossing route of Riau-West Sumatra. A lot of growth magnets and centers of activity will make growth in this area increase. Problems in the infrastructure sector are the main problems in this region. According to data from the

IOP Conf. Series: Earth and Environmental Science **340** (2019) 012014 doi:10.1088/1755-1315/340/1/012014

Tambang District Statistics Agency in Figures 2014 [2], exposure to data on the development of built land and infrastructure development in the suburbs of Pekanbaru shows differences in development with Pekanbaru City, from the land use data of only ± 10,000 Ha of land that is still untapped. And from data on facilities and infrastructure, one of the roads is ± 100 km of roads that are still in a damaged condition. As well as other problems such as slums and so on. Therefore, researchers are interested in conducting research in this area in order to find out what infrastructure is the priority of development in this area in order to develop a sustainable area or more commonly known as "sustainable city". While for the purposes of this study are as follows: (i) Identifying infrastructure developments in the suburban of Pekanbaru; (ii) Analyze the priority level of infrastructure development in the suburbs and Pekanbaru; (iii) Formulating infrastructure development strategies in the suburban of Pekanbaru in order to create sustainable development.

2. Methods

The method used in this research is:

1. Identification of Infrastructure Development in the Suburban of Pekanbaru City The method used is the descriptive quantitative-qualitative method, this research is obtained from the observations and existing secondary data. Then it is explained to describe the conditions and what are the infrastructure in the research area. In addition, at this stage also describes the development of infrastructure with see its development from the period or span of 5 years of development. The assumption of taking this method is obtained in accordance with the development period of an area for 5 years. Infrastructure developments were taken from 2003, 2008 and 2013.
2. Analysis of the Priority Level of Infrastructure Development in the Suburban of Pekanbaru City two methods are used the quantitative-qualitative descriptive method. This stage begins with sampling which is considered expert or there is a relationship with the research area. Offices who are respondents in the Tambang District are: Regent's office, District office, Development Planning Agency, Departement of Public Work, Promotion and Investment Agency and Academics. An analytical hierarchy process (AHP) questionnaire was used distribute to the expert sample. Data from the questionnaire will be used as input for the AHP analysis using expert choice software, then processed to obtain a value that describes the priority level of the infrastructure development and then explained descriptively. AHP is a retrieval technique decisions developed for cases that have various levels of (hierarchy) analysis.
3. Formulating Infrastructure Development Strategies in Pekanbaru Suburban.
   The method used is descriptive qualitative method with SWOT analysis. In formulating these efforts and strategies based on local government policies, infrastructure development priorities, and field conditions. SWOT is a tool that can be used for qualitative analysis. This analysis can be based on logic that can maximize strengths and opportunities, but simultaneously can minimize weaknesses and threats (Rangkuti, 2001 in [3]).

3. Result and discussion

3.1. Identification of infrastructure development

The highest development of infrastructure is economic facilities, which reached 54% from 2003 to 2013, while infrastructure which experienced a decline in development was health advice, where from 2003 to 2013 its development decreased by 17%. The average infrastructure development in Tambang District from 2003 to 2013 was 20.75%.

Table 1. Development of infrasctruture in Tambang District (year 2003-2013)

Infrastructure	2003	2008	2013	Development	
				Average	%
Education Facility	70	86	90	2	8
Economic Facility	105	218	599	49,4	54
Health Facility	25	20	15	-1	-17
Worship Facility	150	152	142	-0,8	-2
Sport Facility	84	86	84	0	0
Road	21	38	40	1,9	19
Electric	2378	7675	16011	1363,3	52
Brigde	141,5	321	435,5	29,4	33
Clean Water	6250	8786	16743	1049,3	33
Housing	5439	8786	16743	1130,4	37
Telecomunication	5238	5240	12478	724	32
				Average	20,75

Source: Analysis, 2015

3.2. Analysis of priority levels of infrastructure development

The first step in this analysis is to determine the structure of the hierarchy. This hierarchical structure consists of five levels.

- *Level 1* is the main objective; the Priority of Infrastructure Development in Tambang District; *Level 2* is a supporting factor for achieving the main objectives; Physical Factors, Social Factors, Economic Factors;
- *Level 3* is a criterion relating to each factor, consisting of; Land Use, Environment, Population, Community Activities, Regional Benefits, Procurement Costs;
- *Level 4* is a sub-criteria which is a derivative of the previous criteria, which consists of; Land Availability, Land Value, Environmental Quality, Environmental Impact, Population Growth, Increased HR, Labor, Community View, Community Behavior, Community Income Level, Regional
- Development, Regional Original Income (PAD), Mutiplier Effect, Government Costs, Investment, Maintenance cost;
- *Level 5* is an alternative that is the choice in the priority of infrastructure development in Tambang District.

For more details can be seen in Figure 1. Furthermore, weighting is carried out on: Factors, Criteria - Sub Criteria, and Alternatives.

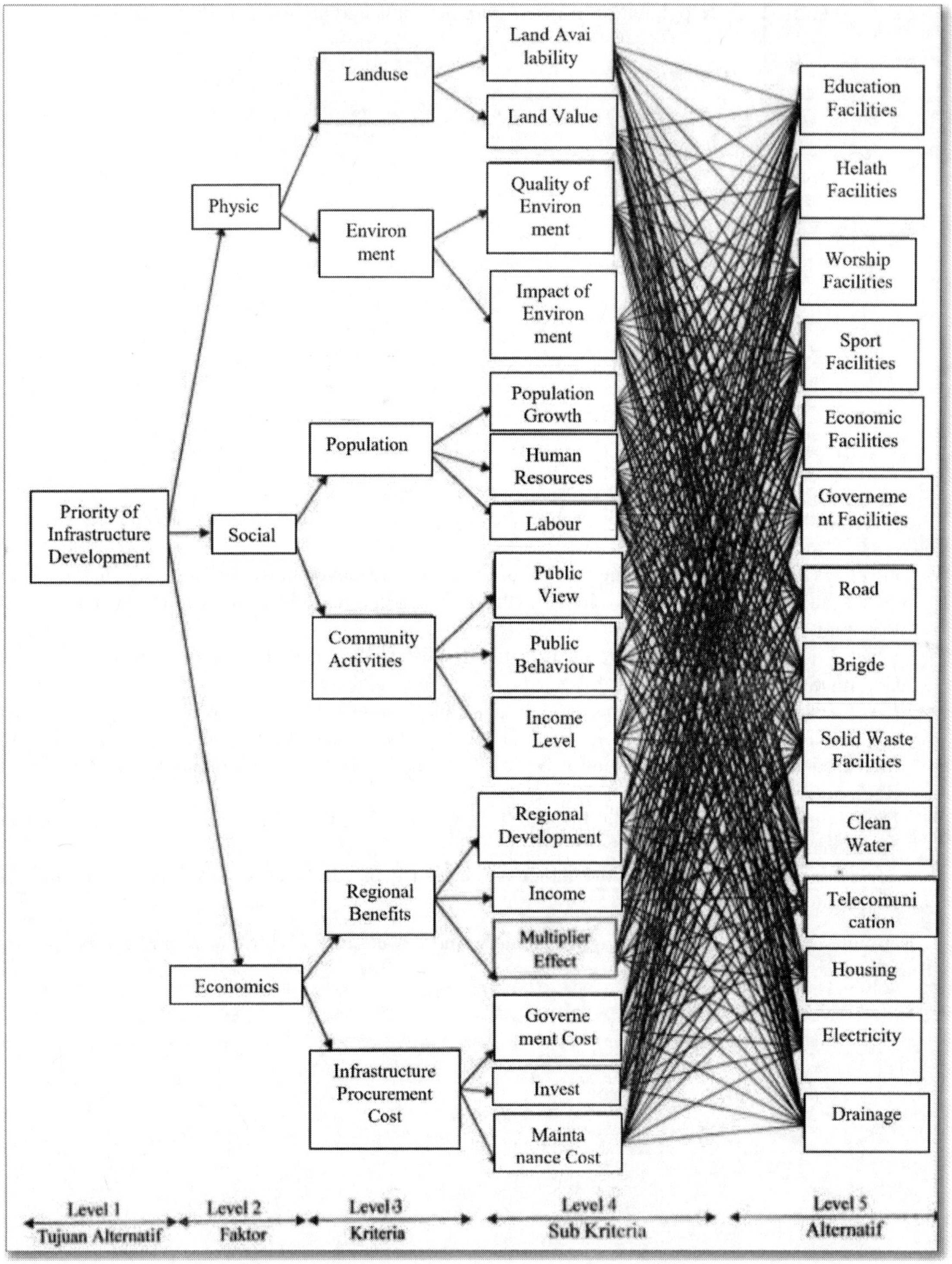

Figure 1. Priority Levels of Infrastructure Development
Source: [4,5,6,7]

The results of the Infrastructure Development Priority Level are: (1) Road, weight of 0.165; (2) Housing, weight of 0.095; (3) Education Facilities, weights of 0.089; (4) Economic Facilities, weight of

0.086; (5) Health facilities, weights of 0.075; (6) Clean water, weight of 0.074; (7) Bridges, weights of 0.067; (8) Electricity, weight of 0.066; (9) Worship facilities, weights of 0.059; (10) Telecommunication, weight of 0.051; (11) Government facilities, weight of 0.049; (12) Solid waste and drainage facilities, weight of 0.042; (13) Sports facilities, weights of 0.041.

3.3. Formulating infrastructure development strategies in Pekanbaru Suburban

The following are formulated Strengths, Weaknesses, Opportunities and Threats:

1) Strength (S):
   1. Tambang District located in the suburban of Pekanbaru,
   2. Land supports broadly,
   3. The existence of a regional cooperation program for Pekansikawan,
   4. The policy of the Kampar Regency government towards the Tambang District,
   5. The potential of extensive agricultural and plantation land,
   6. The development area in the national scale education area,
   7. Development area is located on a national road crossing.

2) Weakness (W):
   1. Land in the development area in the form of peat land,
   2. Quality of human resources is still low,
   3. Land use is low,
   4. The pattern of research area is not organized.

3) Opportunities (O):
   1. Support from the policies of the Kampar Regency government through RP2KP (Settlement Area Development and Development Plans)/Urban Settlement and Infrastructure Development Strategies (SPPIP) and RPJMD,
   2. Regional-scale Regional Cooperation,
   3. Government Transparency in Accepting Investment,
   4. Communities heterogeneous.

4) Threat (T):
   1. Land conflict with the community,
   2. Degradation and deterioration of the quality of the environment,
   3. Accumulation of development on the outskirts of the city,
   4. The rapid use of land tends to ignore the spatial plan.

Based on the results of the analysis that has been carried out and is based on the priority objectives of infrastructure development in the suburban area of Pekanbaru, the suburban area development strategy is described as follows:

1. Utilizing the position of the Tambang District to attract infrastructure development investors (S1,S2,S3,S4-O1).
2. Utilizing large areas of land to support development in the research area (S2-O2).
3. Utilizing support from the government through development policies by promoting the region (S4-O3,O4)
4. Utilizing the potential of the Tambang District to attract domestic and foreign investors for infrastructure development (S1,S2,S5,S6,S7-O2,O3,O4).
5. Utilizing government support to handle land in less optimally by repair and so on (W1-O1)..
6. Enhancing and developing regional cooperation between regions in order to create competent human resources in managing development in the development area (W2-O2,O4).
7. Utilizing the openness of the government in accepting investments to make land use in regional development (W3-O3).
8. Creating an organized pattern of regional space with the support of diverse communities and support from the government (W4-O4).

9. Utilizing community and government cooperation to minimize conflict between government and society (S3-T1).
10. Utilizing infrastructure development to create environment so that degradation and degradation of environmental quality can be overcome (S2,S3,S4 –T2).
11. Use of inter-regional cooperation to create a pattern of space and land use in accordance with the development plan (S2-T4).
12. Promote collaboration between government and society to avoid conflict (W2,W3-T1).
13. Using an environmentally friendly development system to minimize environmental degradation (W3-T2).
14. Utilizing vast land to create infrastructure development that is in accordance with the needs of the community (W3,W4-T3,T4).

4. Conclusion

Based on the results of the discussion on the priority study of infrastructure development in the suburban of Pekanbaru (Case Study:Tambang District, Kampar Regency), it can be concluded :

1. Infrastructure Development in Tambang District
   Most of the infrastructure development in Tambang Subdistrict has always experienced an increase, which is quite significant is the Economic facilities, up to 54%, while the facilities that have decreased are health facilities have decreased by 17% and the facilities of worship decreased by 2%. The average development of infrastructure in Tambang District from 2003 to 2013 was around 20.75%.
2. The Priority Level of infrastructure development in Tambang District is the road, housing, facilities: education, economic, health, clean water, bridges, electricity, worship facilities, telecommunications, government facilities, waste and drainage facilities and sports facilities.
3. The development strategy of the Tambang District which is suitable to be applied in this region is the development of sustainable areas. This is in accordance with the location of the area that is very possible to do this. Sustainable development must be adjusted to the development plan that has been prepared by the government. As has been known this area has been included in the Settlement Development and Development Plan (RP2KP) by the district government Kampar. Of course the plan must be supported by other aspects such as transportation, electricity, clean water, sanitation and so on. This is so that development can be well organized and in accordance with the agreed Regional Spatial Planning (RTRW). Another strategy in the development of this region by utilizing regional cooperation between the areas of the Tambang Subdistrict and the City of Pekanbaru, can be in the form of cooperation in the provision of infrastructure development and so on. In this case both governments must sit together and make an agreement in the form of a Memorandum of Understanding (MOU) so that sustainable and sustainable development will be created.

5. References

[1] Yunus H S 2009 *Dinamika Wilayah Peri-Urban: Determinan Masa Depan Kota.* (Yogyakarta: Pustaka Pelajar)
[2] Kecamatan Tambang Dalam Angka 2003, 2008, 2012, 2013, 2014 (Bangkinag: Badan Pusat Statistik)
[3] Kodoatie Robert J 2005 *Pengantar Manajemen Infrastruktur* (Yogyakarta: Pustaka Pelajar)
[4] Jayadinata Johara T 1999 *Tata Guna Tanah dalam Perencanaan Pedesaan Perkotaan dan Wilayah.* Edisi ketiga. (Bandung: Penerbit ITB)
[5] Ilhami 1988 *Strategi Pembangunan Perkotaan di Indonesia.* (Surabaya: Usaha Nasional)
[6] Khairuddin 2000 *Pembangunan Masyarakat: Tinjauan Aspek Sosiologi, Ekonomi dan Perencanaan* (Yogyakarta: Liberty)
[7] Maqin Abdul 2011 *Pengaruh Kondisi Infrastruktur terhadap Pertumbuhan Ekonomi di Jawa Barat.* Trikonomia, Volume 10, No.1, Juni 2011, Hal 10-18, ISSN 1411-514X

[8] Branch Melville 1955 *Perencanaan Kota Komprehensif: Pengantar dan Penjelasan.* (Yogyakarta: Gadjah Mada *University Press*)
[9] Budimanta Arif 2005 *Pendekatan Pembangunan Kota Berkelanjutan, dalam Bunga Rampai Pembangunan Indonesia dalam Abad 21: Konsep dan Pendekatan Pembangunan Perkotaan di Indonesia*. Buku 1 (Jakarta : Yayasan Sugijanto Soegijoko dan URDI)
[10] Dikun Suyono 2005 *Pendekatan Sektoral Dalam Pembangunan: Pengembangan dan Pengelolaan Infrastruktur, dalam Bunga Rampai Pembangunan Indonesia dalam Abad 21: Konsep dan Pendekatan Pembangunan Perkotaan di Indonesia*. Buku 1.(Jakarta : Yayasan Sugijanto Soegijoko dan URDI)
[11] Herjanto Eddy 2009 *Sains Manajemen: Analisis Kuantitatif untuk Pengambilan Keputusan.* (Jakarta: PT Grasindo)
[12] Kampar Dalam Angka 2003, 2008, 2014 (Bangkinang: Badan Pusat Statistik)
[13] Kuncoro M 2003 *Ekonomi Pembangunan, Teori, Masalah dan Kebijaksanaan* (Yogyakarta: UPP AMP YKPN)
[14] Mitchel B et all 2003 *Pengelolaan Sumberdaya dan Lingkungan* (Yogyakarta: Gadjah Mada University Press)
[15] Nasution S R 2013 *Proses Hirarki Analitik Dengan Expert Choice 2000 Untuk Menentukan Fasilitas Pendidikan Yang Diinginkan Konsumen*. Jurnal Teknik FTUP, Volume 26 Nomor 2 Juni 2013.
[16] Raco J R 2010 *Metode Penelitian Kualitatif: Jenis, Karakteristik dan Keunggulannya* (Jakarta: PT Grasindo)
[17] Rustiadi Ernan et all 2009 *Perencanaan dan Pengembangan Wilayah.*(Jakarta: Yayasan Obor Indonesia)
[18] Sjafrizal 2012 *Ekonomi Wilayah dan Perkotaan* (Jakarta: Raja Grafindo Persada)
[19] Tarigan Robinson 2005 *Perencanaan Pembangunan Wilayah*. Edisi Revisi (Jakarta: PT Bumi Aksara)
[20] Vargas Ricardo 2010 *Using The Analytic Hierarchy Process (AHP) to Select and Prioritize Projects in a Portfolio* (North America: PMI Global Congress 2010)
[21] Yunus H S 2000 *Struktur Tata Ruang Kota* (Yogyakarta: Pustaka Pelajar)
[22] Yunus H S 2005 *Manajemen Kota: Perspektif Spasia* (Yogyakarta. Pustaka Pelajar)

Acknowledgments
This work is partially supported by Urban and Regional Planning Departement, Faculty of Engineering, Islamic University of Riau.

IOP Conf. Series: Earth and Environmental Science **340** (2019) 012015 doi:10.1088/1755-1315/340/1/012015

Neighbourhood space for formal housing based on social cohesion in Jember Region

D J Koesoemawati[1], H Yuswadi[2], A Ratnaningsih[3], R Alfiah[1] and M Firmansyah[1]

[1] Urban and Regional Planning Department, Faculty of Engineering, Universitas Jember
[2] Sociology Department, Faculty of Social and Political Sciences, Universitas Jember
[3] Civil Engineering Department, Faculty of Engineering, Universitas Jember

Email: dewi.teknik@unej.ac.id

Abstract. Nowadays "community without closeness" does exist and makes the closeness to space is no longer important for social relations such as in formal housing in Jember Region. The housing area in Jember Region divided into two types, such as high density housing and low density housing which are influenced by their type of house. Qualitative analysis method that is used in this research are synchronic reading and typology. Synchronic analysis technique based on maps, observation, and interview to several source and used in this research to describe settlement pattern. In the fact, the community who live in formal housing prefer to spend their time in home without paying attention the socialization with their neighbourhood and using public open spaces. It makes segregation and weaken the social cohesion. Therefore, this research is needed to provide a social model of environmental space based on social interaction patterns and location points of interaction in low and high density formal housing in Jember region. Based on the result, it is known that the developer need to designing enclave on formal housing road spaces that are used for social interaction and control each other in housing security, according to the concept of defensible space and neighbourhood space model namely "Enclave Transpatial"

1. Introduction

Phenomena that occur in cities show that the community tendencies are heterogeneous and the social relations do not always account for proximity to space because of the development of communication technology [1]. Based on the phenomena above, there was a research about social space and neighbourhood space which study about the correlation between proximity space and social relations. Nowadays "community without closeness" is possible, it makes the closeness to space is no longer important for social relations [2] in [3]. The area of the environment where the housing exists is very important to maintain the quality of life, so the distance to space becomes important [4] in [3]. Basically, the relationship between the road network and the building that forms the road space as a social space will affect the neighbourhood space model. The neighbourhood patterns play a role in the formation of social interaction [5] .

Jember Region well known with the agriculture and plantation product. In line with its development, nowadays Jember Region has begun to be noticed, especially by developer of formal housing. Housing area has the second biggest land use in the cities area of Jember Region with total housing area are

IOP Conf. Series: Earth and Environmental Science **340** (2019) 012015 doi:10.1088/1755-1315/340/1/012015

3.130 ha. The built area are concentrated in Kaliwates (1.178 ha) and Sumbersari (1.102 ha). It used for residential housing area and commercial area. Many new residences are developed by formal developer or personal resident using agricultural areas which is changed its functions became formal housing.

The residential area in Jember Region divided into two types, such as high density housing and low density housing which are influenced by their type of house. Generally, the housing or residence built by developer (formal housing) has build every building with good facilities such as road and public open space. The arrangement of housing layout such as road network and the width of the road has influence on the intensity of their social interaction. Social interaction is supported by the potential of social cohesion as behavior modal and community activities. Social cohesion covers sense of belonging, social trust, and generalized reciprocity and cooperation, and also social harmony [6]. Public open space in this research associated the road and the open space of housing facilities. The open space, is not only a place for recreation, but also to help for improving social cohesion and give advantages of quantitative environment [7]. Public open space should be used as socialization area with others, or community with other communities. But, the fact that public open space is less used, it made the level community socialization directly become weak. Eventhough, public open space is node and communication facility of social connective to create interaction among community groups [8]. For example, between young generation and older generation, ethnic, religion, social class, economy, and etc. While nowadays the community who live in formal housing prefer to spend their time in home without paying attention the socialization with their neighbourhood. The underutilization of public open spaces has made the level of socialization of the community directly weakened, it makes segregation and weaken the social cohesion. Therefore, this research is done to know the neighbourhood space model in formal housing and the influenced factors are related with potential of social cohesion in that environment. In addition, there were concerns related to security in formal housing so that residents built many portals unplanned. Therefore, a study is needed to provide a social model of environmental space based on social interaction patterns and location points of interaction in low and high density formal housing in Jember region.

2. Studyerea

Administratively, study area in this research is cities area of Jember Region, including two districts which haspotention the growth of formal housing, such as Kaliwates and Sumbersari district. The delineation of its territory can be seen in the figure 1 as follow:

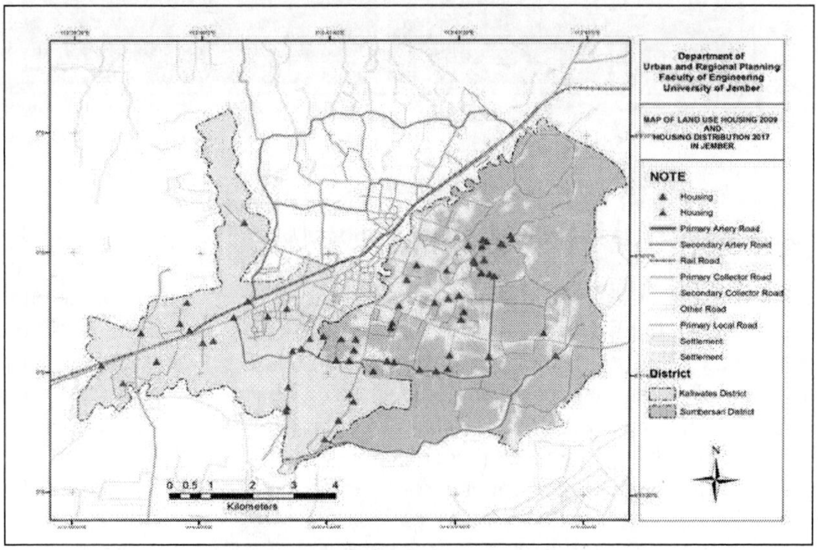

Figure 1. Research Area Deliniation
Source: Digitization Result, 2018

3. Method

This research used qualitative descriptive to describe systematically, factual, and accurately about fact and population characteristic in certain area [9]. Qualitative analysis method that is used in this research are synchronic reading and typology. Synchronic analysis technique based on maps, observation, and interview to several source and used in this research to describe settlement pattern [9]. Synchronic reading model is used to read the potential of community social cohesion related with spatial concepts. The aspect that is seen later, including shape and place, elements comparison spatially (length and width), percentage of environment is limited by time limitation, and opening spatially. Survey technique "static snapshots" is one of the technique or observation method/survey which is used to write people's movement activities and people static activities related with space used [10]. Static snapshots method in this research is used to know the community interaction pattern in formal housing related with space used. Formal housing that is surveyed in Kaliwates and Sumbersari grouped based on same type of road space and availability facilities of open space.

4. Result and discussion

4.1. Housing condition

Based on the survey result, formal housing area (housing that is built by developer) in study area is devided in housing area with density and low density housing that is influenced by plot area andtype of house. Based on the density and house type, it is grouped in 2 groups in each Kaliwates and Sumbersari district which is clustered from survey result on 80 housing. Formal housing characteristics in study area is reflected in the planned arrangement and also completed with good facilities. Each housing parcels has the average of base shear coefficient (BSC) 80-90% with very high density and base shear coefficient (BSC) 70-80% with low density.

Figure 2. Formal settlement
Source: Jember Spatial Document 2015-2035

4.2. Neighbourhood space and social cohesion in settlement formal Jember

Perception of social cohesion is contributed by thė use of public open space. In one of the research is found that 60% from 116 respondents stated that the use of public open space is contact with low intensities, such as see and watch other people, give or information, or give comments casually [11]. However, this contact ass the first steps and fundamental to create the variation of interpersonal and social relationship. Later, this interaction and social relation will meet the social cohesion [12].

Nowadays, the trend of living at housing in Jember raised concerns related with social cohesion in housing area. Based on the survey result in housing, it is known that influenced factors for weakening of social cohesion on formal housing in Jember such as:

1. Communities are limited their social interaction space because of fenced house that limits the house accessibility.
2. Communities taste at modern living, instant, practical, and efficient, make developers provide the housing which is designed to fulfill basic needs of residence without pay attention for social interaction need of housing community.

3. Practical design and highlight interest for each the housing owner make the interaction between them is difficult to build.

Based on the observation on formal housing in Jember with high density, then the points of location for community who live in housing to do their social interaction is at the road space. The social interaction on the certain time, such as at 06.00-08.00 a.m. when mobile vegetable traders and at 15.30-17.00 p.m. when some of women enjoy their leisure time while invite their children to play. The characteristic of community social interaction on housing with high density similar with the social interaction of communities who live in village.

The lack of open space, especially on middle and lower class of housing, but the community of housing is still can do their activities together, so the chance to know, is known and interact with others. Spill market also as generator for social interaction activities, especially in housing with high density. Spill market is farmed because of mobile vegetable traders activity who has many consumes, so their activities in the long time, that is known static activity.

Based on the observation at formal housing in Jember with low density, then many social interaction is done in public open space in this housing, because of housing public facilities are really designed according to it's function.

Housing residents used public open space to do exercises in the morning at 05.00 – 06.00 a.m. The road space is really used as circulation. The social interaction on formal housing in Jember proved that human as social creatures, with limitation of fascilities as area in doing social interaction because of fenced house design and limitation of public open space, then housing residents keep doing the social interaction although with limited time. Spatially, social cohesion in formal housing in Jember Region can be seen in figures 3,4,5 and figure 6.

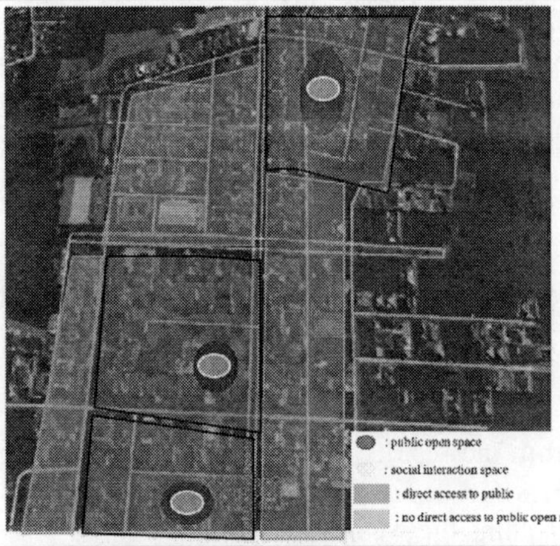

Figure 3. Sempusari Housing

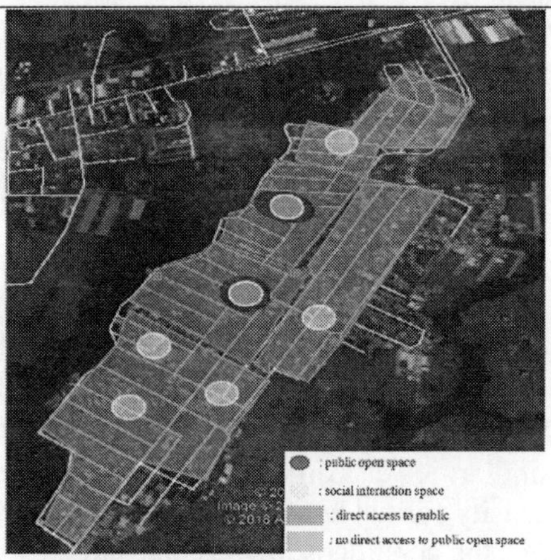

Figure 4. Griya Mangli Housing

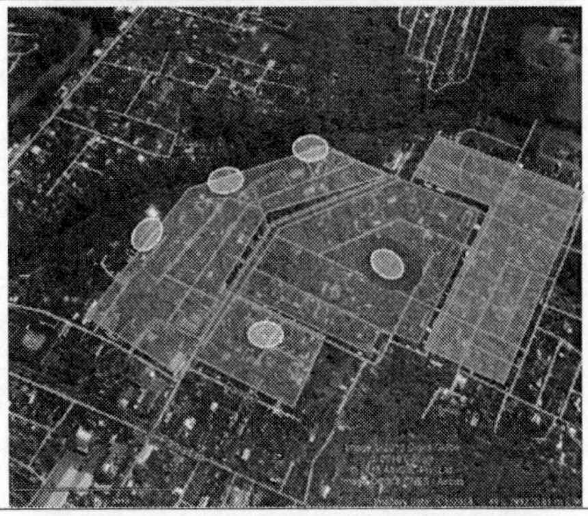

Figure 5. Gunung Batu Housing

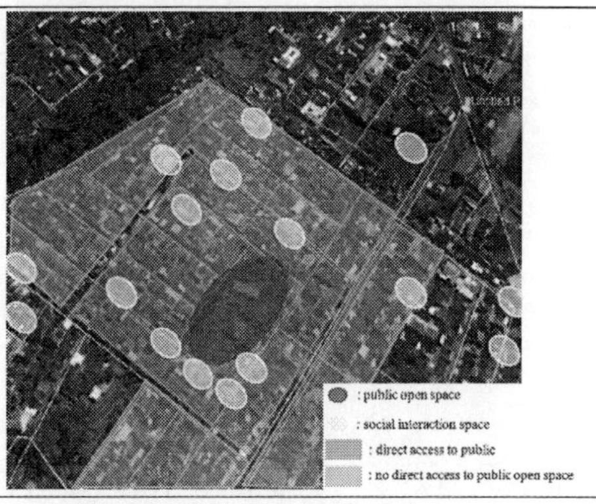

Figure 6. Mastrip Housing

Sempusari housing in Kaliwates Subdistrict and Gunung Batu Housing in Sumbersari Subdistrict is a low density type housing with gate community, because to enter housing is monitored in the guard post and above 21.00 certain roads are closed portal for housing security.

Public open spaces are used housing resident for sports activities as well as social interaction with other housing residents who both take advantage of the function of open public spaces, residents who are far away with the location of open public spaces will be constrained by direct access to public spaces.

The Griya Mangli housing in Kaliwates Subdistrict is a high density type housing that has easy access to enter the housing. The security aspect is low because there is no gate community, so many portals are installed. Public open space is limited, so that it utilizes the road space for social interaction at certain hours, namely the morning dominated by mothers and evening dominated by children. Economic activity as a generator of social interaction. Social interaction activities at the location point of the presence of vegetable vendors around the street space, so that mothers when shopping as well as caring for their children

Theory about open space many put forwand by experts, one of them is definition that open space is social space schome combines social components and physical of environment become a single scheme. Basically, open space give a chance for community to have reaction then the at the end they will create a city community. As definition, open space has different characteristic in accordance with the scale. Neighbourhood space is open space which is developed in housing environment to do public activities, such as playing, doing sport, enjoying the leisure time like the activities in formal housing complex.

Weakening of social cohesion on housing environment can be caused by the developer do not provide an available interaction space/neighbourhood space. Healthy social space is usually not become an important part from planning. Where as in essence, neighbourhood space can help housing communities to meet others.

Modern formal housing tend to eliminate the availability interaction space based on efficiency reason with build the house corncidentally. Housing is also often do not provide on open space that can be used together by communities. The weaknesssess of this social cohesion create a social gaps in formal housing. And it can be solved by good social interaction between housing community. Home design and location, facing each other can help neighbours to watch others, so the orientation of housing residents not only at home.

Economy activities as generator of social interaction. Social interaction activities in the location point is the presence of mobile vegetables traders, then some mothers buy vegetables while they nurturing their children.

4.3. Neighbourhood space model

The similarities and differences in social interactions from the results of research on high density and low density formal housing in the Kaliwates and Sumbersari Subdistricts, explained in Table 1 which explains the conditions of weekend and table 2 which explains the weekday.

Table 1. Social Interaction Activities

Description	High Density Formal Housing	Low Density Formal Housing
Differences		
Social Interaction Activity	Weekend is used by residents to interact among residents, including social interaction when shopping at vegetable sellers, sports activities and at meetings in one of the housing residents.	During the holidays, many residents do activities outside the housing, including recreation out of town, lunch and dinner outside the house, so that only a few residents go shopping at vegetable sellers. On weekdays the interaction between residents is reduced due to their activities at their workplaces.
Static activity	Static activity generated by vegetable sellers is active in a place on a residential	There is no static activity that is raised by the vegetable seller on a holiday because the

Description	High Density Formal Housing	Low Density Formal Housing
	road starting at 05.00 am - 09.30 am and it is carried out every day both weekdays and weekend	residents of low-density formal housing on holidays are more active outside the housing, including recreation. Static activity occurs during weekdays starting at 5:00 a.m. - 9:00 a.m. Static activity generated by vegetable seller active in a place on a residential street starting at 05.00 am - 09.30 am and carried out every day both weekdays and weekend
Social interaction time	The time of social interaction occurs between 05.00 - 09.30 both weekdays and weekend	The social interaction time of some nonMuslim residents in housing occurred at 7:00 a.m. to 9:00 a.m. during activities in the church in the housing. Most of the other residents carry out activities outside the housing.
Distance point of social interaction activity	The distance of the residents' social interaction activity points is an average of 50-75 meters	The distance of the residents' social interaction activity points is more than 100 meters
Housing Security	There are many portals on several ends of the road that are not properly conceptualized.	There are many portals on several ends of a well-planned residential road, so that residents can only get out and enter the housing on the main route which is easily monitored by the housing security guard.
Similarity		
Social Interaction Activity	The types of activities in both housing are the same, namely shopping and sports and children's playground	
Static activity	Both activities are generated by vegetable sellers starting at 05.00 am - 09.00 am, except on Sundays there are no vegetable sellers in low density housing	
Social interaction time	Most of housing residents doing sports A small percentage of housing residents and shopping at the vegetable sellers shop at vegetable sellers at 5:00 to 7:00. At from 5:00 to 7:00 7:00 a lot of people had started walking towards the church in the housing.	
Housing Security	The portal closes from 10:00 p.m. to 5:00 p.m.	

Based on the analysis of empirical conditions, basically social interactions carried out by residents mostly use the road space. only a small percentage of residents live in the vicinity of open spaces that make direct use of open spaces as housing facilities. Based on the analysis, it is necessary to plan a neighbourhood space model both in formal high density and low density housing, namely: 1) the concept of space that accommodates static activity generated by vegetable sellers; 2) the concept of small open space at some points on the residential road that are semi-public open space for the social interaction among housing residents; 3) Applying the concept of defensible space as a concept of space design to improve the housing security. Based on observations, it was seen that there was a gathering point for housing residents during vegetable shopping activities, so a neighbourhood space design that utilizes street space and open space refers to the concept of defensible space described in Figure 7 as follows.

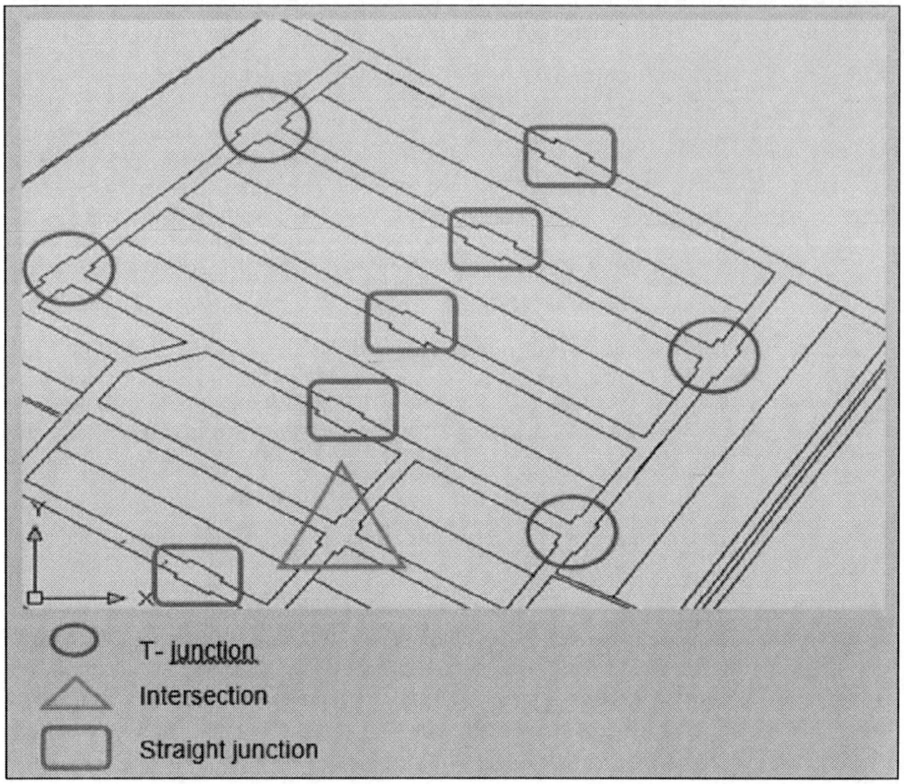

Figure 7. Neighbourhood space model in formal housing

Designing enclosures on formal housing road spaces that are used for social interaction is also a citizen effort to control each other in housing security, according to the concept of defensible space [14] with the main principle, namely territoriality; natural surveillance; image and "milieu" or environment.

5. Conclusion

Based on the research result, there was 3 (three) elements in establishing a neighbourhood space to develop social cohesion. Such as road space, spill market in formal housing that is static activity and public open space. Social cohesion in formal housing is need to be increased, where the housing layout with road pattern type is very influenced to evoke social interaction activities, especially housing with high density. The placement of public open space that is potentially to evoke social interaction activities should be design in the area center of housing that is surrounded by a road, so the housing community who has activity in public open space can supervise the housing condition. The developer need to designing enclave on formal housing road spaces that are used for social interaction and control each other in housing security, according to the concept of defensible space and neighbourhood space model namely "Enclave Transpatial"

6. References

[1] Koesoemawati D J 2016 Social Cohesion Potential of Pendalungan Communities and Urban Space Integration in Jember *International Journal of Indonesian Society and Culture*. Vol 8 (1): 145-154

[2] Webber M M. 1964 *The Urban Place and The Non-Place Urban Realm* in Exploration into Urban Structure, 79-153, Philadelphia PA: University of Pennsylvania Press

[3] Rofe Y 1995 Space and Community - The Spatial Foundations of Urban Neighbourhoods an Evaluation of Three Theories of Urban Form and SocialStructure and Their Relevance to the Issue of Neighbourhoods *Journal of Berkeley Planning*, Vol.10, hal. 107-125

[4] Banerjee T and Baer W C 1984 *Beyond The Neighborhood Unit: Residential Environments And Public Policy*, New York City, NY;London, UK: Plenum Press.

[5] Hapsari H D 2011 *Kaitan Pola Neighbourhood pada Inetraksi Sosial Penghuni* Skripsi. Jakarta: Prodi Arsitektur, FakultasTeknik Universitas Indonesia

[6] Harpham T G E and Thomas E 2002 Measuring social capital within health surveys: key issues. *Health Policy and Planning*, 17 (1), 106-111

[7] Braza M 2003 *Parks, Community Gardens, and Open Space in Urban Neighbourhoods* http://www.neighbourhoodcoalition.org/Smartgrowth/article.a sp?art

[8] Carr S Francis M Rivlin L G and Store A M 1992 *Public Space* Australia: Press Syndicate of University of Cambridge

[9] Darjosanjoto E T S 2006 *Penelitian Arsitektur di Bidang Perumahan dan Permukiman* ITS Press Surabaya

[10] Hillier B 1994 *Space is The Machine: A Configurational Theory of Architecture* Cambridge UK Cambridge Univ Press

[11] Pasaogullari N and Doratly N 2004 Measuring Accessibility and Utilization of Public Spaces in Famagusta, *CITIES 21* (3): 225 – 232.

[12] Fiorillo D and Sabatini F 2011 *Quality and Quantity: The role of Social Interactions* in selfreported individual health Social Science and Medicine. 73. 1644-1652.

[13] Hester D D 1994 *Tentang Teori Intermediasi Keuangan*, De'Economist, 142.

[14] Newman O 1972 *Defensible Space* New York: Macmillan

Smart economy for coastal resource management in Surabaya City

D Rahmawati[1], N I Arvitrida[2], D Lastomo[3], Kusnadi[4] and Rindawati[5]

[1] Department of Urban and Regional Planning, Institut Teknologi Sepuluh Nopember
[2] Department of Industrial Engineering, Institut Teknologi Sepuluh Nopember
[3] Department of Electrical Engineering, Institut Teknologi Sepuluh Nopember
[4] Faculty of Cultural Studies, Universitas Negeri Jember
[5] Faculty of Geography, Universitas Negeri Surabaya

Email: d_rahmawati@urplan.its.ac.id

Abstract. Smart economy as a part of smart city concept has been emerging in recent years to promote urban growth. It infuses digital infrastructure in the economic activities, not only to organize the flow of finance or information but also the production and distribution of goods. This paper provides a review of smart economy concept for coastal resource management in a large coastal city in Indonesia. Surabaya is taken as a case as it is the second largest city in Indonesia with abundant fisheries resources. However, these resources are vulnerable due to the dynamic urban demand and pressures. The approach used in this study considers the supply chain aspects of fishery commodities, involving supply and demand characteristics, the chain of distributions, and the price variation along the supply chain. In addition to the literature review, an evidence-based study is also presented by considering the Indonesian government program in managing the national fisheries logistic system. This paper offers two-fold contributions. It shows the urgency in promoting smart economy concepts in coastal resources in coastal cities and the importance for determining a suitable concept of supply chain management practice and digital prototype design of fisheries resources management as instruments of smart economy concepts.

Keywords: *smart economy, coastal resources, coastal cities*

1. Introduction

The democratization of ICT around the world has led many scholars and policy makers to a discussion on sustainable, resource-conserving, and resilient smart cities. One of the emerging topics is smart city economic development that is appropriate to different cities, countries, and continents. It can be possible that each city in particular country and continent possess differing challenges to smart city economic development. Smart economy as part of smart city concept has been emerging in recent years to promote urban growth. It infuses digital infrastructure in the economic activities, not only to organize around finance or information but also the production and distribution of goods. There is no unanimous definition on smart city or smart economy. The interrelationship of smart economy with smart city is also difficult to decipher. It is not clear whether a city is smart because of its smart economy or smart city is the reason behind the workings of a smart economy. It is also not clear from the literature how spatial system of a smart city can be designed in such a way that a smart economy may start functioning [1,2]. A Smart City System comprises of six key building blocks: (i) smart people, (ii) smart economy, (iii) smart mobility, (iv) smart environment, (v) smart living, and (vi) smart governance. These six building blocks are closely interlinked and contribute to the 'Smart City System'. Smart economy has these emerging characteristics: competitiveness; knowledge-based economy; creativity and innovation; establishment of innovation clusters; innovation through networks; IoT; sharing economy; mutual cooperation between enterprises, research institutions, and citizens; entrepreneurship; job creation; social responsibility; green economy; Triple Helix Model in which governments, businesses, and academia have changing roles; and the use of ICTs. Within the digitally networked society of today, 'sharing economy' has emerged as a new economic or business model, which goes hand in hand with smart cities. The sharing economy can be defined as the economic model in which demand and supply are immediately in contact through an online platform, in order for the supply side to directly provide services and/or products with an underlying aim to improve the use of assets and to reduce transaction costs. Two distinct sharing economy models are in operation: (i) asset hubs where 'a single company owns the goods or assets and sells access to users on a temporary basis' (by hour or day), and (ii) peer-to-peer networks where 'various would-be suppliers are connected with various would-be users' using an online platform [3,5].

Cities on the coast are often associated with major ports, which facilitate cheap sea transport of goods, which in turn attracts major industries. Economic growth provides employment and investment opportunities, coastal cities acting as a magnet for people looking to improve their economic status. In response, many urban areas are being developed or expanded to meet the needs of new coastal residents for housing, sanitation and transport. Since 1970, houses have been built at the rate of 2000 units per day in coastal areas of the US and tourism continues to develop in the Mediterranean, the area could host up to 350 million seasonal tourists every year by 2025 [3,6]. Coastal resource refers to the natural and cultural resources of the coastal zone. Coastal land is usually owned and/or managed by multiple agents of private, communal, corporate and government bodies, whereas coastal waters are usually owned and/or managed solely by governments. The uniqueness of the coast is further enhanced by the value of its resource such as fish and offshore mineral reserves, and more recently recreation, economic development, and aquaculture sites, considered by the population to be common property in high demand, and for subsistence use by coastal dwellers. Exploitation of such resources raises their value, with a consequential demand for equitable resource allocation. Therefore, resource planning often forms an integral part of coastal management programs. Renewable coastal resources are primarily exploited in the fisheries sector by commercial, subsistence and recreational fishers and the aquaculture industry.

Surabaya is the second biggest city in Indonesia, which geographically located as coastal city where coastal resource, both natural and cultural resource are abundant yet vulnerable and facing many threats due to urban pressures. To fulfill the coastal resource demands of its 3.5 million populations and beyond, especially for fish commodities, Surabaya is struggling to manage this resource as daily basis supply from 18 community-scale fish ports and from larger area in East Java. According to the Surabaya Spatial Planning documents, there will be 4 city-scale fish ports established in Surabaya and considering the dynamics of economic situation that Surabaya has, the integration of smart economy concepts is an important need to be applied. The use of smart economy concept has the urgency to balance supply and demand, and mitigate environmental degradation through greater efficiency, also reduce the strain on resources by streamlining the delivery of energy and services as well [4]. The readiness of Surabaya city implementing smart economy has been long recognized as Surabaya has already integrate the use of technology to connect, supervise, and control the resource in the city in effective and efficient way so that the citizen of Surabaya can achieve optimal services. There are many awards that given to Surabaya related to its success implementing Smart City concept; i.e. Indonesia Smart City Index in 2015. So far, the use of technology as instruments toward Smart City has been focused in environment, education, procurement, and citizen administration needs. There is opportunity to develop more, especially for coastal resource management.

This paper provides a review of smart economy concept for coastal resource management in a large coastal city in Indonesia. Surabaya is taken as a case as it is the second largest city in Indonesia with abundant fisheries resources. However, these resources are vulnerable due to the dynamic urban demand and pressures. This paper offers two-fold contributions. Firstly, it shows the reason why we need to promote smart economy concepts in coastal resources in coastal cities. Secondly, this paper is important for further stages, particularly in determining a suitable concept of supply chain management practice and digital prototype design of fisheries resources management as instruments of smart economy concepts.

2. Methods

This research employs two main stages which are described as follows. First, we explore the smart economy concepts that require to consider in making coastal resources to be essential parts of smart city. In doing so, a desk study is conducted, particularly in reviewing the available practices of smart economy in coastal cities in various countries. This theoretical-driven approach provides a foundation that is expected be a basis of understanding the potential of a coastal resources toward smart city. Second, we analyse empirical evidences related to fish resource management in Surabaya city, one of the large coastal city in Indonesia taken as study case. The primary data is collected through observation and interviews, while the secondary data is obtained from the city government. A simple correlation analysis is used to characterize the fishery problems. Hypothesis testing is also performed to infer the significance of the statistical relationship.

There are variables specifically set to this study, reviewed from smart city literatures and coastal management literatures, and synthesized in the context of this study. The variables are mentioned in Table 1 below.

Table 1. The Variables Observed

Indicator	Variables
Smart Economy	Innovation cluster in economic activities
	Digital infrastructure and IoT (Internet of Things) in economic activities
	Mutual cooperation between economy actors
Coastal Management	Fish commodities/ resources
	Facilities in fisheries management

3. Results and Discussions

3.1 Exploring smart economy concept related to coastal resource

Technological needs in managing logistics, such as fisheries resources, smart concepts are important to be carried out in the management of material handling and movement processes. This is because consumers require accurate information and certainty on the availability of goods at affordable prices. Robotics and Automation News Report states that in order to support this, perfect supply chain management is needed and is supported by optimal inventory efficiency and processes in all aspects from production to distribution to end customers. Innovation in terms of logistics is important to do because the more supply chains that can integrate and coordinate activities, the greater the possibility to optimize the flow of goods from suppliers to consumers. The condition of the company in dealing with fluctuations in demand changes and shorter delivery times, thus requiring greater flexibility in managing goods to be more efficient. This is done so that companies can continue to be sustainable and have competitive and good capabilities [8].

Information disclosure or data is an important aspect in realizing smart cities in the aspect of supply chain management because of the many processes and participation involved in it. One of the innovations is the blockchain and IoT that helps all participants to get involved together. Blockchain in the field of logistics has been widely applied in American states, especially in the port sector. However, apart from all existing technologies, human involvement in supply chain management is important for planning and making the right decisions, because not all decisions can be taken automatically needed by human logical thinking. Smart Logistic and Supply Chain is used to make the logistics system more efficient, effective, connected and flexible to meet needs quickly, if viewed further to connect the economy in all aspects and improve real-time economy. Besides that, it is used to change the way of working in managing aspects such as data / information analysis and it requires a skill to do the management in making the right and fast decisions along with changing demand. The following is an overview of smart logistics and supply chains in various fields. In Figure 1 illustrates the evolution of the logistics system and supply chain towards broader and more flexible autonomy, even to the point where there are no more warehouses in the supply chain system [10,11].

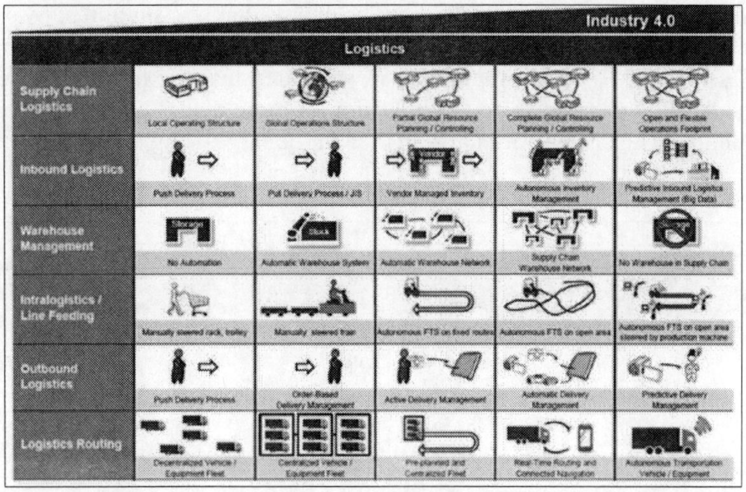

Figure 1. The Evolution of Logistic System and Supply Chain as Smart System
Source: I-Scoop, 2016

Coordination between economic activity actors, especially in the field of fisheries commodity management is very important to do especially the nature of consumers who demand an economical and varied product with an increase in service parts and fast delivery. Changes to business processes are needed such as low costs and fast delivery processes [12,13]. In this case the company must excel in several fields such as innovation in capacity and speed of responding to consumer demand. To deal with these problems, it is necessary to implement efficient supply chain management and adapt it to the needs and objectives of the company with good responsiveness to fluctuations in demand changes. Business competition is not only about competition between companies but also competition that occurs between economic activity actors involved in the supply chain of fish resources. Coordination between economic activity actors or supply chain stakeholders must go well to be able to face business competition and be an effective way of handling all processes and operations involved in making goods to the end customer. At present, supply chain is closely related to smart city because it provides opportunities and constraints in the process of managing collaboration systems between companies, end customers and the government [14,15,16,17,18].

Meanwhile in Indonesia, the policy and programs supporting the implementation of smart economy and coastal management in Indonesia already provided. According to Republic of Indonesia Ministry of Maritime Affairs and Fisheries Regulation No 5 / Permen-KP / 2014 concerning National Fish Logistic System, fisheries are stated as all activities related to the management and utilization of fish resources and their environment ranging from preproduction, production, processing to marketing, which are carried out in a fisheries business system. The National Fish Logistics System, hereinafter abbreviated as SLIN, is a management system for fish and fishery product supply chains, materials and production equipment, and information ranging from procurement, storage, to distribution, as a unity of policies to increase capacity and stabilize upstream fisheries production systems - downstream, controlling price disparity, and for meeting domestic consumption needs [7,9].

3.2 Surabaya indication in Smart Economy integration of coastal resources management

This section is to describe the analysis of phenomenon appears in Surabaya related to fish resource management. The analysis is described as follows:
Identification of fish resource supply characteristics in the city of Surabaya was conducted to determine the number and type of fish resources produced. The analysis is carried out through the compilation of secondary data from various agencies related to the fish resource supply chain. In this case, the Surabaya City Government of Food Security and Agriculture Board and the East Java Province Government of Marine and Fisheries Board became the main provider of data that was supported by some literature and related stakeholder interviews. Fish resource supply in Surabaya City consists of marine waters and public waters. Based on 2017 data compilation, it was noted that the largest catches in the fisheries sector were in the third quarter with a total production of more than 3000 tons and production values reaching Rp 80 billion. The 5 types of marine fish with the largest supply are shown in the following graph (Figure 2).

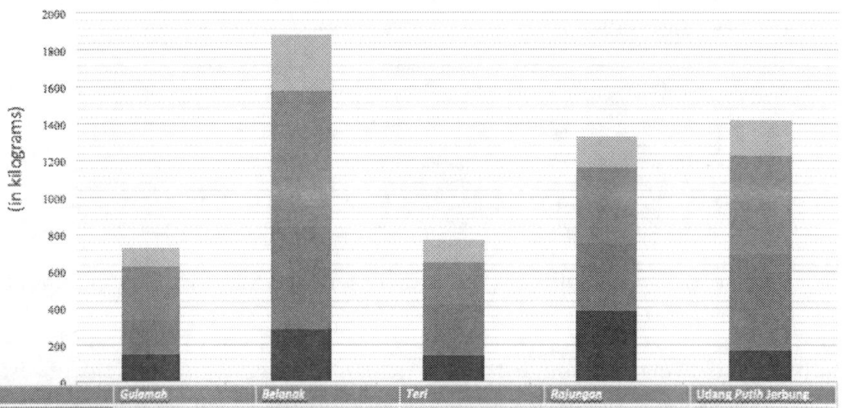

	Gulamah	Belanak	Teri	Rajungan	Udang Putih Jerbung
Oct – Dec	98.6	307.3	121.3	167.1	193.8
Jul – Sept	288.8	740.2	232.5	411.9	527.5
Apr – Jun	184.6	552.6	279.7	367.2	531.2
Jan - Mar	151.8	287.3	139.8	383.1	170.1

Figure 2. Marine Fishery Production and Commodities in Surabaya (top five)

Marine fishery production is dominated by mullet, manyung fish and shelled marine animals such as crabs and white shrimp, while in general fisheries are dominated by cork fish. Variations in the height and yield of fish caught by marine fishermen are influenced by several factors such as climate, fish catch distance, number of hours worked on the sea, rain and wave height. This is reinforced by the results of interviews with the Surabaya Food and Agriculture Security Agency as follows:

> "*Yes, so if the Surabaya fishermen have few catches, especially when there is no fish famine, Ma'am, they are very small. But they are indeed abundant during the certain harvest season, for example grago and the dried small shrimp, ebi.*" – Aris Munandar, Head Division of Fisheries, Surabaya

While the largest ground fishery production is shown in Figure 3. When compared to other regions in East Java Province, Surabaya City has a very low production rate. This is because the area of fish

IOP Conf. Series: Earth and Environmental Science **340** (2019) 012016 doi:10.1088/1755-1315/340/1/012016

catching in the city of Surabaya is too narrow and most of the fishermen in Surabaya only rely on a one day fishing system. When compared with the level of fish consumption of the people of Surabaya City, fish production in a year is still not sufficient for fish consumption needs that reach 43 kg / capita / year and the national target of 47 kg / capita / year. Figure 4 shown the position of Surabaya City based on its fish production compared to other coastal districts / cities in East Java on average over the past 7 years (2010-2016). To meet the fish consumption needs of the people of Surabaya City, fish supplies are imported from other regions. Based on information from one of the stakeholders of the Surabaya city government, most of the fish demands in Surabaya are supplied from various regions such as Pekalongan, Brebes, Tuban, Lamongan, Gresik, Probolinggo and Pasuruan. Fish distribution originating from outside the city of Surabaya enters the *Pasar Pabean*, which is one of the largest fish markets in the city of Surabaya. The function of the Pasar Pabean in the fisheries supply chain roled as collector and distributor of fish supply to a variety of smaller markets in the city of Surabaya and surroundings.

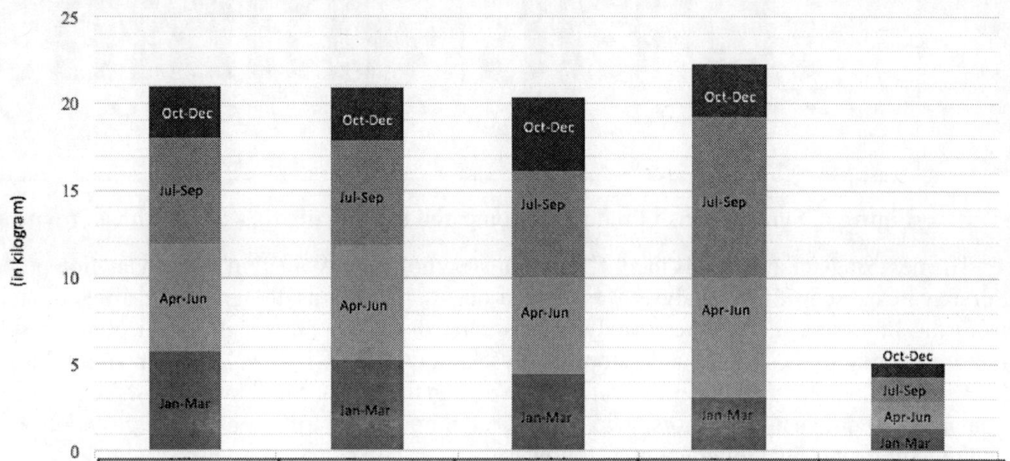

	Nila	*Tawes*	*Mujair*	*Gabus*	*Lele*
Oct-Dec	2.9	2.9	4.2	3	0.7
Jul-Sep	6.1	6.1	6.1	9.2	1.5
Apr-Jun	6.2	6.6	5.6	6.9	1.5
Jan-Mar	5.7	5.2	4.4	3.1	1.3

Figure 3. Ground Fishery Production and Commodities in Surabaya (top five)

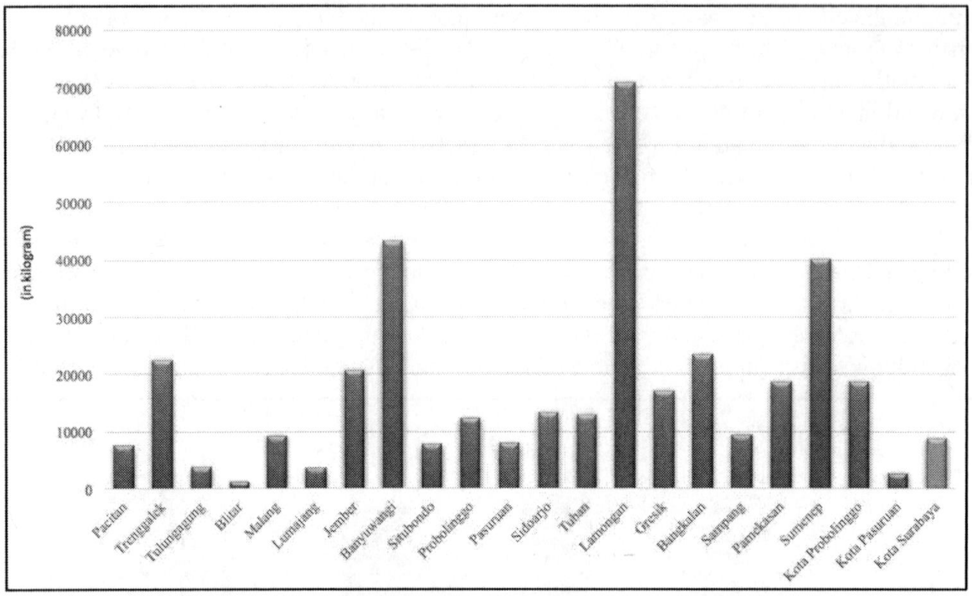

Figure 4. Surabaya Fish Production compared to other districts in East Java Province

The next stage of analysis is identification of stakeholders involved in the fish auction process in the Surabaya City scope. The analysis was carried out by identifying the types and roles of stakeholders from the literature, then content analysis was carried out to confirm the truth of the results of the literature study to those who understood the research context. The results obtained are 7 affected and interested stakeholders in fisheries auction activities in the city of Surabaya. The stakeholders consist of fishermen, middlemen/ collectors, wholesalers, retail traders, distribution service providers, consumers, and the Surabaya City Government as follows:

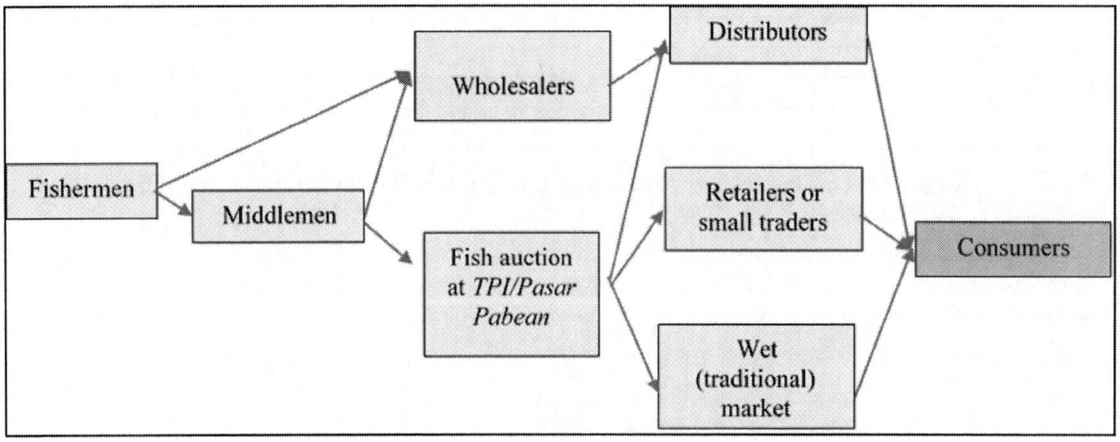

Figure 5. Cluster pattern of economic actors of fisheries activities in Surabaya

Fishermen

There are around 2,600 fishermen that exist in Surabaya city up to 2017 with the potential of capture fisheries along the coastline of 47.4 km2. The existing fishing groups are spread in 9 sub-districts in Surabaya: Gununganyar, Rungkut, Sukolilo, Mulyorejo, Bulak, Kenjeran, Krembangan, Asemrowo, and Benowo Subdistricts. They conduct traditional fishing with daily patterns with the main role are catching fish, selling fish directly in local ports (mostly through middlemen), earning money from the sale of fish, and providing good quality fish.

IOP Conf. Series: Earth and Environmental Science **340** (2019) 012016 doi:10.1088/1755-1315/340/1/012016

Collector/ middlemen
They offer prices to fish buyers, get services from a fish auction and provide information on fish prices appropriately
Wholesalers
Their role is selling fish on a broad scale and / or to other areas. They benefit from the sale of fish and they provide fish needed by consumers
Retail traders
Their role is selling fish on a smaller scale (e.g. in housing or *kampung* area). They benefit from the sale of fish by providing fish needed by consumers
Distribution service provider
Their role is transporting the fish from fish auction/market to other regions, both in the city or outside the city. They have to transport the commodities to destination with good quality, using cold chain.
Consumer
Their role as a final consumer of fish and want to have good quality fish. They spend money according to the agreed price. They also the key of fisheries demand.
The government of Surabaya
The Surabaya City Government for fisheries activities consists of the Food Security and Agriculture Board of Surabaya City (DKPP) and Market Coordinator in Surabaya (PD Pasar Surya). Their role is to regulate the fisheries policy, conduct management of Fish Auction Places (TPI) and fish markets. They have rights in authority of regulating or implementing policies. Their obligation is to handle related issues in the field of fisheries management in accordance with the main tasks of the governmental function.

The last stage of analysis is to find the correlation between the fish resources and the fisheries actors. In dry season, the fish production level is higher than in the wet season. This is because the fishermen should cope with seasonal challenges in the sea during the wet season. To understand whether any correlation exist between the level of production and price, a simple correlation analysis is conducted and shown in Table 2. From the analysis, there is no significance correlation between the level of fishery production and price in Surabaya. This is an interesting finding as in urban areas the production level of fish does not have a significant impact on the market price in the same areas related to fish resource management.

Table 2. The summary output of correlation analysis using linear regression

SUMMARY OUTPUT								
Regression Statistics								
Multiple R	0.116977709							
R Square	0.013683784							
Adjusted R Square	0.000532901							
Standard Error	0.522368142							
Observations	77							
ANOVA								
	df	*SS*	*MS*	*F*	*Significance F*			
Regression	1	0.283925682	0.283925682	1.040522108	0.310979271			
Residual	75	20.46513567	0.272868476					
Total	76	20.74906136						
	Coefficients	*Standard Error*	*t Stat*	*P-value*	*Lower 95%*	*Upper 95%*	*Lower 95.0%*	*Upper 95.0%*
Intercept	0.067793991	0.059781237	1.134034602	0.260391808	-0.05129634	0.186884322	-0.05129634	0.186884322
Percentage of price changes	0.524935836	0.514612778	1.020059855	0.310979271	-0.500225388	1.55009706	-0.500225388	1.55009706

We also test the correlation between the number of fishermen and the price disparity. The price disparity considered is between the fish selling price sold by the fishermen to the collector and the fish price at the consumer level. This test is carried out against the background of the phenomenon of competition, where the more competitors, the price will be set lower than other competitors. As can be seen in Table 3, there is no correlation between these two factors. This indicates that the higher number of fishermen does not cause changes in fish price set by the fishermen.

Table 3. Correlation test between the number of fishermen and price disparity of fishermen and collectors

	Number of fishermen	*The average of price disparity between fishermen and middlemen*
Number of fishermen	1	0.052884284
The average of price disparity between fishermen and middlemen	0.052884	1

4. Conclusion

Surabaya City's fish production which is still not sufficient for consumption needs, the fair trade concept among players in the fisheries supply chain is also not valid. Based on the primary data collected, the fishermen sell their catch or cultivation with relatively high price disparities compared to the prices offered by middlemen. the profit margins obtained by fishermen and are very much different when compared to middlemen, especially for hard-shelled commodities, such as lobsters, crabs, crabs, and white snapper. This can occur due to several factors, such as the waiter not knowing the market price directly, the middleman is the main customer or the only customer so that fishermen do not have other options to sell their catch or cultivation, and also the role of TPI does not function properly. Therefore, smart systems for fisheries resource management can be one solution so that fishermen can interact with potential customers in fisheries supply chains so that fishermen have access to prices and have the freedom to determine prices.

5. Acknowledgedments

Authors would like to acknowledge encouragement from LPPM ITS through the research funding of Pakerti Scheme and the opportunity to cooperate with stakeholders in Surabaya city.

6. References

[1] Herzberg, C. (2017). Smart Cities Digital Nations: How Digital Urban Infrastructure Can Deliver a Better Life in Tomorrow's Crowded World. Roundtree Press, California.

[2] Kumar, V, Dahiya, B. (2017). Smart Economy in Smart Cities. Springer.

[3] Kay, R., Alder, J. (2005). Second Edition: Coastal Planning and Management. Taylor and Francis.

[4] Rahmawati, D., Sulistyarso, H., Ariastita, P.G., Yusuf, M., Paramasatya, D.A. (2018). Smart Kampung for Surabaya Smart City: Criteria Redefined. IOP Conference Series: Earth and Environmental Science Vol. 202/Issue 1.

[5] Woensel, van, T. (2012). Smart logistics. Eindhoven: Technische Universiteit Eindhoven.

[6] Olafsdottir, G, Johannesson, G. (2010). Views of stakeholders in fish supply chains on implementation of electronic systems for traceability and temperature data. Brussels.

[7] Setijadi. (2016). Implementasi Sistem Logistik Ikan Nasional Sebagai Solusi Persoalan Sektor Perikanan (*Implementation of the National Fish Logistics System as a Solution to the Fisheries Sector Problem*). Supply Chain Indonesia, Bandung.

[8] APEC Policy Support Unit. (2009). Logistics: Connectivity for Goods and Services. First Economic Committee Meeting, Singapore.

[9] Peraturan Menteri Kelautan dan Perikanan Republik Indonesia No 5/PERMEN-KP/2014 Sistem Logistik Ikan Nasional (*Indonesian Ocean and Fisheries Ministry Regulation about the National Fish Logistics System*)

[10] Caragliu, A., Del Bo, C., & Nijkamp, P. (2011). Smart cities in Europe. Journal of Urban Technology, 18(2), 65–82.

[11] Cocchia A., Smart and Digital City: A Systematic Literature Review, DOI: 10.1007/978-3-319-06160-3_2

[12] Cohen, B., 2013. What exactly a smart city? http://www.boydcohen.com/smartcities

[13] Diana, A., Balaceanu, C., Cinstatinescu, E.M., 2015, Smart Economy Concept-Fact and Perspective, JEL Classification: I30, I31

[14] Ding, W., 2013. Study of Smart Warehouse Management System Based on the IOT. Intelligence Computation and Evolutionary Computation, Volume 180, pp. 203-207.

[15] Gamero, R., 2012, Why Do We Need Smart Cities? https://www.telefonica.com/es/web/public-policy/blog/articulo/-/blogs/why-do-we-need-smart-cities-, Visited on 30 Januari 2019

[16] I-Scoop, 2016, Logistics 4.0 and smart supply chain management in Industry 4.0, https://www.i-scoop.eu/industry-4-0/supply-chain-management-scm-logistics/, Visited on 27 Januari 2019

[17] Martínez D.R.S., Gracia T.J.H., Muñoz E.M., García A.C., 2017, Smart Cities' Challenge: How to Improve Coordination in the Supply Chain., Innovation, Technology, and Knowledge Management. Springer, Cham

[18] Melanson, A., 2018, Intralogistics is the Secret to a Smart Supply Chain https://aethon.com/intralogistics-is-the-secret-to-a-smart-supply-chain/, Visited on 27 Januari 2019

Spatial interaction between urban and peri-urban regions of *Surabaya* metropolitan area and its impact to the carbon footprint of transportation sector

E B Santoso[1], B U Aulia[1], D N Aninditya[1] and K D M Handayeni[1]

[1] Urban and Regional Planning Department, Institut Teknologi Sepuluh Nopember, Surabaya, 60111, Indonesia

Email: santosoeb@gmail.com

Abstract. Surabaya Metropolitan Area is the dominant urban agglomeration in East Java Province as the center of regional economic development. The population growth and economic activities have significantly affects to the periphery area, one of which is Sidoarjo Regency where is directly border to the southern part of Surabaya City. Sidoarjo Regency as a supporting area has a main function in the provision of residential and industrial areas which causes the expansion of urban areas of Surabaya. The relationship between urban and peri-urban areas has the potential to increase the flow of people, goods and vehicles from Surabaya to Sidoarjo and vice versa. It causes congestion and air pollution problems from vehicles in the corridor of the area with a high level of interaction. The purpose of this study is to estimate the level of spatial interaction between urban and peri-urban areas as well as the environmental impacts, mainly carbon emissions and ecological footprint. The analysis was carried out using the gravity method to measure spatial interactions, while carbon emissions analysis uses quantitative method with data on emission factors and the number of vehicles between Surabaya City and Sidoarjo Regencies, as well as their effects on ecological footprint. The results show that the peri-urban area surrounding border of Surabaya City has the highest level of spatial interaction. The strong interaction between urban areas and peri-urban areas has an impact on increasing carbon emissions between these two regions and also their ecological footprint.

Keywords: *Spatial interaction, Peri-urban region, Carbon footprint, Transportation sector*

1. Introduction

Indonesia as one of the countries that are growing rapidly both in the economic, social and demographic sectors has the central points of strategic growth with the main characteristics of the development is metropolitan area as a centre of economic activity. The development of economic activity centres in the metropolitan area is supported by the strategic location of the area, both geographically and accessibility which is closely related to the transportation of goods and people. The metropolitan area in Indonesia consists of one growth point (city) which is supported by several regencies or surrounding cities as a supporting area for physical, economic and social activities. The development of this region resulted in changes from rural areas to urban areas. This phenomenon is called urban sprawl where changes in the physical appearance of the area become less rural and show more urban characteristics [1]. This area is called a peri-urban area which has the main land use characteristics as an agriculture area which is slowly transforming into a built-up area, rapid population growth with high population density, and a change in the mindset of rural people into urban communities [2]. The central city of growth and the peri-urban area has a close relationship with the fulfillment of consumption needs, where the peri-urban area provides natural resources that the city cannot provide. In the economic field, this relationship can be seen in the supply and demand process.

Surabaya and the surrounding urban areas are referred as a Gerbangkertasusila metropolitan area which causes urban sprawl in the city of Surabaya to other regions, one of which is Sidoarjo District [3]. Based on BPS Sidoarjo Regency in 2016, Waru and Taman districts which are directly adjacent to the City of Surabaya experienced the highest population growth compared to other districts in Sidoarjo with 11.38% and 10.79% respectively. Whereas, in other areas located close to Parks and Waru districts namely Sukodono, Gedangan, and Candi also experienced an increase in population density roughly 7,000 people / km2. Another peri-urban characteristic that appears to occur in Sidoarjo Regency is the change in land use from agricultural land to residential and industrial land, around 5.3 thousand of which have been diverted in land use. The calculation is based on remote sensing analysis from 2002 and 2017. Land use changes into built-in areas, namely settlements and industries will directly increase the number of vehicles and the movement of vehicles from the peri-urban area to urban areas, especially in districts that are directly adjacent to urban areas because in that area lies a residential area that cannot be provided by urban areas. CO^2 emissions derived from burning fossil fuel vehicles contribute 50% of the effects of greenhouse gases [4]. CO^2 gas is an important gas and if in normal circumstances it is not dangerous and useful in the process of photosynthesis of plants, excessive amounts can pollute the air and cause greenhouse gas effects [5]. The increase in population and the number of motor vehicles as a source of CO^2 emission increases its concentration in the air [6,7].

Therefore, a study is needed related to the relationship between urban areas and peri-urban as well as their effect on the increasing number of vehicles and carbon emissions to implement the sustainable sustainable development in the metropolitan area. The objectives of this research were, 1) seeing the pattern of urban sprawl in Sidoarjo regency with normalized prevention of built-up index (NDBI) method by using Landsat satellite image year 2002 and year 2018 with the help of GIS, 2) studying the characteristics of regional calculating by gravity model, and the last 3) calculating carbon emission on Sidoarjo regency.

2. Data and methods

2.1 Study site

Sidoarjo Regency is located on East Java which is directly adjacent to the City of Surabaya. This area is included in the development area of Gerbangkertasusila metropolitan with the main land use of agriculture with a large area of 36.2% of the total land area and 17.5% built land consisting of residential land, commercial area, industrial area, and other land use.

Figure 1. Administration map of Sidoarjo Regency

The population of Sidoarjo Regency reached to 2,223,002 peoples in 2017 according data recorded by Central Bureau of Statistics of Sidoarjo Regency. It is indicated that the most densely population district is Waru with 242,004 inhabitants. Meanwhile. the smallest number of population is in Jabon district which up to 61.015 peoples.

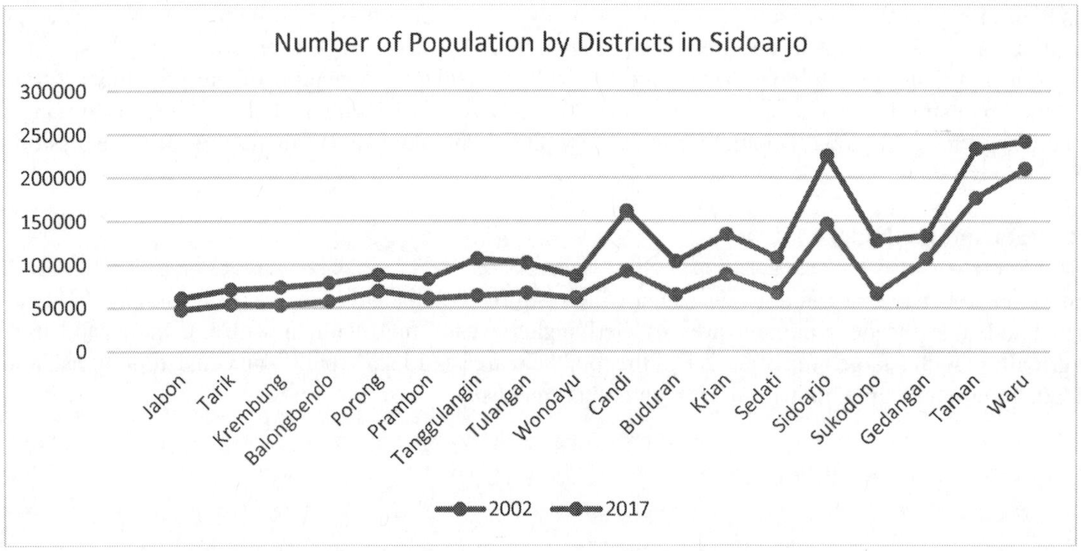

Figure 2. Population number on each districts in Sidoarjo Regency

The table above shows that the largest population is located in the Districts of Waru, Taman, and Sidoarjo with roughly 230 thousand people on each district. Population fluctuations are one of the attributes of urban sprawl where there is a movement of residents from the city of Surabaya who moved to Sidoarjo regency due to the limited land of settlements in the city of Surabaya.

2.2 NDBI

Normalized Difference Built-up Index (NDBI) analysis required landsat imagery to delimitate built-up. Images that were used in this analysis selected from the available imagery the earliest and the latest using satellite captures with clear skies to get more precise result. The images that were selected to reflect urban growth trends in Sidoarjo regency were from 2002 and 2017 imagery from landsat imagery. Based on the calculation of standardized differentiation of Band 5 and Band 4 will result binary image. In which, woodland and farmland pixels have value close to 0 (zero), negative value for waterbodies, and positive values for built-up areas.

2.3 Gravity model

This analysis method is used to calculate the magnitude of inter-regional spatial interaction. Gravity model measures the strength of interaction between 2 different poles (in this case 2 different regions namely Surabaya and Sidoarjo). Calculations was done buy multiplying the number of populations between Surabaya population and districts in Sidoarjo and then divided by the square of distance between those two regions times to constata number (1).

The general formulation above is adopted by Vaz & Nijkamp resulted a new formulation. Mass from the initial formulation is likened to be change of population in a city with the acceleration of built-up changes [8]. With that assumptions, F_p is population changes in the area, and a is changes of built-up area.

$$F = k\frac{m_1.m_2}{{r_{12}}^2} \quad (1)$$

F = interaction forces with Surabaya
F_p = changes of population
a = changes of built-up area
m_1 = mass of urban sprawl $area_1$
m_2 = mass of urban sprawl Surabaya
r_{12} = distance of $area_1$ and Surabaya
k = constanta

2.4 Carbon emission calculation

Emissions released from transportation mode use activities are calculated through fuel consumption per type of transportation mode. This amount of consumption is derived from the average population consumption of BBM from the 2017 Income Bureau of East Java Province data [9].

To find out the number of emissions released from these two activities, the energy consumption is calculated with the emission factor. The amount of CO^2 emissions generated from an activity can be written in the formula below.

$$ECO_2 = A \times FE \quad (2)$$

ECO_2 = CO_2 Emission
A = Energy Consumption
FE = Emission Factor

Based on Suhandi in Srikandi (2009), the emission factor of a vehicle with fossil fuel is 2.5 Kg CO2 /liter [10]. This type of vehicle is only distinguished from the type of fuel, including small vehicles (motorcycles, cars, jeeps, sedans, etc.) namely gasoline and large vehicles (buses, trucks, heavy vehicles, etc.) with diesel fuel. Both types should have different uses of the type of fuel used, namely gasoline for small vehicles and diesel for large vehicles. Fossil fuel consumption per year can be seen on table 2. from SUSENAS 2014 and Ismayanti [11].

Table 1. Fossil Fuel Consumption per-Vehicle

Fossil Fuel Type	Number of Uses (Litre/year/unit)
Gasoline	140.28
Diesel Fuel	3,819.1

3. Result and discussion

3.1 NDBI

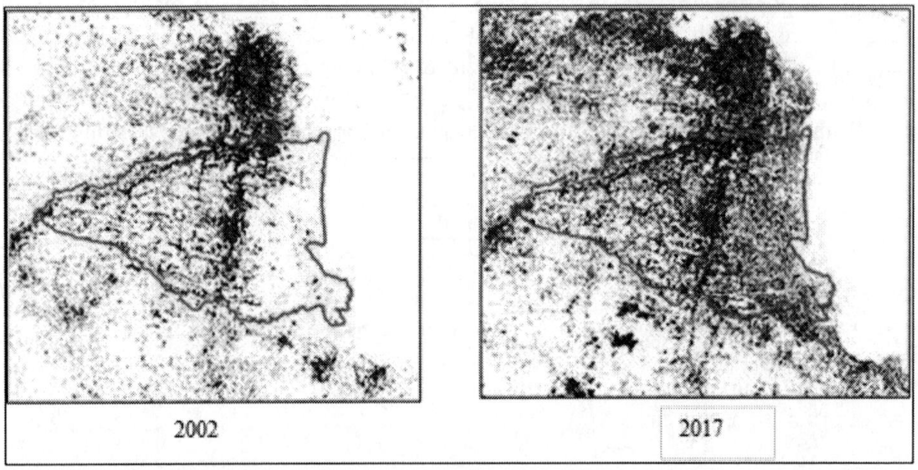

Figure 3. NDBI result of Sidoarjo Regency year 2002 and 2017
Source: [12]

Based on the calculation of the raster data from Landsat imagery, it can be seen that there is an increase in the density of the built-up area as shown in Figure 4. This raster shows also the magnitude of the NDBI index that will be used as an input gravity model.

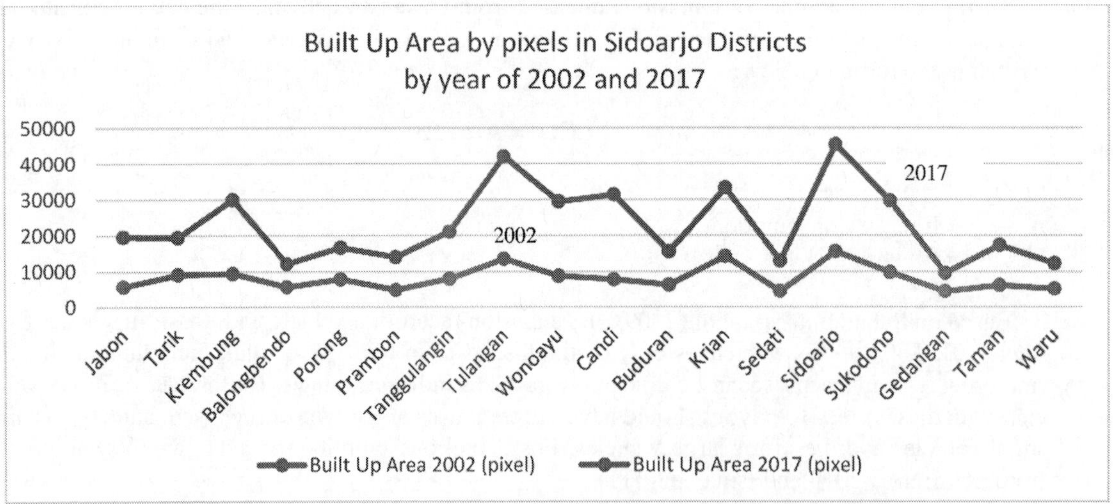

Figure 4. Built-up area transformation number of Sidoarjo in 2002 and 2017
Source: [12]

3.2 Gravity model

This analysis method is used to calculate the magnitude of inter-regional spatial interaction and calculate the strength of interaction between the two regions. The results of the gravity model calculation are shown in Table 2.

Table 2. Results of interaction analysis of Sidoarjo's Districts to Surabaya City

No	District	a	Fp	$m_1 = \frac{Fp}{a}$	r_{12}	F = $k\frac{m_1.m_2}{r_{12}^2}$*
1	Jabon	8096	13332	1,64673913	31,152	0,01372
2	Tarik	1269	17294	13,62805359	29,744	0,12456
3	Krembung	11155	20761	1,861138503	28,198	0,01893
4	Balongbendo	788	21446	27,21573604	26,652	0,30983
5	Porong	985	18854	19,14111675	28,193	0,19474
6	Prambon	4361	22400	5,136436597	25,907	0,06189
7	Tanggulangin	5058	42839	8,469553183	25,785	0,10301
8	Tulangan	15098	35020	2,319512518	23,778	0,03318
9	Wonoayu	11993	25366	2,115067122	19,868	0,04333
10	Candi	15857	69055	4,354859053	23,219	0,06532
11	Buduran	3033	38875	12,81734257	18,254	0,31106
12	Krian	4871	46351	9,515705194	19,643	0,19942
13	Sedati	3997	40745	10,19389542	16,682	0,29623
14	Sidoarjo	14338	78431	5,470149254	21,267	0,09780
15	Sukodono	10124	60998	6,025088898	14,619	0,22799
16	Gedangan	723	26749	36,99723375	12,983	1,77500
17	Taman	4989	56754	11,37582682	11,454	0,70123
18	Waru	1962	31578	16,09480122	10,120	1,27082

*m_2= 8,086478867

The results of these calculations can be seen that each districts have different interactions values. Then the interaction value is classified into 5 classes from highest to lowest.

Table 3. Range and classification of each districts in Sidoarjo Regency

Classification	Range	Districts
Class 1	1,77500 – 1,42275	District Gedangan
Class 2	1,42274 – 1,07048	District Waru
Class 3	1,07047 – 0,71821	-
Class 4	0,71820 – 0,36595	District Taman
Class 5	0,36594 – 0,01368	District Sukodono, Jabon, Tarik, Krembung, Balongbendo, Porong, Prambon, Tanggulangin, Tulangan, Wonoayu. Candi, Buduran, Krian, Sedati, dan Sidoarjo

Based on Table 3 it can be known that the biggest interaction in Sidoarjo Regency towards Surabaya City is Gedangan District, the second strongest namely Waru District which has a direct border and has access levels of national and provincial roads. Whereas the weakest is located far from the border of

IOP Conf. Series: Earth and Environmental Science **340** (2019) 012017 doi:10.1088/1755-1315/340/1/012017

Surabaya City namely Sukodono, Jabon, Tarik, Krembung, Balongbendo, Porong, Prambon, Tanggulangin, Tulangan, Wonoayu. Candi, Buduran, Krian, Sedati, and Sidoarjo.

Based on these results it can also be seen that the closer and more densely populated and the built area of Surabaya City has a strong urban character than districts that are farther from Surabaya. the existence of urban sprawl occurring in Sidoarjo Regency shifts from Waru District and Taman to District Gedangan where there is a rapid development of residential areas.

3.3 Carbon emission calculation

Calculation of carbon emissions is based on the number of vehicles, type of vehicle fuel and emission factors as described in the equation 4. The number of vehicles will be shown in table 4, the total of Carbon Emissions is presented in Gha (Globalhectares).

Table 4. Total Vehicle Emission on Sidoarjo Regency

Total of Small Vehicle	Conversion of uses (lt/year/unit)	Total of Large Vehicle	Conversion of uses (lt/year/unit)	Total	Emission Factor	Total x Emission Factor (Gha)
1.446.358	140.28	46,054	3,819.1	378,779,931.6	2.5	946,949,829.10

Table 5. Total vehicle emission on 3 strongest interaction Districts in Sidoarjo

District Name	Total Carbon Emission (Gha)
Sidoarjo	9,586,463
Buduran	4,431,832
Candi	6,898,798
Porong	3,756,742
Krembung	3,143,717
Tulangan	4,358,947
Tanggulangin	4,563,374
Jabon	2,599,104
Krian	5,747,422
Balongbendo	3,356,834
Wonoayu	3,707,371
Tarik	3,021,845
Prambon	3,549,419
Taman*	9,944,796
Waru*	10,308,837
Gedangan*	5,681,651
Sedati	4,609,678
Sukodono	5428,152

**Main Urban Sprawl Districts*

Based on the calculation of carbon emission in Sidoarjo Regency which is shown in table 6, the total carbon emissions in all Sidoarjo regency is 946,949,829.10 which accounts for nearly 20% of the total number of vehicles in the Metropolitan Gerbangkertasusila area and occupies the second position of the largest number of vehicles after Surabaya City. Based on the data above, it can also be seen that 3 districts with the highest level of interaction with the City of Surabaya contribute to large carbon emissions. Waru and Taman Districts which are directly related to Surabaya City contribute the largest carbon emissions, followed by District Gedangan which has the largest spatial interaction value of Surabaya City with a value of 48,296,953.70 Gha.

4. Conclusion

From the three analyzes that have been carried out in this study (including NDBI analysis, Gravity Model and Carbon Emission Calculation), it can be concluded that the peri-urban area in the southern part of Surabaya City has converted the agricultural land use area to a built-up area of up to 43.95% due to the expansion of urban areas. The increase in spatial interaction in the urban area with the peri-urban area in Sidoarjo occurs due to the movement of people and goods through the transportation sector, which is viewed from the Gravity model and sees the impact on increasing the carbon footprint in the area with the highest spatial interaction. Districts of Gedangan, Waru, and Taman contributed 48,296,953.70 Gha of carbon emissions, 103,088,367.19 Gha, and 99,447,959.65 Gha respectively.

5. References

[1] Yunus, H. S. (2008). Dinamika wilayah peri-urban: determinan masa depan kota. Pustaka Pelajar.

[2] Iaquinta, D. L., & Drescher, A. W. (2000). Defining periurban: understanding rural-urban linkages and their connection to institutional contexts. In Tenth World Congress of the International Rural Sociology Association (Vol. 1, pp. 3-28).

[3] Siswanto, V. K., Routray, J. K., & Dewi, R. (2014). Rural urban disparity in and around Surabaya region, Indonesia. IPTEK, 25(2), 41.

[4] Puslitbangkim (Pusat Penelitian dan Pengembangan Pemukiman) Departemen Pekerjaan Umum. (2005). *Keterkaitan Penyelenggaraan Bangunan dengan Emisi CO2.* Bandung.

[5] Kirby. (2008). *Emisi CO2 pada Sistem Perumahan Pekotaan.* Dalam http://repository.usu.ac.id/bitstream/123456789/33596/3/Chapter%20II.pdf. Diakses pada tanggal 18 Agustus 2018.

[6] Rees, William and Mathis Wackernagel. (1996). Urban Ecological Footprints: Why Cities Cannot Be Sustainable and Why They are A Key to Sustainability. Elsevier Science Inc., https://doi.org/10.1016/S0195-9255(96)00022-4. New York.

[7] Wackernagel, M., Rees, William E. Rees (1996). Revisiting carrying capacity: Area-based indicators of sustainability, Population and Environment, Volume 17, Number 3, Pages 195-215.

[8] Vaz, E., & Nijkamp, P., (2014) Gravitational forces in the spatial impacts of urban sprawl: An investigation of the region of Veneto, Italy, Habitat International.

[9] Ilma, Atina. (2016). Arahan Optimasi Pemanfaatan Lahan di Kabupaten Mojokerto dengan Pendekatan Telapak Ekologis. ITS. Surabaya.

[10] Novianti, Srikandi dan Direjana. (2009). Pengaruh Karakteristik Faktor Emisi terhadap Estimasi Beban Emisi Oksida Nitrogen (NOx) dari Sektor Transportasi (Studi Kasus: Wilayah Karees, Bandung) (Jurnal). Bandung: Fakultas Teknik Sipil dan Lingkungan Intitut Teknologi Bandung.

[11] Ismayanti, dkk. (2008). *Kajian Emisi CO2 Menggunakan Persamaan Mobile 6 dan Mobile Combustion dari Sektor Transportasi di Kota Surabaya.* < http://digilib.its.ac.id/public/ITS-Undergraduate-16090-3307100053-Paper.pdf.> acessed in August 2017

[12] Aulia, B.U., Santoso, E.B., Hapsari, A.D., Safitri, E.W. (2018). Determining the Spatial Connectivity of Surabaya's Peri-urban Area (in Sidoarjo) towards Urban Rural Linkage Framework

The study of potential area for urban farming at Surabaya City

E Umilia[1] and D Saptarini[1]

[1] Urban and Regional Planning Department, Institut Teknologi Sepuluh Nopember, Surabaya, 60111, Indonesia

Email: umilia84@gmail.com

Abstract. Surabaya City is one of the big cities at East Java with rapid area development. This rapid area development, as usual, is followed by the rising of its citizen's growth, which creates a challenge in fulfilling the food necessity. This problem causes the food prices to escalate because of its limited amount, but on the other hand, the requests are kept on increasing. Now on, food security is an important issue for Surabaya City. Thus, efforts of increasing the food security level must be done, one of these efforts is urban farming. This study aims to locate the suitable areas for urban farming activities at Surabaya City. To gain that aim, an analysis method of Wight Overlay is used. From This Weight Overlay Analysis, the result retrieved is in the form of chosen areas' map which can be used as the urban farming activities locations. The chosen areas are distributed over to some locations at four districts of Benowo, Lakarsantri, Pakal, and Sambikerep.

Keywords: *Area's Identification, Urban Farming, Weighted Overlay*

1. Introduction

The food necessity becomes the basic needs for human's life sustainability, health, also as a source of energy for doing activities. The food necessity itself develops according to the number of the population, lifestyle, social and religion, income level, and individual taste [1]. The food security, based on UU No. 18 tahun 2012 (Indonesian regulation, number 18, of the year 2012), is a condition of the food fulfillment for the country to every individual, which reflected on the availability of enough food in amount and quality, save, various, nutritious, thoroughly, and affordable also not against toward religion, belief, and culture of the people, for sustainable healthy, active, and productive life [2].

The problems which related to food security are not only happened in countrysides. The more people live in cities than countrysides leads to a food security issue which should be well maintained. The city of Surabaya is the biggest metropolitan city at East Java, which has rapid regional development. This rapid development raises the also rapid population growth, which affects the foodstuff prices. The consequence is the foodstuff prices become more expensive, because of the limited amount, when the request is kept on escalating. To this day, Surabaya still depends on the continuity of rice supply from nearby areas, such as Sidoarjo, Gresik, and Lamongan for fulfilling its food necessity.

Thus, as explained previously, the food supply's right amount, good quality, and thoroughly distributed to all level of society with affordable price, is become a challenge to all the food stakeholders, such as government, business world from BUMN (country's business entity) and private sector, also other related parties. Thus, an effort must be done for leveling up the food security of Surabaya by creating food sources inside Surabaya City or taking from the outside sources, which one of these efforts is by working on urban farming. In order to maintain this urban farming development at Surabaya City, a corresponding study is needed.

2. Research method

2.1 Research approach and type

The approach which is used in this study is a normative approach. The normative approach in this study is an approach which is used to formulate a strategy based on existing data and information, also is accorded to farming related regulations and laws at Surabaya City.

Meanwhile, the research type of this study is comparative research. This study is done in order to compare the similarity or difference of two or more facts and characteristics of a study's object based on a particular framework.

2.2 Data collection method

The data collecting method in this study is done through primary and secondary surveys. The primary survey is done through interviews with related stakeholders of local government and other parties which are expertise in urban farming plan. The data collected by secondary survey is done through a study of related literature, also by collecting data from public service agency or related institutions, they are Development Planning Agency for Surabaya City, Central Bureau of Statistics, and Food and Agricultural Security Services of Surabaya City. The used data in this study are:

1. The width of the farming areas
2. The availability of primary food
3. The width of crop yield
4. The map of existing areas' usage
5. The existing map of agricultural designation areas distribution
6. The map of agricultural designation areas plan
7. RTRW (Spatial Development Plan) of Surabaya City

Beside using above data from related agency or institution, this study also observing theories and literature in the form of previous studies which is retrieved from periodical published journals from domestic also international sources. To be outlined, the theories' sources are divided into:

1. The general manual, especially from textbooks
2. The specific manual, in the form of reports which are specially derived from professional journals.

2.3 Data analysis method

The data analysis of this study is in the form of a quantitative method. This analysis aims to locate and analyze the priority areas for urban farming at Surabaya City, by using weight overlay analysis technique of ArcGIS 10.1 software, or other versions of it.

Weighted Overlay Analysis Technique is a spatial analysis technique which combines various type of maps in an overlapping way/overlay by the support of Geographic Information System (GIS) Software [4]. In this method, a raster data which has pixel as the smallest unit of measurement is used, in order to have every pixel scored and assessed thoroughly. The following is the framework of the study analysis:

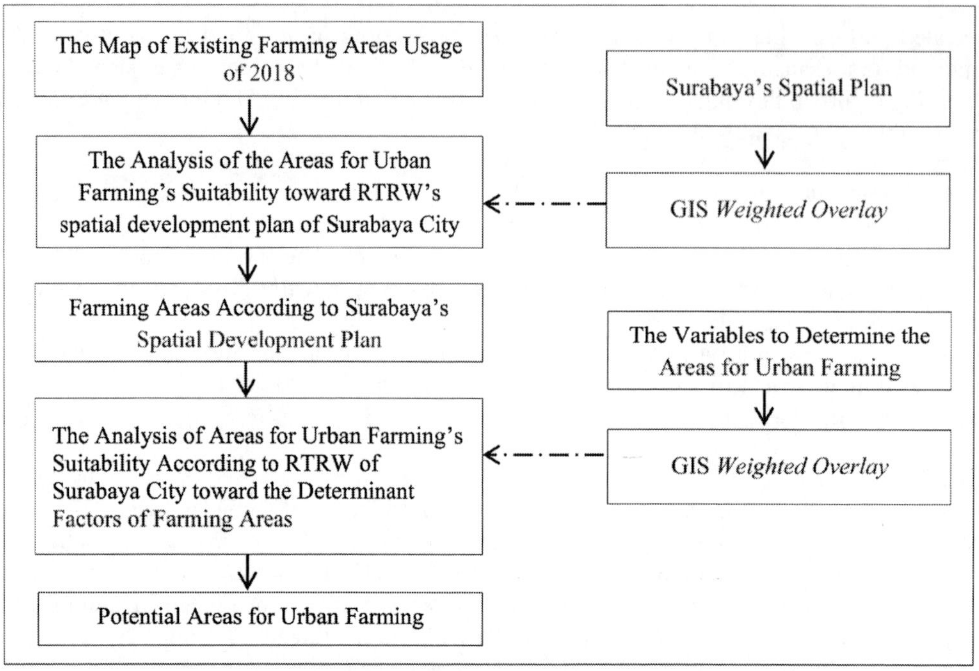

Figure 1. The Framework of Study Analysis

3. Result and discussion

3.1 The analysis of the potential areas determination for urban farming by existing farming approach

The used approach in determining the areas for urban farming is existing farming approach of the year 2018. This approach is used in order to accommodate the necessity zone for farming areas which is not included in the spatial development plan or RTRW's spatial pattern plan of Surabaya City for the year 2018. Besides, based on the interview of the Field Agriculture Instructor of the Food and Agricultural Security Services, there are farming areas which are not accommodated by the spatial development plan of Surabaya City. Thus, by using this approach of existing farming, the scenario resulted by this study can become a suggestion for future RTRW of Surabaya City.

The existing farming areas which will be used are those areas that have already maintained by the Field Agriculture Instructor of the Food and Agricultural Security Services of Surabaya City, which located at the districts of Benowo, Lakarsantri, Pakal, and Sambikerep. Also, some farming areas which located on the land owned by the Surabaya City's government with a status of the former of village's treasury land that the existing condition of the usage type of these farming areas is rice field, fishpond, and moor/field.

Table 1. The distribution of existing farming areas at Districts of Benowo, Lakarsantri, Pakal, and Sambikerep

District	Sub District	Areas' Usage Types (Ha)			
		Rice Field	Fishpond	Moor/Field	Total
	Kandangan	5,73	217,27	0,57	223,57
	Klakah Rejo		15,50		15,50
Benowo	Romokalisari		490,76		490,76
	Sememi	9,09	217,40	8,73	235,23
	Tambak Osowilangun		648,47		648,47
	Total of Benowo District	**14,82**	**1589,40**	**9,30**	**1613,52**
	Bangkingan	116,26		108,02	224,29
Lakarsantri	Jeruk	107,75		2,06	109,81
	Lakarsantri	76,60		6,83	83,43

District	Sub District	Areas' Usage Types (Ha)			
		Rice Field	Fishpond	Moor/Field	Total
	Lidah Kulon	92,52		0,08	92,60
	Lidah Wetan	0,05		20,84	20,89
	Sumurwelut	21,89		30,58	52,47
Total of Lakarsantri District		**415,07**		**168,40**	**583,47**
Pakal	Babat Jerawat		113,38	36,79	150,17
	Benowo		42,52	79,85	122,37
	Pakal		249,19	262,22	511,41
	Sumberejo		309,83	2,37	312,21
	Tambakdono		315,47		315,47
Total of Pakal District			**1030,39**	**381,24**	**1411,63**
Sambikerep	Beringin	0,34	6,15	108,48	114,97
	Lontar			159,71	159,71
	Made		1,42	228,71	230,13
	Sambikerep	0,08	2,74	93,17	95,99
Total of Sambikerep District		**0,42**	**10,31**	**590,07**	**600,80**
Total		**430,32**	**2630,10**	**1149,01**	**4209,43**

Before doing the location determination analysis, an indicator and parameter are needed as the determinant variables [5]. To determine the variables in this study spatial approach are used, thus, only spatial variables that is used in this study. Below is the variables' synthesis of the suitability of urban farming location at Surabaya.

Table 2. The Parameter of Areas' Suitability Variables for Urban Farming at Surabaya City

No.	Variables	Parameter	Attributes 0 : Not Suitable; 1 : Suitable
1.	**Topography**	Below the elevation of 100 MAMSL	0 : elevation >100 1 : elevation <100
2.	**Soil Type**	Alluvial soil type	0 : beside alluvial soil type 1 : alluvial soil type
3.	**Disaster Proneness**	Outside Disaster Prone Areas	0 : on disaster zone 1 : outside disaster zone
4.	**Land Slope**	Below 8% land slope	0 : >8% land slope 1 : <8% land slope
5.	**Irrigation/Drainage**	Inside 1 kilometer of irrigation/drainage's buffer zone	0 : >1 kilometer range to the irrigation system 1 : <1 kilometer range to the irrigation system
6.	**Road Access**	Outside 100 meters road's buffer zone	0 : < 100 meters range to road 1 : > 100 meters range to road

From these six variables, the suitable farming areas toward the Surabaya's spatial pattern plan are evaluated with the result as follows:

1. **Topography**
   The chosen farming areas will be assessed by its topography. The farming areas which located on <100 MAMSL (Meters Above Mean Sea Level) are scored as suitable, however, those areas which located >100 MAMSL are scored as not suitable. This is the result of the areas suitability:

Table 3. The suitability toward topography

Districts	Sub Districts	Suitability (Ha)		Total (Ha)
		Not Suitable	Suitable	
Benowo	Kandangan		223,57	1229
	Klakah Rejo		15,5	22

Districts	Sub Districts	Suitability (Ha)		Total (Ha)
		Not Suitable	Suitable	
	Romokalisari		490,7	3824
	Sememi		235,22	1654
	Tambak Osowilangun		546,63	3856
Total of Benowo District			**1611,62**	**10585**
Lakarsantri	Bangkingan		224.29	381
	Jeruk		109,81	26
	Lakarsantri		83,42	27
	Lidah Kulon		92,6	40
	Lidah Wetan		20,89	10
	Sumurwelut		52,47	647
Total of Lakarsantri District			**593,48**	**1131**
Pakal	Babat Jerawat		150,17	1217
	Benowo		122,38	1084
	Pakal		511,41	4701
	Sumberejo		312.21	1090
	Tambakdono		315,47	1358
Total of Pakal District			**1411,64**	**9450**
Sambikerep	Beringin		144,97	1772
	Lontar		159,71	1257
	Made		230,13	2894
	Sambikerep		95,99	1013
Total of Sambikerep District			**600,8**	**6936**
Total			**4207,51**	**28102**

Based on their elevations, all the farming areas are scored as suitable because all those areas are under 100 MAMSL.

2. **Soil Type**
The chosen farming areas will be assessed by its soil type. The areas with alluvial soil type are scored as suitable. On the other hand, all areas which have other soil type are scored as not suitable. The following table contains the result of the suitability of existing farming areas toward its soil type.

Table 4. The suitability toward soil type

Districts	Sub Districts	Suitability (Ha)		Total (Ha)
		Not Suitable	Suitable	
Benowo	Kandangan	82,1	141,6	223,6
	Klakah Rejo	12,6	2,8	15,5
	Romokalisari		490,7	490,7
	Sememi	284	206,7	253,3
	Tambak Osowilangun		646,6	646,6
Total of Benowo District		**123,1**	**1488,4**	**1611,7**
Lakarsantri	Bangkingan	168,9	55,4	224,2
	Jeruk	38,1	71,7	109,8
	Lakarsantri	81,9	1,4	83,4
	Lidah Kulon	2,5	90,1	92,6
	Lidah Wetan		20,9	20,9
	Sumurwelut	51,4	1,1	52,5
Total of Lakarsantri District		**342,9**	**240,6**	**583,4**

Districts	Sub Districts	Suitability (Ha)		Total (Ha)
		Not Suitable	Suitable	
Pakal	Babat Jerawat	15,1	135,1	150,1
	Benowo	93,4	29	122,3
	Pakal	101,2	410.3	511,4
	Sumberejo	174,2	138	312,2
	Tambakdono		315,4	315,4
Total of Pakal District		**383,8**	**1027,8**	**1411,4**
Sambikerep	Beringin		115	115
	Lontar		159,7	159,7
	Made		230,11	230,1
	Sambikerep		96	96
Total of Sambikerep District			**600,8**	**6936**
Total		**849,8**	**3357,7**	**4207,5**

According to the soil types at those four districts, the majority of areas have alluvial soil type, thus scored as suitable. From a total of 4,207.4 Ha, 3,357.7 Ha areas have suitable soil type. For Benowo District, the suitable soil type areas are 1,488.5 Ha from 1,611.6 Ha. 240.6 Ha from 583.5 Ha areas at Lakarsantri District have suitable soil type. 1,027.8 Ha areas at Pakal District have suitable soil type from a total of 1,411.6 Ha. All 600.8 Ha areas of Sambikerep District have suitable soil type.

3. Disaster Proneness

The chosen farming areas will be assessed by its fire and flood disaster proneness. The areas which located at the save zone from fire and flood disaster-prone areas are scored as suitable. On the other hand, all areas which located at the fire and flood disaster-prone areas are scored as not suitable. The table below is the result of the suitability of existing farming areas toward its disaster proneness.

Table 5. The suitability toward fire disaster proneness

Districts	Sub Districts	Suitability (Ha)		Total (Ha)
		Not Suitable	Suitable	
Benowo	Kandangan		223,6	223,6
	Klakah Rejo		15,5	15,5
	Romokalisari		490,7	490,7
	Sememi		235,3	235,3
	Tambak Osowilangun		646,6	646,6
Total of Benowo District				**1611,6**
Lakarsantri	Bangkingan		224,2	224,2
	Jeruk		109,8	109,8
	Lakarsantri		83,4	83,4
	Lidah Kulon		92,6	92,6
	Lidah Wetan		20,9	20,9
	Sumurwelut		52,5	52,5
Total of Lakarsantri District				**583,5**
Pakal	Babat Jerawat		150,1	150,1
	Benowo		122,3	122,3
	Pakal		511,4	511,4
	Sumberejo		312,2	312,2
	Tambakdono		315,4	315,4
Total of Pakal District				**1411,6**
Sambikerep	Beringin		115	115
	Lontar		159,7	159,7
	Made		230,1	230,1

Districts	Sub Districts	Suitability (Ha)		Total (Ha)
		Not Suitable	Suitable	
	Sambikerep		96	96
Total of Sambikerep District			**600,8**	**600,8**
Total			**4207,511**	**4207,511**

Seen from the fire disaster prone factor, all the farming areas which have reviewed, are scored as suitable because all of those areas are on the outside of the fire disaster-prone zone.

Table 6. The suitability toward flood disaster proneness

Districts	Sub Districts	Suitability (Ha)		Total (Ha)
		Not Suitable	Suitable	
Benowo	Kandangan	157,3	66,2	223,6
	Klakah Rejo		15,5	15,5
	Romokalisari	490,7		490,7
	Sememi	157,1	78,2	235,3
	Tambak Osowilangun	646,6		646,6
Total of Benowo District		**1451,7**	**159,9**	**1611,7**
Lakarsantri	Bangkingan	101,4	122,8	224,2
	Jeruk	37,1	72,8	109,8
	Lakarsantri	5,1	78,4	83,4
	Lidah Kulon	56,2	36,4	92,6
	Lidah Wetan		20,9	20,9
	Sumurwelut	3	49,4	52,5
Total of Lakarsantri District		**202,8**	**380,7**	**583,4**
Pakal	Babat Jerawat	112,8	37,3	150,1
	Benowo		122,3	122,3
	Pakal	219	292,3	511,4
	Sumberejo	3,6	308,6	312,2
	Tambakdono	292,3	23,1	315,4
Total of Pakal District		**627,7**	**783,6**	**1411,4**
Sambikerep	Beringin		115	115
	Lontar		159,7	159,7
	Made		230,1	230,1
	Sambikerep		96	96
Total of Sambikerep District		**0**	**600,8**	**600,8**
Total		**2282,3**	**1925,2**	**4207,5**

And on the other hand, as seen from the flood disaster prone of those areas at four districts, from a total of 4,207.4 Ha, 1,925.2 Ha areas are outside of the flood disaster zone. For Benowo District, the suitable areas are 159.9 Ha from 1,611.6 Ha. 380.7 Ha from 583.5 Ha areas at Lakarsantri District are suitable areas. 783.8 Ha areas at Pakal District are suitable areas from a total of 1,411.6 Ha. All 600.8 Ha of Sambikerep District are suitable areas.

4. **Land Slope**

The chosen farming areas will be assessed based on its land slope. The areas which located at the <8% slope are scored as suitable. On the other hand, all areas which located at the >8% slope are scored as not suitable. The result of the suitability of existing farming areas toward its land slope is shown below.

Table 7. The suitability toward land slope

Districts	Sub Districts	Suitability (Ha)		Total (Ha)
		Not Suitable	Suitable	
Benowo	Kandangan	9,1	214,47	223,57

Districts	Sub Districts	Suitability (Ha)		Total (Ha)
		Not Suitable	Suitable	
	Klakah Rejo	2,88	12,62	15,5
	Romokalisari		490,7	490,7
	Sememi	62,58	172,64	235,22
	Tambak Osowilangun		646,63	646,63
Total of Benowo District		**74,56**	**1537,06**	**1611,62**
Lakarsantri	Bangkingan	224,29		224,29
	Jeruk	109,81		109,81
	Lakarsantri	83,42		83,42
	Lidah Kulon	92,6		92,6
	Lidah Wetan	20,89		20,89
	Sumurwelut	52,47		52,47
Total of Lakarsantri District		**583,48**		**583,48**
Pakal	Babat Jerawat	27,27	122,91	150,17
	Benowo	56,32	66,05	122,38
	Pakal	242,60	268,8	511,41
	Sumberejo		312,21	312,21
	Tambakdono		315,47	315,47
Total of Pakal District		**326,19**	**1085,44**	**1411,64**
Sambikerep	Beringin	112,26	2,71	114,97
	Lontar	159,71		159,71
	Made	230,13		230,13
	Sambikerep	95,99		95,99
Total of Sambikerep District		**598,09**	**2,71**	**600,8**
Total		**1582,3**	**2625,21**	**4207,51**

According to slope at those areas at four district, from a total of 4,207.4 Ha, 2,625.21 Ha areas have a suitable land slope. For Benowo District, the suitable areas are 1,537.07 Ha from 1,611.6 Ha. The areas from all 583.5 Ha at Lakarsantri district are not suitable. 1,085.43 Ha areas at Pakal District are suitable areas from a total of 1,411.6 Ha. 2.71 Ha from 600.8 Ha areas of Sambikerep District are suitable areas.

5. **Irrigation/Drainage**

The chosen farming areas will be assessed by the availability of irrigation. The areas which located inside 1 kilometer radius from the irrigation system are scored as suitable. On the other hand, all areas which located outside 1 kilometer radius from the irrigation system are scored as not suitable. The result of the suitability of existing farming areas toward its land slope is in the next table:

Table 8. The suitability toward irrigation/drainage

Districts	Sub Districts	Suitability (Ha)		Total (Ha)
		Not Suitable	Suitable	
Benowo	Kandangan	50,2	173,3	223,6
	Klakah Rejo		15,5	15,5
	Romokalisari	137,1	353,7	490,7
	Sememi	10,4	224,8	235,3
	Tambak			646,6
	Osowilangun	144	502,6	
Total of Benowo District		**341,7**	**1269,9**	**1611,7**
Lakarsantri	Bangkingan	8,1	216,2	224,2
	Jeruk		109,8	109,8
	Lakarsantri	5,6	77,8	83,4
	Lidah Kulon		92,6	92,6

Districts	Sub Districts	Suitability (Ha)		Total (Ha)
		Not Suitable	Suitable	
	Lidah Wetan		20,9	20,9
	Sumurwelut		52,5	52,5
	Total of Lakarsantri District	**13,7**	**569,8**	**583,4**
Pakal	Babat Jerawat	50,7	99,5	150,1
	Benowo	55,1	67,3	122,3
	Pakal	225,7	285,7	511,4
	Sumberejo		312,2	312,2
	Tambakdono	20	295,4	315,4
	Total of Pakal District	**351,5**	**1060,1**	**1411,4**
Sambikerep	Beringin	30,4	84,5	115
	Lontar	21,7	138	159,7
	Made	75,5	154,6	230,1
	Sambikerep	16,7	79,3	96
	Total of Sambikerep District	**144,3**	**456,4**	**600,8**
	Total	**851,2**	**3356,3**	**4207,5**

Based on the irrigation's availability at those four districts, from a total of 4, 207.4 Ha, 3,356.3 Ha areas have reachable irrigation system. For Benowo District, the suitable areas are 1,269.9 Ha from 1,611.6 Ha. The areas from all 583.5 Ha at Lakarsantri district, 569.7 Ha are suitable. 1,060.2 Ha areas at Pakal District are suitable areas from a total of 1,411.6 Ha. 456.5 Ha from 600.8 Ha areas of Sambikerep District are suitable areas.

6. Road Access

The chosen farming area will be assessed by the road access range. The areas which located outside radius of 100 meters from the road access are scored as suitable. On the other hand, all areas which located inside radius of 100 meter from the road access are scored as not suitable. The result of the suitability of existing farming areas toward its road access range is in the next table.

Table 9. The suitability toward the road access

Districts	Sub Districts	Suitability (Ha)		Total (Ha)
		Not Suitable	Suitable	
Benowo	Kandangan	46,87	176,69	223,57
	Klakah Rejo	11,39	4,11	15,5
	Romokalisari	134,45	356,24	490,7
	Sememi	123,86	111,36	235,22
	Tambak			646,63
	Osowilangun	170,26	476,36	
	Total of Benowo District	**486,83**	**1124,76**	**1611,62**
Lakarsantri	Bangkingan	52,11	172,17	224,29
	Jeruk	9,82	99,99	109,81
	Lakarsantri	42,77	40,65	83,42
	Lidah Kulon	61,11	31,49	92,6
	Lidah Wetan	**19,39**	1,5	20,89
	Sumurwelut	35,55	16,91	52,57
	Total of Lakarsantri District	**220,75**	**362,71**	**583,48**
Pakal	Babat Jerawat	52,42	97,75	150,17
	Benowo	66,13	56,24	122,38
	Pakal	175,64	335,76	511,41
	Sumberejo	114,63	197,58	312,21
	Tambakdono	12,74	302,73	315,47
	Total of Pakal District	**421,56**	**990,06**	**1411,64**
Sambikerep	Beringin	54,58	60,39	114,97

Districts	Sub Districts	Suitability (Ha)		Total (Ha)
		Not Suitable	Suitable	
	Lontar	75,6	84,1	159,7
	Made	124,92	105,21	230,13
	Sambikerep	72,73	23,25	95,99
Total of Sambikerep District		**327,83**	**272,95**	**600,79**
	Total	**1457,08**	**2750,5**	**4207,51**

Based on the road access range at those four districts, from a total of 4, 207.4 Ha, 2,750.5 Ha areas are outside of road range, which is suitable. For Benowo District, the suitable areas are 1,124.77 Ha from 1,611.6 Ha. The areas from all 583.5 Ha at Lakarsantri district, 362.72 Ha are suitable. 990.07 Ha at Pakal District are suitable areas from a total of 1,411.6 Ha. 272.95 Ha areas from 600.8 Ha of Sambikerep District are suitable areas.

3.2 *The analysis of suitability of the urban farming areas toward the existing areas by the farming area determination factors*

After working on the suitability analysis for each variable, the operation of addition is done in the form of "if meeting the variable's criteria is scored 1 and 0 if not". Thus, from seven variables, the areas with the end result of 7 are chosen. The scoring system used is:

Table 10. The scoring system for areas' determination for urban farming

Score	Explanations	Suitability Value
1	Do not meet 6 variables	Not Suitable
2	Do not meet 5 variables	Not Suitable
3	Do not meet 4 variables	Not Suitable
4	Do not meet 3 variables	Not Suitable
5	Do not meet 2 variables	Less Suitable
6	Do not meet 1 variable	Suitable
7	Meet all variables	Suitable

The Weight Overlay Analysis's result (on Picture 2.) is containing some locations of suitable areas for urban farming at Surabaya. The following table is the result of the suitability analysis of potential areas for urban farming.

Table 11. The analysis of suitability of potential areas for urban farming

Districts	Sub Districts	Suitability Result (Ha)			
		Not Suitable	Less Suitable	Suitable	Total
Benowo	Kandangan	10	94,6	119	223,6
	Klakah Rejo	0	8,5	6,9	15,4
	Romokalisari	29,4	212,7	248,6	490,7
	Sememi	3,1	92,9	139,3	235,3
	Tambak Osowilangun	28,5	257,2	360,9	646,6
Total of Benowo District		**71**	**665,9**	**874,7**	**1611,6**
Lakarsantri	Bangkingan	28,4	138,3	57,6	224,3
	Jeruk	2,2	9,1	98,4	109,7
	Lakarsantri	4,5	43,7	35,2	83,4
	Lidah Kulon	2,3	30,8	59,5	92,6
	Lidah Wetan	0	0	20,9	20,9
	Sumurwelut	14,1	25,9	12,4	52,4
Total of Lakarsantri District		**51,5**	**247,8**	**284**	**583,3**
Pakal	Babat Jerawat	0,8	79,4	70,1	150,3
	Benowo	12,3	68,6	41,6	122,5
	Pakal	38,3	191,2	282	511,5
	Sumberejo	0	81,2	231	312,2

Districts	Sub Districts	Suitability Result (Ha)			
		Not Suitable	Less Suitable	Suitable	Total
	Tambakdono	0	29,9	285,6	315,5
	Total of Pakal District	**51,4**	**450,3**	**910,3**	**1412**
Sambikerep	Beringin	0	12,9	102,1	115
	Lontar	0	17,8	141,9	159,7
	Made	0	40,3	189,8	230,1
	Sambikerep	0	16,5	79,5	96
Total of Sambikerep District		**0**	**87,5**	**513,3**	**600,8**
	Total	**173,9**	**1451,5**	**2582,3**	**4207,7**

Based on the suitability analysis of potential areas for urban farming at those four districts, from total 4,207.4 Ha, 61.4% of it are suitable, 34.5% are less suitable, and the not suitable areas are 4.1%.

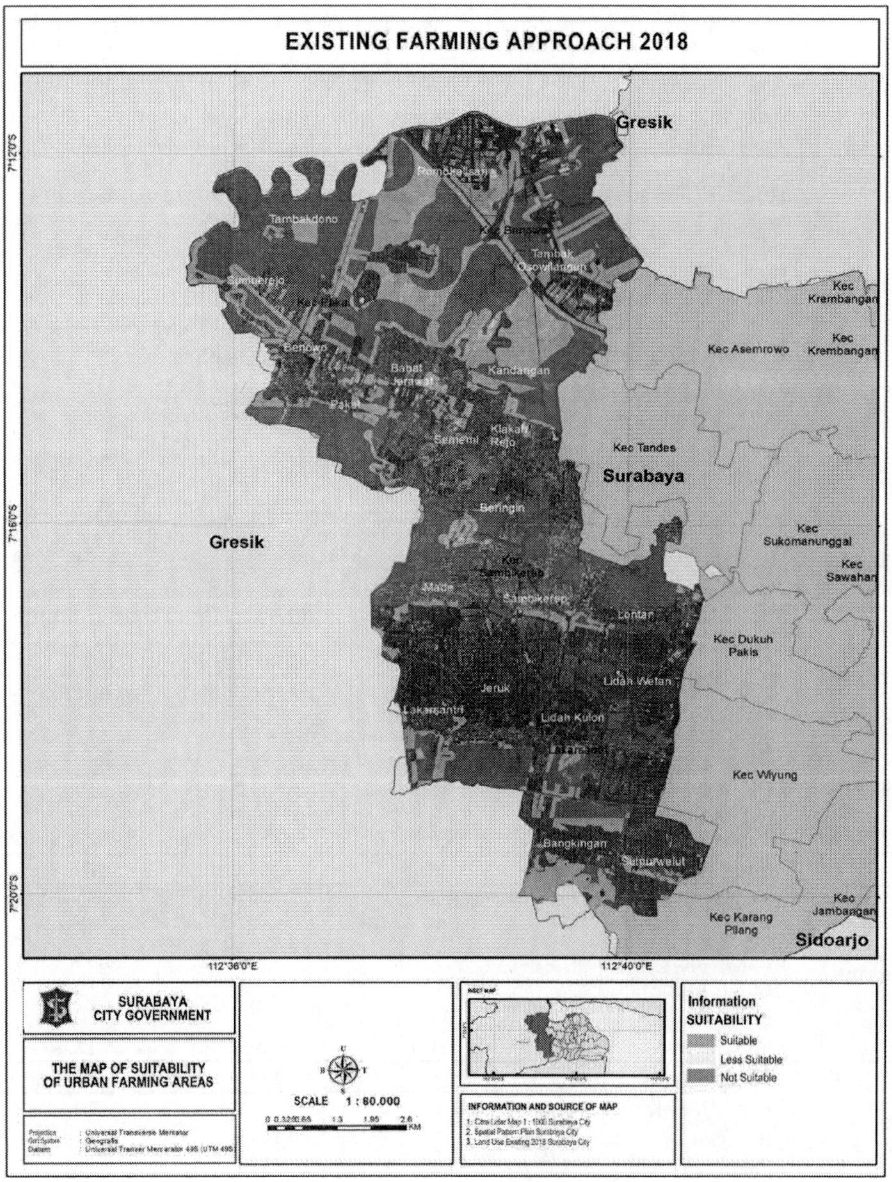

Figure 2. The map of suitability of urban farming areas at Surabaya City

Below are the percentage of the suitability of each districts areas:

a. Benowo District
   - Suitable: 64.5%
   - Less Suitable: 41.3%
   - Not Suitable: 4.4%
b. Lakarsantri District
   - Suitable: 48.7%
   - Less Suitable: 42.5%
   - Not Suitable: 8.9%
c. Pakal District
   - Suitable: 54.3%
   - Less Suitable: 31.9%
   - Not Suitable: 3.6%
d. Sambikerep District
   - Suitable: 85.4%
   - Less Suitable:14.6%
   - Not Suitable: 0%

4. Conclusion

According to the previous result and discussion, it can be concluded that in determining potential areas for urban farming activities, there are six variables which should be considered. They are topography, soil type, disaster proneness, land slope, irrigation/drainage, and road access. From the suitability analysis by using weighted overlay method, the areas which have suitability factor for urban farming at Surabaya City are those areas at four districts of Benowo, Lakarsantri, Pakal, and Sambikerep, with the width for each district's areas are 874.7 Ha, 284 Ha, 910.3 Ha, and 513.3 Ha.

5. References

[1] Zaroni. 2017. *Supply Chain – Logistik Pangan.*

[2] Syaifullah, Yunan. (2013). *Ketahanan Pangan dan Pola Distribusi Beras di Propinsi Jawa Timur*. **6** 103-11.

[3] Sihgiyanti, Vika Jessy. 2016. *Evaluasi Implementasi Program Urban Farming Oleh Dinas Pertanian di Kota Surabaya*. **4** 264-72

[4] Karana, R.C, & Suprihardjo, R. (2013). *Mitigasi Bencana Banjir Rob di Jakarta Utara*. **2** C31-6.

[5] Subroto, Gatot. 2016. *Pemodelan Spasial Lahan Pertanian Pangan Berkelanjutan di Kabupaten Jombang Jawa Timur* (Jurusan Perencanaan Wilayah dan Kota ITS Surabaya) p 62-63

IOP Conf. Series: Earth and Environmental Science **340** (2019) 012019 doi:10.1088/1755-1315/340/1/012019

Economical valuation for measuring renewable energy efficiency as the implementation of smart city concept at Surabaya City

E Umilia[1] and E B Santoso[1]

[1]Urban and Regional Planning Department, Institut Teknologi Sepuluh Nopember, Surabaya, 60111, Indonesia

Email: umilia84@gmail.com

Abstract. Energy is the most important part of supporting human necessity. For this while, the crude oil energy which is mostly needed at Indonesia are fuel oil and electricity. Unfortunately, the domestic crude oil reserves are diminishing and global crude oil's price has always fluctuated. Thus, the development of renewable energy is necessary. However, the alternative energy development in Indonesia still relies on fossil energy, which is not environmental friendly yet. Smart energy is one of the existed approaches in the concept implementation of smart city. Surabaya city has already had a number of innovations in smart energy implementation of alternative uses in utilizing renewable energy for particularly public services, such as solar cell, garbage power plant, and the application of wind turbine. Some of the mentioned alternative energy have been tried at government's buildings, other government's facilities, and public street lighting. There are some occurrences of problems such as the lack of readiness of the supporting facility, the energy production made a few trial project slowed down to be implemented. This study is trying to economically valuate about the efficiency level of renewable energies which have been tried. This study aims to offer an advantage of choosing the best alternative energy to be mass implemented. Thus, the smart city concept, especially smart energy, can be thoroughly implemented using the most efficient type of energy.

Keywords: *Smart city, Smart energy,Economical valuation, Energy efficiency*

1. Introduction

Energy is the most important part of human life because almost all of the human activity always need energy. Most energy used in Indonesia comes from fossil energy in the form of crude oil and natural gas. If it was examined from development segment, the energy system in Indonesia since shows that the fossil energy source still be the main source in fulfilling the domestic energy necessity. The favorable fossil energy is crude oil, natural gas, and charcoal. For many years, crude oil dominates the energy supply and utility domestically as fuel oil and electricity.

The most useful for crude oil is gasoline and diesel fuel, meanwhile for household necessity, people tend to choose kerosene. However, because of the program of kerosene to LPG (Liquid Petroleum Gas) conversion, kerosene's price becomes too high and rarely existed which makes people substitute it with LPG for their energy needs. Unfortunately, the particular program is not solving the fuel problem in society. It is because the society's worriedness of a severe blown potential which caused by the package's leaking, also the distribution of LPG is not thoroughly yet which cause rarity in some areas.

Fuel oil is a form of energy which has an important role in the activities of industrial sector, transportation sector, also household sector. The diminishing of domestic crude oil and the increasing of domestic fuel oil consumption leads to the rarity of fuel oil in some areas in Indonesia. This condition forces Indonesia to imports fuel oil from other countries and unable to rely on the global crude oil source because of its fluctuated price, which will deplete country's foreign exchange and threatening national energy security.

The diminishing of domestic crude oil reserves, the fluctuated price of global crude oil, and the availability of various domestic alternative energy potencies become a background of consideration that the alternative energy development is necessary. However, according to Presidential Regulation No. 5 of 2006 concerning National Energy Policy, this day the developing portion of alternative energy still rely on fossil energy which aims in increasing the supply and utility of natural gas and charcoal. Meanwhile, the renewable alternative energy development which is environmentally friendly still get relatively limited portion even though it has already increased.

Energy has ever been separated from the environmental issue. The most globally trending issue of environmental nowadays is global warming and climate changes. Since before, fossil fuel oil is issued as the cause of global warming [1]. Fossil fuel oil which is incompletely combusted will cause the CO_2 gas rising to the earth's surface and obstructing the earth heat's reflection. This occurrence leads to high temperature on the earth surface.

The development of renewable and environmentally friendly alternative energy are relevant matters to issues about energy and environment nowadays. This happens because of the energy sector has a strong relation toward the environment, in which the energy sector is affecting the environment while on the energy production period until its utility by changing the environment state.

The development of renewable alternative energy which is environmentally friendly and available for local structure can be a multi-advantages instrument, as a substitute in order to make people will reduce their dependence on fossil energy, realizing the sustainability of the environment, and supplying energy which is accessible to local people for its quantity, quality, and value. There are a lot of kind of renewable alternative energy which is environmentally friendly and locally available which can be developed.

Surabaya city is the second biggest city in Indonesia and is a trade and service city by becoming the central interconnection to the east part of Indonesia. This fact means that the economic growth and development of Surabaya City are affecting its citizen's mobility. The high mobility leads to the citizen's consumption of fuel oil escalate too. Besides that matter, the centralized economic activities of Indonesia's east part at Surabaya City is increasing the electricity consumption.

Thus, it is necessary to conduct a study about the potential of alternative energy development for public necessity at Surabaya City. This should be done in order to prevent the rapid activities which are happening to not interrupt the existing supply of energy source.

2. Research method

2.1 Research type

This study is included to comparative study which is done in order to compare the similarity and difference of two or more facts and characteristics of a study's object based on particular framework. There are few main steps of comparative study, which are: 1) to formulate and define the problem; 2) to explore and analyze the existing literature 3) to formulate the theoretical framework and hypotheses, also any other assumptions which are used; 4) to create a research design by determining the subject using the chosen data collection technique, then categorized the characteristics, attributes, or other matters which is corresponding with the study's problem in order to ease the analysis of cause and effect; 5) to test the chosen hypotheses, create an interpretation about the correlation using suitable statistic

IOP Conf. Series: Earth and Environmental Science **340** (2019) 012019 doi:10.1088/1755-1315/340/1/012019

technique; 6) to put through generalization, conclusion, also policy implication; and 7) to form a report in a scientific writing. [2]

The method used in this study is quantitative and qualitative method. The quantitative method is calculating the data using formulas, meanwhile the qualitative method is analyzing the calculating result and comparing them to the existing standards.

2.2 Data collection technique

The data collection technique in this study is using secondary survey. The data collected by secondary survey is done through a study of related literature, also by collecting data from public service agency or related institutions, they are Bappeda Kota Surabaya (development planning agency for Surabaya City) and BPS (central bureau of statistics) in this study. The used data in this study are:

1. The general pictorial of Surabaya City, including administration profile, geography, also the physical condition of the city base
2. The electricity usage for Surabaya City's government buildings
3. The electricity usage for PJU (Penerangan Jalan Umum, public street lighting)
4. The electricity usage for buildings or areas which are financed by Surabaya City's government
5. The examples of utilization of alternative energy source at Surabaya City.

Beside using above data from related agency or institution, this study also observing theories and literature in the form of previous studies which is retrieved from periodical published journals from domestic also international sources. To be outlined, the theories' source is divided into:

1. The general manual, especially from textbooks
2. The specific manual, in the form of report which is specially derived from professional journals.

2.3 Data analysis

The data analysis in this study is done in the form of qualitative and quantitative analysis. The qualitative analysis is implemented to gain the pictorial of the potential of alternative energy development at Surabaya City by using comparative analysis.

Meanwhile, the quantitative analysis is implemented for determining the priority in choosing the later to be developed alternative energy source at Surabaya City by using Benefit Cost Ratio method. Benefit Cost Ratio is one of the investment feasibility methods to evaluate a project or investment by comparing its economic benefits toward its economic cost [3]. As for the criteria to determine the single decision is by examining the value of B/C which will show resulting values are greater or equal to one, or less than one. The following are the valuation criteria of benefit cost ratio [4]:

If $B/C \geq 1$, thus the alternative investment or project is feasible, approved (1)

If $B/C < 1$, thus the alternative investment or project is not feasible (2)

3. Result and discussion

3.1 PLTS (Solar System Power Plant)

3.1.1 PLTS Analysis for Building Utility

In evaluating the efficiency of Solar System Power Plant usage for building, three schemas of valuation are done, as follows:

A. Valuation Schema of the Usage of 100% from PLN (State Electricity Company)

This valuation's aim is to understand the public facility's electricity charge if only using power from PLN without using solar system power at all.

Below is the result of valuation of the electricity charge annually, using 100% power from PLN. This valuation is retrieved from invoice data from PLN monthly for several government building which is chosen as study's sample.

Table 1. The valuation schema of using 100% power from PLN

No	Public Facility	Annual Charge	
A	KANTOR JIMERTO		
1	Pemda Tk. II / Puskesmas Ketabang	Rp	65.682.018,00
2	Kota Praja / Rumah Dinas Walikota	Rp	214.216.317,00
3	YYS Kas Pembangunan / Rumah Dinas Wawali	Rp	44.527.800,00
4	Kota Praja / Balai Kota Surabaya	Rp	55.419.520,36
5	Gedung Keuangan T2 KMS / Kantor Pemkot Surabaya (6 lantai)	Rp	2.046.115.791,00
6	Kota Madya Dati II Sby / Komplek Kantor Pemkot Surabaya (kantor balaikota)	Rp	1.862.063.949,82
B	KANTOR BAPPEKO	Rp	269.812.923,00

B. Valuation schema of the usage of 50% power from PLN and 50% power from PLTS

This valuation is used to understand the amount of savings made by government if reducing the power usage from PLN to 50% and simultaneously increasing the power usage from solar system to 50%. The investment valuation is included in the following schema.

The following is the valuation result of electricity charge annually by utilizing both powers from PLN and solar system in 50% part of each. By using PLTS, the expenses of investment, operational, and maintenance should be considerate.

Table 2. The valuation schema of using 50% power from PLN and 50% power from PLTS

No	Public Facilities	MIXED (50% PLN + 50% PLTS)	
		Annual Charge	
A	KANTOR JIMERTO		
1	Pemda Tk. II / Puskesmas Ketabang	Rp	260.898.275,67
2	Kota Praja / Rumah Dinas Walikota	Rp	124.031.847,39
3	YYS Kas Pembangunan / Rumah Dinas Wawali	Rp	260.904.611,11
4	Kota Praja / Balai Kota Surabaya	Rp	220.716.426,85
5	Gedung Keuangan T2 KMS / Kantor Pemkot Surabaya (6 lantai)	Rp	1.031.367.495,50
6	Kota Madya Dati II Sby / Komplek Kantor Pemkot Surabaya (kantor balaikota)	Rp	940.263.077,94
B	KANTOR BAPPEKO	Rp	158.391.394,83

The cost from the city government's electricity charge is able to be diminished to 97.15%. In this valuation, the cost of the PLTS procurement investment is omitted. This valuation's goal is to understand the PLTS role which is aiming in reducing the electricity charge from using the power from PLN, as an existing or projection matter.

Table 3. The saving percentage of the valuation schema of using 50% power from PLN and 50% power from PLTS

No	Public Facility	Saving Percentage (without investment cost)
A	KANTOR JIMERTO	
1	Pemda Tk. II / Puskesmas Ketabang	94,44%
2	Kota Praja / Rumah Dinas Walikota	99,87%
3	YYS Kas Pembangunan / Rumah Dinas Wawali	91,43%

IOP Conf. Series: Earth and Environmental Science **340** (2019) 012019 doi:10.1088/1755-1315/340/1/012019

No	Public Facility	Saving Percentage (without investment cost)
4	Kota Praja / Balai Kota Surabaya	94,43%
5	Gedung Keuangan T2 KMS / Kantor Pemkot Surabaya (6 lantai)	99,99%
6	Kota Madya Dati II Sby / Komplek Kantor Pemkot Surabaya (kantor balaikota)	99,99%
B	KANTOR BAPPEKO	99,86%
		97,15%

If considering investment cost, thus the city government will reach BEP (Break Even Point) in 17 years, when the percentage will be on the positive value.

C. Valuation Schema of the Usage of 100% Power from PLTS

This valuation is used to understand the amount of savings made by government if only utilize the power usage from PLTS 100%. The required energy to have the electricity is retrieved 100% from the sun. The investment valuation is included in the following schema.

The following is the valuation result of electricity charge annually by utilizing 100% power from PLTS. By using PLTS, the expenses of investment, operational, and maintenance should be considerate. Using 100% power from PLTS means not using any power from PLN at all.

Table 4. The valuation schema of using 100% power from PLTS

No	Public Facility	SOLAR CELL	
		Investment Cost	Maintenance/year
A	KANTOR JIMERTO		
1	Pemda Tk. II / Puskesmas Ketabang	Rp 456.114.533	Rp 7.297.833
2	Kota Praja / Rumah Dinas Walikota	Rp 33.847.378	Rp 541.558
3	YYS Kas Pembangunan / Rumah Dinas Wawali	Rp 477.281.422	Rp 7.636.503
4	Kota Praja / Balai Kota Surabaya	Rp 386.013.333	Rp 6.176.213
5	Gedung Keuangan T2 KMS / Kantor Pemkot Surabaya (6 lantai)	Rp 16.619.200	Rp 265.907
6	Kota Madya Dati II Sby / Komplek Kantor Pemkot Surabaya (kantor balaikota)	Rp 18.462.206	Rp 295.395
B	KANTOR BAPPEKO	Rp 46.969.867	Rp 751.518

The able-to-diminished cost from the existing electricity charge of city government is 94.29%. In this valuation, the cost of the PLTS procurement investment is omitted.

Table 5. The Saving Percentage of the Valuation Schema of Using 100% Power from PLTS

No	Public Facility	Saving Percentage (without investment cost)
A	KANTOR JIMERTO	
1	Pemda Tk. II / Puskesmas Ketabang	88,89%
2	Kota Praja / Rumah Dinas Walikota	99,75%
3	YYS Kas Pembangunan / Rumah Dinas Wawali	82,85%
4	Kota Praja / Balai Kota Surabaya	88,86%
5	Gedung Keuangan T2 KMS / Kantor Pemkot Surabaya (6 lantai)	99,99%

No	Public Facility	Saving Percentage (without investment cost)
6	Kota Madya Dati II Sby / Komplek Kantor Pemkot Surabaya (kantor balaikota)	99,98%
B	KANTOR BAPPEKO	99,72%
		94,29%

Considering investment cost, thus the city government will reach BEP (Break Even Point) in 17 years, when the percentage will be on the positive value.

3.1.2 PLTS analysis for PJU (Public Street Lighting) usage
The usage of public street lighting is costed by the PLN and should be paid by local government. For this payment interest, the PJU will be listed as a must be paid component in monthly electricity invoice. PLTS at Surabaya City is utilized for PJU. The PJUs which have been using power from PLTS are: PJU Jembatan Surabaya dan Air Mancur, PJU Jalan Raya, PJU Tanggul, PJU THP Kenjeran, PJU THP Kenjeran Parkir, Lampu Taman THP Kenjeran. The following is the analysis of investment cost and the savings percentage by utilizing PLTS for PJU.

Table 6. The investment cost and saving rate

No	PJU (Public Street Lighting)	Solar Module Needs		Battery Needs	Investment Cost	Probabi lity	Savings
		5 hours/ day	120 WP	100 AH	kurs 13.600,00		
1	PJU Jembatan Surabaya dan Air Mancur	1.875,90 WP	16 buah	36 buah	Rp 255.122.400,00		99%
2	PJU Jalan Raya	0,06 WP	1 buah	1 buah	Rp 8.012,97	Too Small	
3	PJU Tanggul	258,15 WP	3 buah	1 buah	Rp 35.108.587,16		55%
4	PJU THP Kenjeran	0,06 WP	1 buah	1 buah	Rp 8.547,17	Too Small	
5	PJU THP Kenjeran Parkir	951,93 WP	8 buah	1 buah	Rp 129.462.915,14		57%
7	Lampu Taman THP Kenjeran	0,03 WP	1 buah	1 buah	Rp 4.056,28	Too Small	
Average Savings							**70%**

Based on the above valuation analysis, thus the government is able to save 70% off from the electricity charge by utilizing the solar system power for powering PJU.

3.2 PLTB (Wind Power Plant)
The development of Wind Power Plant at Surabaya City is adopting the Wind Power Plant on Nusa Penida Island because the average wind speed in this area is lower than 5m/s. A WES80 turbine is chosen accordingly as its used for low rate of 3.0m/s wind speed. Each WES80 turbines gives average electricity power as much as 135,000 kWh (kilowatt an hour) annually. 1 billion rupiahs is needed to build up each turbine, which means every kWh, it will cost US$ 10-15 cents. The valuation of the electricity usage by utilizing power from PLTB at Surabaya City shown in the table below:

Table 7. The valuation of electricity usage by utilizing PLTB at Surabaya City

Description	Amount
Average Electricity Requirement per Month	3.436.025,65 kWh
Average Electricity Needs per Year	41.232.307,78 kWh

IOP Conf. Series: Earth and Environmental Science **340** (2019) 012019 doi:10.1088/1755-1315/340/1/012019

Description	Amount
Bill of PLN Electricity Account per Month (TDL Rp 1.467,- per kWh)	Rp 5.040.649.626
Bill of PLN Electricity Account per Year (TDL Rp 1.467,- per kWh)	Rp 60.487.795.514
Average city wind speed in Surabaya	17 knot = 9,26 m/s
The need for wind turbines (1 wind turbine generates 135,000 kWh of electricity per year)	305 unit
Investment cost for PLTB (1 turbine Rp. 1 billion)	Rp 305.000.000.000
Bill of PLTB Electricity Account per Year (TDL US $ 10 cents per kWh = Rp. 1,360 per kWh)	Rp 56.075.938.581
Amount of Saving by Using PLTB	Rp 4.411.856.932 7,29%

From the previous valuation, it is known that for supply the public electricity needs, it will require 305 units of WES80 turbines with the buildup investment cost is 305 billion rupiahs. The total charge of electricity usage by utilizing PLTB for public affair is 56,075,938,581.00 rupiahs. This amount is more efficient 4,411,856,932.00 rupiahs than using power from PLN, or can save around 7.29%..

3.3 PLTSa (Garbage Power Plant)

According to the existing data, it can be understood that the production result of PLTSa Benowo for the year 2017 is 6,349,298 kWh and the selling potential to PLN is 15,847,847,808.00 rupiahs, which means can be counted that the savings in garbage maintenance for the year 2017 is 16%. The complete information is shown in the table below:

Table 8. The cost saving of garbage maintenance of 2017

No.	Month	Garbage Maintenance Cost 2017		Production Result of PLTSa Benowo 2017		Saving Cost of Garbage Maintenance 2017
		Tonage	Total	KWh	Selling Potential to PLN	
1	Desember	46500	Rp 7.253.302.500,00	672.148,00	Rp 1.677.681.408,00	23%
2	Januari	127600	Rp 19.903.686.000,00	692.164,00	Rp 1.727.641.344,00	9%
3	Februari	44800	Rp 6.988.128.000,00	873.550,00	Rp 2.180.380.800,00	31%
4	Maret	49600	Rp 7.736.856.000,00		Rp -	0%
5	April	48000	Rp 7.487.280.000,00		Rp -	0%
6	Mei	49600	Rp 7.736.856.000,00	1.028.666,00	Rp 2.567.550.336,00	33%
7	Juni	48000	Rp 7.487.280.000,00	1.101.736,00	Rp 2.749.933.056,00	37%
8	Juli	49600	Rp 7.736.856.000,00	1.142.240,00	Rp 2.851.031.040,00	37%
9	Agustus	49600	Rp 7.736.856.000,00	838.794,00	Rp 2.093.629.824,00	27%
10	September	48000	Rp 7.487.280.000,00		Rp -	0%
11	Oktober	49600	Rp 7.894.089.600,00		Rp -	0%
12	November	48100	Rp 8.028.082.400,00		Rp -	0%
TOTAL		**659000**	**Rp 103.476.552.500,00**	**6349298**	**Rp 15.847.847.808,00**	

And by understanding the above analysis table, the average savings in the garbage maintenance cost for the year 2017 is 16%.

In order to know the savings for the year 2018, a projection of production result at PLTSa Benowo is valuated. Thus, the projection of production result of PLTSa Benowo for the year 2018 is 10,884.511 kWh with the selling potential to PLN reach 27,167,739,099,43 rupiahs. And as for its saving amount, this projection for the year 2018 is 26%. Further information can be seen on the next table:

Table 9. The projection of garbage maintenance cost of 2018

No	Month	Projection of Garbage Maintenance of 2018Sampah 2018		Projection of Production Result of PLTSa Benowo 2018		Saving Cost of Garbage Maintenance 2018
		Tonage	Total	KWh	Selling Potential to PLN	
1	December	52700	Rp 8.795.840.800,00	907.042,57	Rp 2.263.978.258,29	26%
2	January	52700	Rp 8.795.840.800,00	907.042,57	Rp 2.263.978.258,29	26%
3	February	49300	Rp 8.228.367.200,00	907.042,57	Rp 2.263.978.258,29	28%
4	March	52700		907.042,57	Rp 2.263.978.258,29	26%
5	April	51000	Rp 8.512.104.000,00	907.042,57	Rp 2.263.978.258,29	27%
6	May	52700	Rp 8.795.840.800,00	907.042,57	Rp 2.263.978.258,29	26%
7	June	51000	Rp 8.512.104.000,00	907.042,57	Rp 2.263.978.258,29	27%
8	July	52700	Rp 8.795.840.800,00	907.042,57	Rp 2.263.978.258,29	26%
9	August	52700	Rp 8.795.840.800,00	907.042,57	Rp 2.263.978.258,29	26%
10	September	51000	Rp 8.512.104.000,00	907.042,57	Rp 2.263.978.258,29	27%
11	October	52700	Rp 8.974.590.700,00	907.042,57	Rp 2.263.978.258,29	25%
12	November	51000	Rp 9.107.937.000,00	907.042,57	Rp 2.263.978.258,29	25%
TOTAL		**622200**	**Rp104.622.251.700,00**	**10884511**	**Rp 27.167.739.099,43**	

4. Conclusion

According to this study, thus it can be concluded that by utilizing the renewable energy, Surabaya City's government are able to manage some amount of cost savings. Meanwhile, the saving rate of each type of alternative energy is concluded as follows:

a. PLTS for buildings, the saving rate of this project reaches 97.15% (50% power from PLN and 50% power from PLTS), 94.29% (100% power from PLTS)
b. PLTS for PJU, the saving rate of this project reaches 70%
c. PLTSa, the saving rate of this project reaches 26%
d. PLTB, the saving rate of this project reaches 7.29%.

5. References

[1] Wahyuni, S. 2013. *Biogas Energi Alternatif Pengganti BBM, Gas, dan Listrik* (Jakarta Selatan : PT. Agro Media Pustaka) p 3-4

[2] Nazir, Mohammad. 1988. *Metode Penelitian* (Jakarta : Ghalia Indonesia) p 69-70

[3] Shively, Gerald. 2012. *An Overview of Benefit-Cost Analysis.*

[4] Subagyo, Ahmad. 2007. *Studi Kelayakan Teori dan Aplikasi* (Jakarta: PT. Elex Media Komputindo) p 33

IOP Conf. Series: Earth and Environmental Science **340** (2019) 012020 doi:10.1088/1755-1315/340/1/012020

Ecosystem services-based land use suitability analysis in East Java Province

F Firmansyah[1], M Yusuf[1], P R Setiawan[1] and C Susetyo[1]

[1] Urban and Regional Planning Department, Institut Teknologi Sepuluh Nopember, Surabaya, 60111, Indonesia

Email: ff.fendy.firmansyah@gmail.com

Abstract. Land Use planning method nowadays has been developed from many perspectives. One of the considered approaches is ecosystem based which is integrating ecosystem services in constructing land use plan scenario. The concept of ecosystem services is strictly linked to changes in land use and land cover, which affect ecosystem services, moreover this determine a decrease in global environmental and biodiversity loss. In this paper we aim to construct an ecosystem services-based industrial suitability map as an input for land use planning. The construction of ecosystem services map process involves expert judgement from stakeholders in order to know which ecosystem services classification should be taken into account in determining spatial allocation for industrial use. Furthermore, the carrying capacity map is constructed based on physical aspects such as morphology, water availability, disaster vulnerability, etc. In the construction process, stakeholders are also asked to give weight to the physical aspects, this is important since the data are then merged using weighted overlay method. The suitability map is resulted from the overlay of carrying capacity map and ecosystem services map in East Java Province scope.

Keywords: *land use, carrying capacity, ecosystem services, industrial land suitability*

1. Introduction

The industrial sector is one of the main sectors driving the regional economy. Therefore, the plan to develop industrial estates is one of the main focuses in the preparation of spatial plans both at the national, provincial and regency/city levels. While it is discussed in the national spatial plan, the industrial sector plan is also regulated in the National Industrial Development Master Plan (RIPIN) for 2015-2035. In the RIPIN, one of the most important points of the mission, targets, policies and industrial development strategies is to develop the central areas of industrial growth, industrial estates, and small and medium industrial centers [1]. This encourages the government both at the provincial and regency/city levels to develop industrial development plans in their respective regions.

In East Java Province, the Provincial Industrial Development Plan is regulated in the provincial regulation draft on the East Java Provincial Industrial Development Plan (RPIP) for 2018-2035. In the regulation, it was explained that the focus of the RPIP was on the East Java industrial development area. Furthermore, the plan explained that industrial development in East Java is divided into 4 stages, the first stage is to build synergy between industrial development areas and between sectors. Where the purpose of this synergy is to improve production efficiency and to create environmentally friendly industries. The second stage is to build synergy between the scale of production to strengthen the competitiveness of the industries. The third stage is industrial development directed at expanding market share, especially the expansion of exports and the fourth stage is achieving the vision of industrialization which is a world-quality industrial province [2] [3].

At present, industries in East Java Province are growing faster than national industries [2]. Industrial growth in East Java has implications for improving the economy in East Java. Based on GRDP data in 2017, it can be seen clearly that the industrial sector is the main sector contributing the largest GRDP in East Java Province amounting to 586,258.6 billion rupiahs, followed by the trade sector of 367,185.2 billion rupiahs, and the construction sector of 197,699 billion rupiahs [4].

Based on the development and contribution of the industrial sector in East Java Province, as well as the vision of the East Java Government who wants to realize the development of environmentally friendly industrial estates [5], an analysis is needed regarding the suitability of the environment-based industrial area in East Java Province. The result of this study is expected to provide input for the East Java Provincial Government in preparing their spatial plan revision, especially for the industrial sector in East Java Province. In spatial planning, the formulated and and agreed strategies are run by institutions in the government. Any decision made by them are influencing the change of spatial distribution of activities. Environmental services are important elements in land use management, especially in determining the future change of land use for development purpose [6].

The ecosystems can provide a range of direct and indirect services for people and other living organisms. Moreover, climate change and human activity can influence these services, particularly land-use changes in human activity. The influence of climate variations and land-use change influence the ecosystem services, such as carbon sequestration, water flow regulation, and food and fiber production, at a variety of scales [7]. Recent issue about the use of ecosystem services is related to its dynamics in value. The use of a decentralized technology to monetise the gaining of knowledge about biodiversity is considered to be a practical way to enable a currency backed by biodiversity to play a role in the financial system. Taxonomists and ecological expertise would play important role in valuing this biodiversity and ecosystem. Furthermore, this would put biodiversity and ecosystems at the center of a financed system that fully reflects its value [8].

Consideration of environmental aspects in planning is important as an initial step in realizing sustainable development. In addition, consideration of environmental aspects in planning can create environmental sustainability in the development process. So far, there have been many environmental assessment techniques that can be applied, one of which is the assessment of ecosystem-based environmental services. The ecosystem services assessment is intended to know the spatial distribution and quality patterns of each ecosystem service in each ecoregion assessed by means of land cover data. Therefore, this analysis technique is considered appropriate to be included in the suitability analysis of industrial estates in East Java Province.

2. Method

The methods used in this study include the literature study, and policy review. Literature study is used to examine theories and concepts regarding ecosystem services, capabilities and land suitability, while policy review is used to review the spatial documents of East Java Province for the need to determine the location of industries in East Java. In addition, policy review is used to implement technical guidelines related to analysis techniques for ecosystem services and techniques for analyzing physical and environmental aspects in spatial planning.

The analysis technique carried out in this study is divided into several stages of analysis in accordance with the objectives to be achieved. First is land capability analysis, it is carried out by overlaying several land capability indicators obtained from technical guidelines for analyzing physical and environmental aspects in spatial planning regulated in Minister of Public Works Regulation No. 20/PRT/M/2007, this analysis obtained a map of land capability in East Java Province. Second is the analysis of ecosystem services for industrial estates, this analysis is carried out through several stages, namely the determination of the index value of ecosystem services obtained from expert assessment, where the

IOP Conf. Series: Earth and Environmental Science **340** (2019) 012020 doi:10.1088/1755-1315/340/1/012020

ecosystem services themselves are formed from overlaying ecoregion and land cover maps in East Java Province [9]. The overlay results expert judgment to provide an index of environmental services for the 23 ecosystem services. After that, several ecosystem services were selected from the 23 ecosystem services to determine the value of industrial estate ecosystem services through Focus Group Discussion (FGD) by relevant government agencies and academics. After obtaining several ecosystem services for industrial estates, the overlay technique was carried out to produce a map of the value of ecosystem services for industrial estates.

The final stage of this research is suitability analysis of industrial areas in based on ecosystem services. In this stage, overlay analysis of land capability maps and ecosystem service value map for industrial estates is carried out. The result of this analysis is land suitability map for industrial zones based on ecosystem services. The analysis stages of this study can be explained through the following diagram:

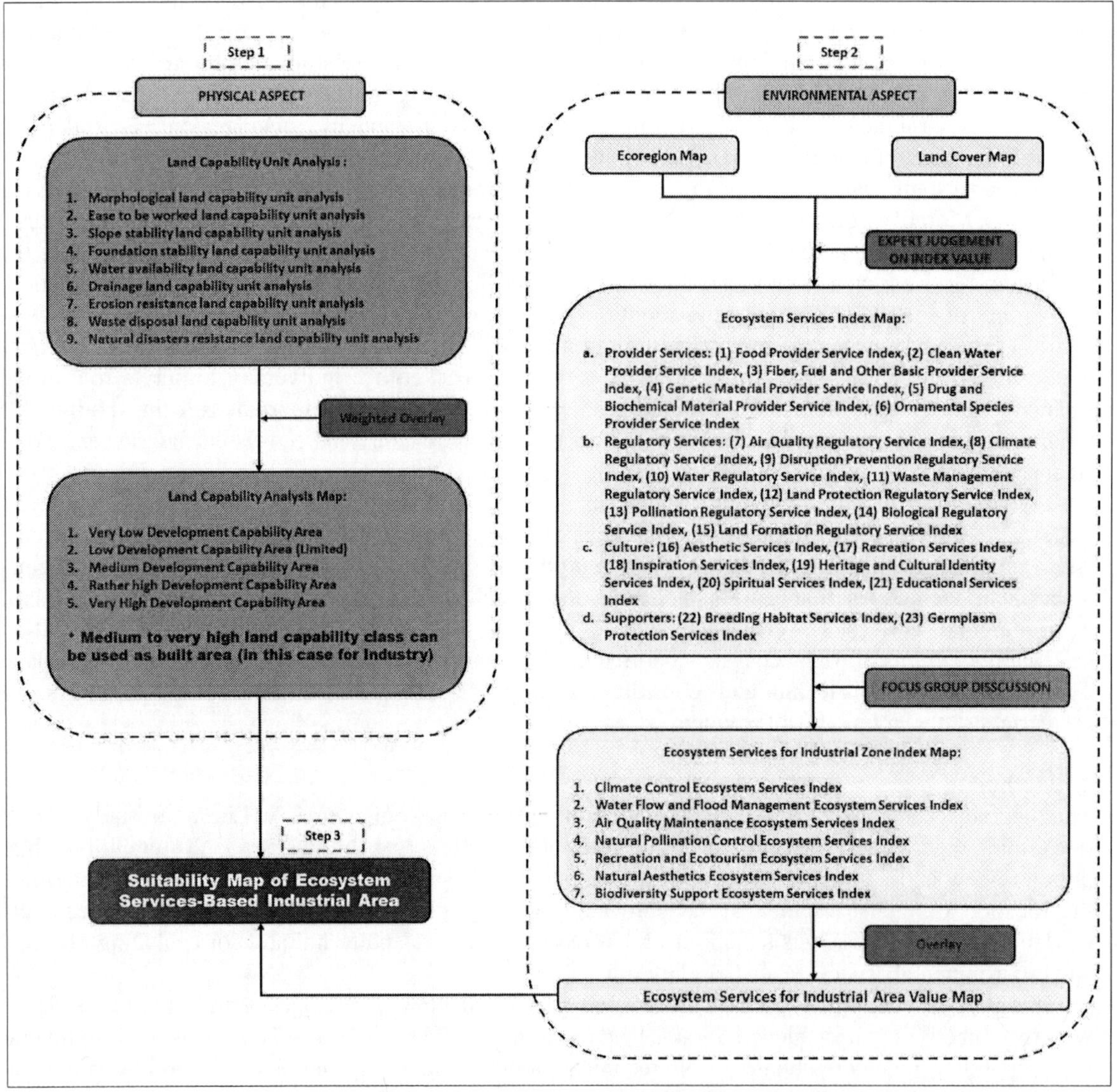

Figure 1. Research analysis steps
Source: Analysis, 2018

3. Result

3.1. Land capability analysis in East Java Province

Analysis of land capability is used to determine the value of land capability of a region. The indicators used in this analysis are mofological land capability units, ease of work, slope stability, foundation stability, water availability, drainage, erosion, waste disposal, and natural disasters. Each unit of land capability has its own land physical parameters within each unit analysis, and each unit of land capability has its own weight influence in influencing the results of the land capability analysis. The physical parameters used for each unit of land capability and the influence weight of each unit of land capability used are based on technical guidelines for analyzing physical and environmental aspects in the preparation of spatial plan issued by Minister of Public Works Regulation No.20/PRT/M/2007. The result of the conducted analysis can be explained as follows.

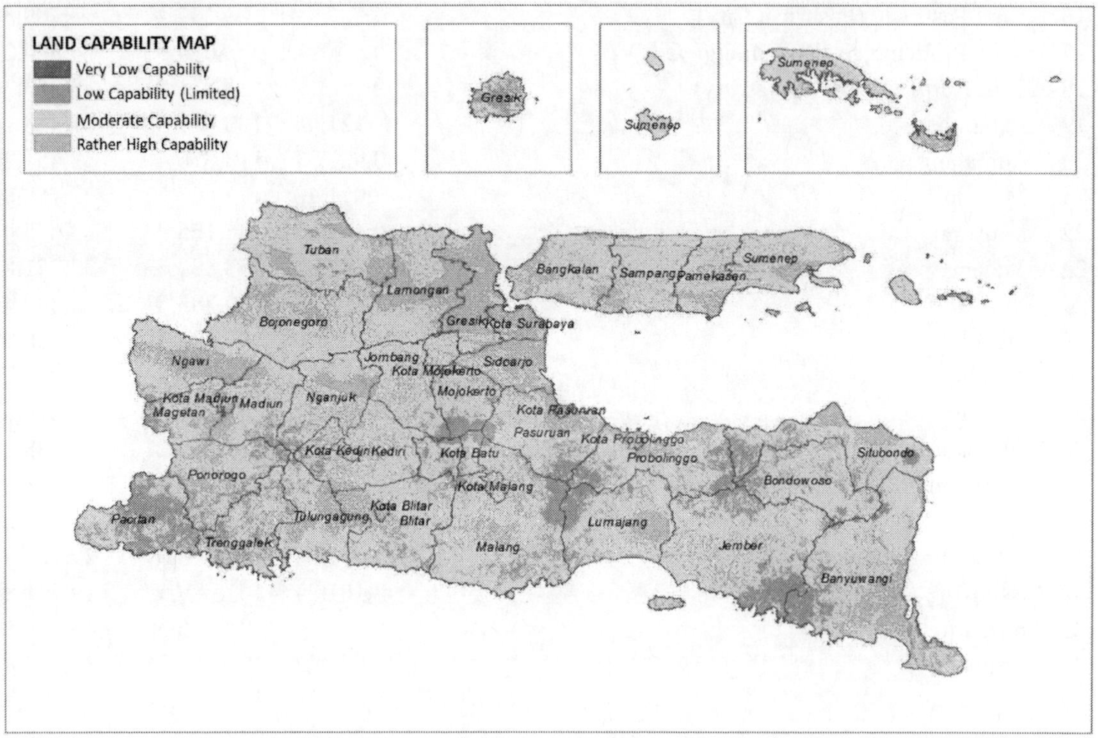

Figure 2. Land capability analysis of East Java Province
Source: Analysis, 2018

Table 1. Land capability of each Regency/City in East Java Province

No.	Regency/City	Land Capability (Ha)			
		Very Low	Low (Limited)	Moderate	Rather High
1	Bangkalan	0.52	1,716.29	94,188.00	36,171.74
2	Banyuwangi	6.45	52,837.06	240,864.72	73,260.29
3	Blitar	5.08	16,826.79	136,330.30	25,681.40
4	Bojonegoro	-	730.30	182,005.67	52,349.81
5	Bondowoso	-	27,687.13	118,784.44	11,848.65
6	Gresik	9.12	-	33,621.64	83,987.60
7	Jember	24.93	71,049.97	237,564.37	28,985.31

No.	Regency/City	Land Capability (Ha)			
		Very Low	Low (Limited)	Moderate	Rather High
8	Jombang	-	5,420.27	87,034.03	20,608.08
9	Kediri	-	12,571.43	113,361.46	29,430.72
10	Kota Batu (Batu City)	-	6,596.84	12,779.58	406.83
11	Kota Blitar (Blitar City)	-	-	1,191.89	2,186.63
12	Kota Kediri (Kediri City)	-	218.67	4,475.30	2,087.18
13	Kota Madiun (Madiun City)	-	-	1,316.12	2,361.46
14	Kota Malang (Malang City)	-	39.17	6,950.18	4,335.03
15	Kota Mojokerto (Mojokerto City)	-	-	1,340.61	715.65
16	Kota Pasuruan (Pasuruan City)	-	-	232.22	3,709.88
17	Kota Probolinggo (Probolinggo City)	-	-	684.53	4,884.24
18	Kota Surabaya (Surabaya City)	-	-	5,821.51	28,263.36
19	Lamongan	-	121.44	100,417.50	77,418.16
20	Lumajang	-	30,921.57	118,976.40	33,451.20
21	Madiun	-	7,911.10	77,663.57	27,768.80
22	Magetan	-	10,165.58	48,165.93	13,695.47
23	Malang	19.37	63,196.99	257,809.43	32,351.83
24	Mojokerto	-	15,473.59	70,283.41	14,399.48
25	Nganjuk	-	8,709.10	87,932.50	34,526.95
26	Ngawi	-	3,637.41	83,262.54	54,932.72
27	Pacitan	6.05	80,639.87	64,346.55	1,270.33
28	Pamekasan	-	4,133.40	58,657.21	17,938.24
29	Pasuruan	-	23,052.24	97,669.29	31,178.46
30	Ponorogo	-	30,048.35	92,997.43	21,612.21
31	Probolinggo	-	35,671.64	110,946.69	29,724.46
32	Sampang	-	3,101.24	101,689.53	19,901.86
33	Sidoarjo	-	-	24,346.30	49,568.19
34	Situbondo	-	34,957.47	100,161.71	33,572.80
35	Sumenep	75.20	4,578.36	155,646.14	51,196.91
36	Trenggalek	47.47	34,860.84	88,570.15	3,978.04
37	Tuban	-	2,795.71	140,897.78	56,514.11
38	Tulungagung	2.38	9,243.74	74,879.04	32,628.62

Source: Analysis, 2018

Based on the analysis, it can be explained that in general the ability of land in East Java Province is very supportive to be used as a built-up area. From the total land area in East Java, 87.73% or about 4,282,768 hectares have moderate land capability - rather high, which means that the land capability class can be used as a built area, while the remaining 12.27% has very low land capability classes - low (limited), which means the land capability class is very suitable for non-built land or protected areas.

In more detail, the results of the analysis of land capabilities for each region (regency/city) in East Java show that almost all regencies/cities in East Java have a land capabilities class that can be used as built-in areas. For example, the Blitar City, Mojokerto City, Pasuruan City, Probolinggo City, Surabaya City, and Sidoarjo Regency, 100% of the area has a moderate-to-rather high land capability class that is suitable to be developed as a built area. Whereas for regencies/cities that have land capability classes very low - limited (limited) are, Pacitan Regency with a proportion of 55.14% of the total area, Batu

IOP Conf. Series: Earth and Environmental Science **340** (2019) 012020 doi:10.1088/1755-1315/340/1/012020

City with a proportion of 33.35% of the total area, and Trenggalek Regency with proportion of 27.39% of the total area. This means they are not suitable to be used as a built area. Spatially the distribution of land capabilities in East Java can be explained in Figure 1.

3.2. Ecosystem services value analysis of industrial area in East Java Province

To map ecosystem services in an ecoregion, a method of valuing ecosystem services is carried out using proxy landuse. Based on the land cover map, the ecosystem service index of each class the total ecosystem service index can be calculated and (Mashita, 2012). Moreover, the pattern of spatial distribution and quality of each ecosystem service in each ecoregion discussed by means of land cover.

In this study, the process of analyzing the value of ecosystem services for industrial area is carried out through several stages, first is to assess each ecosystem service through expert judgment, from the expert's assessment spatial distribution and quality pattern of each ecosystem service can be mapped through the ecosystem services index. the second step is to select several ecosystem services that affect the carrying capacity and capacity of the industrial area. In this stage, FGDs are conducted with the government and academics to determine the carrying capacity of the capacity of the ecosystem service-based industrial area. Based on the results of the FGD, there are 7 ecosystem services that have influence on industrial areas, which are: (1) climate controlling ecosystem services, (2) water flow and flood management ecosystem services, (3) air quality maintenance ecosystem services, (4) natural pollination controlling ecosystem services; (5) recreation and ecotourism ecosystem services; (6) natural aesthetics ecosystem services; (7) biodiversity-supporting ecosystem services.

After preparing data on 7 ecosystem services that affect the carrying capacity and capacity of the industrial area, then an overlay analysis is conducted to obtain the overall value of ecosystem services for the industrial area. The results of the overlay analysis for the value of industrial area ecosystem services can be explained as follows.

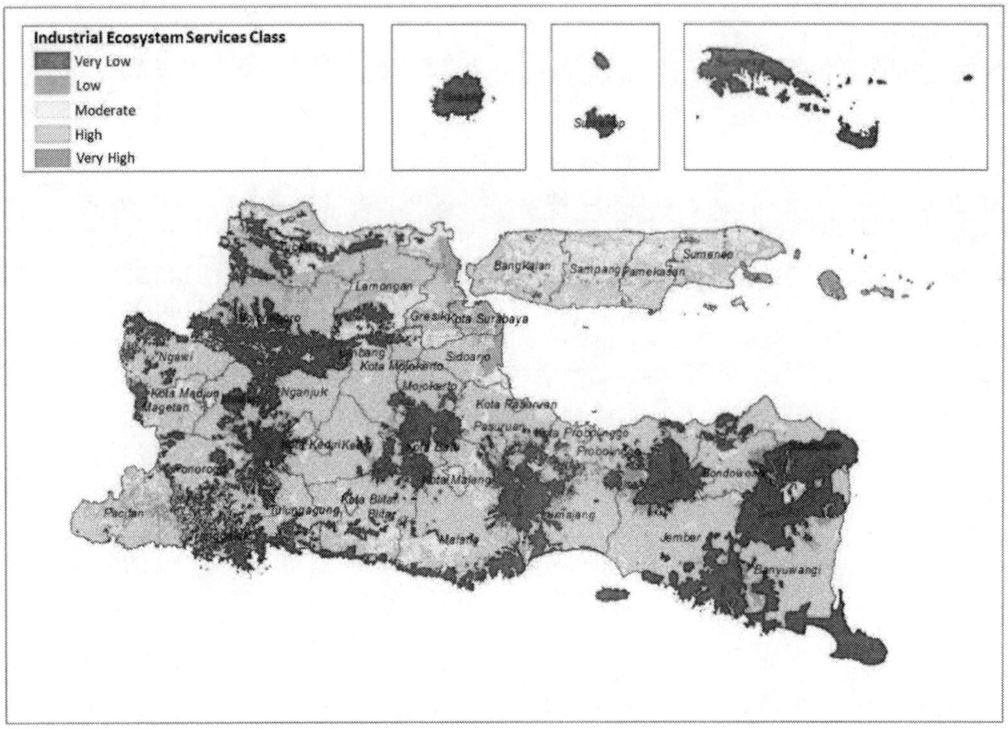

Figure 3. Map of *ecosystem services for industrial estates* in East Java Province
Source: Analysis, 2018

Table 2. Industrial ecosystem services distribution of each Regency/City in East Java Province

No.	Regency/City	Industry Area Ecosystem Services Class (Ha)				
		Very Low	**Low**	**Moderate**	**High**	**Very High**
1	Bangkalan	3,390.79	10,205.74	70,609.75	47,545.36	255.20
2	Banyuwangi	179,571.47	571.01	29,105.72	137,962.03	19,716.75
3	Blitar	37,252.07	3,980.71	54,010.71	83,026.47	568.08
4	Bojonegoro	97,546.97	21,465.58	20,087.65	95,887.72	50.30
5	Bondowoso	63,003.65	134.25	23,239.78	63,807.69	8,134.85
6	Gresik	21,646.36	21,177.93	44,487.95	39,748.01	-
7	Jember	123,537.41	1,229.72	46,252.23	154,692.91	11,800.32
8	Jombang	25,730.41	714.55	23,090.37	63,349.33	177.72
9	Kediri	23,109.91	-	36,749.06	93,865.57	1,639.08
10	Kota Batu (Batu City)	10,900.17	-	2,796.52	6,041.63	44.93
11	Kota Blitar (Blitar City)	-	-	2,002.28	1,376.24	-
12	Kota Kediri (Kediri City)	301.51	-	3,163.00	3,316.64	-
13	Kota Madiun (Madiun City)	-	-	2,373.64	1,303.93	-
14	Kota Malang (Malang City)	-	172.29	8,189.50	2,962.60	-
15	Kota Mojokerto (Mojokerto City)	-	-	1,189.54	866.71	-
16	Kota Pasuruan (Pasuruan City)	6.94	1,555.77	1,248.52	1,130.87	-
17	Kota Probolinggo (Probolinggo City)	-	-	2,963.14	2,604.89	-
18	Kota Surabaya (Surabaya City)	45.56	20,554.46	10,908.83	2,562.22	-
19	Lamongan	32,204.87	13,267.30	55,585.77	76,801.85	95.71
20	Lumajang	59,987.41	1,263.21	20,674.34	93,712.47	7,708.90
21	Madiun	47,614.17	339.29	17,870.04	46,998.62	521.36
22	Magetan	7,604.08	228.69	26,414.24	37,086.76	251.14
23	Malang	111,631.63	5,623.76	82,461.70	146,239.21	7,390.00
24	Mojokerto	28,011.89	1,371.12	21,008.26	49,205.51	559.68
25	Nganjuk	51,302.12	265.90	21,377.90	58,187.09	35.55
26	Ngawi	45,960.23	3,108.50	29,284.51	62,919.44	33.48
27	Pacitan	2,341.20	15,830.79	28,528.30	82,371.40	14,779.88
28	Pamekasan	1,042.52	11,539.19	54,246.83	13,830.24	-
29	Pasuruan	31,821.08	2,042.91	34,361.48	79,508.37	4,166.16
30	Ponorogo	50,046.25	4,206.39	36,771.11	52,099.49	1,092.20
31	Probolinggo	58,403.57	284.33	24,774.38	82,096.04	10,784.47
32	Sampang	792.43	10,704.65	75,064.78	37,959.04	119.30
33	Sidoarjo	302.75	15,040.09	27,550.01	31,021.65	-
34	Situbondo	84,833.15	315.15	17,663.97	59,227.19	6,652.40
35	Sumenep	77,168.15	26,104.29	68,609.10	38,473.35	120.68
36	Trenggalek	63,188.77	12,027.68	19,704.07	31,018.12	1,470.42
37	Tuban	56,733.85	11,386.96	64,902.06	67,002.60	-
38	Tulungagung	41,149.05	3,540.31	26,997.60	44,631.76	431.48

Source: Analysis, 2018

Based on the analysis, it can be explained that in general the condition of ecosystem services for industrial area in East Java is distributed in each class starting from the very low level of ecosystem services - low, medium, and high - very high. Ecosystem services class very low - low has a proportion of 33.94% of the total area of East Java or around 1,438,182 hectares, moderate ecosystem services class has a proportion of 23.26% of the total area of East Java or about 1,136,318 hectares, and high ecosystem services class - very high has a proportion of 42.80% of the total area of East Java or around 2,091,041 hectares.

Based on the analysis, it can be explained that in general the condition of ecosystem services for industrial area in East Java is distributed in each class starting from the very low level of ecosystem services - low, medium, and high - very high. Ecosystem services class very low - low has a proportion of 33.94% of the total area of East Java or around 1,438,182 hectares, moderate ecosystem services class has a proportion of 23.26% of the total area of East Java or about 1,136,318 hectares, and high ecosystem services class - very high has a proportion of 42.80% of the total area of East Java or around 2,091,041 hectares.

More detail, regencies/cities that have a very low - low ecosystem service class with a large proportion, among others are the Surabaya City with a proportion of 60.46% of the total area, Trenggalek Regency with 59.03% of the total area, and Bojonegoro Regency with 50.63% of the total area. Whereas for regencies/cities that have high - very high ecosystem services with a significant proportion, among others are Pacitan Regency with a proportion of 67.54% of the total area, Kediri Regency with a proportion of 61.47% of the total area, and Jombang Regency with proportions 56.19% of the total area. Spatially, the distribution of industrial area ecosystem services in East Java can be explained in Figure 3.

3.3. Ecosystem services-based industrial area suitability analysis in East Java Province

In this study, industrial area suitability analysis is obtained from the results of overlaying land capability map and ecosystem service value map for industrial area in East Java Province. Technically, the two maps will be given weight for each class. For land capability maps, each class starts from very low - very high, will be given influence weight 1 - 5, while for ecosystem services maps in industrial area will be given a weight of 5 - 1 for each class of ecosystem services ranging from very low classes - very high. For ecosystem services map, the assumption of the highest (score 1) to lowest (score 5) weight is given because regions with high levels of ecosystem services have high environmental service value, so that the utilization is more suitable for the protection of the area not to be used as a built-up area, especially as industrial area. So, the higher the value of ecosystem services in an area the smaller the influence to be used as a built-up area. The results of the suitability analysis based on ecosystem services in East Java can be explained as follows.

Table 3. Distribution of industrial area suitability in each Regency/City in East Java Province

No.	Regency/City	Industry Area Ecosystem Services Class (Ha)				
		Very Inappropriate	Not Suitable	Moderate Suitability	Suitable	Very Suitable
1	Bangkalan	106.88	4,213.58	105,783.51	21,902.87	-
2	Banyuwangi	37,338.73	170,083.97	143,978.41	15,525.86	-
3	Blitar	5,111.47	40,358.31	119,369.91	13,998.36	-
4	Bojonegoro	270.78	97,474.03	111,177.10	26,116.29	-
5	Bondowoso	14,780.92	63,818.74	76,424.48	3,296.08	-
6	Gresik	9,357.37	11,776.78	54,630.83	51,023.15	272.11
7	Jember	53,062.44	93,299.28	175,564.76	15,586.13	-
8	Jombang	3,695.95	23,718.81	74,123.83	11,523.79	-
9	Kediri	5,334.57	24,767.56	110,276.23	14,985.26	-
10	Kota Batu (Batu City)	5,534.19	6,274.86	7,757.91	216.28	-

No.	Regency/City	Industry Area Ecosystem Services Class (Ha)				
		Very Inappropriate	Not Suitable	Moderate Suitability	Suitable	Very Suitable
11	Kota Blitar (Blitar City)	-	-	2,032.24	1,346.27	-
12	Kota Kediri (Kediri City)	149.05	204.29	4,794.73	1,633.09	-
13	Kota Madiun (Madiun City)	-	-	1,864.04	1,813.54	-
14	Kota Malang (Malang City)	-	0.51	7,376.34	3,947.54	-
15	Kota Mojokerto (Mojokerto City)	-	-	1,695.59	360.67	-
16	Kota Pasuruan (Pasuruan City)	-	-	1,217.69	2,717.47	6.94
17	Kota Probolinggo (Probolinggo City)	-	-	2,866.36	2,701.67	-
18	Kota Surabaya (Surabaya City)	-	-	4,811.93	29,219.38	39.78
19	Lamongan	4.47	32,322.11	111,600.03	34,028.87	-
20	Lumajang	21,726.57	51,456.20	101,551.83	8,611.74	-
21	Madiun	1,958.72	51,317.10	49,328.27	10,739.38	-
22	Magetan	3,033.07	7,986.41	52,782.02	7,783.40	-
23	Malang	36,513.24	100,238.40	198,368.06	18,226.61	-
24	Mojokerto	13,419.44	16,789.04	62,011.66	7,936.34	-
25	Nganjuk	4,871.13	49,855.23	64,520.09	11,922.11	-
26	Ngawi	1,880.63	45,433.86	76,813.40	17,178.27	-
27	Pacitan	579.00	62,426.66	71,182.11	9,663.79	-
28	Pamekasan	16.93	1,216.05	59,281.77	20,144.02	-
29	Pasuruan	11,403.22	32,319.53	94,785.20	13,369.91	22.13
30	Ponorogo	13,308.69	46,610.29	70,188.06	14,108.41	-
31	Probolinggo	19,551.78	59,110.01	89,487.15	8,193.85	-
32	Sampang	105.64	1,941.46	103,673.83	18,829.43	89.85
33	Sidoarjo	-	-	38,204.35	35,476.26	233.89
34	Situbondo	19,289.55	83,707.83	58,492.27	7,202.21	-
35	Sumenep	780.41	74,289.61	93,925.16	40,402.61	1,077.77
36	Trenggalek	13,101.03	62,553.97	43,992.01	7,762.06	-
37	Tuban	243.30	56,863.55	119,831.73	23,086.90	-
38	Tulungagung	3,898.19	40,968.06	54,122.41	17,761.55	-

Source: Analysis, 2018

Based on the analysis, it can be explained that in general the suitability of ecosystem service-based industrial area in East Java is dominated by the moderate class with a proportion of 53.62% or 2,619,887 hectares, meaning that this suitability class can still be used as an industrial area with certain amount of engineering techniques and accompanied by a study of environmental impacts. Industrial area suitability with very inappropriate - not suitable class have a proportion of 35.08% of the total area of East Java. This can be interpreted that the land suitability class is not suitable to be used as an industrial area because it has many physical boundaries such as high slope, low soil structure, susceptibility to disasters, and others, and the area in this suitability class has a high value of ecosystem services that must be protected. Industrial area suitability with suitable - very suitable class have a proportion of 11.30% of the total area of East Java or around 552,083 hectares. This means that the land suitability class has low physical and environmental services limitations to be developed as industrial area.

In more detail, the results of the suitability analysis of ecosystem service-based industrial land that has the appropriate proportion of industrial area suitability classes - very suitable is Surabaya City with 85.88% of the total area, Pasuruan City with 69.11%, of the total area, Madiun City 49.31% of the total area, Probolinggo City 48.52% of the total area, Sidoarjo Regency 48.31% of the total area, Gresik Regency 40.37% of the total area, Blitar City 39.85% of the total area, and Malang City 34.86% of the total area. Whereas for regencies/cities dominated by very low - low industrial land suitability, among others are Situbondo Regency with proportion 61.06% of the total area, Batu City with proportion 59.69% of the total area, and Trenggalek Regency with proportion 59.38 % of the total area. Spatial distribution of the ecosystem service based industrial land suitability class for each Regency/City in East Java can be explained through Figure 4.

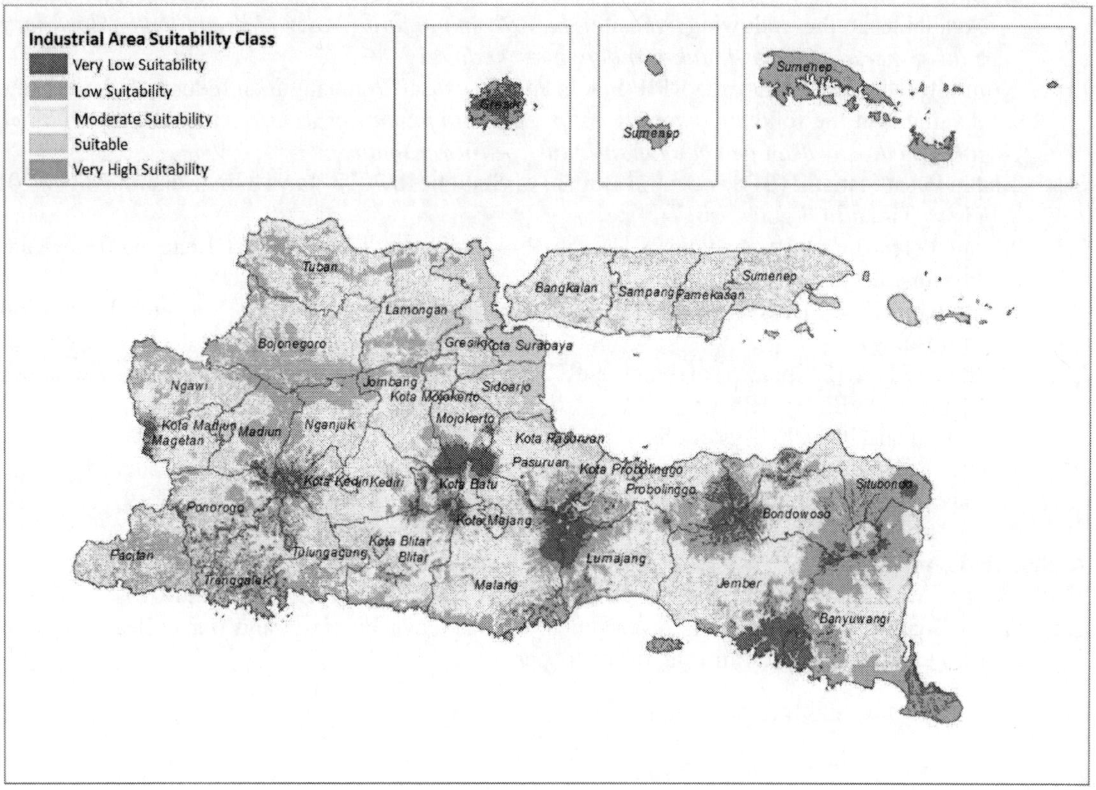

Figure 4. Suitability Map of Ecosystem Services-Based Industrial Land in East Java Province
Source: Analysis, 2018

4. Conclusion

In principle, many methods and approaches can be used in determining the suitability of industrial area, one of which is the ecosystem services approach. The suitability of industrial services based on ecosystem services is considered very good to be used in determining the suitability of industrial land because with this approach we can examine environmental aspects both physically and based on ecosystem services deeper in determining the industrial area.

Based on the results of the suitability analysis of ecosystem service-based industries in East Java Province, the result is that almost all regencies/cities in East Java Province have land suitability potential to be developed as industrial area. Spatially, it can be explained that almost all East Java Province have industrial area suitability with moderate class suitability. Furthermore, for high to very high industrial land suitability classes tend to be in the North East Java Region such as Gresik Regency, Surabaya City, Sidoarjo Regency, Pasuruan City and Probolinggo City. In addition, several other regencies/cities which are also dominated by high industrial land suitability classes to very high levels are Madiun City, Blitar City and Malang City.

Finally, the results of this study are expected to be input for the Government in determining the industrial area in East Java Province. In addition, it is expected that a study of economic, infrastructure, social and cultural aspects can be collaborated with the results of this study in order to obtain a more accurate suitability industrial area determination.

5. References

[1] Pusat Komunikasi Publik Kementerian Perindustrian. 2015. Rencana Induk Pembangunan Industri Nasional 2015-2035

[2] Kominfo Jatim. 2018. DPRD Jatim Sahkan Perda Rencana Induk Pembangunan Industri. Available at the following website: *http://kominfo.jatimprov.go.id/read/umum/dprd-jatim-sahkan-perda-rencana-induk-pembangunan-industri*

[3] Kominfo Jatim. 2018. Raperda RPIP Fokus Pada Kawasan Pembangunan Industri di Jawa Timur. Available at the following website: *http://kominfo.jatimprov.go.id/read/umum/raperda-rpip-fokus-pada-kawasan-pembangunan-industri-di-jawa-timur*

[4] Badan Pusat Statistik (BPS – Central Bureau of Statictics). 2017. Jawa Timur dalam Angka 2017 (Jawa Timur in Figures, 2017)

[5] Badan Perencanaan Pembangunan Daerah Provinsi Jawa Timur. 2011. Rencana Tata Ruang Provinsi Jawa Timur Tahun 2011-2031

[6] S. Ronchi. 2018. Ecosystem Services for Spatial Planning. (Cham: Springer International Publishing)

[7] J. Zhan, Ed. 2015. Impacts of Land-use Change on Ecosystem Services. (Berlin, Heidelberg: Springer Berlin Heidelberg)

[8] S. Muddiman. 2019. Ecosystem Services. (Cham: Springer International Publishing)

[9] Deputi Bidang Tata Lingkungan Kementerian Lingkungan Hidup. 2014. Pedoman Penentuan Daya Dukung dan Daya Tampung Lingkungan Hidup

Acknowledgements

This research was supported by Institut Teknologi Sepuluh Nopember. We would like to say our gratitude to Board for Planning and Development of East Java Province and our colleagues for the support during preparation and finishing of this paper.

The impact of fiscal decentralization on economic performance in Indonesia

F Setiawan[1] and A F Aritenang[1]

[1] Urban and Regional Planning Department, Institut Teknologi Bandung, 40132, Indonesia

Email: setyawanfreda@gmail.com

Abstract. Regional studies literature argues that neighboring regions could impact each others' economic performance trough trade, technology diffusion, capital flow, and political stability. However, the emergence of decentralization has put the above argument into a new context. This study aims to explain the impact of fiscal decentralization on economic. The study analyse the impact of fiscal decentralization using lag value and ignores the existence of spatial dependency between regions. Our study concludes that the fiscal decentralization has a significant effect on economic performance at a lag value of 3 years, implying that public budgeting will have a significant impact on increasing economic performance on 3 years later. In addition, the paper provides evidence that regions with similar economic performance are located nearby suggesting the presence of spatial dependence.

1. Introduction

In general, decentralization is the transfer of power or authority from the central government to local governments [1,2,3,4]. One of the important dimensions of decentralization for local governments is fiscal decentralization. This policy is related to the submission of fiscal authority including revenue and expenditure of local governments [5,6,7]. The implementation of this policy is expected to increase the allocation and efficiency of the provision of public goods [2,8]. This efficiency can drive local and national economic growth [4].

The view of classical fiscal federalist theory does not explicitly explain the relationship between fiscal decentralization and economic performance [7]. However, research on this field has become very relevant after large-scale decentralization in socialist countries such as the Soviet Union, China, Eastern Europe in the 1980-1990s. As studies found inconlcusive findings on the relations that theoretical and empirical justification is required to examine the relationship between decentralization and economic performance [6,7,9,10].

Studies at both regional and state perspectives show contradictory results. For instance, Xie and Zou [9] and Stansel [11] found that fiscal decentralization promotes economic growth. The paper by Zhang and Zou [9] answers this by showing that fiscal decentralization could effect to economic performance through the efficiency of resources allocation in the public sector. In addition, since the local government as a key actor in providing public goods and services to the community, this has made it increasingly important to know the impact of fiscal decentralization on the economic, social and political aspects [7]. Meanwhile, other studies negative and insignifcant effects [7, 12,13].

So far the research that analyzes the impact of fiscal decentralization still ignores the impact of other regions. The paper by Oates recognized that the assumption of decentralization ignored the potential impact of cross-regional public goods supply [8]. Another assumption was no mobility of cross-regional economic units, implying that individuals do not move between regions. However, there should still be a possibility for someone to move locations. This may influence the selection of fiscal variables in this study. In addition, Ramirez and Loboguerrero argue that in the literature on economic growth still ignores spatial dependence between regions [14]. The existence of spillover effects between regions is important for economic growth. Thus, spatial linkages cannot be ignored in the analysis of regional economic performance because modeling can be biased.

In the Indonesian context, there a are several studies that examine the empirical relationships between fiscal decentralization and economic performance [15,16,17,18,19] and the study on spatial dependences of economic performances [20,21,22,23,24,25]. However, these studies ignored the potential impact of spatial dependence on fiscal decentralization of neighboring regions. Hence, this study attempts to examine to which extent the spatial dependence impact of fiscal decentralization on economic performance in districts in Indonesia. The paper uses econometricts analysis and compares traditional OLS and spatial models.

The paper is organized as follows; the following section discuss the data and research methodologies and followed by the analysis results and discussion section. The last section provides research conclusion and policy implications.

2. Data and research methods

This study uses a quantitative approach using secondary data. These data include locally-generated revenue (PAD), general allocation funds (DAU), special allocation funds (DAK), revenue sharing funds (DBH), capital expenditure, percentage of people in poverty. The period of these data from 2008 until 2011. These data were obtained from the websites of Ministry of Finance of (MoF), the Indodapoer (Indonesia Database for Policy and Economic Research), and the Central Bureau of Statistic (*Badan Pusat Statistik*/BPS).

The method of data analysis uses spatial regression by weighting in the form of a matrix that describes the closeness of the relationship between observations or known as spatial weight matrix. In his seminal book on spatial regression, Anselin (2006) explains that spatial regression analysis is related to specifications, estimates and examinations of regression models that combine spatial effects [26]. According to Matthews (2006) the existence of spatial dependencies has 2 (two) important implications [27]. First, spatial autocorrelation can be disruptive. Second, if it is intended to find spatial interaction, the spillover nature of spatial dependencies can be considered as a substantive part. Therefore, ignoring spatial correlation can cause estimates to be inefficient [27,28]. In the analysis, the data is processed every year so that it is cross section.

The spatial weighting into 3 (three) types, namely intersection, distance, general. In the analysis of Spatial Lag Model (SLM) and Spatial Error Model (SEM) it only works for spatial weights that correspond to continuity relationships symmetrically [26]. In other words, the spatial weighting that can be used is rook, queen, bishop, distance band contiguity but not for k-nearest neighbors. This study uses a spatial weighting matrix based on distance. This is because observations in Indonesia consist of several scattered islands so that there are limitations when using contiguity.

Spatial regression analysis is related to specifications, estimates and examinations of regression models that combine spatial effects. These effects are affected by measurements at other locations. This causes spatial data to be less suitable for the smallest regression analysis. First, Spatial Lag Model (SLM) is a spatial regression model whose dependent variable is spatial correlation, meaning that this model has a dependency between one observation in an area with other observations in its neighboring region [29]. SLM does not only discuss the effect of uncountable independent variables but also considers the value of the dependent variable on other objects [27]. Second, the Spatial Error Model (SEM) shows the relationship of errors. The spatial error model evaluates the extent to which grouping of outcome variables is not explained by measurable independent variables. This model captures the influence of independent variables that are not measurable or errors [29].

IOP Conf. Series: Earth and Environmental Science **340** (2019) 012021 doi:10.1088/1755-1315/340/1/012021

3. Result and discussions

This section provides the research analysis of relationship between fiscal decentralization to economic performance, spatial dependency model and a discussion of the results. The analysis is displayed in the following table (Table 1). The Locally-generated revenue (*Pendapatan Asli Daerah*/PAD) is a regional income derived from the ability of the region to multiply its potential. The sources of PAD consist of local taxes, regional retributions, the results of the separated management of regional assets, and other legitimate sources. While General Allocation Fund (*Dana Alokasi Umum*/DAU) is a balancing fund to cover regional financial capacity inequality measured by basic allocation and fiscal gap. In principle, the regions with the largest spending and the smallest taxes will receive the largest transfer funds. Because the function of this fund is to cover the gap horizontally.

The Special Allocation Fund (Dana Alokasi Khusus/DAK) is an state revenue and expenditure budget allocated to regions to fund specific activities that are in line with national priorities. DAK is usually to finance the needs of basic service facilities and infrastructure that have not reached certain standards to encourage the acceleration of regional development. With the allocation of this balance, the areas that are still left behind can do development. This development will later be able to meet basic needs as well as the needs of supporting economic activities so that it can help increase regional fiscal capacity in order to reduce the degree of underdevelopment.

The Revenue Sharing (*Dana Bagi Hasil*/DBH) is a fund from the state revenue and expenditure budget that is distributed based on a certain percentage according to the applicable provisions. This fund consists of tax components and natural resources. While the capital expenditure is a form of investment expenditure related to procurement, improvement of regional assets. This asset procurement plays an important role in helping community activities, especially those related to economic activities.

First, the table shows evidence of spatial autocorrelation as indicated by the Moran's I test of residual (residual or error) regression. The Lagrange Multiplier test for Lag and Error also shows indications of spatial autocorrelation. The Ordinary Least Square (OLS) regression using lag value of 1 year, 2 years, 3 years. The results show that all the main components of fiscal decentralization consisting of PAD, DAU, DAK, DBH have a significant impact on alpha <0.05 with a lag value of 3 years. This finding implies that fiscal allocation in year t will have a significant impact on t + 3. The discussion of each component of fiscal decentralization is as follows.

Table 1. The Comparison Result of OLS and Spatial Regression Model

Dependent Components	Gross Domestic Regional Product (GRDP)					
	OLS			Spatial Regression		
	Lag 1	*Lag 2*	*Lag 3*	*Lag 1*	*Lag 2*	*Lag 3*
Moran'I	0.0495	0.01913	0.06908	-	-	-
Robust (LM-Lag)	0.00155	0.00041	0.05067	-	-	-
Robust (LM-Error)	0.4804	0.00029	0.48339	-	-	-
PAD	0.000026* (33.66)	0.0000335* (30.32)	0.00003* (39.26)	0.0000263* (34.61)	0.0000338* (31.26)	0.0000037* (39.88)
DAU	0.00001* (5.17)	0.000002* (2.87)	0.000016* (9.61)	0.000011* (5.69)	0.0000019* (2.86)	0.0000016* (9.16)
DAK	-0.000015*** (-1.09)	0.0000062*** (0.71)	-0.00006* (-5.54)	-0.000016*** (-1.184)	0.00000628 (0.73)	-0.000006* (-5.32)
DBH	-0.0000199** (-1.69)	0.00000004 (0.327)	0.000004* (2.67)	-0.0000002** (-1.917)	0.0000001 (0.11)	0.000004* (2.50)
BM	0.00002* (7.47)	0.000012* (4.95)	-0.000005** (-1.75)	0.0000201* (7.52)	0.000012* (4.99)	-0.0000052** (-1.65)
PR	-88073.1*	-50335.4*	-29804.5***	-76004.3*	-38485.3***	24234***

Dependent Components	Gross Domestic Regional Product (GRDP)					
	OLS			Spatial Regression		
	Lag 1	*Lag 2*	*Lag 3*	*Lag 1*	*Lag 2*	*Lag 3*
	(-3.34)	(-1.96)	(-1.23)	(-2.92)	(-1.53)	(-1.00)
Dummy Metropolitan	202427* (3.26)	298488* (4.43)	164238* (2.43)	187895* (3.09)	277565* (4.21)	159524* (2.39)
Dummy Java	137131* (2.26)	320787* (5.519)	175494 (0.27)	762617*** (1.23)	237318* (3.87)	-160851 (-0.24)
Lag Coefficient	-	-	-	0.109* (3.45)	0.127* (3.69)	0.07* (2.13)
Lambda Coefficient	-	-	-	-	-	-
R-Square	0.95	0.95	0.95	0.96	0.9519	0.95
Observation	450	450	450	450	450	450

Note: *PAD: Locally-Generated Revenue, DAU: General Allocation Fund, DAK: Special Allocation Fund, DBH: Total Revenue Sharing, BM: Capital expenditure, PR: Poverty Rate*
** significant at $\alpha < 0.05$, ** significant at $\alpha < 0.1$, *** significant at $\alpha < 0.5$*
t-statistics are reported in parentheses

First, the PAD variables tend to have a consistent impact on lag-1, lag-2, and lag-3. But when compared to the coefficient and t-statistic values, PAD contributes the most to economic performance at a lag of 3 years. This is presumably because it takes a span of time for the construction of public services in advance so that it can later move the economy. The PAD variable have a positive and significant impact on economic performance. The result suggest that PAD contributes positively as increased of local taxes and levies through the authority of taxing power. This local revenues act as an incentive instrument for development allocation. The greater the revenue size, the higher economic growth found in the region.

Another component that might be the source of PAD is public savings in banking. This deposit consists of demand deposits, time deposits, savings, and foreign exchange exchanges. The source of income is included in other legal income. This shows that the increasing public savings in the form of demand deposits, savings accounts, foreign exchange, the more legitimate other income components. The role of banking can be realized as an institution that brings together debtors and creditors. With deposits in banks can provide capital assistance so that it can move the economy.

Second the DAU shows the direction of positive and significant relationships in each testing year. However, after comparing the coefficient value, the test on the lag of 3 years contributes (in terms of coefficient value) which is greater than other year lags. This shows that at 3 years lag, DAU contribution is very important for regions with low fiscal capacity. After economic activity develops, the DAU contribution begins to decrease because the regional fiscal capacity begins to develop. This is indicated by the reduced coefficient of DAU and on t-statistics.

The DAU component has a positive effect allegedly due to budget allocations for employees that affect the level of employee consumption. Employee spending is also possible to increase regional productivity. For example, an additional amount of the teacher salary expectedly improve teachers' wellbeing and thus, improve teaching capacities that may lead to student competencies and regional productivity.

Third, DAK variable indicates the direction of an inconsistent relationship. However, the role of DAK is very significant in the 3 years lag. This shows that it plays an important role for regions with low fiscal capacity. This means that regions with low economic performance tend to receive greater DAK allocations. This DAK contribution turned out to be significant in driving the economy in the area. After an increase in fiscal capacity, the role of DAK began to decrease in significance. This can be shown in the lags of 1 year and 2 years which tend to be insignificant towards economic performance, which implies that regions are increasingly autonomously fiscal.

While the DBH variable has a significant impact on the lag of the 3rd year. In lag 1, and 2 years, this DBH contribution is not significant enough to economic performance. This means that the DBH allocation in 2008 had a significant impact on the regional economy in 2011. DBH components show a significant influence on economic improvement. This fund has a significant impact on GRDP after 3 years because this fund can be allocated for development which does not directly impact the increase in GRDP. This DBH plays a role in covering the fiscal gap vertically, but not horizontally. If an area is economically developed, it is suspected that there are regions that are still lagging behind so that the role of DBH is less significant to improve economic performance.

The capital expenditure component has a very significant role in economic performance at a lag of 1 year and 2 years. However, if the review of the variable coefficient shows a greater coefficient value at 1 year lag. The capital expenditure component plays a significant role in the 1 year lag. This means that the role of spending on lags 2 and 3 years is less significant. This is presumably because the regional fiscal capacity is still low so that the allocation for development expenditure is still low. After fiscal capacity begins to grow, the contribution of capital expenditure is increasingly significant because of the greater budget allocation for development.

This means that the capital expenditure component plays an important role in regional economic performance because it provides faster benefits. This capital expenditure can also be a form of government consumption so that it directly impacts the increase in GRDP.

3.1. Control variables: Poverty Rate, Dummy of Metropolitan, and Dummy Java

The relationship of control variables to economic performance also shows different results. Variable percentage of poor people has a significant negative relationship only at year lag. This means that regions with a high percentage of poor people can have implications for the decline in economic performance. Poor people tend to have lower levels of education and skills which have implications for regional productivity and low economic performance.

Metropolitan dummy variables tend to show a significant positive relationship. This means that regions in the metropolitan area tend to have higher economic performance. The metropolitan area has a rapid economy due to industrial concentration. This is because there are information flow factors and knowledge exchange. Economic performance is also influenced by the greater fiscal capacity of the metropolitan area because of the potential tax base.

Java dummy variable shows a positive and significant relationship at 1 year and 2 years lag on economic performance. This means that the regions in Java tend to have a higher GRDP compared to other islands. This is because the condition of Java Island is more advanced when viewed from accessibility, availability of infrastructure that supports productivity of economic activity.

4. Discussion

Taking into account the effect of spatial dependencies causes the model to be more specific. The best spatial regression model produced is Spatial Lag Model. This model shows that dependencies occur on the dependent variable, meaning that the performance of a region's GRDP is also influenced by the performance of fiscal decentralization of the surrounding area. The use of spatial regression analysis influences model specifications. In testing with a lag of 1 year, several coefficients from independent variables showed an increase such as PAD, Capital Expenditures, although this increase was not significant.

Similary with a lag of 2 years, there was a change in the coefficient on PAD, DAK, percentage of poverty. Whereas in the 3-year lag test, shows that there is a change in the PAD coefficient, Capital Expenditures. In table 1, the lag coefficient for each test shows positive and significant. This means that there is a grouping of regions with similar characteristics. An area with high GRDP is surrounded by areas with high GRDP, and vice versa. This positive relationship occurs spatial dependence meaning that an area also needs other areas to complement each other.

Another important finding is the importance of DBH as the fund potentially causes regional inequality due to differences in resources. The study from Waluyo (2007) suggest that DBH is considered only benefits resource-rich regions, following higher tax revenue sharing, and large urban districts that earns higher revenue from property tax. Allocation of revenue sharing funds for key sectors can increase regional economic growth. The greater the DBH, the greater the allocation for the development of public services. For that reason, if an area wants to receive a high DBH, it must be able to explore the potential of regional wealth.

The existence of a spatial dependency phenomenon shows that in planning it is necessary to consider the externalities generated in other regions. For example, tax policy in an area will also affect the surrounding area. Although local taxes are needed to boost PAD, the determination also considers the surrounding area. explains that the authority to collect taxes in the regions potentially leads to unfair competition between regions that may cause a poor investment climate [30]. This can later affect the regional economic performance. Therefore, Ramirez and Loboguerrero (2002) suggest collaborating that provides economic benefits across regions [14].

5. Conclusion

In general, fiscal decentralization has an impact on the economic performance of districts/cities in Indonesia. In terms of revenue, fiscal decentralization has the most significant impact on year 3 lag. This means that government investment and consumption of public services will have an impact on economic improvement in the span of 3 years after its establishment. This implies that investment in financing is needed in order to improve economic performance in order to encourage the creation of fiscal independence.

This study also shows that there are spatial dependencies in the relationship of fiscal decentralization to economic performance. This provides a lesson that analysis with panel data is not enough to analyze economic performance due to the phenomenon of spatial dependence. The results of this study support the statement of Oates (2006) which reveals an effect of fiscal decentralization on neighboring area [8]. This provides an important lesson that development in an area also has implications for adjacent areas.

The implication of the finding is that that the regional autonomy policy has changed the paradigm of local governments to act more inward looking [2,31]. This is a challenge when dealing with cross-regional development that has spatial dependency, thus development need to be carried out in an integrated manner to have more effective impacts through cooperation between regions. Accordingly, decentralization could lead to the need to work together or establish partnerships with neighboring cities/districts [2] and shaping metropolitan industrial landscapes that reflect the socio-political and economic dynamics at national levels [32].

6. References

[1] Hammond, G W. Tosun, M S. (2009). *The Impact of Local Decentralization on Economic Growth: Evidence from U.S Countries*. IZA: Germany

[2] Firman, T. (2014). *Inter-local-government partnership for urban management in decentralizing Indonesia: from below or above? Karmantul (Greater Yogyakarta) and Jabodetabek (Greater Jakarta) compared*. Space and Polity. Vol. 18, No. 3, 215-232

[3] Devkota. 2014. *Impact of Fiscal Decentralization on Economic Growth in the Districts of Nepal*. International Center for Public Policy. Andrew Young School of Policy Studies. USA

[4] Martinez-Vazquez, et.al. (2015). *The Impact of Fiscal Decentralization: A Survey*. Governance and Economic research Network (GEN).

[5] Zhang, Tao and Zou, Heng-fu. (1998). *Fiscal Decentralization, Public Spending, and Economic Growth in China*. Journal of Public Economics.

[6] Schneider, Aaron. (2003). *Who gets what from whom? the impact of decentralisation on tax capacity and pro-poor policy*. IDS Working Paper 179

[7] Yushkov, Andrey. (2015). *Fiscal decentralization and regional economic growth: theory, empirics, and the russian experience*. Russian Journal of Economics 404-418.

[8] Oates, Wallace E. (2006). *On the Theory and Practice of Fiscal Decentralization*. IFIR Working Papers. College Park: University of Maryland.

[9] Xie, D and Zou, Heng-fu. (1999). *Fiscal Decentralization and Economic Growth in the United States*. Journal of Urban Economics 45, 228-239.

[10] Martinez-Vazquez. (2011). *The Impact of Fiscal Decentralization: Issues in Theory and Challenges in Practice*. Phillipines: Asian Development Bank.

[11] Stansel, Dean. (2005). *Local Decentralization and Local Economic Growth: A cross-sectional examination of US metropolitan areas*. Journal of Urban Economics 55-72

[12] Rodriguez-Pose, Andres & Ezcurra, Roberto. (2010). *Is fiscal decentralization harmful for economic growth? Evidence from the OECD countries*. Journal of Economic Geography No.11, 619-643.

[13] Thiessen, Ulrich. (2001). *Fiscal Decentralization & Economic Growth in High-Income OECD Countries*. European Network of Economic Policy Research Institutes.

[14] Ramirez, M. T. and Loboguerrero, A.M. (2002). *Spatial Dependence and Economic Growth: Evidence from a Panel of Countries*

[15] Booth, A, 2005, *The Evolving Role of the Central Government in Economic Planning and Policy Making in Indonesia*. Bulletin of Indonesian Economic Studies, 41 (2). pp. 197-219.

[16] Brodjonegoro, B., 2003, *The Indonesian Decentralization after law revision: Toward a better future?*, Department of Economics, University of Indonesia.

[17] Brodjonegoro, B., 2009. *Fiscal Decentralizatioj and Its Impact on Regional Economic Development and Fiscal Sustainability*. In: C. J. Holtzappel & M. Ramstedt, eds. Decentralization and Regional Autonomy in Indonesia: Implementation and Challenges. Singapore: ISEAS, pp. 196-221.

[18] Dartanto, T. & Brodjonegoro, B. P., 2003. *Dampak Desentralisasi Fiskal di Indonesia Terhadap Pertumbuhan Ekonomi dan Disparitas Antar Daerah: Analisa Model Makro Ekonometrik Simultan*. Jurnal Ekonomi dan Pembangunan Indonesia, 4(1), pp. 17-38.

[19] McCulloch, N. & Sjahrir, B. S., 2008. *Endowments, location or luck ? evaluating the determinants of sub-national growth in decentralized Indonesia*, Jakarta: World Bank.

[20] Aritenang, A.F., 2014., *The spatial effect of fiscal decentralisation on regional disparities: the case from Indonesia*. Indonesian Journal of Geography, 4 (1). pp 1-11

[21] Aritenang, A.F. (2016), *The Impact of Decentralisation and ASEAN FTA on Indonesia's Regional Economic Convergence*. Institute of Southeast Asian Studies Publications, Singapore

[22] Aritenang, A. and Sonn, J.W. (2018). *The Effect of Decentralization and Free Trade Agreements on Regional Disparity in a Developing Economy: The Case of Indonesia*, 1993– 2005. International Journal of Urban Sciences

[23] Karmaji, (2010), *Analisis Spasial untuk Identifikasi Lokasi Pemusatan Potensi Pertanian*, in Reinvensi Pembangunan Ekonom Daerah, Penerbit Esensi

[24] Puspitasari, M., Nurmalasari, V., and Sjafii, A., (2010), *Investigating the Economic Growth Impact on Poverty Reduction in East Java,* in Priyarsono, D.S, and Ernan, R. Regional Development in Indonesia. Crestpent Press: Bogor, Indonesia

[25] Syabri, I. (2003), *Explanatory Spatial Data Analysis for Flow Data: Exploring the Error Term of Spatial Interaction Models*, Jurnal Perencanaan Wilayah dan Kota 14(2): 49-68.

[26] Anselin, Luc. (2006). *Spatial Regression*. Spatial Analysis Laboratory, Department Geography, University of Illinois

[27] Matthews, Stephen A. (2006). *GeoDa and Spatial Regression Modeling*. Population Research Institute

[28] Di Rienzo, Cassandra. Fackler, Paul. Goodwin, Barry K. (2000). *Modeling Spatial Dependence and Spatial Heterogeneity in County Yield Forecasting Models*. AAEA Conference, Tampa Summer 2000. Session 8

[29] Anselin, Luc. (2003). *Spatial Weights*. Urbana-Champaign, University of Illinois

[30] Seknas Fitra (2012). *Kupas Tuntas Hubungan Keuangan Pusat Daerah.* Jakarta Selatan

[31] Tarigan, A. (2009). *Kerja sama Antar Daerah (KAD) untuk Peningkatan Penyelenggaraan Pelayanan Publik dan Daya Saing Wilayah*. Jakarta: Bappenas

IOP Conf. Series: Earth and Environmental Science **340** (2019) 012021 doi:10.1088/1755-1315/340/1/012021

[32] Hudalah, D. and Aritenang, A. (2017). *Industrial economies on the edge of Southeast Asian metropoles: from gated to resilient economies* in Andrew McGregor, Lisa Law, Fiona Miller (Eds). Routledge Handbook of Southeast Asian Development. London: Routledge.

[33] Karlinda, Elizabeth. (2015). *Optimalisasi Fiskal bagi Pertumbuhan Ekonomi: Peran Belanja Operasional dan Belanja Modal dalam APBD*. Laporan Penelitian. KPPOD

IOP Conf. Series: Earth and Environmental Science **340** (2019) 012022 doi:10.1088/1755-1315/340/1/012022

Determination of location criteria of PKL Center as tool for success relocation

H Idajati[1], C Susetyo[1] and E W Safitri[1]

[1] Urban and Regional Planning Department, Institut Teknologi Sepuluh Nopember

Email: ide_archits@yahoo.com

Abstract. Street vendor is an informal sector that performs its economic activity by using an area which is not in accordance with the government's spatial planning and considered as an additional to the urban issues, however the street vendor is an essential aspect in Indonesian urban cities. The relocation of street vendors is aiming for location management in accordance to carry out training program about the services activities of the informal sector by the government. At 2016, Surabaya is actively on beautifying the urban area by relocated the street vendors, including all street vendors along Arif Rahman Hakim Street. Those street vendors have been relocated to several Street Vendor Centers along the same street.The effort of minimizing any the changes of those street vendors to return to the side of the road to vend is by measuring the effectiveness of the street vendor's relocation, targeted on these points: 1) identifying the street vendor's characteristic, 2) measuring the effectiveness level of the previous relocation, and 3) formulate a location's criteria which is suitable for the street vendors of Arif Rahman Hakim Street in order to improve the effectiveness of the relocation.Thus, it can be concluded that these street vendors is in good term. And relatively concluded, the relocation of Arif Rahman Hakim Street's street vendors is quite effective.

1. Introduction

The street vendor is an informal sector which performs their economic activity using side walk area which caused by inexistence of a respected place from the government. Meanwhile, as stated by McGee and Young [1], the description of street vendor, or "hawkers", is people who sell their goods in the form of goodies or services at a public area, especially on side walk or road side. The street vendor is the fastest solution that comes in handy to cope with the life expectations by vending, with low funding and skills. This kind of attempt is considered as a salvation which is able to provide jobs for workers who do not accepted to the formal sector [2]. Street vendor as the addition of urban issues, in fact, has the potential of increasing jobs opportunity, implementation of small investment flows, relatively more absorbing of wage component by beverages production, and is on the job training site for the prospective entrepreneurs. Street vendor is indeed an essential part of Indonesian cities.

The problems arise when the street vendor is developing and occupying the strategic public area within a city. From McGee and Young [1], the strategic location is one of the street vendor characteristics. The existence of uneven expanding and spreading of the street vendor leads to the necessity of an area that is able to accommodate. The unregulated and shabby location of the expanded street vendor can withdraw the image of a particular area. Street vendor also considered as the reason of public facility disruption and the reason of public anxiety of traffic jam occurrences.

IOP Conf. Series: Earth and Environmental Science **340** (2019) 012022 doi:10.1088/1755-1315/340/1/012022

The reason of these problems is the unprepared location that should be provided by the government. The existence of location is the key of street vendor problems. Surabaya government is gradually carrying out a relocation program for the street vendors to be moved to available Street Vendor Centers.

This relocation of street vendors is aiming for location management in accordance to carry out training program about the services activities of the informal sector by the government. In accordance to Peraturan Daerah Kota Surabaya No. 9 Tahun 2014 about providing locations for street vendors, this relocation is decided by the Surabaya government to organize the street vendors.

In 2016, Surabaya government is actively beautifying the urban area by relocating street vendors around the city. Arif Rahman Hakim Street is one of the street vendor control and eviction program targets. This street is a strategic location because it is included in Keputih Village region which is close to ITS circle and ITS's students settlements. The street vendors along the street accommodate the necessity of the local residents and the ITS's students. But they also give a negative effect of declining environment quality along Arif Rahman Hakim Street. The occurrence of the still increasing number of street vendors makes the street become narrower and causes traffic jam. Thus, arranging and controlling the street vendors through relocation is carried out.

These street vendors are relocated to several places. The prepared areas are at Convention Hall and Deles region, which both are prepared by the government and located on the same street. However, to accommodate in limited areas, the vendors should pass a selection stage. Those vendors who do not pass the selection are relocated to personal or private maintained Street Vendors Center areas. Some of those centers are successfully increasing the street vendor's income, even though most location are not, such as both relocation areas on Arif Rahman Hakim Street. The consequence is some vendors returning to the road side or side walk.

The effort of minimizing any the changes of those street vendors to return to the side of the road to vend is by measuring the effectiveness of the street vendor's relocation, targeted on these points: 1) identifying the street vendors characteristic, 2) measuring the effectiveness level of the previous relocation, and 3) formulate a location's criteria which is suitable for the street vendors of Arif Rahman Hakim Street in order to improve the effectiveness of the relocation. This research is very beneficial for the government for its suggestion in urban area arrangement in accordance to urban spatial panning without disrupting the economic development of the informal sector.

2. Methods

This community service activity based research is going through these stages of:

2.1. The stages of literatur study

Literature study is done to understanding the related theories of informal sector and street vendor topic. Thus, the affected variables and sub variables of the determination of Street Vendor Center's location are obtained as seen in Table 1 and the variable of relocation effectiveness is shown in Table 2.

2.2. Stages of data collection

The data collection process divided to secondary and primary data. The secondary data mentioned is an implementation report of the street vendor relocation and the relocation planning, especially the street vendors along Arif Rahman Street. The primary data is the observation result and in-depth interviews with vendors and related agencies which is done in the research location. Meanwhile, for the samples, there are 4 (four) location as the research's observation object, they are:

a. Street Vendor Center cluster of Deles (A),
b. Street Vendor Center cluster at Convention Hall (B),
c. Street Vendor Center cluster next to BRI building (C), and
d. Street Vendor Center cluster at Wisata Kampus.

Table 1. Variable and Sub-Variable in determining Vendor Street Center's Location

No	Variable and Sub-Variable	Code
1	Strategic Location	-
	- near the people's activity	A1
	- near the bus or railway station	A2
	- near the housing	A3
	- near the settlements	A4
	- near the public transportation	A5
	- near the activity center, CBD, etc.	A6
2	Cluster Area	B
3	Visibility	C
4	Avability of parking lot	D
5	Avability of electricity	E
6	Avability of clean water	F
7	Avability of drainage system	G
8	Avability of trash container	H
9	Avability of a place to pray	I
10	Avability of toilet	J
11	Avability of menu	K
12	Avability of tables and chairs	L
13	Avability of wash basin	M
14	Facility of goods storage	N
15	Facility of centered cashier	O
16	Queue Number	P
17	Fire extinguishers tool	Q
18	Stands Pattern/Layout inside the Vendors Center	R

Table 2. Indicator and Variable in Measuring Relocation Effectiveness

No	Variable and Sub-Variable
Adaptation Indicators	
1	Sales and Marketing Adaptation
2	Way of Sell Adaptation
3	New Place Adaptation
4	New Costumer Adaptation
5	Way of Serving the New Customer Adaptation
6	New Way of Cooking Adaptation
Productivity Indicators	
1	Sales Improvement
2	Profit
3	Number of Customer
4	Expenses
5	Number of Goods
6	Number of Variation
7	Goods Inovation
8	Ease of Shopping Raw Materials
Job Satisfactory Indicators	
1	Chances of Improving

No	Variable and Sub-Variable
2	Security
3	Incomes Improvement
4	Customer Improvement
5	Improvement of Sales Environment Quality
6	Job Daily Condition
7	Facilities

2.3. The stage of analysis
The stage of analysis is based on the divided of the research targets, that including:

i. **First Target**: Identification of the area of relocation's characteristic using quantitative descriptive analysis,
ii. **Second Target**: Identification of the street vendor's characteristic through in-depth interview which is analysed using content analysis,
iii. **Third Target**: Measurement of the effectiveness level of street vendor's relocation through in-depth interviews which analysed with content analysis, and
iv. **Fourth Target**: Formula of the location's criteria which is suitable for street vendors from Arif Rahman Hakim Street in order to improve the relocation's effectiveness using descriptive comparative.

Those analysis stages are expected in resulting a recommendation of street vendor's location criteria which is in accordance to the area's spatial planning, aiming on the improvement of street vendor's relocation's effectiveness level.

3. Results and discussion

3.1. First target: identification of the relocation area's characteristic
Based on observation object, from 4 (four) locations there is only one location which is managed by the government and built on government's land. It is the Street Vendor Center of Convention Hall cluster. However, all the existed Street Vendor Center is under Dinas Koperasi's management. Based on the analysis result, it was obtained that there are characteristics of a street vendor's relocation area which is seen from variables and sub variables, derived from the determining process of Street Vendor Center's location which are shown in all four centers. The characteristics are:

a. The building has the fore and aft, out spread, and in the form of U letter,
b. The location is around the resident's activity places, far from transportation station, reachable from the vendor's settlement, near from the residents' settlement, accessible by public transportation, and close to public activities centers,
c. The building expans around105 - 1200 m^2,
d. The building is highly visible,
e. The cluster provides parking lot,
f. The building supported by electricity and clean water,
g. The building does not have internal drainage channel,
h. The building has dump stall, musholla, and toilet,
i. The cluster does not have communal menu list,
j. The building provided with tables and chairs,
k. The cluster does not including centered cashier and queue number, also
l. Only one cluster has APAR (lightweight fire extinguisher)

3.2. Second target: identification of street vendor center
According to the survey's result, it is occurred that:

a. The vendors dominated by 40 years old and up people,
b. Most of the vendors are citizen of Surabaya,

c. The vendors are residing around Arif Rahman Hakim Street, such as Gebang, Keputih, Klampis, and Semolowaru areas,
d. The business ownership are personal, not franchise,
e. Most of the vendor'ss latest educational degree are high school,
f. In majority, vendors sell foods and baverages, also
g. Vendors earn more than 300,000 rupiahs a day.

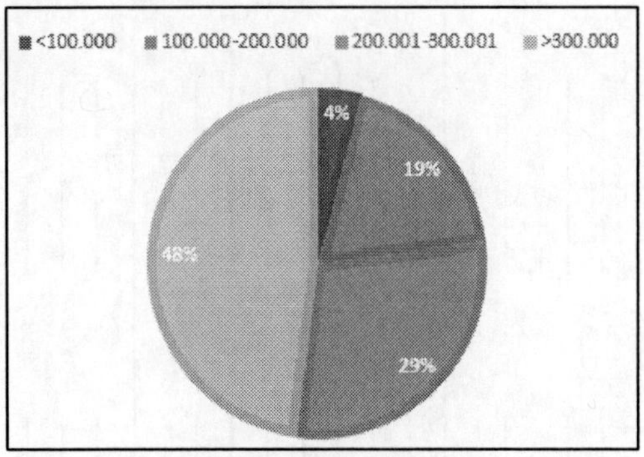

Figure 1. Street Vendor's Income per Day

3.3. Third target: measurement of the relocation effectiveness level of the street vendors center
The measurement take on by indicator's scoring about adaptation, productivity, and job satisfactory, also from every variable on table 2. From the scoring result on table 3, it can be seen that in every locations, the effectiveness value of every variables are different.

For adaptation variable, street vendor center near BRI building is the most effective center with 3.96 score from 1 to 5 scale. As explained by respondent of the survey, this street vendor center's vendors are able to adapt in a new place by adapting to a new selling system, new trading method, and new service ways.

In productivity variable, Deles Street Vendor Center and Street Vendor Center next to BRI building have the best productivity level compared to the other two. The productivity values of each are 3.39 and 3.38. The most distinguished values are profit increasing, customer number increasing, and the easiness of purchasing raw materials. Meanwhile, other street vendor centers are unable to increase their productivity.

The last variable is related to job satisfactory. For this variable, Street Vendor Center at Convention Hall has the best value of 4.08 from 5 scale. The vendors of this center consider that their relocation resulting in the increasing of safety working, opportunity to develop, also better environment and facilities quality. Other street vendor centers reach 3.5 or above in this variable.

Overall, the highest effectiveness level is gained by Deles Street Vendor Center with the average value of 3.6. The Street Vendor Center next to BRI building reachs 3.59. And for both of Street Vendor Center at Convention Hall and at Wisata Kampus get the average values of 3.44 and 3.31.

Table 3. Scoring Result of the Relocation Effectiveness Assessment

Indicator	Variable	Deles		Convention Hall		Nest to BRI		Wisata Kampus		Total NRV	Total NRI	Total Average
		NRV	NRI	NRV	NRI	NRV	NRI	NRV	NRI			
Adaptation	Sales and Marketing Adaptation	3.59		3.38		4		2.5		3.53		
	Way of Sell Adaptation	3.64		3.13		4		2.5		3.5		
	New Place Adaptation	3.82		3.38		3.5		2.5		3.61		
	New Costumer Adaptation	3.68	3.68	3.75	3.67	3.75	3.96	4	3.25	3.72	3.64	
	Way of Serving the New Customer Adaptation	3.68		4		4.25		4		3.83		
	New Way of Cooking Adaptation	3.68		4.38		4.25		4		3.92		
Productivity	Sales Improvement	3.14		2.38		3.25		3		2.97		
	Profit	3.05		2.38		3.5		3		2.94		
	Number of Customer	3.27		2.38		3.75		3		3.11		
	Expenses	3.27		2.38		3		2.5		3		3.46
	Number of Goods	3.36	3.39	2.38	2.58	3.5	3.38	3.5	3.19	3.17	3.13	
	Number of Variation	3.64		2.5		3		3.5		3.31		
	Goods Inovation	3.68		2.75		3.25		3.5		3.42		
	Ease of Shopping Raw Materials	3.68		3.5		3.75		3.5		3.64		
Job Satisfactory	Chances of Improving	3.5		2.63		3.75		2.5		3.28		
	Security	3.32		4.88		3		4.5		3.69		
	Customer Improvement	3.91	3.76	3.88	4.08	3.5	3.45	4	3.50	3.86	3.70	
	Improvement of Sales Environment Quality	4		4.13		3.75		4		4		
	Facilities	4.09		4.88		3.25		2.5		4.08		
	Average of Each Cluster NRI		3.61		3.44		3.59		3.31			

Description:
NRV = Average Value of Variabel
NRI = Average Value of Indicator

3.4. Fourth target: determining the suitable location for the street vendors

In this research target, criteria of a location which is suitable for street vendors at Arif Rahman Hakim Street will be formulated. The determination of the criteria is done by comparing the result of the first, second, and third research targets which are shown on table 3 and table 4. The comparison between effectiveness value and location characteristic are:

a. It is occurred that the highest value as many as 17 variables are accomplished by the location's characteristics of Deles Street Vendor Center. If it is analyzed by the relocation effectiveness value, it is shown that thc adaptation value reach 3.68, which is the most dominant among other effectiveness values. There for, in that particular location of street vendor center which has adequate facilities, the vendors are very adaptable,
b. The comparison result of the Street Vendor Center next to BRI building has 16 achieved variables with the most dominant effectiveness adaptation value of 3.95. It is proven that the vendor s of this center is very adaptable,
c. The Street Vendor Center cluster at Convention Hall fulfills about 15 variables with the most dominant effectiveness adaptation value of 3.67. In this center with adequate facilities, the vendors are adaptable, and
d. The lowest effectiveness adaptation level of Street Vendor Center with the most dominant value of 3.28 is gained by Street Vendor Center of Wisata Kampus. This center also has adequate facility with satisfy vendors.

As for the comparison between effectiveness value and the street vendors characteristics are described below:

a. It is occurred that the vendors at Deles Street Vendor Center has most of the positive values, such as has a lot of vendors, moderate rent cost, high income, and has appropriate kiosk space, which resulting on the highest productivity value of 3.38 and job satisfactory value of 3.60,
b. The cluster of Street Vendor Center next to BRI building has only few vendors, high renting price, moderate income, which give 3.37 score for the high productivity value and 3.46 for the job satisfactory value,
c. With low productivity value of 2.57 and the highest job satisfactory value of 3.64, Street Vendor Center at Convention Hall has the most vendors with the moderate income, and
d. Fewer vendor numbers with the highest income and the highest rent cost, the Street Vendor Center at Wisata Kampus get moderate productivity value of 3.18 and 3.28 for moderate job satisfactory value.

The comparison between the effectiveness value and the spreading pattern/diagram are as follows:

a. It is occurred that the vendors at Deles Street Vendor Center has most of the positive values, such as has a lot of vendors, moderate rent cost, high income, and has appropriate kiosk space, which resulting on the highest productivity value of 3.38 and job satisfactory value of 3.60,
b. The cluster of Street Vendor Center next to BRI building has only few vendors, high renting price, moderate income, which give 3.37 score for the high productivity value and 3.46 for the job satisfactory value,
c. With low productivity value of 2.57 and the highest job satisfactory value of 3.64, Street Vendor Center at Convention Hall has the most vendors with the moderate income, and
d. Fewer vendor numbers with the highest income and the highest rent cost, the Street Vendor Center at Wisata Kampus get moderate productivity value of 3.18 and 3.28 for moderate job satisfactory value.

Based on the disscussion above, it can be concluded that the location center of street vendors relocation must meet these following criteria:

- **In order to improve the adaptation in the new street vendor's center, then:**
  a. Have a strategic location
  b. The new street vendors center must not far from the old location

IOP Conf. Series: Earth and Environmental Science **340** (2019) 012022 doi:10.1088/1755-1315/340/1/012022

c. *Have a minimum space 3x3,5 meter square for each stand*
d. *Have a spacious stand for cooking*

- **In order to improve the productivity of the vendors, then:**
  a. *Have a maximum rental fee Rp 600.000,00 per month*
  b. *Pay the retribution for a maximum amount Rp 20.000,00 per month*
  c. *Have a LONG or CIRCULAR layout of vendors center*
  d. *Have many variations of be sold goods (foods, beverages, others)*
  e. *Have less than 30 vendors in one vendors center*
  f. *The new street vendors center must not far from the old location*

- **In order to improve the job satisfactory, then:**
  a. *Have adequate facilities*
  b. *Have adequate utilities*
  c. *Have a high visibility of the street vendors center's building*
  d. *Have an independent street vendors center's manager*
  e. *Have a clear and licenced street vendors center's land*

4. Conclusions

As the result of the analysis and the activities'achievment target, it can be concluded that:

1) The characteristic of the relocation area in general are has the fore-and-halt building type, out spreaded, and in the form of U letter pattern, is near from strategic locations but far from transportation station, expanding around 105 - 1200 m^2, good visibility, has parking lot, has electricity and clean water access, and has some other facilities in the building. However, it maintained without internal drainage, without comunal menu list, wothout central cashier, and only one cluster with APAR (a light-weight fire extinguiser),
2) The majority vendor's characteristic are people on their 40's or above, citizen of Surabaya, reside around Arif Rahman Hakim Street such as Gebang, Keputih, Klampis, and Semolowaru, has previous job as street vendors, own a personal vending non franchise bussiness, with the latest educational degree of high school, vend food and baverages, and earn more than 300,000 rupiahs a day,
3) The relocation of street vendorsa along Arif Rahman Hakim Street is quite effective with the scoring value as high as 3.46 (1- 5 scale), which can improve the vendor's productivity by providing facilities for the Street Vendor Centers. Even though in detailed analysis, every centers need special adjustments and treatments which are different from one another and according to the relocataion effectiveness value for every variables in the indicator, in order to improve the adaptation, productivity, and job satisfactory aspect.
4) The criteria that must be fulfilled in order to improve the adaptation in the new street vendor's center are: have a strategic location, the new street vendors center must not far from the old location, have a minimum space 3x3,5 meter square for each stand, and have a spacious stand for cooking.
5) The criteria that must be fulfilled in order to improve the productivity of the vendors are: have a maximum rental fee Rp 600.000,00 per month, paying the retribution for a maximum amount Rp 20.000,00 per month, have a Long or Circular layout of vendors center, have many variations of be sold goods (foods, beverages, others), have less than 30 vendors in one vendors center and the new street vendors center must not far from the old location.
6) The criteria that must be fulfilled in order to improve t the job satisfactory are: have adequate facilities, have adequate utilities, have a high visibility of the street vendors center's building, have an independent street vendors center's manager, and have a clear and licenced street vendors center's land.
7) The necessary thing to do for street vendor's relocation are:

- Carrying on the market research upon the street vendor's location. This is the first step of determining the Street Vendor Center cluster location without affecting the vendor's customers,
- Performing socialization and simulation activities for vendors, also actively communicate with the vendors. The obligated contents in the socialization, despite the regulation, the retribution cost, and provided facilities,
- Training the vendors after the relocation, thus the vendors are adapted to the new location, and
- Creating promotion medias about the new relocation centers to prefent losing customers and maintaining the vendor's productivity.

The main challenges of utilization of those technologies are the security aspect of the technologies itself because the implementation of applications to increase a city's adaptability will also increase in its vulnerability, because technologies, especially in form of computer systems, are prone to malicious attacks, i.e. from hackers and computer viruses. Therefore, implementation of new technologies in promoting a resilient city must be taken very cautiously.

This paper shows the possibility constructing a System dynamics-based flood modelling based on input both from raw data and 1D flood modelling. The visualization funcionality of System dynamics should give an added value to the flood modelling practice, because planners and decision makers can evaluate the existing condition of flood hazard and generate scenarios to various possible conditions. Even though the 2D and 1D2D hydrologic modelling is not possible to be transformed into a System dynamics, the visualization of the water network can give a lot of advantages.

For our future work, we planned to construct a complete model of System dynamic-based flood modelling, based on the existing 1D flood models for a particular city. By contructing a multiple System dynamics-based flood modelling from various 1D flood, we hope that we can evaluate the accuracy of the flood modelling if it is constructed using a System dynamic approach. An integrated model of System dynamics and 1D flood modelling is also possible, that would bring a complete new approach in flood modelling for urban environment.

5. Recommendations

Based on the result of the research's analysis, it can be recommended to the government that the effort of improving the effectiveness of the street vendor's relocation in accordance to the spatial planning through:

- Direct involving of the vendors in determining the relocation place including facilities providance, reachable vendors, and accessibility for the customers. This recommendation is to anticipate any losing customers from vendor's previous location also prevent the vendor's to find other location outside the provided are by the goverment, in the worse scenario is returnal of the vendors to the previous location,
- The spreading pattern of the Street Vendor Center cluster built as in relocation area's criteria, at least as shown in the following picture:

IOP Conf. Series: Earth and Environmental Science **340** (2019) 012022 doi:10.1088/1755-1315/340/1/012022

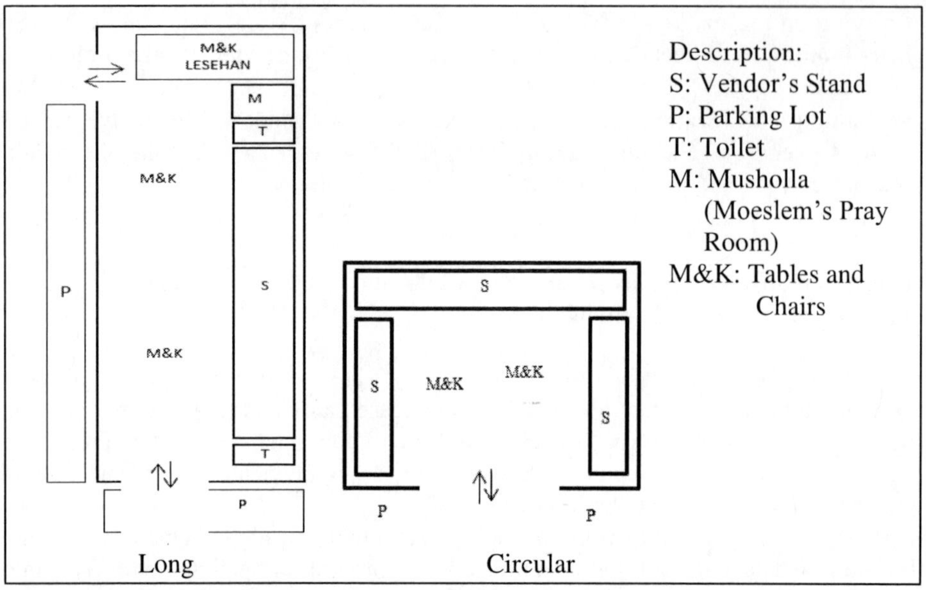

Figure 2. The Visualization of Proposed Floor Plan in Future Vendor Street Center

- Maximize the number of the vendors and the variation of the to be sold goods, around 30 - 40 vendors with the facility of 3 x 3.5 meters standard kiosk which sell various foods and beverages,
- Offers the standard rent cost around 15,000 - 20,000 rupiahs a day fo a kiosk, or 400,000 - 500,000 rupiahs a month,
- Set the parking cost of 2,000 rupiahs for two wheelers and 5,000 rupiahs for four wheelers, to attract customers using affordable price, also
- Street Vendor Center's managers have targets of getting more visitors by profiding facilities as shown in the criteria table, maintaining the foods and baverage's quantity, the facility's cleanliness and, the security of the center's area, also carry on the promotional activities.

6. References

[1] McGee, T.G. and Yeung, Y.M. (1977) *Hawkers in Southeast Asian Cities: Planning for The Bazaar Economy*. Ottawa International Development Research Centre.
[2] Usman, Sunyoto. (2006) *Malioboro*. PT Mitra Tata Persada.
[3] Alisjahbana. (2004) *Public Policy of Informal Sector.* ITS Press.
[4] Handoko, Hani. (2001) *Personnel and Human Resource Management.* BPFE Jogja.
[5] Indonesian Ministry of Agrarian and Spatial Planning Law No. 2 of 2016 about Street Vendor.
[6] Steers, R.M. (1977) *Organizational Effectiveness: A Behavioral View.* Goodyear Publishing.
[7] Tjiptono, Fandy. (2002) *Marketing Strategy.* Andi.

IOP Conf. Series: Earth and Environmental Science **340** (2019) 012023 doi:10.1088/1755-1315/340/1/012023

Creating cultural and heritage tourism route as tool for development tourism strategy (Case study: Surabaya Kalimas River Area)

H Idajati[1] and F E Nugroho[1]

[1] Urban and Regional Planning Department, Institut Teknologi Sepuluh Nopember

Email: ide_archits@yahoo.com

Abstract. Tourism routes have gained prominence in recent years and have greater pulling over potential to entice visitors to spend time and money. Route development in Indonesia has not grown enough, which makes visitor confused about tourism objects as well as the accessibility. However, Surabaya as the second biggest city in Indonesia attracts domestic and foreign tourists to cultural and heritage objects, especially in Kalimas River area. However, tourist in Surabaya have a limited time to stay which less than a day then visits another attraction outsite the city. Poor accessibility within the city also contributes to the problem. Based on the potential and problems above, the research aim is to identify the potential route connecting tourists with cultural and historical attractions through increased the tourist mobility. This qualitative research used the theoretical descriptive analysis to find specific criteria and the formation of pathways, then descriptive analysis to determine the implementation strategy. The results were that six route groups identified based on relationship of each tourism object criteria and the accessibility between them. To implement the route, strongly need to improve the public transportation and its supporting facilities.

1. Introduction

Routes were one of the important parts of the tourist product long before tourism was defined [1]. Routes can determine from model or pattern of product and services tourism identification showing specific location map [2]. The route in ancient time means that there are waterways along the coastline that transport people by boats before building the railways. The invention of private transportation meant new possibilities of travel, even roads mainly built for other than tourism and recreation purposes. The developments of tourism routes as a tourist attraction has gained prominence in recent years and has greater pulling over potential to entice visitors to spend time and money [3]. It is why in this study focusing on the ease of accessibility that playing an important role in tourism.

Surabaya City, located in East Java, has approximately 46 cultural and heritage tourist attractions along Kalimas River. In the past, The Kalimas River has a vital role in raising growth of Surabaya as a port town and trade center as well as being one of the most crowded water transportation [4]. Therefore, Arabian, Chinese and European cultures enter through trade indirectly making Surabaya an old city that has an old city character. Furthermore, there are many kinds of cultural and historical Travel Attractions such as, Surabaya City Hall, Tunjungan Area, Gedung Grahadi, Tugu Pahlawan Monument, Monumen Jenderal Soedirman, Submarine Monument and Pemuda Hall [5]

However, the data gained from the Surabaya City Culture and Tourism Agency in Surabaya (2017), showed an imbalance in the number of visitors amid the objects. In 2016, The Submarine Monument was 323,875 tourists and The Djoko Dolog, which were only 952 tourists. Inequality was more apparent on foreign tourist visits seen on House of Sampoerna reached 34,450 visitors, while Loka Jala Crana had no visitors. Some of them had decreased visitors because of unreachable location and inappropriate accessibility.

Futhermore, the characteristics of tourists in Surabaya, especially foreign tourists, just as a transit place before continuing their tour to Bromo, Tengger [5]. This is because of the Juanda airport and Tanjung Perak Port as the main entrance to East Java. It made Surabaya have no attraction in the view of the visitors. Whereas according to the Victoria University research journal on Tourism Behavior, tourists have the desire to visit many tourist destinations and various attractions in a place and limited time (Miah, 2017). Surabaya City Government revitalized various tourism objects but they are difficult to reach by the tourists because of the low access [6]. The mode and road conditions are in poor conditions. In another word, the objects accessibility has a significant effect on the number of tourists [7].

In addition, many cultures and heritage objects have no bus stop and lack access to public transportation, thus people prefer private transportation. The low accessibility and lack of tourist bus networks has resulted in people having to use public transportation and make several transportation changes from the starting point to the destination [8]. Then the accessibility of tourists to public transportation in the form of Lyn in Surabaya also hampers because it is inadequate in terms of quantity and poor quality of service seen from the length of the waiting time [9]. Implementation strategies needed to make sure the development to the tourism route. Moreover, given the accessible tourism activity to cultural and heritage objects that scattered from north to south in Surabaya.

Given the constraints, the developing of tourist routes is important to facilitate access to tourism object especially to the culture and heritage. The necessity of tourist routes in term of to facilitate tourists in access the activities independently, understand the whole flow of tourist trips within the city, to connect the tourists to local wisdom, and to increase either interest or knowledge in culture. With the development of tourist routes, several objects assembled in easily accessible paths so that tourists can easily determine what objects they want to enjoy within the limited time [10].

2. Methods for route development

The research is going through these stages of:

2.1 The stages of literatur study

Literature studies have done to understanding the related theories of route development. Thus, chosen variables and sub variables as the determination of Route Development Table 1. and the accessibility criteria is shown in Table 2.

2.2 Stages of data collection

The data collection processes are with primary and secondary surveys. The primary survey done with collecting data directly with in-depth-interviews conducted with key stakeholders. In addition, five key respondents who influence and importance in the development of cultural and historical tourism routes in Surabaya, such as: the Surabaya City Planning Agency, Culture and Tourism Agency, Travel Agents, one of the manager in oldest tourism object (House of Sampoerna), and Professor Johan Silas as an academic and a member of the Surabaya cultural heritage organization. While secondary surveys conducted through survey to the relevant agencies, regulatory and policy documents, planning, and prior research as theoretical descriptive analysis. The tourism object lists around the Kalimas River in the city of Surabaya are on the following map.

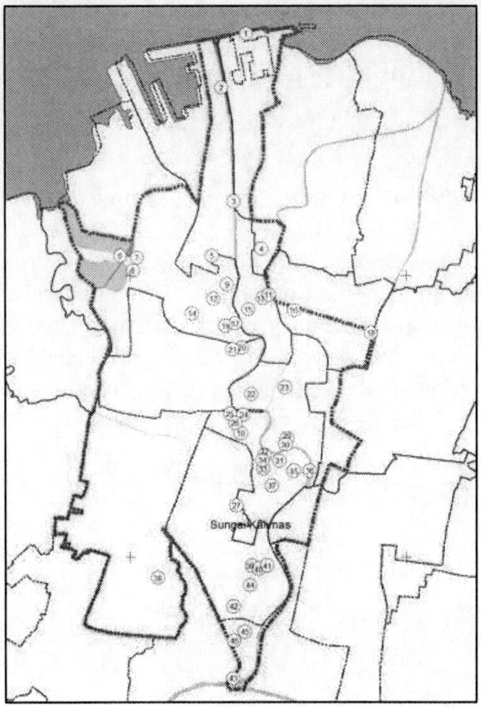

Figure 1. Cultural and heritage objects in Surabaya along Kalimas River

Table 1. Variable and criteria in determining tourism route

No	Variable dan Sub-Variable
1	Proximity Distance (between tourism destinations)
2	Road Condition
3	Transportation Mode
4	Transportation Facility Availability

Former final project research has done to determine the route variables ("Development of Cultural and Historical Tourism Route based in Accessibility"). The variables were to find out the hierarchy in which indicators and variables become the priority as input in the process of combining individual maps. In developing the tourist route, the need to elaborate tourism object characteristics and the criteria compiled. In terms of implementation, stakeholder opinions needed to formulate the right strategy in implementing tourist routes in Surabaya.

Theoretical descriptive analysis use to analyse the process in the form of specific criteria to identify the cultural and heritage tourist route with the priority indicators and variables in the previous analysis process. The theoretical descriptive analysis itself is an analysis that compares existing conditions in the field with policies, theories, and previous research related to each variable (Gabor, 2010). Determination of explanatory criteria from this analysis process is to strengthen the existing conditions in the field to form "object groups".

2.3 The stage of analysis

The stage of analysis based on the divided of the research targets, that including:

i. **First Target**: Tourism object similarity and significant in cultural and historical theme identification,

ii. **Second Target**: Identification of the accessibility to establish the routes between tourism objects to accommodate tourist mobility effectively,

iii. **Third Target**: Formulate the strategy of implementation to develop tourism routes. Those analysis stages will result strategy of tourism route implementation.

3. Results and Discussion

3.1 First target: tourism object similarity and significant in cultural and historical theme identification

The priority criteria compiled in the previous paper [12], therefore the adjustments to identify route through tourism objects group must form between the existing tourism attraction conditions and the criteria. The adjustments are as a basic data for overlay the maps to identify which tourism objects are in the same area or linked to the public transportation. Based on the criteria for the tourism object (tourism attraction, proximity, road condition, transportation mode and facility availability), it was found that there is six cultural and heritage tourism group. It identified from the distance between the object, the attraction similarity, and the facility availability.

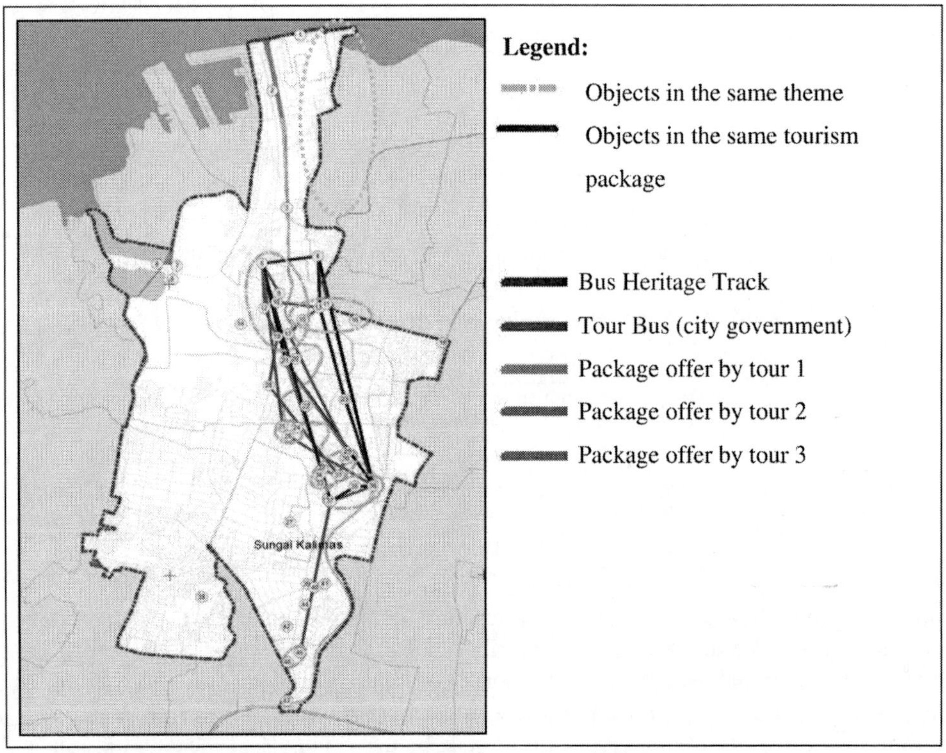

Figure 2. Connection between the objects along Kalimas River of Surabaya

The similarity between objects based on culture and heritage object analysed from the tourism object theme, the availability of tour packages by several agents, and objects that became the governments reference when there were guests from outside the city and abroad. This similarity will be closely related to the significance of the object's itself as part of history in the city of Surabaya. One of history that significant to Surabaya is a strong memory about of the struggle to be independence in the City of Heroes, which symbolize a thick story. Moreover, the collecting tour package information available from every tourism agent and government were to found that there was priority of objects in each group: starting from the well-known of themes, places then to differences in experience offered to the tourists. The groups are:

1. Group 1: Jalesveva Jayamahe Monument, Kalimas Traditional Port, and Pethekan Bridge

2. Group 2: House of Sampoerna, Internatio Building, Bank Niaga Building, PTPN XXII Building, Health Museum, Pertamina Building, Governor's Office, Kepanjen Church, and Tugu Pahlawan.
3. Group 3: Saint Louis Building, the Cathedral Church, the National Police Monument, Darmo Hospital, Santa Maria Building, Ki Ajeng Bungkul Tomb, Bank BI Library, and Wira Surya Monument.
4. Group 4: Ampel Area, Hok Tiek Han Temple, Kya-kya and Jembatan Merah
5. Group 5: Siola, Tunjungan Area, Peneleh Tomb, Cheng Ho Mosque, and Indonesian National Building.
6. Group 6: City Hall, Jendral Sudirman Monument, and Balai Pemuda, then Grahadi Building, Joko Dolog Statue, Governor Suryo Monument, RRI Building, Submarine Monument, and Bamboo Monument.

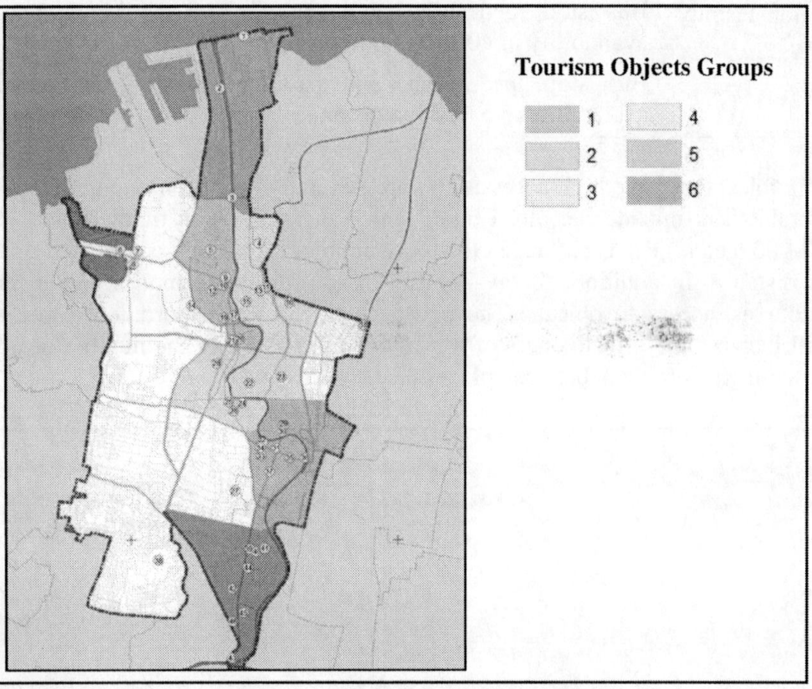

Figure 3. Six group of tourism object through analysis based on criteria in observation

3.2 Second target: identification of the accessibility to establish the routes between tourism objects to accommodate tourist mobility effectively

The establishment of routes is highly needed to consider access between objects to accommodate tourist mobility. The establishment of tourist routes based on accessibility begins with an analysis of the proximity of the distance between the object. The criteria obtained previously for the proximity variable between the object based on the length of the connecting road that reached by pedestrians with a maximum range of 2 km. Therefore, to identify tourist route of the city of Surabaya, pedestrians can reach the range between object distances are from 0.05 to 2 km as the maximum limit that.

Table 2. The accessibility criteria and its typology

Accessibility Criteria	Typology	
Proximity Distance (between tourism destinations)	<1 km	High accessibility
	1-1.5 km	Medium accessibility
	1.5-2 km	Low accessibility

Accessibility Criteria	Typology	
Road Condition	No traffic jam and road in good condition	High accessibility
	Congestion and side road obstacle (street vendors or on-street parking)	Low accessibility
Transportation Mode	>1 type of mode transportation and connecting line	High accessibility
	< 1 type of mode transportation	Low accessibility
Transportation Facility Availability	Bus stop, pedestrian, and parking area availability in 0 – 400 m distance.	High accessibility
	Bus stop, pedestrian, and parking area availability in 400 – 1000 m distance	Medium accessibility
	Bus stop, pedestrian, and parking area availability in > 1000 m distance	Low accessibility

Tourism object that have high accessibility are generally located in the middle of Surabaya City, besides several others outside the city center, which make them difficult to access. However, the tourists (100-700 meters) can easily reach the tourism object with high accessibility on foot within the comfortable distance. In addition, the availability of transportation modes has an important role in facilitating tourists' access to objects. The existence of mass transportation, especially for tourism object, it will increase the effectiveness of tourist time in achieving the number of object visited and increase the willingness to walk between places [13].

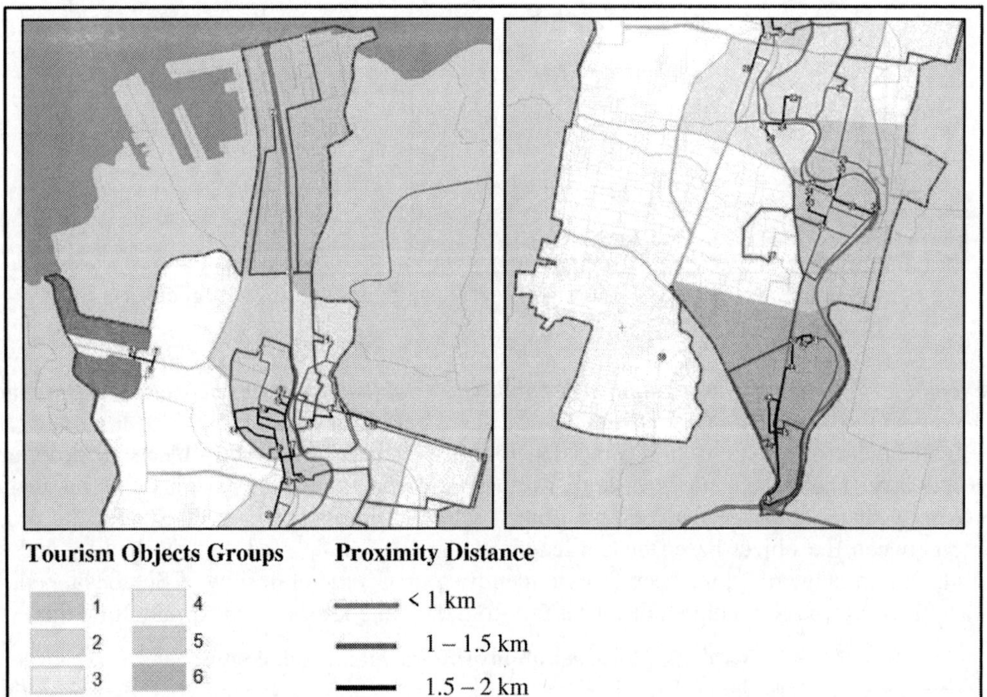

Figure 4. Map analysis proximity distance between attractions

The analysis continued to determine the availability of modes for each group of route, which have formed into six groups from the previous analysis. The criteria for forming route based on transportation modes, firstly the types of modes that pass more than one tourism object and have connecting routes. In addition, the easiness when the object may not in the same bus line, but connected to other type of transportation therefore tourist can determine their route, in other words, get off at any stop.

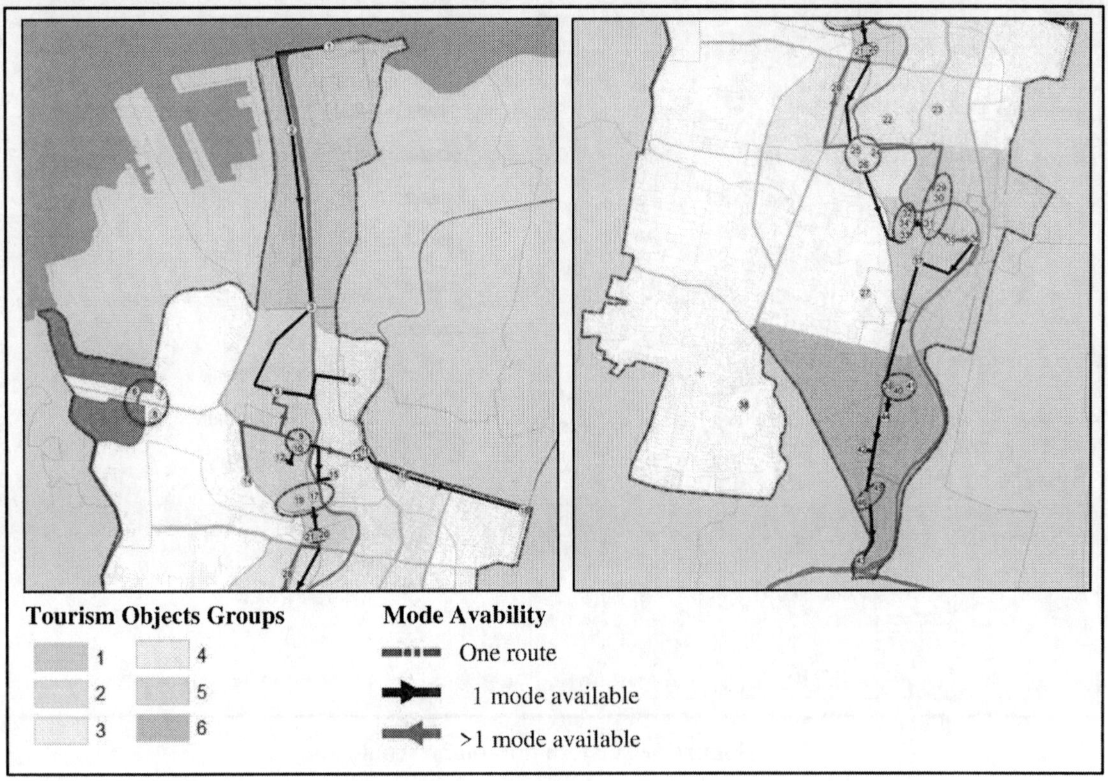

Figure 5. The availability modes in Surabaya

The analysis results combined between tourism objects ease of the access, and the availability of facilities in the tourist routes. Route analysis obtained the distance between tourism objects with high, medium and low accessibility; the availability of modes produced which meets the criteria if the modes of transportation available are more than three. The results of the analysis of the road access conditions are non-accessible due to poor road conditions, congestion, and side obstacles in the form of on-street parking, street vendors and one-way street. However, the availability of transportation facilities criteria must available around the route therefore it can be a more point to the route itself and comfort the tourist. The identification results showed that the high accessibility between route groups was 2, 3, 4, 5, 6, in sequence and low accessibility in group 1.

Identifying the accessibility process between tourism object was combine several individual maps of each variable. The process uses the overlaying technique that performed on ArcGIS Software to produce a combined map containing all the information / attributes of the maps of each of these variables. The following are the results of merging maps:

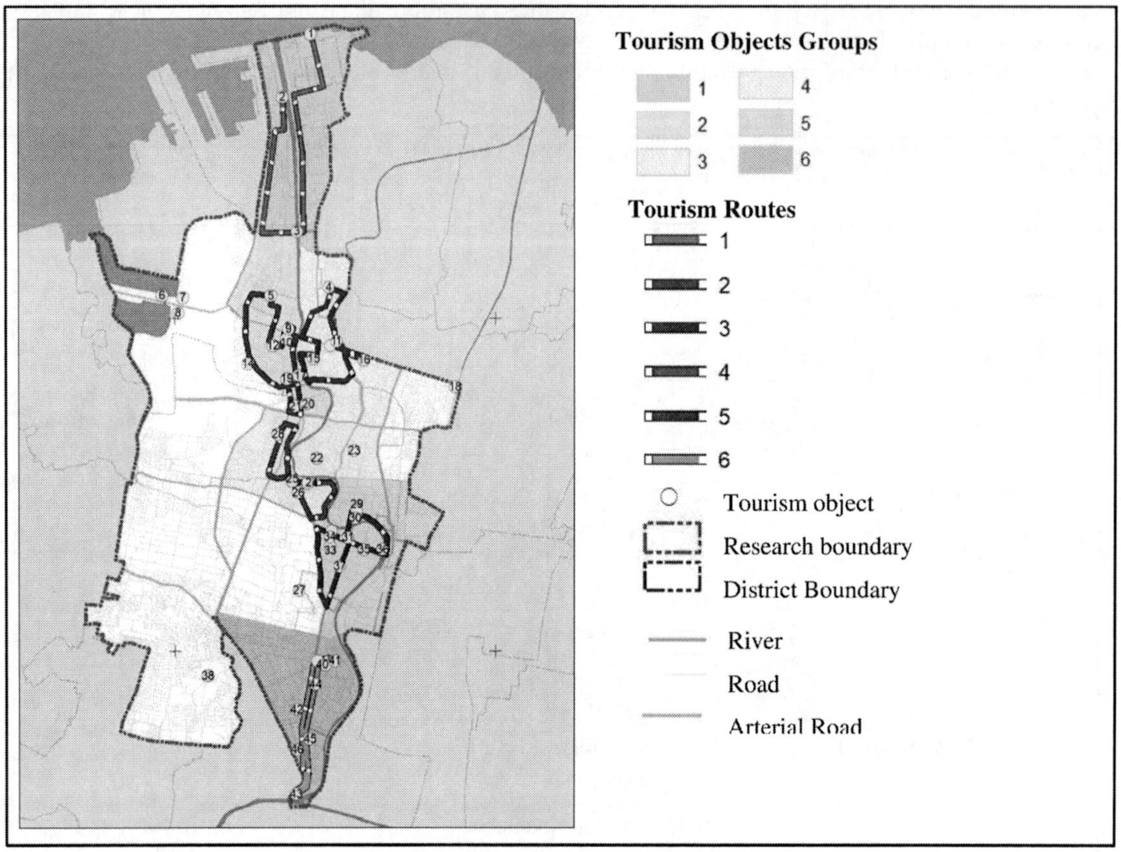

Figure 6. The result of tourism route

3.3 Third target: formulate the strategy of implementation to develop tourism routes

The results of the formation of the group of cultural and historical tourism routes in Surabaya City continued to the strategy to develop in terms of implementation. The formation of tourism groups that develop into routes have been analyzed previously based on priority indicators, the first one was the priority of tourism object itself, second the accessibility between the objects, and the last one was the facilities in the object. Based on that, the six group of tourism object which have the same characteristics found with a variety of attractions. Then, the accessibility of each group is analyzed, so it was found the proximity of the distance, the availability of modes, the conditions of road access, and the availability of transportation facilities. Finally, an analysis of the availability of accommodation, restaurants, and souvenir shops conduct to inform which object has that amenities reached at a close distance.

The analysis above combined with results of interviews that have been conducted with five different stakeholders. The stakeholders consisted of the government agencies (City Planning and Development Agency and Tourism Agency), private stakeholders represented by one of the travel agents and one of the tourism object managers (House of Sampoerna), and Prof. Johan Silas as an academic, which also part of Cultural Heritage, team of Surabaya. The following below is a summary of the results of the analysis:

1. Route 1

Route 1 consists of Jalesveva Jayamahe Monument, Kalimas Traditional Port, and Pethekan Bridge. The third objects above connect with maritime theme offering different attractions. However, the accessibility of this group is poor, as seen from the proximity of the results of the analysis including the low accessibility. This is because the distance between the objects is approximately 1.5-2 km and must ride their private transport to reach other objects. Apart from the availability of modes, only some local public transportation is through this area. Furthermore, in terms of availability of transportation facilities, the bus stop is located more than 400 m and there is no pedertrian connecting the objects. From providing facilities, accommodations and restaurants, it is difficult to find. While the souvenir shop is located on the North Quay (outside study area), so the conclution is that Group 1 is not accessible for tourists.

Based on the opinions of stakeholder's academics and managers of tourism object, conclusions and suggestions for group 1 are:

i. Add more objects for tourist attraction, especially with the coastal attraction around North Kalimas area, so that it is attraction that is more diverse.
ii. Increase the access between Jalesveva Jayamahe Monument and Kalimas Traditional Port with public transportation or build crossing bridges of the Kalimas River. Moreover, providing shelter facility in 400 m distance and adding sidewalks.
iii. Improve the facilities in northern Surabaya to provide tourism because the facilities are mainly lodging accommodations.

2. Route 2

House of Sampoerna, Internatio Building, Bank Niaga Building, PTPN XXII Building, Health Museum, Pertamina Building, Governor's Office, Kepanjen Church, and Tugu Pahlawan are the tourism object in route 2 and it part of the cultural heritage triangle area in Surabaya, which Veteran-Rajawali- Kepanjen Street. The results of the analysis in this route are the accessibility were easily connected between the objects, then there is a strong relationship in terms of distance proximity (100-600 m). Whereas there is the availability of modes, such as buses, public car, and even tour buses provided by the House of Sampoerna. There are no obstacles to road conditions and transportation facilities as well as the distances. Then the supporting (accommodation and restaurants) are available in the neighbourhood, moreover the souvenir shops located in the House of Sampoerna.

Based on stakeholder opinions at the interview, conclusions for group 2 summarized:

i. The tourism objects are located in a triangular area that has cultural heritage buildings. Increase the activities or attraction by revitalizing buildings on Jalan Kepanjen, Jalan Veteran and Jalan Rajawali.
ii. Improve the access with the availability of more than one mode of public transportation, then additional bridges between objects
iii. Improve the quality and the quantity transportation facilities such as, the shelters in close distance (approximately 400 m).
iv. Improve the pedestian path to tourism objects.

3. Route 3

The tourism objects around Darmo street have good accessibility because it in the downtown area of Surabaya; it can be seen from the proximity of the distance between the objects, the availability of modes, adequate road conditions, and transportation facilities such as the maintained sidewalks. Route 3 consists of the Saint Louis Building, the Cathedral Church, the National Police Monument, Darmo Hospital, Santa Maria Building, Ki Ajeng Bungkul Tomb, Bank BI Library, and Wira Surya Monument. However, in terms of providing facilities, there are only affordable accommodations, while the number of small restaurants and souvenir shops is not available.

Directions for group 3 based on stakeholders are:

i. Increase access with the public transportation development that passes through Jalan Darmo, especially for tourism due to the large number of passengers.
ii. Transportation facilities must improve in bus stops facility and parking spaces to connect users to public transportation.
iii. Increase the standard of tourism facilities, especially for restaurants, such as hygiene and health standards.

iv. Build crossing bridges on Jalan Darmo for faster access of tourist.

4. Route 4

There are some of the tourism area such as Ampel Area, Hok Tiek Han Temple, Kya-kya and Jembatan Merah Area which rich in cultural traditions as well as the attractions because it is housing area with different ethnicities. For example, Kampung Ampel, Kampung Keraton, and a village in the Chinatown are the cultural village that integrated to other villages which already developed by the Surabaya city government. However, accessibility in this group is difficult (the distance between the areas are more than 1000 m), there are no bus stop, no pedestrian path, and parking tends to be on street, which can disrupt road traffic.

Based on two opinions of respondents, some of the suggestions from the stakeholders are:

i. Integrate the Tourism object with the cultural village, which was the beginning of the formation of cultures in Surabaya, such as Arab Village, Keraton Village, and villages in the Chinatown.
ii. Upgrade the quality and the quantity of public transportation, especially in the area of Ampel because many tourists go there.
iii. Add more souvenir shops around this area, especially in the around Jembatan Merah area which can be combined with restaurants.
iv. Improve transportation facilities such as bus stops in the Ampel and Jembatan Merah (Chinatown) areas, then pedestrian restoration (from the parking lot of the bus / car to Ampel) so that tourists get comfort when walking to the tourism spot.
v. Provide parking and street vendors space, especially in front of the Ampel Gate, so it would not to disturb the flow of traffic.

5. Route 5

Tourism object in-group 5 are Siola, Tunjungan Area, Peneleh Tomb, Cheng Ho Mosque, and Indonesian National Building. This area is among the most restored and promoted, because it is located in the middle of the city and regional identity as a history of Surabaya is very thick. Analysis in terms of accessibility shows that this area is high accessibility because the distance between tourism objects is close, moreover reached on foot. Then the transportation mode is in good condition with the new SuroBoyo bus. However, the road conditions tend to congested due to the volume of vehicles at certain hours and there are no visible facilities for bus stops and rides. In terms of facilities, it is very adequate with a variety of accommodation and restaurant choices, while the souvenir shop is available at Siola.

Based on several stakeholder opinions and the results of the analysis, the directions for group 4 are:

i. Increase the activities attraction by revitalizing the functions of buildings on Jalan Tunjungan are commercial in order to attract tourists.
ii. Improve the transportation facilities such as bus stops on Tunjungan Street and improving the quality of sidewalks on Bubutan Street so that tourists get comfort when using transportation modes if there is congestion.
iii. Add number of souvenir shops, especially in the Indonesian National Building.

6. Route 6

Route 6 consists of City Hall, Jendral Sudirman Monument, and Balai Pemuda, then Grahadi Building, Joko Dolog Statue, Governor Suryo Monument, RRI Building, Submarine Monument, and Bamboo Monument. The objects connect with the theme of the Independence Day History in Surabaya and have a good accessibility. Then the road conditions are well maintenance even though there is congestion at some points. The availability of sidewalks is adequate, although in this region there inadequate modes of transportation. In terms of facilities, accommodations and restaurants are affordable with Plaza Surabaya as a shopping center, while the souvenir shop is at City Hall and Surabaya Plaza.

Based on several stakeholder opinions, here are the suggestions to implement route 6:

i. Improve the quality and quantity of public transportation that passes through the area because it is the center of the office and government of the City of Surabaya (City Hall, Grahadi, and Balai Pemuda)
ii. Improve the number of bus stops at Jalan Pemuda (in front of Plaza Surabaya or Balai Pemuda) as the main people attraction or meeting point.
iii. Provide parking space on the several area.

4. Results

Here are the results of possible access for tourist and its strategy to implement tourism route effectively:

Table 3. The routes and the possible access of each tourism object

No.	Routes	Tourism Objects	Strategy
1	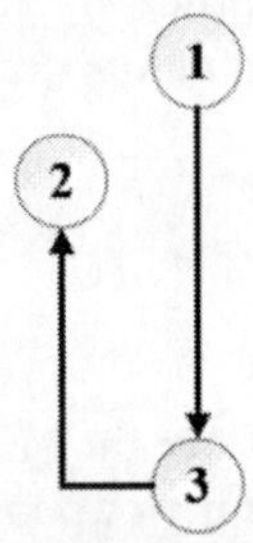	1: Monumen Jalesveva Jayamahe 2: Pelabuhan Kalimas 3: Jembatan Pethekan	1) Add more objects for tourist attraction 2) Increase the access 3) Improve the facilities in northern Surabaya
2	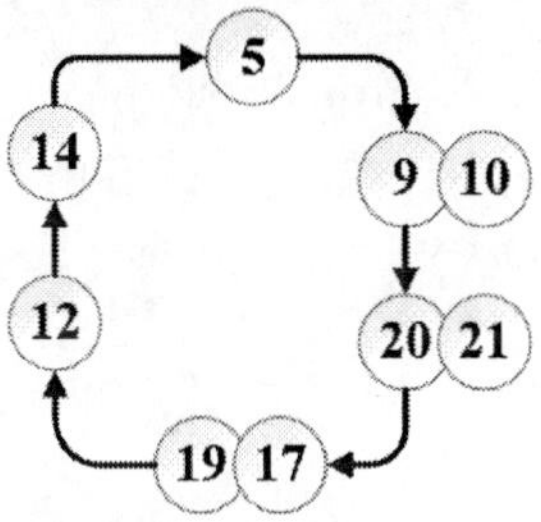	5: House of Sampoerna 9: Gedung Internatio 10: Gedung Bank Niaga 12: Gedung PTPN XI 14: Museum Kesehatan 17: Gedung Pertamina 19: Gereja Kepanjen 20: Kantor Gubernur 21: Monumen Tugu Pahlawan	1) Increase the activities or attraction by revitalizing buildings 2) Improve the access with the availability of more than one mode of public transportation 3) Improve the quality and the quantity transportation facilities 4) Improve the pedestrian path to tourism objects
3	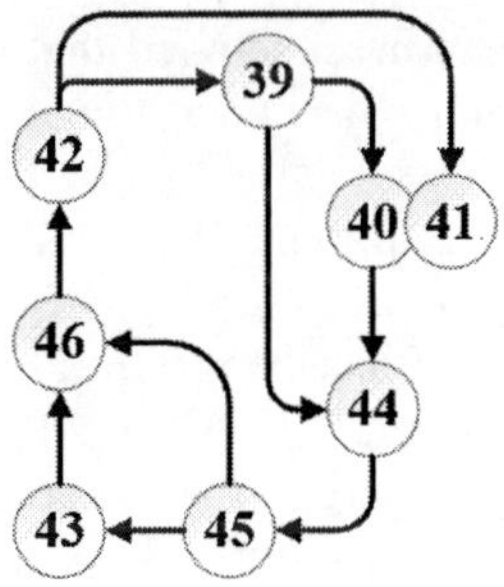	39: Monumen Perjuangan POLRI 40: Gereja Kathedral 41: Gedung Saint Louis 42: Rumah Sakit Darmo 43: Monumen Wira Surya 44: Gedung Santa Maria 45: Makam Ki Ageng Bungkul 46: Perpustakaan Bank BI dan Monumen Bahari	1) Increase access with the public transportation development 2) Transportation facilities improvement such as bus stops and parking spaces 3) Increase the standard of tourism facilities 4) Build crossing bridges on Jalan Darmo
4	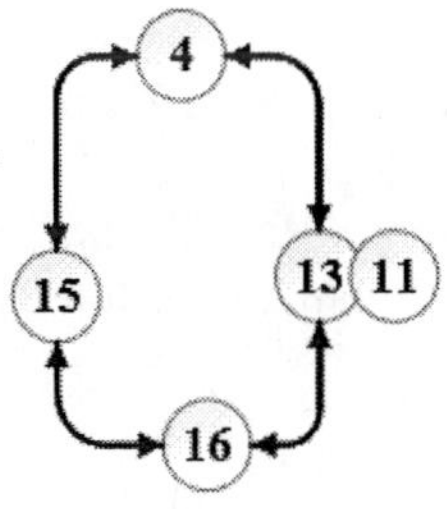	4: Makam Sunan Ampel 11: Klenteng Hong Tiek Hian	1) Integrate the Tourism object with the cultural village 2) Upgrade the quality and the

No.	Routes	Tourism Objects	Strategy
		13: Kawasan Jembatan Merah dan Kya-Kya 15: Klenteng Hok An Kiong 16: Klenteng Boen Bio	quantity of public transportation 3) Add more souvenir shops around this area 4) Improve transportation facilities such as bus stops 5) Provide parking and street vendors space
5	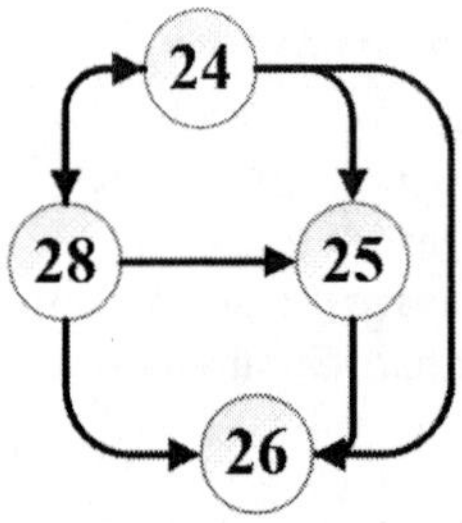	24: Gedung Cak Durasim 25: Siola / Museum Surabaya 26: Kawasan Tunjungan 28: Gedung Nasional Indonesia	1) Increase the activities attraction 2) Improve the transportation facilities 3) Add number of souvenir shops
6	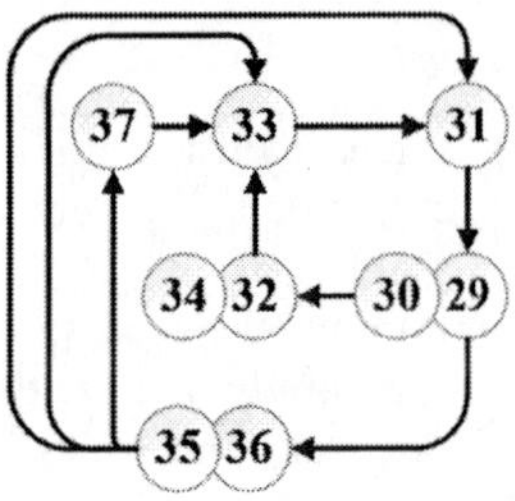	29: Balai Kota 30: Monumen Jendral Soedirman 31: Balai Pemuda 32: Gedung Grahadi 33: Patung Joko Dolog 34: Monumen Gubernur Suryo 35: Gedung RRI 36: Monumen Kapal Selam	1) Improve the quality and quantity of public transportation 2) Improve the number of bus stops 3) Provide parking space

Source: Analysis, 2018

5. Conclusions

As the result of the analysis and the activities'achievment target, it can be concluded that this study identified six groups of route based on the relationship criteria between objects and the accessibility for each variable. To conclude, each the routes classified:

1. Route 1 consists of Jalesveva Jayamahe Monument, Kalimas Traditional Port, and Pethekan Bridge.
2. Route 2 consists of the House of Sampoerna, Internatio Building, Bank Niaga Building, PTPN XXII Building, Health Museum, Kepanjen Church, and Tugu Pahlawan.
3. Route 3 consists of the Saint Louis Building, the Cathedral Church, the National Police Monument, Darmo Hospital, Santa Maria Building, Ki Ajeng Bungkul Tomb, Bank BI Library, and Wira Surya Monument.
4. Route 4 consists of Ampel area, Hok Tiek Han temple, Hok An Kiong temple, Boen Bio temple, Kya-kya and Jembatan Merah area.
5. Route 5 consists of Siola, Tunjungan Area, Peneleh Tomb, and Cheng Ho Mosque, and Indonesian National Building.
6. Route 6 consists of City Hall, Jendral Sudirman Monument, and Pemuda Hall, Grahadi Building, Joko Dolog Statue, Governor Suryo Monument, RRI Building, Submarine Monument, and Bamboo Monument.

6. Recommendations

Based on the result of the research's analysis, it recommend implementing tourist routes in Surabaya to improve the quality and quantity of public transportation in Surabaya, increased activities offered by the cultural and heritage objects by revitalizing the functions, and provide more transportation facilities such as bus stops and pedestrian path.

7. References

[1] Council of Europe, 2016. *Cultural Routes of the Council of Europe*, Madrid.

[2] Association of Southeast Asian Nation (ASIAN), 2018, *Guideline of ASEAN Cultural Hertage Tourism Travel Pattern*

[3] Meyer, D., 2004. *Tourism routes and gateways : key issues for the development of tourism routes and gateways and their potential for Pro-Poor Tourism.* , (April), pp.1–31. Available at: http://www.odi.org.uk/resources/details.asp?id=3100&title=routes-gateways-tourism.

[4] Idajati, Hertiari. 2014. *Cultural And Tourism Planning As Tool For City Revitalization The Case Study Of Kalimas River, Surabaya-Indonesia* , Procedia - Social and Behavioral Sciences 135 (2014) 136 – 141

[5] Development Planning Agency of Surabaya, 2017, Review Rencana Induk Pariwisata Daerah Surabaya (Review RIPDA)

[6] Development Planning Agency of Surabaya, 2010, RDTRK UP. Tunjungan,

[7] Winaya, P.P.J., 1995. *Pengaruh Tingkat Aksesibilitas terhadap Jumlah Wisatawan pada Rute Perjalanan Pariwisata di Bali*. Institut Teknologi Bandung.

[8] Aminah, S., 2007. Transportasi Publik dan Aksesibilitas Masyarakat Perkotaan. *Media Masyarakat Kebudayaan dan Politik*, 10(1).

[9] Development Planning Agency of Surabaya, 2010, RDTRK UP Wonokromo, 2010

[10] Csapó, J., 2012. The Role and Importance of Cultural Tourism in Modern Tourism Industry, Strategies for Tourism Industry. *University of Pecs, Institute of Geography. In Tech: Hungary.*

[11] Bungin, Burhan. 2005. Metodologi Penelitian Kuantitatif. Kencana Prenada Media Group: Jakarta.

[12] Nugroho, Felicia and Idajati, Hertiari. 2017. Heritage Tourism Route Criteria Based on Accessibility (Case Study: Surabaya Kalimas River Area).

[13] Thériault, M. & Des Rosiers, F.D., 2011. Modeling Urban Dynamics: Mobility, Accessibility and Real Estate Value. *Modeling Urban Dynamics*, (Hermes Science/Lavoisier).

Urban settlement growth factors through ekistics element approach (Case study: Jember City)

I A Farizkha[1], RR D J Koesoemawati[1], R A Suprobo[1], R N Listyawati[1] and N N Hayati[1]

[1] Urban and Regional Planning Department, Universitas Jember

Email: ivanagustafariska@gmail.com

Abstract. Urban Sprawl phenomenon occurs because of the uncontrolled development of urban settlements. The uncontrolled development of these settlements has resulted in a decline in environmental quality due to changes in certain land functions. Ecistic is a tool that is considered appropriate for understanding the settlement. Research uses ecistic elements to see a settlement in one large frame, namely total settlements patterns with a classification scheme related to elements of settlements, namely nature, anthropos, society, shells, networks. This research was conducted to understand the conditions of residential growth to find out the growth factors that occur. Several stages of analysis carried out in this study include identification of settlement characteristics, trend analysis using periodic regional image analysis, as well as analysis of factors that influence settlement development using AHP techniques. The results showed that settlement development began to occur sporadically in the period 2006-2018 to the south with the highest density concentration remaining in the city center. In the physical organization of space, the development of settlements forms an octopus pattern (Octopus/Star Shape Cities). The results of the AHP analysis strengthen these conditions where the location of settlements and activities is the main factor that causes the growth of settlements.

Keywords: *Ecistic, Urban Sprawl, Housing Development.*

1. Introduction

Urban areas always develop dynamically along with the increase in population due to growth and migration. The dynamic development of urban areas will emerge urban problems. One of the problems that often take place in urban areas is the uncontrolled growth of settlements, without the planning process, and tends to form a dirty environment. The phenomenon of the development of settlements like this is called urban sprawl [1].

In fact, Urban Sprawl has many negative effects on the physical development of the city [2]. Urban sprawl contributes a lot in decreasing the quality of settlements because most settlements that grow due to urban sprawl are not efficiently served by infrastructure, reduce catchment areas, increase pollution (land, water and air) and are the beginning of the formation of slums if the development continues to be left without the right planning process [3].

One of the areas in East Java that experienced the development process of the region from rural to urban areas and the increase in land conversion was Jember. Land conversion that occurs is in the area around the urban center and is spread. The high level of activity in the urban areas of Jember, has led to a pull of human movement towards the city. Pulling human movement coupled with increasing population growth, will also increase the needs of the house. However, when viewed from the price of land, prices in the city are relatively high, causing housing providers to prefer to build housing in an area that is a little further from the city center to get affordable land prices. The fertile land conversion occurs in many of these conditions.

Jember is a potential area in housing development [4]. This is based on various factors including the area, location, population, activity attraction, home ownership status and the number of developers who have grown up. Demographically, Jember has a population of 2.3 million with a growth rate of 0.31% (BPS, 2012). However, when viewed from the status of home ownership there was a decrease from 88.93% in 2010 to 88.37% in 2011. REI (Real Estate Indonesia) is a formal institution engaged in housing. According to REI (2013), the number of housing developers in Jember Regency reached 23. Through the number of formal housing developer institutions, the potential for housing development can be ascertained to continue to increase if not offset by the Jember government control measures.

A control measure for the development of urban settlements is an urgency that must be implemented to minimize the phenomenon of urban sprawl. The focus of this study is focused on settlements because the starting point of the development of a city is settlement. Settlement growth will determine how a good city structuring quality can be realized. In a process of drafting a control concept, it is necessary to formulate the right concepts and strategies based on a comprehensive understanding of the settlement growth process as well as what factors influence the development of settlements so that a concept of control is targeted.

The approach to understanding the condition of settlements is done by looking at the aspect of forming a settlement. Aspects of the formation of these settlements are explained by the understanding of the ecological theory proposed by Doxiadis (1976). According to Doxiadis, understanding the characteristics of settlements must be carried out comprehensively in one large frame, namely "total settlements patterns". Economics studies settlements in two-way classification schemes. The first classification relates to the hierarchy of settlements based on scale while the second classification relates to elements of settlements, namely nature, anthropos, society, shells, networks [5]. In the process of controlling the development of settlements, in addition to understanding the characteristics of settlements, it is also necessary to pay attention to the time dimension in the process of understanding settlements, namely the historical and approximate how these settlements will be in the future [6].

Referring to the understanding of the problems above, the objectives of this study are:

1. Identification of the characteristics of settlements in the urban areas of Jember
2. Analysis of the spatial pattern of settlement development
3. Analysis of the factors causing the growth of settlements in Jember urban areas through an ecological element approach.

The results of this journal discussion are expected to be able to contribute ideas to the determination of urban structuring policies. Jember urban settlement is an appropriate study area for understanding the growth pattern of urban settlements because Jember urban is one part of the Besuki Raya region which has the potential to develop into a metropolitan area such as Gerbangkertasusila and Jabodetabekpunjur. Unanticipated developments will cause problems in the future such as those occurring in big cities in Indonesia such as Jakarta or Bandung. In the context of controlling the development of urban settlements, it is necessary to formulate the right concepts and strategies based on a thorough understanding of the settlement growth process and the causes of the development of these settlements so that the concept of proper control can be formulated to anticipate the challenges of these developments.

2. Study of literature

2.1 Definition of urban sprawl
In general, urban sprawl is uncontrolled settlement growth as a result of the development of a city. The process of urban sprawl occurs due to the need to persuade residential facilities due to urban activities that use the population to find a place to live. According to Setioko [7] sprawl can be described as unplanned, diffuse, low density and unstructured development in the periphery. The suburbs are suspected as the center of urban sprawl growth because the city center area has become saturated with the existence of various facilities and services centers.

According to Yeates and Garner [8], in addition to being caused by an increase in space requirements, sprawl is also caused by changes in the ability of the transportation system, housing construction and infrastructure. There are three types of urban area expansion processes (urban sprawl), namely [9]:

1. Concentric development is characterized by a uniform spillage to all existing urban areas and slow types of perforations.
2. The elongation of the ribbon (ribbon development) is characterized by the development of cities that follow the existing transportation network so that the role of the transportation network plays a very important role in the process of trapping this type of city.
3. A jumping leap (leap frog development/checker-board development) is characterized by irregular city traps or jumping from the parent city. This type of trapping is a city that is ineffective and inefficient.

2.2 Ekistics theory
Ekistics theory is a theory to explain settlement (human settlement) developed by doxiadis [10]. Human settlement is a place that is inhabited by humans that includes content elements and container elements. Elements of ekistics are divided into five parts, namely nature (human), human (antrophos), society (society), reflection (shells) and networks (networks).

a. Natural. Nature is the first element of settlement due to several reasons, namely nature is the oldest element compared to settlement, after that settlements are formed by nature which produces a system [10]. Theoretically, it can be said that settlements are part of nature. The research was carried out by looking at natural elements related to physical appearance of the earth, availability and ability of the environment.
b. Human. Humans are the second element of the element ekistic, because this element follows nature but escorts three other elements [10]. Humans initially started by changing nature by building huts. After that, began to have expertise in the agricultural revolution which subsequently created various types of houses. In research, the human element examined is related to the conditions, numbers, and relationships between individuals.
c. Community. Society is a kinship system among many people who form a network [10]. To understand society as an element of the settlement, first look at its relationship with the physical conditions of the environment such as nature, protection and networks. This study examines the elements of society through sub-variables in the form of social conditions that exist in the scope of settlements (such as livelihoods, income, education level, economic conditions and cultural characteristics that exist in the community in the scope of the research room)
d. Protection. Protection or shells in this case are interpreted as conditions, types and home services. Protection is divided into several categories, namely education, health, administration, security, industry, storage and others [10].
e. Network

   The network is a utility aspect which consists of transportation, electricity, water facilities etc. All types of networks are used and utilized by humans as a residential support facility. The choice of a comfortable dwelling will take into account the availability of the network, especially the transportation network that contributes to the access of occupancy to centers of activities and clean water networks as a means of basic human needs.

2.3 *Settlement development theory*

Settlements are part of the environment in protected areas, both in the form of urban and rural areas that function as neighborhood units of residence or residential neighborhoods and places of activities that support per life and livelihood [11]. The development of settlements can be interpreted as an increase in the size of the residential environment caused by an increase in the population and an increase in the economy of the community so it will increase the demand for housing needs. The choice of settlement location is essentially through various considerations such as location factors, completeness of facilities and infrastructure, and the environmental conditions of the community.

2.4 *City morphology and settlement growth*

Literally, morphology means the science of form. In urban contexts, morphology is the study of forms and shapes from neighborhoods. Form means a form that can be observed and is a configuration of several objects, while a shape is a geometric feature or an external form and an outline of an object. The settlement environment becomes an important keyword, because in the science of planning and designing the city it is stated that civilization starts from settling activities. The complexity in the growth of settlements then forms larger environmental units, namely cities. So the city environment cannot be separated from the neighborhood [12]. So it can be concluded that the development of a good city must begin with good settlement planning. If settlement growth control can be carried out with certain treatments, the formation of an urban area will run well and meet the principles of sustainable development.

The most influential element of city morphology is the road pattern [12]. Where there are 3 (three) known types of road pattern systems, namely: (1) irregular road pattern system (irregular system); (2) a concentric radial road pattern system (radial concentric system); (3) a system of elbow angles or grids (rectangular or grid systems).

The growth and development of the city can be understood by observing morphological components. Functionally and economically, regional growth is affected by land use, buildings, plots and road networks. Urban areas are formed from a system of activities that are complexly linked by movement networks. The interaction between these two systems, the activity system and the movement system makes urban areas have economic value or property values whose distribution is strongly influenced by natural physical characteristics and the support of the two systems. City growth can be observed geographically assisted by map science (cartography). Using maps, the distribution of natural and artificial physical potential can be easily observed and analyzed. For land use, building density, land size and control and the road network can be mapped and logically explained in relation to each other.

3. Method

The research variables used in this study are derived from the synthesis of literature studies related to the exploration of ecological theory and sustainable development approaches. The variables in this study can be seen in table 1.

The research mindset is the process of research stages starting from the background of the research carried out to the stages of the analysis process. The research mindset is discussed to facilitate understanding of the research flow. For more details, the research flow can be seen in Figure 1.

Table 1. Research Variables

Variables	Sub Variables	Indicators
Physical • Nature • House / housing • network	Natural conditions	- Availability of natural resources - Natural conditions
	housing conditions	- Type of housing (permanent or non- permanent) - Housing Model - Location of settlements - Density of residential areas

Variables	Sub Variables	Indicators
		- Land use
	Infrastructure and settlement	- Water resources - electricity infrastructure - Road infrastructure - waste processing infrastructure - Telecommunication infrastructure
Non-physical : • Humans • Community	Social conditions of the community	- number of population - Livelihood - Level of education - The existence of social institutions - Human growth
	economic conditions of the community	- Distribution of community income levels - The existence of economic institutions
	Cultural characteristics of society	- Habits - Activities

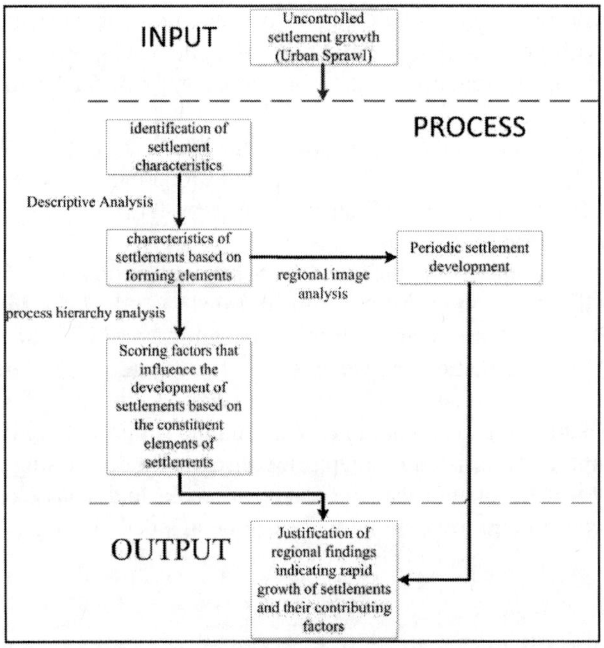

Figure 1. Flow of research process

4. Discussion

4.1 Identification of the characteristics of settlements in the urban areas of Jember based on the variables determined, observations of the characteristics of settlements

a. Nature

- The growth of settlements in the study area can occur evenly basically, except in conditional terms on some areas that have land slopes of 10-25% which are in part of Kaliwates, Patrang and Sukorambi Sub-Districts. Whereas the criteria are not suitable, there are in the Patrang and Sukorambi Sub-Districts which have a slope of> 25%.
- The attraction of natural resources which is the driving force that causes the growth of settlements is fertile soil conditions (argopuro bran and argopuro tuff).

b. Home / housing

- In the area of research in the model of the house does not significantly influence the development of settlements. The model of the house in the average research area is in the form of a modern one, the majority are in the Kaliwates, Patrang and Sumbersari Sub-Districts.
- In terms of settlement location, the fastest growth is in activity centers, especially in areas including urban areas (Kaliwates, Patrang and Sumbersari sub-districts). The development in the next stage is more dominating with the opening of new lands in the suburbs of urban areas. Whereas in Ajung and Sukorambi sub-districts the majority of the intensity of growth follows the area that has natural resources in the form of fertile land.
- At the density level, the study area has different levels of density. In areas including urban areas, Jember has a complex level of density and land use. While in areas not included in the Jember urban area it has a smaller density level.

c. Network
- In terms of the electricity network, the research area has been fulfilled well in various sub-districts. There are households that do not use electricity as lighting does not reach 1%.
- For the road network, residential growth follows the road network in the beginning. Whereas in the future the development of the settlement is more in the suburbs by opening new land.
- Based on solid waste infrastructure, the research area has been fulfilled properly with the presence of officers carrying garbage from home. Whereas in the sanitation system, Ajung Sub-district has a number of households that do not have the highest defecation facilities.
- For the telecommunications network has been fulfilled.

d. Community Social Conditions
- Based on the population and density of Kaliwates and Sumbersari sub-Districts which have the highest population density. Meanwhile, in terms of density, Kaliwates Subdistrict and Ajung have the highest density ratings. Population and high density indicate that the area has a high number of residential needs.
- The livelihoods that affect the growth of settlements are agriculture and trade.
- The existence of social institutions that have the most influence on community activities are the Jember City square and the Jember Sport Garden area. This identifies the Kaliwates, Patrang, Ajung and Sumbersari sub-district have better quality of community compared to Sukorambi Sub-district because of the fulfillment of community facilities to interact.

e. Economic conditions of the community
- The risk of improper settlement formation can be calculated based on the number of families according to their class (pre-prosperous and prosperous I), namely: Kaliwates sub-District 20.7%. Patrang sub-District 40.1%, Sumbersari sub-District 30%, Ajung sub-District 38% and Sukorambi sub-District 55.7%.

f. Characteristics of community culture
- Habits of living in the study area specifically do not have certain habits. Community living habits tend to develop at first by following the pattern of the road with a linear pattern and growing by opening new land in the urban suburbs.
- The culture found in Ajung District is the habit of parents to buy a house that is majority in one district.

The descriptions of the existing settlements in the case study are:

a. Formal and non-formal settlements in urban areas
Formal settlements in urban areas are scattered in various regions including the sub-districts of Jember (Kaliwates, Patrang and Sumbersari). Middle-class formal settlements and above are located in the suburban strategic areas included in certain housing. Whereas for the middle and lower housing areas is located in the suburbs (Urban periphery), namely Ajung and Sukorambi districts. The tendency of settlement conditions in the urban periphery region is related to the intensity and density which tend to be lower. Non-formal settlements or settlements are residential areas that grow without the existence of a developer or not included in a particular

residential area. Village settlements are scattered in various sub-districts that are urban areas. The location of non-formal settlements tends to follow the road network and activity center.

b. Urban Slums

There are slums in the city of Jember which are usually illegal for the establishment of settlements. The slum settlements of Jember City are found in part of the border area of the railroad in Patrang sub-districts, Bedadung River in Sumbersari sub-District. As according to the Regent's Decree regarding the slum area of Jember City which is included in the study area, it is spread in the Jember Kidul Village (Kaliwates Sub-district), Karangrejo (Sumbersari sub-District), Baratan, Bintoro and Jumerto (Patrang sub-District). Slums in rural areas in the study area (Ajung dann Sukorambi) are located in areas close to agricultural land. Slums in rural areas do not form a large area but are mostly scattered dots.

Figure 2. Urban Formal and Non-Formal Settlements

Figure 3. Urban Slums

4.2 Spatial analysis of settlement development patterns

Spatial development patterns can be seen on the periodic map from 1997 to 2018 below.

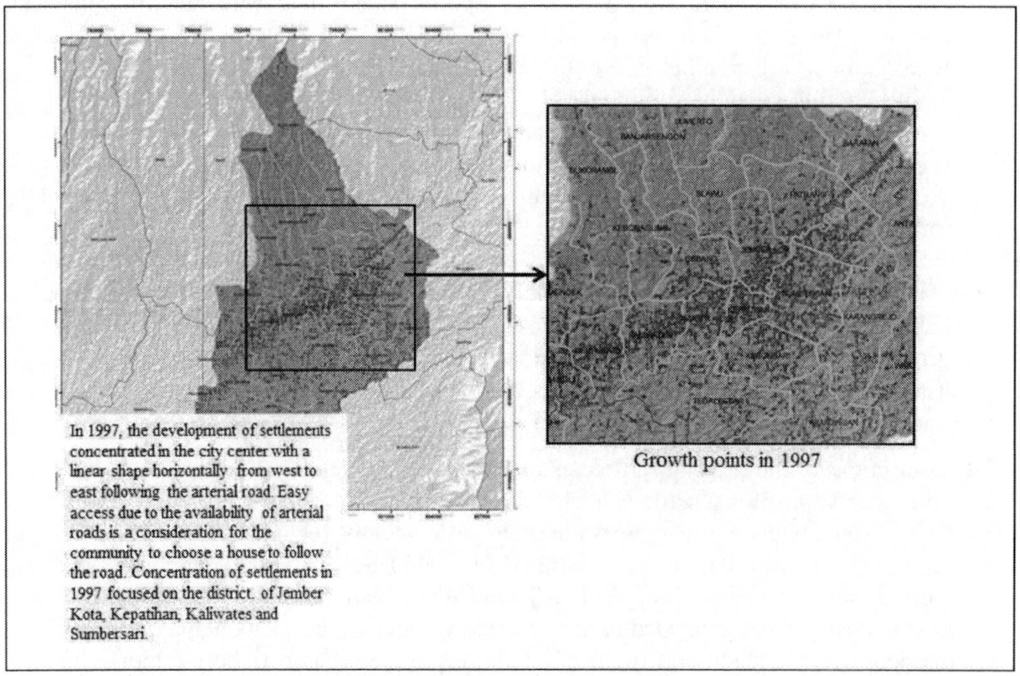

Figure 4. Growth Points in 1997
Source: Analysis, 2018

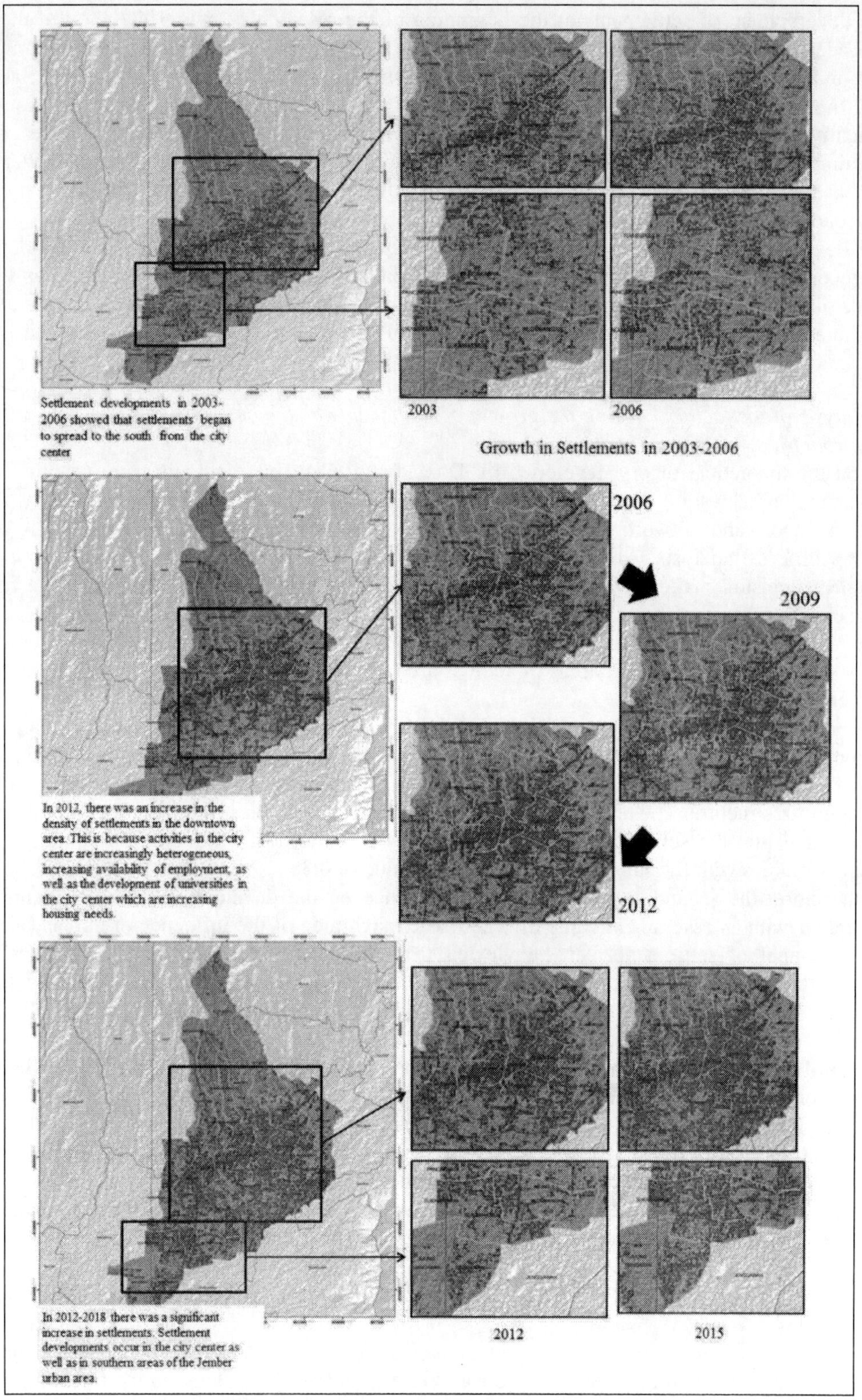

Figure 5. Growth in Settlements in 2003-2015
Source: Analysis, 2018

The development of settlements at the beginning of the research year in 1997 showed that there were concentrations of settlements in the city center, namely Sumbersari sub-District, Patrang sub-District and Kaliwates Sub-district with horizontal forms from west to east following arterial roads. But in the period 2006 to 2018 the development of settlements began to spread southward sporadically, but the highest density concentrations remained in the city center. In the physical organization of space, the development of settlements forms the octopus pattern (Octopus / Star Shape Cities) where in this octopus pattern there are several dominant transportation routes, namely the arterial road that connects Jember-Bondowoso and Jember-Lumajang.

The development of Jember Airport (Notohadinegoro Airport) and Jember Sport Garden starts where according to the 2015-2035 Jember Regency Spatial Plan both regions are planned as strategic areas for the benefit of economic growth. The planned development of the two regions as a strategic economic area can be a trigger for the development of settlements to the south (Ajung sub-District) which began since 2012.

4.3 Analysis of the factors causing the growth of settlements in Jember (through an ekistic element approach)

Based on the theoretical theory developed by Doxiadis (1968), five elements were proposed which later became variables in the study, namely natural factors, human factors, community factors, protection factors and network factors. The five variables are then analyzed using AHP (Analytic Hierarchy Process) analysis. AHP is a process that has been used to assist numerous corporate and government decision makers. The following are the results of the AHP analysis on five research variables based on 15 experts.

The results of the AHP analysis state that the factors that most influence the development of spatially settlements in the urban areas of Jember are factors of housing / housing conditions with an assessment value of 0.261

The condition of housing / housing is a factor that has the highest influence in the development of settlements. This is because people who want to own a house have the greatest consideration of the factors of settlement location, type of housing, location of settlements, level of density and availability of housing infrastructure. The results of the analysis of the sub-variables of the condition of the house / housing stated that the sub-location variable settlement is the most influential sub-variable among other sub-variables with the sub-variable assessment value of 0.385.

Furthermore, the second factor that has an influence on the development of settlements is the human factor with an assessment value of 0.260. The magnitude of the influence of human factors on the development of settlements can be viewed from four supporting sub-variables, namely age, population, human growth and activity. In addition, the assessment value of the largest sub variable is human activity of 0.412. This value means that the higher density of human activity, the more acceleration of the settlement growth will increase.

The third factor that has an influence on the development of settlements in the urban area of Jember is the network condition factor with an assessment value of 0.235 and the sub-variables in it, namely the natural human transfer network, transportation network (land, water, air), the goods transfer network and the energy transfer network. Judging from the sub-variables, the sub-variables with the highest assessment values are land, sea and air transportation networks of 0.415. This means that land, water and air transportation networks have an important role to accelerate settlement growth. One aspect of community support in determining the location of the place of residence they will choose is the completeness of the infrastructure network, especially the easy access to the housing location. This is also the reason why the development of settlements in Jember urban areas tends to follow the road network.

The fourth factor that influences the development of settlements in Jember urban areas is the community factor (eco-culture) with an assessment value of 0.198. Community factors have sub-variables including community quality, income distribution, livelihoods, history and culture. The four variables are then carried out AHP analysis again which results in the distribution of income as a sub variable with the highest assessment value of 0.422. The community aspect is also included in one aspect that influences the choice of settlement location by the community according to the character of the community for convenience. In addition, the sub-variables of culture, livelihoods and income distribution also have an influence on the choice of location of residence by the community.

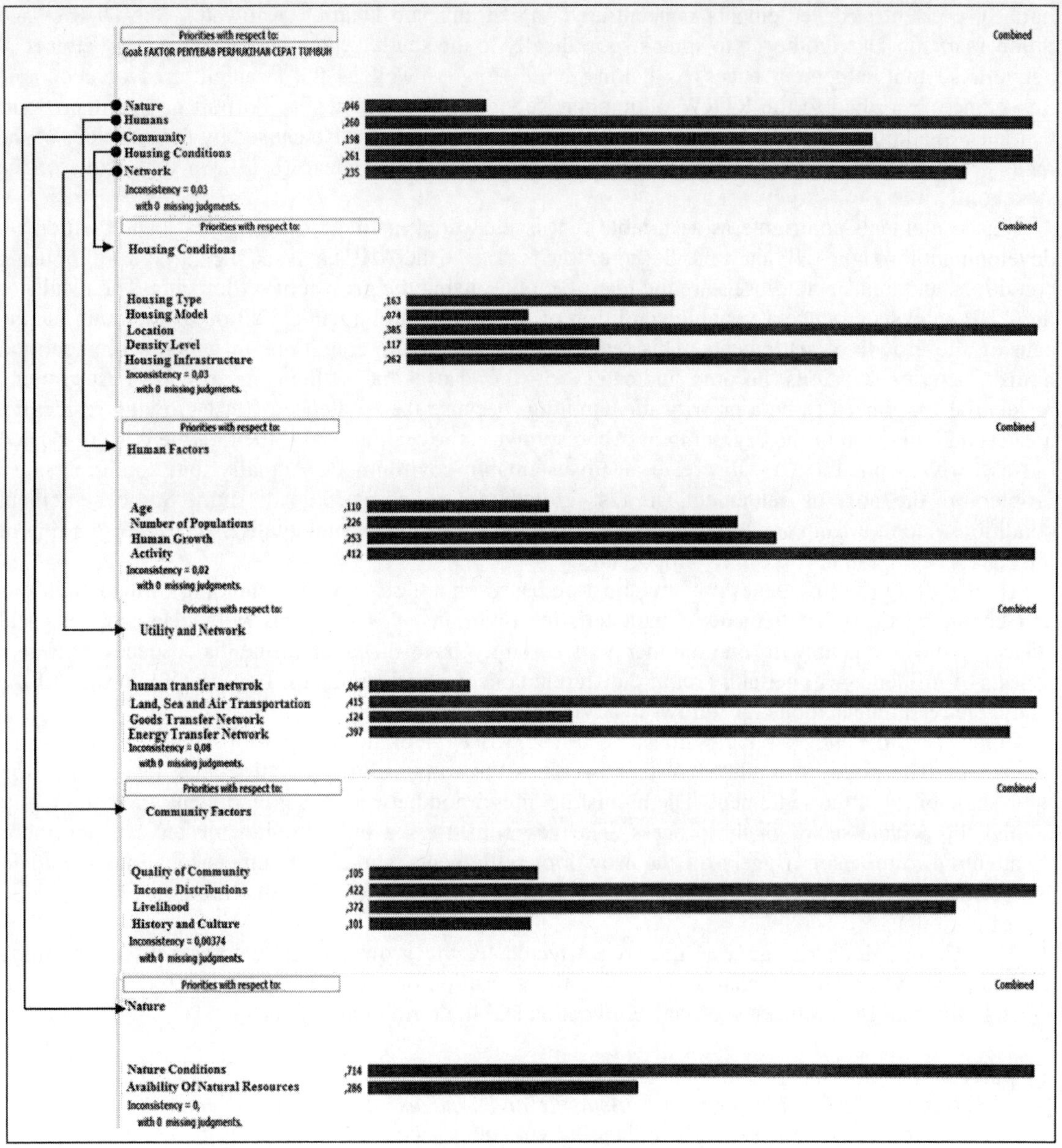

Figure 6. AHP Results
Source: Analysis, 2018

The smallest factor that influences the development of settlements in Jember urban areas is the natural factor with an assessment value of 0.046. This is because natural factors with sub-variables in them are natural conditions and the availability of natural resources is not an important factor in the development of settlements or the last aspects of the community's final consideration in determining the location of settlements for housing.

4. Conclusion

The growth of settlement in Jember happen in peri-urban areas. Based on the results of the analysis, the characterization of urban sprawl occurs in the urban areas of Jember, namely in the Ajung sub-District, precisely in the Ajung Village. These conditions began to appear in 2012 to 2015 where initially concentrated settlements concentrated in Patrang sub-District, Kaliwates Sub-District and Sumbersari sub-District, began to spread sporadically to the south (Ajung District) due to a better road network so that settlement access is getting better too, as well as the strategic area of economic importance according to the RTRW of Jember Regency (Notohadinegoro Airport and Jember Sport Garden). In addition, the development of settlements to the south is also caused by the activities of the factory which become a drag, and the price of land is relatively cheap (based on the results of the stakeholder interview survey).

Ekistic elements contribute as a variable that is measured in finding factors that cause settlement development through AHP analysis. Based on the results of the AHP analysis, the variables of housing conditions and human conditions are the main factors causing the growth of settlements. The results of the AHP analysis produce a variable condition of the house and human condition as the main factors causing the growth of settlements. This result is justified by the conditions in which settlement and human activity locations become indicators of the causes of settlement growth. Fast-growing residential areas need to be a priority for handling, because the existence of fast-growing residential areas is an indication of the beginning of urban sprawl in the region. And if the treatment is not carried out as early as possible, it will create an unsustainable environment. Actually that condition not a problem if the area of settlement (in fast growth area) doesnt indicate slum. Some settlement conditions due to urban sprawl indicate a deterioration of environmental quality. and understanding of the causes of growth becomes very important.

The handling of settlement problems must be based on a thorough understanding of the conditions and characteristics of settlements. Characteristic conditions of settlements with classifications and stages of development with one another will certainly have different influential aspects. Different aspects of influence will certainly require different concepts and strategies. Based on the results of the study, the recommendations for follow-up advice provided are:

1. Concepts and policy strategies must be developed by looking at the root causes of settlement growth. The policy formulated further becomes the responsibility of all parties involved in the sustainability of the settlement. There must be integration between the government, the community and the private sector in the process. The government as a public policy maker can pay more attention to the main triggers for the growth of settlements, namely settlement locations (strategic location of settlement locations) and activities (location of the distribution of economic institutions).
2. The Ajung sub-district area as the area affected by the growth of downtown settlements must immediately formulate a plan to determine the allocation of spatial use in order to control growth and minimize the occurrence of land conversion and the formation of slums.

5. References

[1] Salvatia, L. 2012. *Low-density settlements and land use changes in a Mediterranean urban region.* Landscape and Urban Planning vol **105**.

[2] Ardiwijaya, Vevin S. 2014. *Bandung Urban Sprawl and Idle Land: Spatial Environmental Perspectives.* APCBEE Procedia 10 (2014) 208-213.

[3] Moitra, M. K. 1991. *Environmental improvement of slums The Calcutta experience..* Pergamon press **Vol 26** (Building and Environment): 253-257.

[4] Pratomo, Gigih and Sony Kristiyanto. 2013. *Analisis Sistem dan Peranan Kelembagaan Sektor Perumahan di Kabupaten Jember: Paradigma New Institutional Economics (nie).* Jurnal Equilibrium, **Volume 11** Nomor 1, page 42-56

[5] Doxiadis, C. A. 1976. *Ekistic Elements, Action for Human Settlements***:** 11-27.

[6] Quarterly, T. 1963. *ekistic and traffic*. 439-457

[7] Setioko, Bambang. 2009. *Growth of Urban in Finger Areas (Case Study: Semarang City). Sustainable Slum Upgrading in Urban Area.Informant Settlement and Affordable Housing.* Unit of Research and Empowerment of Housing and Human Settlements Resources, Center for Information and Regional Development, Universitas Sebelas Maret, Surakarta PIPW LPPM UNS. p79- 88

[8] Yeates, Maurice; Garner, Barry. 1980. The North American City. New York: Harper & Row Publishers.

[9] Yunus, Hadi Sabari. 1999. *Struktur Tata Ruang Kota*. Penerbit Pustaka Pelajar: Yogyakarta.

[10] Doxiadis, C. A. 1968. *Ekistics : an introduction to the science of human settlements.* New York : Oxford University Press

[11] Undang-Undang Nomor 4 Tahun 1992. Pemerintah Republik Indonesia.

[12] Pontoh, N. K. 2009. *Pengantar Perencanaan Perkotaan*. Bandung: ITB.

[13] Yunus, Hadi S. 2005. *Megapolitan Konsep, Problematika, dan Prospek*. Yogyakarta: Pustaka Pelajar

IOP Conf. Series: Earth and Environmental Science **340** (2019) 012025 doi:10.1088/1755-1315/340/1/012025

Heritage and the change of meaning: Understanding the urban heritage in Yogyakarta, Indonesia

J P Siregar[1,2]

[1] Department of Regional and Urban Planning, Brawijaya University, Indonesia
[2] School of Design, Faculty of Creative Industry, Queensland University of Technology, Australia

Email: johannes@ub.ac.id

Abstracts. The heritage of Yogyakarta is always situated in a dynamic environment. The conflicting situation is reflected by the ideological tensions during the creation of historical object in the past and the use of heritage in the present day. This issue leads to the underrated appreciation for the historical meaning. This article reveals the ideological process in the creation of heritage object and the recognition of meaning in the present day. The study uses literature study and in-depth interview to reveal the past and currently perceived meaning. This study provides a basis for understanding the change of meaning in heritage discourse.

1. Introduction

Heritage is meaningful not only because of physical uniqueness, but also the narratives embedded on it. The ideological conflicts in the past between the traditional and colonial rulers in Indonesia produce the narratives. The physical representations and narratives become a complete package of heritage.

In the pre-independence period, Yogyakarta grew as a traditional and colonial city [1]. At that time, the traditional and colonial authorities represented hegemonies through the city design. In the present day, the monumental objects created in the past are appreciated as heritage, or the legacy of traditional court and colonial. The new meanings are embedded as the element of identity and economic commodity. The use in tourism industry becomes a government priority [2,3]. This situation reflects the change of meaning is unavoidable. The former representations of authoritarian ideologies have become the current identity and pleasure attraction.

The change of meaning can be threatened with the over-commercialisation of heritage. In the contemporary society, heritage is used as a tourism commodity indicating the change of meaning. The heritage is not only meaningful for local people who understand the narratives, but also for visitors who seek the uniqueness. This phenomenon implies that the narratives are continuously created. The understanding of meaning change is important to anticipating the diminishing of historical and cultural narratives as warned by Fauque [4]. He argues that the domination of economic utility can erode and loose the symbolical meanings of heritage.

Other issues also come as the consequence of urbanisation. Industrialisation and urbanisation is a typical conservation issue in an urban environment [5]. Economic growth is also a potential problem that can erode heritage [6]. Particularly in Yogyakarta, the conservation of heritage competes with the construction of commercial facilities [7,8]. These situations take place because the society and the uses of urban space continuously change. The people always connect heritage to the city function. Consequently, there is a bargain in maintaining the authentic meanings of heritage as the source of historical knowledge and the economic functions.

These issues raise a question about the meaning change of heritage in association with people's perception. This research question is essential because heritage value is a form of social construction [9]. Therefore, this study used in-depth interviews to collect opinions about people's collective memory and culture. This paper aims at introducing a new perspective in recognising heritage by exploring the historical and cultural meanings created in the past and present days.

This paper uses the city of Yogyakarta as the research location with a focus on the historic area of philosophical axis. This case study is located along Malioboro and Mangkubumi Streets. Yogyakarta is relevant case study because this city reflects the mix of traditional and colonial cities. The heritage sites are currently present side by side with the new growing area of the urban centre. Therefore, the area reflects the complexity of historical and cultural narratives, and the attachment of contemporary meanings.

This study offers an approach to the change of meanings by investigating the historical narrative from the past and the current perception of meanings. The first aspect is associated with the creation of memorable objects, or heritage, in the pre-independence period and the second aspect corresponds to people's response to the presence of heritage. This study provides evidence that the current social milieu has changed the meanings; in term of the ideological intention of creating the city is currently perceived as history, identity and economic commodity. This new understanding is important to anticipate the ongoing process of meaning change and to preserve the authentic meanings of heritage.

2. Methodology

The purpose of this paper is to explore the meaning change of historic are in Yogyakarta urban centre. The study contrasts the symbolic meanings of heritage in association with the pre-independence and post-independence situations. The study focus is the influence of socio-political situations on the meanings and how the current society recognised heritage.

The researcher used socio-semiotic as a relevant approach in connecting the meaning to the socio-political context of the society as argued by Gottdiener [10]. The use of this approach is appropriate to the investigation in a dynamic urban situation from the perspective of space users [4]. The use of people's perception in exploring collective memories and experiences is also endorsed by Tweed and Sutherland [11], and Vecco [12]. As a qualitative research, the investigation focused on understanding the meaning constructed in the society as also argued by Fossey *et al.* [13]. The participant's perception is essential because it reflects the interaction between participants and heritage sites. Therefore, the study used in-depth interview to investigate people's perception regarding cultural meaning in the present day. The semi-structured interview was used to maintain flexibility as proposed by Legard *et al.* [14].

The researcher investigated the case study using a time span comprised of the pre-independence or before 1945 and post-independence periods. The situation in the pre-independence period was analysed using data from historical document and research reports. The situation in the post-independence, or the time of research, was investigated using in-depth interviews. Through this way, the researcher is able to compare the historical and cultural narratives created in the two periods.

Regarding in-depth interviews, the study involved 19 participants including 3 persons from expert group and 16 from general public. The expert group was comprised of academics and culturist. Opinion from general public was used to validate the general comprehension about heritage and traditional philosophy. Although the general public were met accidently in public space, the researcher also used criterion sampling to recruit them, based on criteria:

- Adults (from different cohort, such as 20-30, 30-40 and above 40)
- Originally live in Yogyakarta (the participants have experienced the urban space for a long time and has been exposed to local tradition)
- The participants have particular interest in local culture and the city identity.

During interviews, two main questions were asked to trigger in-depth conversation. The questions are:

- How this site (or the heritage sites) means to you?
- How do you remember or believe the stories associated with this site (or the heritage sites)?

The answers revealed themes relevant to the topic. This paper explored narratives developed from the themes regarding the contemporary meanings of heritage in Yogyakarta.

3. Historical context

The city of Yogyakarta experienced the coincidence of classical and colonial periods. This city was founded through the initiation of Gayanti treaty by VOC [15]. This treaty aimed at separating conflicting parties in Mataram Kingdom. As the result, the Kingdom was divided into two smaller kingdoms: Ngayogyakarto Hadiningrat (the current Yogyakarta) and Surakarta (the current Solo). This situation suggests that the colonialist has contributed to the foundation of the kingdom since in the very beginning of Yogyakarta.

At that time the King of Yogyakarta had two political positions: as a traditional ruler and an opponent of colonialist. The role as a traditional ruler was represented through the cosmological concept of the city. The first King of Yogyakarta designed the city using two traditional cosmologies called philosophical axis and *Catursagatra* [16,17]. The physical representations of these cosmologies are depicted in Figure 1. The manifestation of the first cosmology was the alignment of a number of urban objects following the south-north axis. The objects were Panggung Krapyak monument, the Palace complex (*Kraton*) and Tugu Monument. These objects were located in the south, centre and north respectively. This cosmology symbolised a Javanese philosophy known as *Manunggaling Kawula Gusti*. This philosophy encouraged a unity between the leader and people. The position of *Kraton* in the central area emphasised this teaching. Behrend (1989) argued that this central position strengthened the role of *kraton* as *imago mundi* or the centre of universe. Therefore, the King or Sultan was a supreme leader in spiritual and political life.

The representation of *Catursagatra* was the configuration of *Kraton*, the Great Mosque, the northern traditional square or *Alun-Alun Lor* and Beringharjo Market (see Figure 1). This traditional configuration represented Memayu Hayuning Bawono [17]. This philosophy reflected a teaching about leadership in Javanese society on the basis of spirituality, culture and economy.

The antagonistic relationship between the court and colonialist was represented by colonial structures. The ancient Yogyakarta represented the “new Hindia city” in which the office of colonial authority was located near the court palace [1]. This spatial arrangement indicated the polarity of political power in the city. In 1765, the colonialist built a military fort across the colonial office [16]. Sultan reacted this construction by erecting forts surrounding *Kraton*. This situation made political tension between the court and colonialist more obvious.

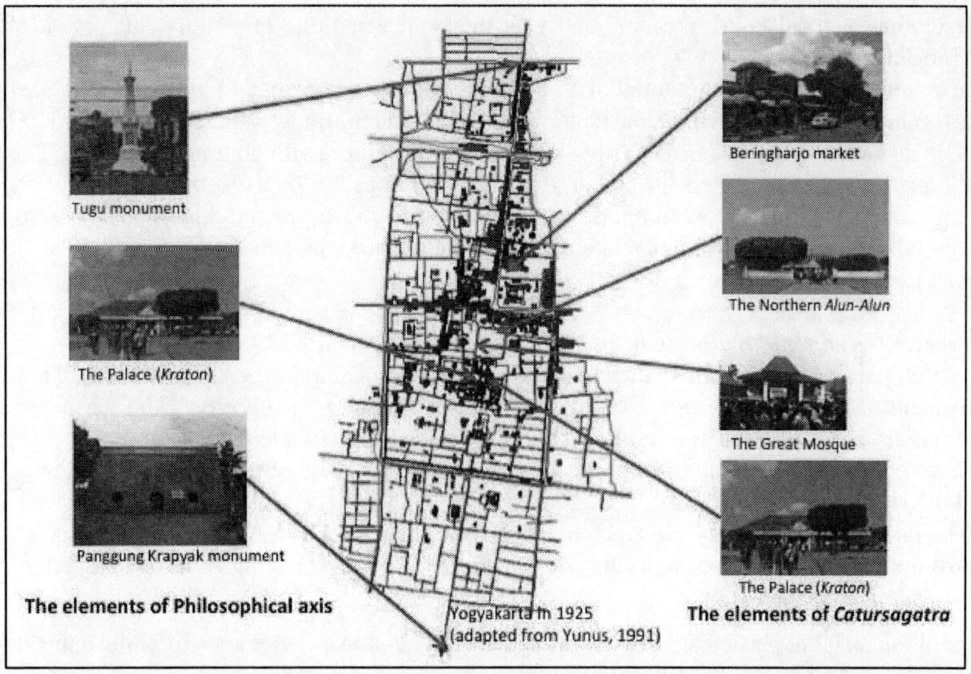

Figure 1. The representations of traditional cosmology in Yogyakarta

The colonialist also introduced industrialisation and modernisation by constructing railways [19]. In 1887, the colonialist built railways crossing Malioboro Street. Kusno [20] argued that the position of railways symbolically neglected the wisdom of philosophical axis. Another form of modernisation was the present of *Indische* architecture [21]. A number of Dutch architects introduced this movement as an effort to combine the Western concept and local tradition [22]. This movement aimed at supporting colonialist economy through a good cooperation with the colonised society [23]. In other words, this movement represented the cultural strategy of colonialist to strengthen the imperialist supremacy.

4. Contemporary context and meanings

4.1 The initiation of cultural heritage and tourism

After the Republic of Indonesia gained its Independence in 1945, the new government developed the city. The urban planning was managed by the central government that initiated road extensions, new infrastructure and the development of educational area [19]. Hence, the city was projected as an educational centre. After 1985, the government introduced modern land uses in the city with Malioboro as commercial area surrounded by residential areas. *Kraton* was designated as cultural heritage area.

Currently, the urban planning of Yogyakarta is regulated by the Municipal Regulation number 1/2015. This regulation designates Malioboro and Mangkubumi areas as commercial zone, while *Kraton* and Vredeburg fort as tourism areas. In addition, the government also protects traditional and colonial buildings using the Heritage Act number 11/2010. Especially the philosophical axis, the government protects this heritage using the Provincial Regulation number 6/2012. Following the protection, the government also promotes the uses of heritage in education and economy. The government also endorses this axis as an international heritage by registered philosophical axis in the tentative list of UNESCO World Heritage in 2017 [24].

This situation reflects the role of government in endorsing local heritage in different platforms. In this sense, the government designates the memorable objects from the past as heritage and uses it for various purposes: the source of education, economy and identity [25]. The study found that the perceived meanings of heritage were associated with the uses of heritage. Currently, people recognise the physical

IOP Conf. Series: Earth and Environmental Science **340** (2019) 012025 doi:10.1088/1755-1315/340/1/012025

manifestations of local court and colonialist as the representations of culture, history, identity and economic commodity.

The cultural and historical potential of Yogyakarta has made the city as a popular tourist destination. The government designates *Kraton* and a number of colonial buildings as tourist sites [26,27]. However, the image as a tourist city has made Yogyakarta in a contestation with commercialisation. Participants disliked a situation when the traditional and historic buildings are overwhelmed by high-rised hotels, shopping centres and other modern buildings. Participant EM02 expressed this issue by asserting:

> "The tall and modern buildings make us can no longer recognise the uniqueness of the city".

4.2 Perceived traditional philosophies

All research participants recognise the elements of philosophical axis and *Catursagatra*. They are also able to mention the wisdoms. For example, the philosophical axis represents the harmony between human, nature and God; and *Catursagatra* symbolises kinship (*kekeluargaan*) and peaceful in social life. These philosophies are the cultural identity of Yogyakarta. The research participant (code GPF03) argued that:

> "Yogyakarta as the city of philosophy has a philosophical axis. There is an imaginary axis, starting from the South Ocean to the Mount Merapi. The axis is comprised of Panggung Krapyak, *Kraton* and Tugu".

In addition, another participant (GPM10) argued that the physical elements of philosophical axis and *Catursagatra* are a part of the cultural existence. It is the heart of Javanese culture (*jantung budaya Jawa*). Therefore, the presence of these symbols is essential to sustaining the culture. Participant GPF06 asserted that:

> "If the historical and cultural symbols had disappeared, replaced by modernity, we might not have the images" ...... "we would be oriented to foreign culture, forget the richness of Yogyakarta local culture".

4.3 Collective memories as social value

Heritage also evokes memories about history associated with the classical and colonial periods. The majority of participants suggest that the presence of *Kraton* and other traditional buildings create an image about the ancient kingdom. On the other hand, the colonial structures remind people about the colonisation. This memory particularly evokes nationalism as argued by participant GPF03:

> "The buildings (colonial buildings) represented our historical value that we had ever been colonised"...... " The buildings reflected the Indonesian fighting spirit (*semangat juang*) gaining the liberty. Moreover, Yogya had been the first capital of Indonesia. This made Yogyakarta special".

These cultural and historical narratives correspond to the heritage values. Participants argue that they are proud of these values and resemble the identity of Yogyakarta and its society. The participants use some words to express the identity and pride, such as *jati diri* (identity), *ciri khas* (particularity), *imej* (image) and *kebanggaan* (pride). Participant GPM10 obviously asserted that:

> "I feel proud of having an identity or particularity. This is my city, this is my culture, and this is the tradition inherited in a very long time across our generations. They (the outsiders / visitors) should know this".

In addition to the historical and cultural values, the participants also mention the role of heritage in the inheritance. Participant GPF02 stated that:

> "Cultural heritage (*cagar budaya*) is a concrete evidence of history".

This statement emphasises the capacity of heritage to contain historical and cultural narratives, and deliver the meanings across generations. Without the heritage, people will gradually forget history.

The in-depth interviews suggest that the meaning is perceived differently than in the past. The next section discusses the change process of meanings.

5. Discussion

The study found that the local society still appreciates philosophical axis and *Catursagatra* as a part of traditional belief and mythology. This belief is fixed in the society and does not change through time. Assmann and Czaplicka [28] suggests this phenomenon as "cultural memory". This kind belief and mythology connects different realities, the cosmos and city urban space, and keeps the metaphor alive [29]. This situation strengthens the symbols of traditional cosmologies as the main media of cultural teaching. These symbols are essential to local identity, as well as the spirituality and moral teaching in Javanese society [30] and across Indonesia archipelago [31]. The construction of identity runs in an illogical way constituting the recognition and implementation of moral and spiritual values of cultural symbols [32]. This leads to what is identified as Javanese and the territory of Java. Therefore, the urban space is meaningful for local people or the believers.

The study also found that the change of social milieu affects the meanings. There is a significant shift of political and social situations between the pre-independence and post-independence periods. In the earlier period, the court and local society are struggling in suppression of colonialist. The polarity of power produces antagonistic symbols. As the time changes, a new government rules the nation. In the new era, the central government recognises the monarchy as the particularity of Yogyakarta side by side with the democratic system of the Republic [33]. There is no more political tensions like in the past. The society currently recognises events in the past as history and the objects associated with the memories as heritage. The change of social agents contributes to social process that gives different ways in recognising and appreciating culture and history [34]. In this term, the government plays a central role in confirming the history and heritage. This action consequently influences the representations of heritage, promotion and the use of heritage. Although the study finds that the role of informal leaders also crucial for disseminating cultural values and traditional philosophy, the government's campaign of heritage gives a wider effects through public policies.

A strong connection between heritage, culture and historical memories is evidenced through in-depth interviews. The recognition of heritage as identity makes the memories popular in the society. The local people of Yogyakarta cannot be separated from the symbol of traditional cosmologies and colonial structures. In this term, the historic sites connect the "chain of popular memories" that makes the city meaningful [35]. The fact that heritage is perceived as a "concrete evidence of history", makes heritage a "historical urban reminders" [36]. The heritage become the medium of sharing collective history and common destiny [37]. In this sense, the physical representations of traditional cosmology and colonisation bind the society to the monarchy, tradition and the independence war. This kind of awareness gives a cultural perimeter to the city and society.

The image as tourist destination is another form of contemporary identity. The use of heritage as an economic potential is not a new idea as suggested by some scholars [38–40]. The current campaign of heritage tourism encourages the heritage conservation and associated development, such as tourist facilities and public space along Malioboro Street. This area currently becomes a melting point between local people, visitors, locality and globalisation (see Figure 2). Hayllar, Griffin and Edwards [41] argues this kind of place as "tourism precinct".

Figure 2. Malioboro as historical street and tourism precinct

The street corridors become a unique place through the sense of memory and traditional belief as suggested by Scannel and Gifford [42]. The traditional cosmologies and colonial heritage assemble a complete package of Yogyakarta heritage. The first requires the living tradition and belief in the society, while the latter needs a commemoration. These representations become the source of cultural and historical narratives that make heritage meaningful. The contemporary use of heritage as tourism commodity is unavoidable. Shepherd [43] warns this action can lead to the "cultural commodification". It is a phenomenon when the economic transaction grows greater that the original interest in local tradition and history. It is also a situation when the locality is defeated by the globalisation represented by the mass production of heritage tourism.

The study endorses the recognition of the "genuine source" of locality as a principal strategy in maintaining the cultural and historical narratives, as well as the use of heritage in tourism industry. The study also proposes that the people's perception is a potential tool in gauging the change of meaning. It is evidenced that the social and political situations contribute to the stability and change of meanings. Therefore, I suggest the current government and society to be aware of this phenomenon, especially in dealing with the intrusion of global symbols or modernity in a traditional and historic urban environment.

6. Conclusion

The historical and cultural narratives are primarily created in the pre-independence periods. The narratives are associated with political tensions between the traditional court and colonial authority. In this period, the two authorities use urban design and architecture to represent their political and cultural influences.

In the contemporary period, the physical representations of historical and cultural narratives are recognised as heritage. The findings from in-depth interviews reflect that the participants still recognise the historical and cultural narratives. The narrative about traditional philosophy is recognised as local wisdom and another is appreciated as history. In this term, the ideological tensions in the past have currently changed into history. The narratives contribute to the identity and nationalism. The interviews also suggest that the physical representation of heritage is important to the succession of historical and cultural narratives. The meaning change is made possible by the change of social agents. The case study demonstrates the authorities as a central agent in creating the symbols as the representations of their political intentions.

7. References

[1] Handinoto. Perkembangan kota di Jawa abad XVIII sampai dengan pertengahan abad XX. Yogyakarta: Ombak Publisher; 2015.

[2] JogjaDaily.com. Tujuh Prioritas Pembangunan Kota Yogyakarta 2016 [Internet]. 2015. Available

from: http://jogjadaily.com/
[3] Yogyakarta TMG of. Heritage city action plan of Yogyakarta. The Republic of Indonesia; 2012.
[4] Fauque R. For a new semiological approach to the city. In: Gottdiener M, Lagopoulos A, editors. The city and the sign: An introduction to urban semiotics. New York: Columbia University Press; 1986. p. 137–59.
[5] Tempo.com. 37 negara bahas kota pusaka di Yogyakarta [Internet]. 2012. Available from: https://m.tempo.co/read/news/2012/09/18/079430217/37-negara-bahas-kota-pusaka-di-yogyakarta
[6] Bisniswisata.co.id. Tiga kota pusaka di Indonesia diharapkan diakui UNESCO [Internet]. 2015. Available from: http://bisniswisata.co.id/tiga-kota-pusaka-di-indonesia-diharapkan-diakui-unesco/
[7] IntisariOnline. Yogyakarta: Keseimbangan dalam pelestarian budaya [Internet]. 2013. Available from: http://intisari.grid.id/index.php/Techno/Science/Yogyakarta-Keseimbangan-Dalam-Pelestarian-Budaya
[8] Tempo.com. Diprotes, bangunan tua di Jogja dijadikan resto cepat saji [Internet]. 2010. Available from: http://properti.kompas.com/read/2010/06/16/17240549/Diprotes..Bangunan.Tua.di.Jogja.Dijadikan.Resto.Cepat.Saji
[9] Avrami EC, Mason R, De la Torre M. Values and heritage conservation: research report. 2000;
[10] Gottdiener M. Postmodern semiotics: Material culture and the forms of postmodern life. USA: Blackwell; 1995. 262 p.
[11] Tweed C, Sutherland M. Built cultural heritage and sustainable urban development. Landsc Urban Plan. 2007;**83(1)**:62–9.
[12] Vecco M. A definition of cultural heritage: From the tangible to the intangible. J Cult Herit. 2010;**11(3)**:321–4.
[13] Fossey E, Harvey C, McDermott F, Davidson L. Understanding and evaluating qualitative research. Aust N Z J Psychiatry. 2002;**36(6)**:717–32.
[14] Legard R, Keagan J, Ward K. In-depth Interviews. In: Ritchie J, Lewis J, editors. Qualitative research practice: A guide for social science students and researchers. Sage; 2003. p. 138–69.
[15] Luthfi, Nazir, Tohari, Winda, Tristiawan. Keistimewaan Yogyakarta yang diingat dan yang dilupakan. Yogyakarta: Ombak Publisher; 2014. 301 p.
[16] Karsono B, Wahid J. Imaginary axis as a basic morphology in the city of Yogyakarta-Indonesia. 2nd International Conference on Build Environment in Developing Countries. Penang, Malaysia; 2008.
[17] Suryanto, Ahmad D, Sudaryono. Aspek budaya dalam keistimewaan tata ruang Kota Yogyakarta. J Perenc Wil dan Kota [Internet]. 2015;**26(3)**:230–52. Available from: http://journals.itb.ac.id/index.php/jpwk/article/view/1354
[18] Behrend T. Kraton and cosmos in traditional Java. Archipel. 1989;**37(1)**:173–87.
[19] Yunus HS. The evolving urban planning: The case of the city of Yogyakarta. Indones J Geogr. 1991;**21(61)**:1–14.
[20] Kusno A. Zaman baru generasi modernis. Yogyakarta: Ombak Publisher; 2012. 163 p.
[21] Veenendaal AM, Knaap PDG. Building modernity. Indische architecture and colonial autonomy, 1920-1940. **Vol.** Bachelor, Faculty of Humanities. [the Netherland]: Utrecht University; 2015.
[22] Cobban JL. Exporting planning: The work of Thomas Karsten in colonial Indonesia. Plan Perspect. 1992;**7(3)**:329–44.
[23] Kusno A. Behind the Postcolonial: Architecture, Urban Space and Political Cultures in Indonesia. Florence: Taylor and Francis; 2014. Available from: http://qut.eblib.com.au/patron/
[24] UNESCO. Historical city centre of Yogyakarta [Internet]. 2017. Available from: http://whc.unesco.org/en/tentativelists/6206/
[25] Indonesia TR of. Act number 11 / 2010: Benda cagar budaya. The republic of Indonesia; 2010.
[26] Yogyakarta T municipal government of. Rencana Aksi Kota Pusaka Yogyakarta. Yogyakarta T

municipal government of, editor. Yogyakarta: The municipal government of Yogyakarta; 2012.
[27] Yogyakarta TMG of. Municipal Regulation number 1 / 2015: The detailed spatial plan and zoning regulation. The Republic of Indonesia; 2015.
[28] Assmann J, Czaplicka J. Collective memory and cultural identity. New Ger Crit. 1995;**(65)**:125–33.
[29] Schultz EA, Lavenda RH. Cultural anthropology: A perspective on the human condition. Vol. 1. Oxford University Press, USA; 2005.
[30] Herusatoto B. Simbolisme dalam budaya Jawa. Yogyakarta, Indonesia: Hanindita Graha Widia; 2001. 126 p.
[31] Pangarsa GW. Merah putih arsitektur nusantara. Yogyakarta: Andi Publisher; 2006. 137 p.
[32] Soedigdo D, Harysakti A, Usop TB. Elemen-elemen pendorong kearifan lokal pada arsitektur nusantara. J Perspekt Arsit. 2014;**9(1)**.
[33] Kurniadi BD. Yogyakarta in decentralized Indonesia: Integrating traditional institution in democratic transitions. J Ilmu Sos dan Ilmu Polit [Internet]. 2009;**13(2)**:190–203. Available from: https://jurnal.ugm.ac.id/jsp/article/view/10961
[34] Zancheti SM, Loretto RP. Dynamic integrity: A concept to historic urban landscape. J Cult Herit Manag Sustain Dev [Internet]. 2015;5(1):82–94. Available from: https://www.emeraldinsight.com/doi/abs/10.1108/JCHMSD-03-2014-0009
[35] Harvey DC. Heritage pasts and heritage presents: Temporality, meaning and the scope of heritage studies. Int J Herit Stud [Internet]. 2001;7(4):319–38. Available from: https://www.tandfonline.com/doi/abs/10.1080/13581650120105534
[36] Lewicka M. Place attachment, place identity, and place memory: Restoring the forgotten city past. J Environ Psychol [Internet]. 2008;28(3):209–31. Available from: https://www.sciencedirect.com/science/article/pii/S0272494408000157
[37] Smith AD. Towards a global culture? Theory, culture and society. In: Steger MB, editor. Globalization and Culture. USA: Edward Elgar Publishing Limited; 2012. p. 349–69.
[38] Graham B, Ashworth GJ, Tunbridge JE. A geography of heritage: Power, culture, and economy. Arnold; Oxford University Press; 2000.
[39] Loulanski T. Cultural heritage in socio-economic development: Local and global perspectives. Environments. 2006;**34(2)**:51.
[40] Salazar NB, Zhu Y. Heritage and tourism. In: Meskell L, editor. Global heritage: A reader. UK: Wiley Blackwell; 2015. p. 240–54.
[41] Hayllar B, Griffin T, Edwards D. Urban tourism precincts. In: Hayllar B, Griffin T, Edwards D, editors. City spaces-Tourist places. USA: Routledge; 2008. p. 3–18.
[42] Scannell L, Gifford R. Defining place attachment: A tripartite organizing framework. J Environ Psychol. 2010;**30(1)**:1–10.
[43] Shepherd R. Commodification, culture and tourism. Tour Stud [Internet]. 2002;**2(2)**:183–201. Available from: http://journals.sagepub.com/doi/abs/10.1177/

Acknowledgements
This work was carried out through a scholarship from the Indonesia Endowment Fund for Education (LPDP) administered by the Government of the Republic of Indonesia.

Reclaiming public space in Avenida dos Aliados: an ethnographic approach in the way society perceptualize their ideal public space

K P Tucunan[1,3] and J T Lopes[2,4]

[1] Urban and Regional Planning Department, Faculty of Architecture, Design and Planning
[2] Sociology Department, Faculty of Arts
[3] Institut Teknologi Sepuluh Nopember Surabaya
[4] Universidade Do Porto

Email: kp.tucunan@gmail.com

Abstract. The public space in Europe has been subjected to the major revitalizations in attempt to lifting the socio-economic value of the city centre as well as Avenida dos Aliados. In the urban planning concepts the interventions to achieve public space as medium has an specific terminology which is " sense of place". It is a design concepts for a place to ensure that every public space able to have identity and a distinctive character to the area and igniting the interactions among the people. Using ethnography as a method, This study give insights to the the definition of the public space according to the people perceptions, comparison of the quality of an *Avenida dos Aliados* as a public space after the revitalization and also highlighting what nowaday's society expected for a public space in Avenida dos Aliados. In order to reclaimed the public space, the society highlighting there is need to include art components (like local artist and creative ornaments), historical information about the city, developing green space, involving technology to stimulate the interaction and prolonged stay, creating inclusive facilities (like games, facility for kids, etc.), and also develop local market in the Avenida dos Aliados.

Keywords: *Public space, Ethnography, Avenida dos Aliados*

1. Background

The concepts of space is defined as many things in many theories. Space is seen as a capital object instead of "pure space" according to Lefebvre as defined in *The Production of Space* [1]. In another definition, the geographer defined the "pure space" as a *terra nulius*[2] or boundless three-dimensional extent in which objects and events have relative position and direction [3]. This approach to views space as a container or only in the physical dimensions, which is mostly debatable, due to its dynamics and inability to control the space as an empty land or nobody's land. Supporting the idea that there are no *terra nulius*, Lefebvre stated that no space is empty, as it is a physical embodiment of a social pattern or otherwise. In an interdisciplinary study involving an anthropologist and geographer about the sea, Jackson proposed a theory in the famous publication "*The Water is Not an Empty Space*" about *mare nullius* which is the term of the space is not empty[4]. The most important highlight of Jackson findings is that even in the most extreme case like the ocean which space is seems to be a *terra nulius*, in fact it is not. Jackson, found the that water is capitalized legally and illegally. There are conflicts in the water that indicating that the space is way more than a container [4].

While there are debates regarding the concepts of space, Habermas stated that public space is considered to be that space to which the citizens of a polity have access and enjoy free right of use [5]. The term of "public" and "free right of use" lead to the sense that public space is a *terra nullius* objects, where public can use or claim the public space as their belonging. Some issues such as question "who owns the public space? "[6] arise as a response of the mis-use of the public space by the private sector. Other findings in Indonesia's case the society misuse the public space to be more semi public space or semi private space [7] [8]. This is the reflection of mismanagement in public spaces that cause by the common mistakes of people perceiving the public spaces.

Most common mistake including: 1). seeing spaces as just a place in the urban area and as "*terra nulius*" 2). believing that space are statics or as an object without any relations with the environment and social developments of its citizen. While on the other hand, public spaces have to be seen as a basis of social, economic and politic representatif of of the city. This mean that space is not static, conflict may arise due to stakeholder differences point of view in seeing the space. This also has a meaning that space is produced by, and continually reproduce, a set concrete social and spatial relations particular to a given region as Habermas define that public space has a definition as "any medium, occasion or event which prompts open communication among strangers" [9]

In the urban planning concepts the interventions to achieve public space as medium has an specific terminology which is " sense of place". It is a design concepts for a place to ensure that every public space able to have identity and a distinctive character to the area and igniting the interactions among the people[10]. Sense of place is a symbolic relationship with the place which is formed by giving emotional or affective meaning to a particular place. This provides the basis of individual's and group's understanding to place. The concept of sense place is composed of three components, namely activities, setting of place, and human experience [10][11] These interplays is signified by Cross, as the mutual relationship and harmony between inhabitant activity in a place that gives them the opportunity to experience the community. Hence, gradually, it constructs the attachment to the place and to the community belong to them [10][11][12].

The public space in Europe has been subjected to the major revitalizations in attempt to lifting the socio-economic value of the city centre as well as Avenida dos Aliados [13]. This study aimed to overlook the definition of the public space according to the people perceptions, quality of an *Avenida dos Aliados* as a public space after the revitalization and also highlighting what nowaday's society expected for a public space in *Avenida dos Aliados.*

2. Managing public spaces: an attempt to revitalizing the public space as a social medium

2.1. Sense of place as a social interactions triggers

There a definite need of an public space to bridging the social interactions and communications and this is comes in many terminology from sense of place to the terms such as place attachment [11][12][14] [15] and also community attachment [16] These terminologies have peculiar similarity in a way to perceive the relation between the individual and their surrounding environment. Sense of place means valuable and meaningful place where people are physically and socially engaged to that place. The important point to build the sense of place is physical and emotional connection, sense of belonging that will encourage people's behaviours to protect their place [17].

The concept sense of place is also closely related to psychological and physical concepts [18] The sentiment toward place were created based on three types of relationship: cognitive, behavioral, and emotional dimension. Cognitive aspect are attributed to the spatial perception (geometry and form) of the space. Behavioral aspect refers to the functional relationship between space and activities. This relationship defines the utility of space that able to accommodate the people needs. Emotional aspect are attribute to the meaning of place to a person and this can be different from one person to others, it depends on experiences, motivations, intelectual background and physical characteristics itself [11].

Furthermore, there are factors in forming sense of place. It is created from at least three different entities, setting of space, activity and interaction, and experience or emotion of individuals as a result of interacting with the surrounding environment, so it is obvious that social relations is crucial. Those factors are listed below:

1. Experience or emotion of individuals:

- Social dimension in space is a generic prescription of an individual expression towards a positive place [19].
- Sense of place relies upon people engagement, community network, and environmental connection [20] These interdependent elements, if they are occurring in a particular setting of place, will create a vibrant and liveable place, which can be easily noticed and perceived by others.
- Cultural values (norm, ethnic group, families, society member) are common in inducing the sense of place by giving a distinctive character towards a place.

2. Interaction and activity features:

The diverse activity features like festival, celebration, cultural event, and gathering is a collective component of the society which have a substantial impact in attracting people, creating emotional feelings to a place and helping the construction of a sense of place [19].

3. Setting of Space:

The physical arrangement/ layout setting of elements can be divided into different aspects of positioning and layout-putting elements in particular places to maximize people's interaction among the people. Whyte have catalogued, in colorful, intricate detail the effects that the layouts and features of built environments have on people's behavior-why some areas become popular, others not so, with whom, and why, with recommendations for how to improve things, including some things below [21];

- Public furniture: including the facilities in the public space (bench, lamp street, sitting arrangement, etc.);
- Path settings and traffic flows: including streets, pedestrians, etc;
- Proximity to the other area: Whyte particularly mentioned that maximum from three block of the center of activities (like: offices, campus, etc);
- Layouts that maximize the natural elements to be enjoyed: sun light, wind, temperature, etc.

Putting all the elements physical into spatial arrangements will creating greater sense of place in some particular public space.

2.2. The public space we idealized: an integrated approach in elaborating the revitalization elements of public space

Places are public spaces if they have some qualities that expected to be present, like:

1) Public interactive qualities, that mean the ideal public space would expect to have the quality to inspire and also dynamic quality;
2) Serve amplifies sociability, that mean the public space expected to be medium to connecting people. Community to community, community to strangers and strangers to strangers, allow convergence between space and action as minimum aggregate of social arrangements and;
3) Public territorialities, which mean the public space has: a.) pre-defined consensus, that it is own by the community b) convergence results of social process that create territorial possibilities of understanding c). express desire and mobilization (social and cultural link being together) [9].

3. Methods

This article is based on ethnography approach, which consist of observations, photography and videography and in depth interview. Ethnography is the recording and analysis of a culture or society, usually based on participant-observation and resulting in a written account of a people, place or institution, Traditionally, ethnographies have focused in depth on a bounded and definable group of people; such as the Dayan, Madurese or other ethnics in the world, that mostly isolated or uncommon to the modern society. Today, they are just as likely to focus on a particular aspect of contemporary social life; such as new reproductive technologies, the meanings of the veil or virtual communication. The concept of ethnography has been developed within social anthropology; but the term is now sometimes used in a looser way in for example opinion and market research. [5]

The rationalistic approach use in these research instead of naturalistic approach, due to the short of the research. Focus on rationalistic allowed researcher to build deduction variable and also indicator to observe and as well as putting the induction results from the field to build local explanations. The main differentiation using the naturalistic is building the theme from the scratch that only viable if the research has much time. The data need to be captured including the demography, user behavioral, physical setting as well as quality of the public space in the perceptions of the user that elaborated in the Table -1.

Table 1. Data capturing, strategy and sampling judgment

Data Capturing	Stategy and Sampling Judgment	Tools
The user of *Avenida dos Aliados*	Participant Observation (weekend – weekdays separtion and also day and night seperation)	Ethnography diary; Photography.
Perceptions and behavior toward public space	Participant Observation (weekend – weekdays separtion and also day and night; In – Depth Interview (differentiate between local and tourist based on early observation judgments).	Ethnography diary; Photography; Videography.
Infrastructure of public space (capturing the physical layout and also facilities for people in society perceptions)	In – Depth Interview (differentiate between local and tourist based on purposeful sampling: maximum variation).	Semi structured open ended questionnaire; Videography.
Accessibility of public space (capturing the proximity and ease to access the public space)	In – Depth Interview (differentiate between local and tourist based on purposeful sampling: maximum variation).	Semi structured open ended questionnaire; Videography.
Dynamic quality of public space (capturing public interactive qualities and the ability to create sense of places)	In – Depth Interview (differentiate between local and tourist based on purposeful sampling: maximum variation).	Semi structured open ended questionnaire; Videography.
Public space's quality to inspire (serve amplify sociability and public territorialities)	In – Depth Interview (differentiate between local and tourist based on purposeful sampling: maximum variation).	Semi structured open ended questionnaire; Videography.

4. Discussion and findings

4.1. Space configuration in Aliados: physical element, human and activities dialogue in Aliados
Based on the preliminary survey and interview, researcher found the distinctive demographic characteristic, which are between tourist and local. The activities of users showing similar characteristic on young people and couple, but show contrast in family and groups. The following table resumes the user demography in the Avenida dos Aliados.

IOP Conf. Series: Earth and Environmental Science **340** (2019) 012026 doi:10.1088/1755-1315/340/1/012026

Table 2. User demography in Avenida dos Aliados

User Tourist Demography	User Local Demography
– Single (20-50 years old) – Couple (25-40 years old) – Family (parents and kids, sometimes with grandparents) – Small Group (less than 5 people) – Large group (5 -30 people)	– Young people (15-30 years old) usually in couple or group – Family (30-60 years old) – Old People (More than 60 years old) – Group: Tuna (coordination, practices), Worker (protest)

The behaviors between the tourists and local have some differentiation. They are including the differentiation: 1). The traffic flows, pit stops (between 3 areas of Aliados), 2). The length of stays, 3). Activities differentiation. The relationship between the human, activities and traffic flows can be seen in Table 2.

Table 3. Relations between user demography activities and traffic flows in Avenida dos Aliados

User Demography	**Activities**	**Traffic Flows, Pit Stops and Length of Stay**
1. Tourists – Single (20-50 years old) – Couple (25-40 years old) – Family (parents and kids, sometimes with grandparents) – Small Group (less than 5 people) – Large group (5 -30 people)	- Meeting points; - Information points (place to get the tourist information); - Seeking directions (from map or other sources); - Selfies/ take photograph;	*Traffic flow*: Tourist tend to have direction from the Hotel Interconintal leaving to Camara Municipal. On the opposite, local mostly use scatter directions that can be seen in the Figure -1; *Pit stops:* The pit stops is not different from the local, although mostly like to use the pit stops in the horse statue or near Camara Municipal. The pit stops usually, near fir the object atrraction like a statue or public furniture. This can be visualized in Figure – 1.
2. Local – Young people (15-30 years old) usually in couple or group – Family (30-60 years old) – Old People (More than 60 years old) – Group: Tuna (coordination, practices), Worker (protest)	- Meeting points; - Enjoying the sun; - Playing card/ family gathering; - Selfies; - Video gaming; - Break/ lunch, etc; - Group show (practices); - Protest.	*Length of Stay:* It is quite obvious that the local have a longer length of stay comparing to the tourist. The local however still resemble Avenida dos Aliados a public space that they usually know, like having a green space, place to enjoying the sun, place to meet and greet a friend, while the tourist on the other hand, mostly do not consider Aliados as a place to stay longer than just opening the map, meeting point and enjoying the scenery. This can be visualized in Figure – 2.

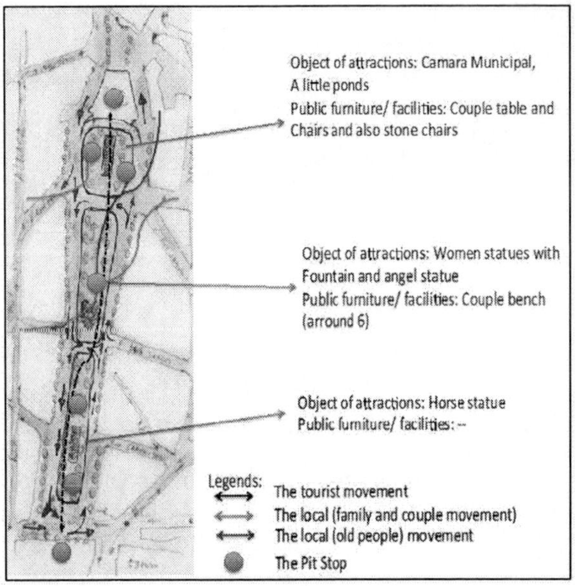

Figure 1. Perceptual spatial movement, objects of attraction and public furniture in Aliados

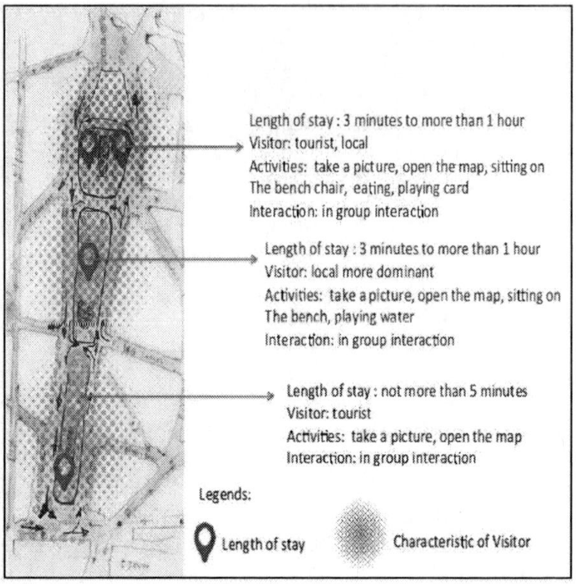

Figure 2. Spatial behavioral of visitor in Avenida dos Aliados

4.2. The public space: public perceptions, preferences and idealism

The way "people" or inhabitants see the public space is the most important thing in the terms of reclaiming the public space in Aliados. There are some perceptions where there is a need to be present in the every good public space, including in Avenida dos Aliados. The criteria are mentioned in the quotation below:

- Place that to be historical/ sacred places of the city
  "Some place where it is like yours and simultaneously to be respectful the culture and the historic area or spirit of the space (Respondent-2, Group, Tourist, 22 years old)
- Place where everybody can going there and spend sometimes and do something
  "In Germany, people in public space doing a lot of activity; read newspaper, doing some job, not like in here, everybody like just starring something like "eerghh"...." (Respondent -3, Group, Tourist, 21 years old)
- Parks or other green area
  "Public spaces is a place for people, like a city center where you can find people, every one can enjoy their time, like green place" (Respondent-6, Single, Tourist, 29 years old)
- Place to socializing
  "Public space is a place where you are comfortable to be there, to meet people there, to comfortable a lot of things, to be relaxed and it has to be free to everyone and the accessibility has to be easy by metro or by bus and something that connect people themselves and also people with the city (Respondent-8, Family, Local, 45 years old)
- Someplace that feel alive
  "In my mind ideal public space is somewhere I can feel alive" (*Respondent-4, Single, Tourist,* 20 years old)

To advancing the experiences of the user, author use place association to help to recall the memories of the participant about ideal public space, as explained below:

- *The Kur (somewhere in the Algeria) with a coffee shop, playground, family come there*

"Well in my country there is this place name The Kur, the big place with some green places and it is beautiful, complete with some place to drink something, to buy something, and it is in city center" (Respondent-6, Single, Tourist, 29 y.)

- Puerta de Sol (Spain), where there is street performances, street market
  Puerta de Sol is might be you can imagine and feel like public, people somehow just drawn there and want to stay longer because the place feel vibrant" (Respondent-7, Single, Tourist, 23 years old)
- Palacio da crystal
  "Crystal palace garden for me is very beautiful, you have a lot of things will go there and stay there a little longer and I think Aliados need that (Respondent-8, Family, Local, 45 years old)"
- Plaza in Paris
  "Well Aliados is beautiful place with a beautiful building but somehow I did not want to stay longer, I see more beautiful dry plaza in Paris because the view is more beautiful"(Respondent-1,Group,Tourist, 22 years old)"

Comparing the theoretical board perspective about how public space define ideal and the idealization of the society, there are slightly differentiation in both perception but strongly affected the criteria. Most people will idealize the public space due to the "green" qualities of its public space. The respondents assumed "green quality" will bridge and igniting the social interactions in the public space due to the extended length of stay it might be created. There also strong value that people will have perceptions that all the historical space will be public space. This is because there a need to connect to the "human" memories, values that space been created in the past. The comparison between the theoretical and perceptional serves in table 4.

Table 4. Theoretical vs perception of public space

Aspects	Theoritical of Public Space	Perceptional of Public Space
Public interactive qualities	VV	VV
Serve amplify sociability	VV	VV
Public territorialities	VV	V
Green place	--	VV
Historical	--	VV

4.3. Aliados as an public space

There are certain expectations for the people for adequate public space. One of the outputs expected in this research is measuring the needs and perceptional quality of an Aliados as a public space comparing to the ideal public space. The assumption is, the ideal public space have the 5 scores for all the criteria (dynamic, social interaction, public facilities and quality to inspire). The figure 3 shows the comparison the score of Aliados as public space (perceptional) and the ideal public space. Overall, most respondents gave between 2-4 and mostly 3 point for Aliados.

The emphasis point is Aliados has no consistency as a good public space. Some events help Aliados to be great Aliados like festival Sao Joao, valentine day, Camellia festival and some concerts, but respondents complaining that the quality they expected only happening in short and temporal time. Lost of green space also been tremendous issue, especially for local who has expected the "greener" Aliados like in the past. While this is being a local mindset, the tourist turn out seeking the same quality, they expected they can stay longer, connect with the residents and also want to remove perceptional that Aliados is just a street or an empty space.

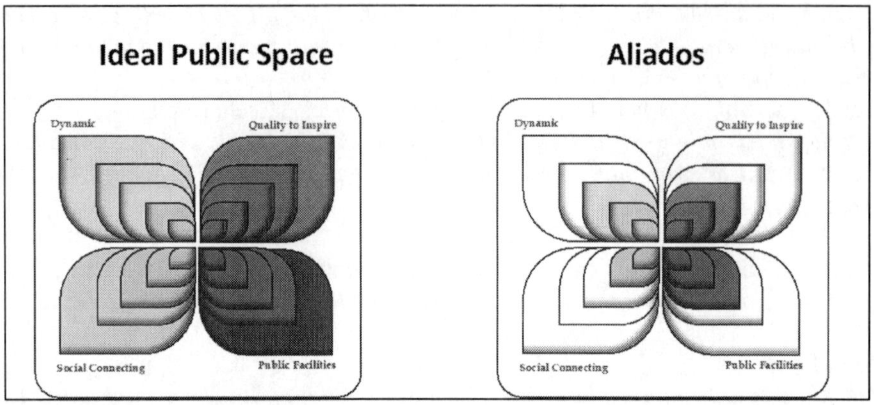

Figure 3. The comparison between ideal public space and *Avenida dos Aliados*

In the qualitative research, the quotation or also known as a verbatim has been used to describe and capture the diction/ vocabulary that has been use by the respondents to prevent misuse of the research and also bring the respondents "as the way it is". This also been used to check the validity and also credibility of the findings of the research (triangulation). Table 4 explaining that respondents give certain scores to the qualities that been measured by researcher and the reason why. Since they did not give the perfect score, the researcher elaborates the positive and negative perceptions toward Aliados.

Aliados dynamic quality mostly advantage by the dynamic of a Porto as a tourist destinations, events and also the historical building that has been perceived by respondents as a nice view and building, while on the contrary the negative qualities cause by the low activity that can be done in the place in regular day. The quality to inspire mostly advantage by the beautiful façade of the building and also connection with the history of the Oporto itself, while the negative quality comes from lack of the inspiring or maybe furthermore connection, information and any other else that can connect between visitor and the place. Most social interactions in Aliados are in-group connections, people judge that Aliados did not bridging the communication between the community and tourist, between strangers and also between group except for the special occasion, on the other hand it is quite good strategy for the government putting some efforts that left the good memories to connecting the people. Public furniture quality is one of the criteria that need to be considering in creating good public space. The respondents responding to the government efforts in putting some public furniture during some events, like box and ornaments during valentine or swinging chair for camellia festival and PORTO letters in some creative space, but on the regular day, most of the respondents think it is quite basic or regular.

Table 5. Positive and negative quality of Avenida dos Aliados as a public space

The Dynamic Quality in Aliados	***The Quality to Inspire in Aliados***	***The Social interactions in Aliados***	***The Public Furniture Quality in Aliados***
Positive (+) quality	**Positive (+) quality**	**Positive (+) quality**	**Positive (+) quality**
City Dynamics	***Beautiful Building***	***Occasional Event in Alidos***	***Occasional Event in Alidos***
"The city of Porto now more open to the people (tourist), Aliados right now more dynamic and vibrant just because the foreigner and the impact of Porto tourism development,	*"Aliados is surrounding by extremely beautiful architecture building, and this is that people don't know, they like to see the beautiful façade of Aliados" (Respondent 2, Group,*	*"3, there is a event, new year eve, football team winning, concert, on so on.. And at some point of the year that connecting local people and tourist (Respondent-8, Family,*	*"if there were a local event, there will be a lot of public furniture in the Aliados, like I told you, swings, box, and other things (Respondent-9, Single, Local, 25 years old)*

The Dynamic Quality in Aliados	***The Quality to Inspire in Aliados***	***The Social interactions in Aliados***	***The Public Furniture Quality in Aliados***
that is way I give Aliados 4" (Respondent-8, Family, Local, 45 years old); **Event in Alidos** *"When there was even in Aliados I Think Aliados could have dynamic quality. I will give 4, like I said previously, the Government held a lot of event in Aliados, even the stage" (Respondent-9, Single, Local, 25 years old)* **Nice View and Building** *"It has it's charm, the building very nice, it is a public space but I don't think it is that kind of public space, because I don't find a lot of people there, maybe it is just some place to have festival or national party, etc (Respondent-6, Single, Tourist, 29 years old)* **Negative (-) quality** **Not a Place To Stay** *"Because it is not a place to stay, Aliados is street for me" (Respondent-7, Single, Tourist, 23 years old)* **Nothing in Aliados** *"Because there is nothing in Aliados, it is like an empty space" (Respondent -5, Group, Tourist, 21 years old)*	*Tourist, 20 Years old)* *"I think I like the building, it is because I don't see the place like this" (Respondent-6, Single, Tourist, 29)* **Historical Story** *"I think Aliados, is the emm.. The most historical park of the city" (Respondent-8, Family, Local, 45 years old)* **Negative (-) quality** ***Don't give any inspiration*** *"No, Aliados don't inspired me a lot" (Respondent-9, Single, Local, 25 years old)* *"I don't know why, but I get no inspiration at all in Aliados" (Respondent 4, Single, Tourist, 20 Years old)*	*Local, 45 years old)* **Negative (-) quality** ***Nothings to Connect With Community*** *"I give Aliados score 3 because Aliados don't gave a lot of things to connect with the community (Respondent 1, Group, Tourist, 20 years old)* *"2, I didn't see Aliados as a place that can connecting between people and community, I told there is nothing here" (Respondent-2, Group, Tourist, 22 years old, tourist)* *"You know, when I was arrived here in the Aliados from the metro I was lost and confused, no one speak English and I need information to go someplace, I have to go to the bar, buy some drinks and than get the Wi-Fi. Its not connected with local community (Respondent-7, Single, Tourist, 23 years old)*	**Negative (-) quality** ***Basic Public Facilities*** *"I give Aliados score 1 because Aliados only have basic public furniture (Respondent-8, Family, Local, 45 years old)"* *"I give 4 because I think is it enough, you know, sometimes the Aliados mean to be empty because it is like some place to walk through" (Respondent-7, Single, Tourist, 23 years old)*

4.4. Reclaiming public space in Avenida dos Aliados

Reclaiming public space in Avenida dos Aliados mean there is a need to optimize the Aliados to be a better public space. The questions asked to the respondent is "what the improvements needed in Aliados so you can give score 5?" The Table 5 giving the verbatim of the each aspects that need to be optimize. Putting something in a continuous period will be massive efforts, especially when the initiative could be from the community itself, like local artists, historical information about what was happened in the Avenida dos Aliados. The quality to inspire can be dramatically increase if the information about Oporto history can be available for public and green space to prolong the people to understand the value of its historical area itself. Local market, map, stuff about the city and also increasing quality of public furniture will benefit the social intearactions. Spesific infrastructure might be recommended like kids stuff and stimulating game, protections to kids during playing in Aliados while there are traffics, another canopy or something that can be use to avoid excess sun light also needed by the respondents.

Table 6. What to improve in Avenida dos Aliados

The Dynamic Quality in Aliados	***The Quality to Inspire in Aliados***	***The Social interactions in Aliados***	***The Public Furniture Quality in Aliados***
Put Something! *"Why not 5? Because Aliados should have something, you know what I mean? When they don't have an event, they should put more in Aliados you know, like swinging chair. During the Christmas they use have swings and put some box in Christmas and that kind of things that attract people, I remember"* (Respondent-9, Single, Local, 25 years old) ***Street Artist*** *"Street artist, it can be example to prolonged my stay, there was a street artist and they play a music and I was watching the street artist and I stand a long time more than 20 minute because they were really good"(*Respondent-7, Single, Tourist, 23 years old)	***Porto Historical Information*** *"Some interfaces, digital information about Porto or the history* (Respondent-9, Single, Local, 25 years old) *"I don't know, maybe some stuff like Porto history, they can put photography or Porto, map or something like that so people can take a look especially the tourist (Respondent-4, Single, Tourist, 20* years old ***Green Space*** *"I think green places will make me give 5 score, because green space make you stay longer and maybe inspire you something (Respondent-1, Group, Tourist, 22 years old)"*	***Interfaces, Information Stuff About The City*** *"Some interfaces, digital information about Porto or the history (Ana Torres, Local)* *"You have put information post and also staff to guide the lost visitor like me (Respondent-7, Single, Tourist, 23 years old)* ***Green Space*** *"Again. Green spaces, something for the kids, Garden maze, something peculiar for the kids will be helpful, public exhibition. Cross in my mind in some European cities. Like public art, painter and artist to add to Aliados" (Respondent-5, Local 23 years old)* ***Local Market*** *"Market will improve a lot, biological market/ local agriculture market, accidental*	***Art, Kids Stuff, Stimulating things*** *"Game, fountain, chairs to relax (lazy chairs) why not some tables, but I don't know small table made by stone, just to add something to add for picnic"(Respondent-8, Family, Local, 45 years old)* ***Tech Things*** *"A lot of things that stimulated grown up and child like hologram illustration, peculiar furniture with something very unique"→ score 1 for now (Respondent-8, Family, Local, 45 years old)* ***Green Space*** *"I think green spaces is more important and the public furniture, like chair, length chair, coffee shop in the middle of the public square" (Respondent-6, Single, Tourist, 29*

The Dynamic Quality in Aliados	*The Quality to Inspire in Aliados*	*The Social interactions in Aliados*	*The Public Furniture Quality in Aliados*
Green Space, Place To Buy *"Make it 5 if you put a green space, there is something for kids, stop the car because it is not calm, mini ice cream too" (Respondent-6, Single, Tourist, 29 years old)* *"With a green space like before I could spend an hour just laying there and bit talk with a friend, eating some sandwich (Respondent-8, Family, Local, 45 years old)*		*market, hand craft, there are a lot of things to be done, frequent market every week or every month will help the connection between local and tourist" (Respondent-8, Family, Local, 45 years old)*	*years old)* ***Some Building Area/roof*** *"The city hall made this choice, they hoping more empty space for the people probably they want to get places where the people meet and greet there. But in the practice they did it wrong because without green spaces, garden and tree we don't have shadow in the summer and tree to sit there in the summer, talk with friend and so on. I think now it is more difficult there just meet people and relaxing (Respondent-8, Family, Local, 45 years old)*

5. Conclusion

As what this paper discuss the human perceptions and need tendencies are crucial for the realistic perspectives of a sustainable public space managements. The public space will be sustainable if there are some qualities that meet the visitor needs. The needs is also evolving from time to time, especially the need to connects. Build up connection between people to people, community to community, strangers to strangers, community to strangers that created the "human" sparks, that created the attachment and also build up the identity of the city it self. *Avenida dos Aliados* need to be re-connect not as an a"temporal space" but also as a place that has value, memories, legacy that need to be hand over generation to generation and community to the traveler.

6. References

[1] Lefebvre, H. (1991). *The Production of Space*, Blackwell, ISBN 0-631-18177-6."Definition of *terra nullius*- English Dictionary". Allwords.com. Retrieved 15 June 2010

[2] Allwords.com. Definition of *terra nulius.*

[3] Space – Physics and Metaphysics". *Encyclopedia Britannica*

[4] Jackson, Sue. (1995). The water is not empty: cross-cultural issues in conceptualizing sea space, May 1995Australian Geographer 26(1): 87-96 DOI: 10.1080/00049189508703133

[5] Jürgen Habermas (1991). *The Structural Transformation of the Public Sphere: An Inquiry into a category of Bourgeois Society*. Cambridge, Mass.: MIT Press, 1991.Who's own the public space

[6] Banarjee, T (2001). The Future of Public Space: Beyond Invented Streets and Reinvented Place. Journal of The American Planning Association. Volume 67. Available from: https://doi.org/10.1080/01944360108976352

[7] Pradinie K, Navastara AM, Martha KE. (2016). Who's Own the Public Space?: The Adaptation of Limited Space in Arabic Kampong. Procedia-Social and Behavioral Sciences. 2016 Jul 14; https://doi.org/10.1016/j.sbspro.2016.06.134

[8] Navastara AM, Pradinie K, Martha KE. (2016). Ambiguous public space towards social sustainability in madurese urban Kampong. Procedia-Social and Behavioral Sciences. 2016 Jul 14;227:507-14. https://doi.org/10.1016/j.sbspro.2016.06.108

[9] Lopes, JT, Hutchkinson R. Public Spaces (2016): Times of Crisis and Change. Emerald Publishing Limited. November 2016.

[10] Altman, I., & Low, S. M. (1992). Place attachment. (I. Altman & S. M. Low, Eds.). New York: Plenum Press. http://doi.org/10.1007/978-1-4684-8753-4

[11] Montgomery, J. (1998). Making a city: Urbanity, vitality and urban design. Journal of Urban Design, 3 (1), 93–116. http://doi.org/10.1080/13574809808724418

[12] Cross, E. J. (2001). What is Sense of Place? In 12th Headwaters Conference (p. 14). Colorado

[13] Carlos J. L. Balsas (2007) City Centre Revitalization in Portugal: A Study of Lisbon and Porto, Journal of Urban Design, 12:2, 231-259, DOI: 10.1080/13574800701306328

[14] Tuan, Y.-F. (1974). Topophilia: A Study of Environmental, Attitutes, and Values. New York: Prentice-Hall Inc.

[15] Stedman, R. C. (2002). Toward a Social Psychology of Place: Predicting Behavior from Place-Based Cognitions, Attitude, and Identity. Environment and Behavior,34(5), 561–581Hummon, D. M. (1992). Community Attachment: Local Sentiment and Sense of Place. In I. Altman & S. M. Low (Eds.), Place Attachment(pp. 253–278). New York: Plenum Press.

[16] Hummon, D. M. (1992). Community attachment: Local sentiment and sense of place. *Human Behavior & Environment: Advances in Theory & Research, 12,* 253-278.

[17] Rogers, Z & Bragg, E (2012), The Power of Connection: Sustainable Lifestyles and Sense of Place. Ecopsychology.

[18] Hashem, H., Abbas, Y. S., Akbar, H. A., & Nazgol, B. (2013). Comparison the Concept of Sense of Place and Attachment to Place in Architectural Studies. Malaysia Journal of Society and Space, 9(1).

[19] Gehl, J. (2006). Life Between Building: Using Public Space. (Sixth Edition). The Danish Architectural Press.

[20] Raymond, C. M., Brown, G., & Weber, D. (2010). The measurement of place attachment: Personal, community, and environmental connections. Journal of Environmental Psychology, 30(4), 422–434.

[21] White WH (1980). The Social Life of Small Urban Spaces. Washington DC: The Conservation Foundation.

Acknowledgment

GRANT from IMPAKT ERASMUS MUNDUS supported this research February 2017 between Institut Teknologi Sepuluh Nopember Indonesia and Universidade do Porto. The authors thanks the respondents that willingly to cooperate and also willingly to be videography. We thank to our colleagues for full support related with preparation and finishing this paper.

The application of environmental management system based on ISO 14001 in Brawijaya Malang University

K E Sari[1] **and S Kamalia**[1]

[1] Urban Regional and Planning Department, Universitas Brawijaya

Email: kartika_plano@yahoo.co.id

Abstract. Universitas Brawijaya (UB) Malang is one of the universities in Indonesia that having the issue of decreasing environmental quality prompted the Rector of UB to put the problem into the University's agenda. To solve this problem, UB plans to apply international standards to support the sustainability of the University, which includes implementing an environmental management system known as ISO 14001 by using Importance-Performance Analysis (IPA) analysis, AHP analysis (Analitycal Hierarcy Process). The first objective in this study is to find out the management's condition on the readiness of the academic community to evaluate the success of the application of ISO 1400 in UB and the second objectives is to produce the extent of preparedness and progress in planning related to ISO 14001 management systems in UB. The results shows that it can be concluded that the current implementation of the environmental management system at UB still lacks the principles of ISO 14001. One of the principles in ISO 14001 that has not been applied at the University indicates lack of preparedness in a University can be seen in the lack of documentation and communication of management policies and commitments on environmental management systems.

Keywords: *environmental; management; university; brawijaya*

1. Introduction

Globalization is demanded for eco-friendly which has a huge impact on the environment. With the existing demands and regulations, the environmental impact of globalization is increasingly minimized. The standard SNI of ISO 14001 wherein this document contains guidance from what aspects fall into the criteria of the environmental management system. In this study will discuss whether ISO 14001 has become part of environmental management efforts in Brawijaya Malang University.

Based on the results of an interview by Dr. Endang Arisoesilaningsih (Chairperson of Brawijaya Malang University Quality Assurance Center) 2017 from the background above, there are several activities that are the focus of this research including:

1. There is no implementation of special programs for the institutional level in applying and ISO 14001 certification at Brawijaya Malang University.
2. There are no program priorities that are in accordance with ISO 14001 requirements, but several criteria for green campus (which are "aligned" with ISO 14001) have begun to be implemented.
3. Provision of a limited budget, so that the current budget priorities are not for implementing ISO 14001.

IOP Conf. Series: Earth and Environmental Science **340** (2019) 012027 doi:10.1088/1755-1315/340/1/012027

4. There is still a lack of readiness of human resources in implementing ISO 14001 requirements in the University of Brawijaya environment.
5. Brawijaya University still has not prepared facilities and infrastructure that are in accordance with ISO 14001 requirements.

The ISO 14001 International Standard is a vehicle for ensuring the performance of the environmental management system. The ISO 14001 standard actually arises as a result of several environmental issues that are often discussed in society. Work units are assessed based on the planning, implementation and evaluation process (PDCA-Plan, Do, Check and Action). The size used by this international institution towards Brawijaya University lies in the implementation of quality management that is adjusted to the requirements of ISO 9001: 2008.

2. Methodology of research

This type of research is qualitative in that the survey method in this study was conducted by distributing questionnaires and interviewing respondents. The following are the research variables used.

This type of research is qualitative in that the survey method in this study was conducted by distributing questionnaires and interviewing respondents. The following are the research variables used.

Table 1. Variable Of The Research

Variable	Sub of Variable
Environmental Policy	Environmental Impact Characteristic of activities
Environmental Planning	Environmental Aspect Action Plan
Implementation And Operation	Structure and Responsibility Training and Control Communication
Emergency Preparedness and Response	Evaluate the need to carry out actions Review the effectiveness of corrective actions Identify and implement corrections
Management Review	Evaluation of program achievements

Determination of respondents is done in a non-probability sampling method where AHP analysis is needed ≤ 5 respondents and for IPA analysis method ≥ 30 respondents.

The analytical tool used from the evaluastive analysis method in this study is the AHP and IPA analysis method with output to evaluate the management conditions that have been achieved based on the ISO 14001 environmental management system, and PDCA (plan -do-check- act) analysis with output to assess performance academic community in the environmental management system which will later become a recommendation for the application of ISO 14001 in Brawijaya University Malang.
The stages of the AHP analysis method in this study are 2 stages:

1. Structuring Stage, where this stage determines the purpose of AHP, the criteria and sub criteria used and what alternatives are available. This stage has the purpose of making decisions in a structured manner based on the existing components.
2. Assessment Phase, which aims to provide weighting values of criteria and sub-criteria as well as alternatives. Giving weight at this stage is given based on the perception of the level of importance directly and also with the graph.

The criteria set for the selection of priorities from the ISO 14001 environmental management system evaluation in Brawijaya Malang University include the 5 variables used in this study is environmental policy (1), environmental planning (2), implementation and operation (3), emergency preparedness and response (4) and management review (5).

3. Results and discussion

In general, ISO 1400 in Brawijaya Malang University is still in the form of a discourse. However, since 2014 Brawijaya Malang University has participated in a green metric program where some of its requirements are the principles of ISO 14001. For example, the convenience of the university environment, electricity energy savings, the application of injection wells for saving water, administrative governance that minimizes paper use, etc. For now Brawijaya Malang University still has no program priorities related to ISO 14001 requirements where environmental policies, environmental planning, implementation and operation, emergency preparedness and response and management reviews are elements contained in ISO 14001 regarding environmental management systems. However, several green metric criteria (which are in line with ISO 14001) have already begun to be implemented in Brawijaya Malang University.

3.1 Environmental policy

Environmental policy in Brawijaya Malang University is still limited to an implementation form where in the environmental policy in Brawijaya Malang University this has not yet been formed as a policy document. The contents of environmental policy in Brawijaya Malang University still do not include a commitment to comply with environmental laws and regulations that are in accordance with other requirements. Meanwhile, Brawijaya Malang University still does not have documents related to environmental policy in SML 14001 and the extent to which the environmental policy in Brawijaya Malang University has been implemented is still in the form of minutes, but UB refers to the green metric UI where indicators of UI green metric can be reference in the application of ISO 14001 in Brawijaya Malang University.

3.2 Enviromental planning

Masterplan making for waste treatment in Brawijaya Malang University is triggered by the number of laboratories, especially in each Faculty in the exact field where some of these laboratories produce B3 waste. Considering that all B3 waste in Brawijaya Malang University, processing is carried out by means of all laboratory activities that produce B3 waste are collected first into one and then for processing the B3 waste is sent and processed in Malang Hospital. The documentation system and requirements related to the environmental management system in Brawijaya Malang University are good enough, where Brawijaya Malang University has planned several aspects to go to ISO 14001 which are among others:

1. The master plan for waste disposal has been carried out since 2017 which was compiled by a team from the planning division of the BAK.
2. It has been planned to make a WWTP which will be in place under the gazebo of the Faculty of Medicine, which will be implemented in 2018.
3. Planning the construction and laying of water catchments planned to be carried out in 2018, one of which is located in the area around the ATM center near the entrance gate of Brawijaya Malang University on Jalan Veteran this year.

The application of external communication that can be done in Brawijaya Malang University includes related agencies and communication with the academic community. By applying these two types of communication, Brawijaya Malang University will receive, document and respond to relevant complaints or complaints from external parties. Environmental communication refers to external and internal communication procedures. The procedure includes:

1. Internal communication between various organizational functions and levels.
2. Acceptance, documentation and responses to relevant communications from related external parties (ISO 14001, 2004).

According to the interview results from Prof.Ir. Sumeru Ashari, M.Agr.Sc., Ph.D. who served as chairman of the 2017 Green metric from the Agriculture Department, Brawijaya Malang University still has not developed and run a good communication system. The communication system (Kabiro with Faculties) in Brawijaya Malang University related to the cleanliness of the environment of the

Finance Admin Bureau is still not fully running smoothly. If the officers are able to do it, then environmental cleanliness in the area of Brawijaya University will be maintained.

3.4 Emergency preparedness and response
Examination actions on physical conditions in the Brawijaya Malang University environment include conditions of several buildings/faculties that have several trash bins for every corner of the building. Some conditions in the trash can in each building/faculty are different. Some still have no sorting at disposal, some have already applied sorting. A lot of trash that has been sorted is found in several buildings/faculties that have applied sorting, among them are Rector, Faculty of Medicine, Faculty of Economics and Business, UB Polyclinic, Samanta Krida Building, Faculty of Agriculture and Law Faculty. The sorting of waste will then be brought to the final disposal site (TPA) near the Faculty of Mathematics and Natural Sciences. This janitor who is within the scope of Brawijaya Malang University is a combination of several internal janitors and external cleaning officers. The internal cleaning officer who is intended is a janitor who carries out cleaning activities in dry waste in the University of Brawijaya environment. Whereas for external officers is a janitor from outside Brawijaya Malang University where the officer carries out his duties in the form of a contract system. This external janitor is in charge of cleaning and sorting dry waste and wet waste in Brawijaya Malang University.

3.5 Management review
Relatively more complete data collection can be seen from 2012 to 2017 where those who participate in the AIM work unit are not only limited to faculties/programs and departments, but also study programs, laboratories and supporting units located in Brawijaya Malang University. However, when viewed from the composition of the number of participation from year to year which shows an increasing trend in the participation of work units in AIM at Brawijaya Malang University is at the level of study programs. In the beginning of 2012 there were still 72 participating study programs, but by the end of 2017 there were 167 participating study programs. From the beginning of 2014 to 2016, it was pointed out that the department was not much involved in the data collection process. Then for the study program which shows changes and variations in data, which shows the trend of decreasing data, namely from its supporting units. If in 2014 is the highest year supporting units, but in 2017 it has decreased to 11 supporting units participating.

Table 2. Conclusions on the Results of Each Variable

Variable	Indicator	Score
Environmental Policy	Brawijaya Malang University still has no official documents related to environmental policy.	3
Environmental Planning	Brawijaya Malang University has planned several aspects related to environmental management systems based on ISO 14001.	5
Implementation and Operation	The closest plan to implement some aspects of the existing master plan is to be implemented in 2018	2
Emergency Preparedness and Response	For now, Brawijaya Malang University has not yet done and scheduled	1

Variable	Indicator	Score
	schedule related to inspection and repair actions	
Management Review	For now, Brawijaya Malang University is still not doing and scheduled schedule related to management progress from ISO 14001	1

The total score for the implementation and operation variable is 2, where Brawijaya Malang University still does not fully schedule according to the planning agenda. For the nearest time, the construction of the parent WWTP will be built which will be built under the gazebo of the Medical Faculty of Brawijaya Malang University in 2018. According to observations that have been made, for examination and repair action variables and management review variables have a total score of 1 which Brawijaya Malang University is still not at both stages.

Table 3. Matrix of Condition Comparison with Brawijaya Malang University Masterplan

Aspect	Masterplan	Existing
Environmental Policy	There is still no masterplan related to ISO 14001 environmental policy documents	Referring to the green metric UI where several aspects of this green metric have led to ISO14001
Environmental Planning	1. UB Zoning Plans for the next few years 2. Plan for the construction of the main WWTP in UB	1. Some development has been carried out 2. Still in the planning process
Implementation and Operation	-	There is no implementation and operation related to ISO 14001 SML in UB
Emergency Preparedness and Response	-	Checking some supporting facilities for several aspects of ISO 14001 related to waste processing at Brawijaya Malang University
Management Review	-	There is no management review related to ISO 14001 SML in Brawijaya Malang University

Some of the problems that hamper the improvement of the ISO 14001 certification program in Universitas Brawijaya, one of which is communication between Kabiro and the Faculty in relation to environmental cleanliness. Here are a few inhibiting factors in improving the ISO 14001 program in Universitas Brawijaya Malang:

1. Brawijaya Malang University there is no implementation of special programs for the institutional level in applying and ISO 14001 certification.
2. Provision of a limited budget, so that the current budget priorities are not for implementing ISO 14001.
3. Still lack of readiness of human resources in implementing ISO 14001 requirements within the Brawijaya Malang University environment.
4. Brawijaya Malang University still has not prepared facilities and infrastructure that are in accordance with ISO 14001 requirements.

According to the interview results from Prof. Ir. Sumeru Ashari, M.Agr.Sc., Ph.D for the future application of ISO 14001 in Brawijaya University must be realized immediately because, the

IOP Conf. Series: Earth and Environmental Science **340** (2019) 012027 doi:10.1088/1755-1315/340/1/012027

application of ISO 14001 must be the face of one of the work programs in Brawijaya Malang University.

3.6 Importance analysis based on Analitycal Hierarchy Process (AHP)
Determination of the score weight in the AHP environmental management system based on ISO 14001 in Brawijaya Malang University uses the current scale, namely the scale of interests with a range of 1 to 9 with the following definition.

Table 4. Scale of Quality Value Criteria

Identification Importance	Definition
1	Both criteria are equally important
3	Criteria A is a bit more important than criterion B
5	A criterion is more important than criterion B
7	Criteria A is more important than criterion B
9	Criteria A is absolutely more important than criterion B
2,4,6,8	Value given if there are two criteria there is a compromise so that the middle value is taken between two consideration values that are close together

Based on the results of AHP calculations using expert choice it can be seen that in the three-step calculation is:

1. Calculating the results of criteria with criteria
2. Calculate the results of criteria with the criteria of each respondent
3. The second calculation is sub criteria with sub criteria
4. Overall calculation of sub-criteria values with sub-criteria

Sub criteria are calculated separately because the limitations of expert choice cannot contain all the sub criteria simultaneously which are too many and have an impact on the inconsistency of sub criteria with sub criteria. Based on the assessment of the AHP questionnaire 5 respondents who have been selected and the calculation of CI can be seen that the results of each respondent can be seen in the table as follows.

Table 5. Assessment Results of Each Respondent's Criteria AHP

Name of Respondent	Score Result					CI
	Environmental Policy	Environmental Planning	Implementation and Operation	Emergency Preparedness and Response	Management Review	
Dr. Ir. Moch Sasmito Djati M	0,597	0,136	0,096	0,096	0,085	0,04
Prof. Ir. Sumeru Ashari, M.Agr.Sc., Ph.D	0,442	0,220	0,159	0,146	0,032	0,09
Dr. Endang Arisoesilaningsih	0,264	0,197	0,197	0,171	0,171	0,02
Dr. Eng. Riyanto Haribowo, ST., MT	0,469	0,225	0,153	0,109	0,044	0,08
Dr. Ir. Sri Utami, MT.	0,264	0,171	0,197	0,197	0,171	0,02

The final calculation of sub criteria and sub criteria is combining all sub criteria so that the sub sub priority results are obtained and knowing the results of the consistency index of all sub criteria with the following score results.

Table 6. Overall Sub-Criteria Results of 5 Respondents

No.	Sub Criteria	Score
1.	Organizational environmental policy	0.233
2.	Communication	0.190
3.	Evaluation to carry out actions	0.111
4.	Identify and correct	0.090
5.	Review the effectiveness of corrective actions	0.076
6.	Evaluation of program achievements	0.065
7.	Training and control	0.065
8.	Action plan	0.054
9.	Structure and responsibility	0.046
10.	Characteristic of activities	0.037
11.	Environmental aspects	0.036
	Consistensies Index	**0.07**

3.7 Importance Perfomane Analysis (IPA)

The average perception of each attribute is the basis for determining whether each attribute on the performance at Brawijaya Malang University is good or not, namely by comparing all the averages on the attributes. The average expectation of each attribute is the basis for determining whether the attribute is important or not.

Table 7. Average Results of Performance Levels

No.	Atribut	Average of Perfomance
1.	Environmental Aspects	3.97
2.	Legislation	3.91
3.	Goals, target and programs	3.88
4.	Resources, roles, responsibilities and authority	3.84
5.	Communication	3.84
6.	Documentation	3.88
7.	Document control	3.88
8.	Emergency preparedness and response	3.91
9.	Monitoring and measurement	3.78
10.	Structuring evaluation	3.84
11.	Record control	3.84
	Total	**11.46**
	Average	**1.04**

To answer the first hypothesis by using measurements from AHP and IPA analysis, the conditions that describe the achievement of the first problem formulation are to find out the management's condition on the readiness of the academic community to evaluate the success in implementing ISO 14001 in Brawijaya Malang University. For this achievement, a calculation process is carried out in which the interests of the AHP results will be adjusted to the interests of the IPA scale by conducting the multiplication process. Furthermore, the results of the multiplication of the AHP interests are also multiplied by the performance in the Natural Sciences, so that the final results can be categorized into three categories: high, medium or low, calculated from the calculation process:

$$\frac{\textbf{highest score – lowest score}}{\textbf{category}}$$

Conclusions related to AHP and IPA multiplication the largest value is found in the multiplication between organizational environmental policies and the control of documents with a score of **0.90** which can be said as a priority for assessing the readiness of the progress of the academic community in Brawijaya Malang University in running ISO 14001 certification related to environmental management systems.

Table 8. Determination of Final Conditions AHP and IPA

AHP		**IPA**		**Total Score**
Sub Criteria	**Score**	**Attribute**	**Score**	
Organizational environmental policy	0.233	Document control	3.88	0.90
Communication	0.190	Communication	3.84	0.73
Evaluation to carry out actions	0.111	Monitoring and measurement	3.78	0.42
Identify and correct	0.090	Record control	3.84	0.35
Review the effectiveness of corrective actions	0.076	Goals, target and programs	3.88	0.29
Evaluation of program Achievements	0.065	Documentation	3.88	0.26
Training and control	0.065	Record control	3.84	0.25
Action plan	0.054	Legislation	3.91	0.21
Structure and responsibility	0.046	Resources, roles, responsibilities and authority	3.84	0.18
Characteristic of activities	0.037	Structuring evaluation	3.84	0.15
Environmental aspects	0.036	Environmental Aspects	3.97	0.14

3.8 PDCA (Plan-Do-Check-Act) Analysis

PDCA analysis is conducted to find out what are the recommendations for the application of ISO 14001 in Brawijaya Malang University related to the environmental management system. It can be seen that for the variables of implementation and operation, inspection actions and improvements as well as management reviews related to the environmental management system based on ISO 14001 in Brawijaya Malang University there is still no action on these three variables because it has not arrived at that stage and there is still no action physical condition planning. So that the stages of management from implementation and operation, inspection actions and improvements and management reviews are still not carried out. It can be concluded that the environmental management system based on ISO 14001 in Brawijaya Malang University is still at the stage of environmental policy and environmental planning.

4. Conclusion

Explained that this research can be concluded that:

1. The implication of this research is that the attention to the environment is getting higher, even in several universities in Indonesia where the academic community is only willing to use products whose manufacturing processes are guaranteed not to damage the environment. Realizing this, the implementation of an environmental management system based on ISO 14001 is a key step of the management system to achieve improvements in the environmental field.
2. Universitas Brawijaya has a commitment that all activities carried out must be environmentally sound. The right way is to implement an environmental management system.
3. The application of the environmental management system based on ISO 14001 in UB has gradually been implemented. This was demonstrated by the University of Indonesia's green metric program which is one of the references for obtaining ISO 14001 certification.

4. The application of objectives and targets, environmental impact management and waste minimization, as well as several environmental management system plans have been carried out in accordance with established procedures.
5. The application of an environmental management system based on ISO 14001 in Brawijaya University, namely by identifying environmental risks and impacts, as well as identifying the level of conformity and the relevant laws and other requirements.
6. The obstacles faced in the implementation of the environmental management system in Brawijaya University include the lack of understanding of the academic community about the importance of environmental problems which considers that this is the responsibility of top management. This causes a lack of awareness of the academic community to implement environmental management system programs that have been implemented.

5. Reference

[1] Imam Munandar. Juni 2017. Pengaruh Sertifikasi Sistem Manajemen Lingkungan Iso 14001 Pada Kinerja Perusahaan. Pascasarjana Sekolah Ilmu Lingkungan Universitas Indonesia

[2] Bima Dirgantara, Harya, Aryo Tri Sambodo. Februari 2015. Penerapan Model Importance Performance Analysis Dalam Studi Kasus: Analysis Kepuasan Konsumen bhineka.com. Jakarta

[3] Chafid Fandeli, Retno Nur Utami, Sofiudin Nurmansyah. April 2008. Audit Lingkungan. Universitas Gajah Mada. ISBN 979-420-6229

[4] Darminto Pujotomo, ST.MT., Agus Yulianto Subekhi. Analisa SML di PT. Janata Marina Indah Semarang Berdasarkan ISO 14001. Program Studi Teknik Industri, Universitas Diponegoro

[5] Dian Anggraini, Lulu, Panji Deoranto, Dhinta Morita Ikasari. Analisis Persepsi Konsumen Menggunakan Metode Importance Performance Analysis Dan Customer Satisfaction Index. Jurusan Teknologi Industri Pertanian Universitas Brawijaya. Malang

[6] Ryan Natasaputra, Muhammad, Parswoto, Yudi Chadirin. November 2015. Evaluasi Efektivitas Penerapan Sistem Manajemen Lingkungan Iso 14001 Di Pabrik Ban XYZ-Jawa Barat. Progran Studi Teknik Sipil dan Lingkungan FTP. Bogor

[7] Setyo Utomo, Budi, Syamsul Maarif, Surjono H.S., Sumardjo. Desember 2011. Aplikasi Analitychal Hierarchy Process (Ahp) Dalam Penentuan Alternatif Pengelolaan Lingkungan Industri Komponen Alat Berat Berbasis Partisipasi Dan Kemitraan Masyarakat. IPB. Bogor

[8] Shoumil Ilhami, Rizka, Dino Rimantho. 27 September 2017. Penilaian Kinerja Karyawan Dengan Metode AHP dan Ratting Scale. Jurusan Teknik Industri FT Universitas Pancasila. Jakarta

[9] Sokovic, M, D. Pavietic, K. Kern Pipan. November 2010. Quality Improvement Methodologies-PDCA Cycle, RADAR Matrix, DMAIC and DFSS. Metrology Institute of the Republic of Slovenia (MIRS). Slovenia

[10] Sueb, Memed, Maria Nety Indramayu Keraf. Maret 2012. Relasi Sistem Manajemen Lingkungan ISO 14001 dan Kinerja Keuangan. FEB Universitas Padjajaran. Bandung

IOP Conf. Series: Earth and Environmental Science **340** (2019) 012028 doi:10.1088/1755-1315/340/1/012028

Impacts of inter-urban transportation railway to regional development (Case study: Sukaraja District - Bogor Regency - West Java Province)

K M Kasikoen[1], Suprajaka[1] and E Martini[1]

[1] Universitas Esa Unggul, Jakarta, Indonesia

Email: ken.martina@esaunggul.ac.id

Abstract. Inter-urban transportation railway becomes the chosen mass transportation mode by the people for daily commuting. The tariff is relatively cheaper and affordable. The travel times are shorter as compared to other modes of transportation. This study aimed to understand the development surrounding Jakarta – Bogor Inter-urban transportation railway station - Cilebut Station located in Sukaraja District. Correlation between passengers and population is 0.99 that shows great impacts to development area surrounding it. Analysis of development aspects are land use, population, and economic change. Using Geographic Information System analysis from the year 2010 to 2015, residential area increased by 18.11% and plantation area decreased by 32.9%. Population was analyzed using average rate of growth from the year 2011 to 2015, the result is 4.6%. Population density increased from 3809 people/sq.km to 4694 people/sq.km. Facilities of industry and grocery store increased, likewise labor structure of business and hotel increased by 11350 persons and services by 3460 persons. But agriculture labor decreased by 894 persons. Based on this research, local government should be alerted. In the case a region has a railway station and is stopped-by inter-urban transportation trains, the area development will take place rapidly, intervention needs to be done.

Keywords: *Inter-urban transportation railway, Regional development, Land use change, Population change, Economic Change.*

1. Introduction

Inter-urban transportation railway becomes one of mass transportation modes preferred by the people for daily commuting. Besides relatively cheaper and affordable, the travel times to destinations are shorter as compared to using other modes of transportation. Republika online newspaper February 1st 2018 [1], according to the report by PT. Kereta Commuter Indonesia as a company which develops inter-urban transportation railway, stated that the passengers of *Commuter Line* trains which serve Jabodetabek routes in 2017 reached 315,849,947, or an average volume of 993.804 passengers per day. The highest record of passengers in a day was 1,076,274. Those numbers are relatively high compared to the users of other mass transportation modes in Jabodetabek. According to Central of Statistic Bureau (BPS), cumulative number of passengers of train in 2017 was 393.3 million or increased by 11.78% as compared to the same period in 2016. The existence of the Commuter Line trains as mode of inter-urban transportationmay elevate the volume of passengers using mass transportation and becomes the most favored transportation mode by the urban people living in Jabodetabek.

The existence of train as one of inter-urban transportation modes save the costs and the travel time urban dweller working in the central of the city. Locations in the vicinity of the stopped by railway stations becomes the chosen alternatives for their dwelling. Although some of the locations may be far away (from the central of the city), since the price of the land is cheaper, easy access to the central of the city and affordable travel cost, they become the deciding factors for choosing those areas for dwelling. In turn, railway inter-urban transportation impacts the development of areas surrounding the stoppėd by railway stations.

Gordana Stefancic, et al [2] stated that the best location for dwelling is the point where the dwelling location is affordable and having cheap transportation cost, in such that low income family will be less living in the area close to the central of the city. The higher income family will live in area farther away from the central of the city. As a result, the total volume of land use for dwelling will increase and city peripheral will expand.

The high number of higher income family dwelling in locations farther away from the central of the city influences the characteristic of commuters actively using inter-urban transportation for their trip. Mackenbach, et al [3] stated that commuters who actively make trips are individuals having high income where active trip frequency is higher at the end of the commuting (working place) and lower at the start of the commuting. Ministry of Transport New Zealand [4], stated that lower costs and enhanced accessibility, due to better transport links and services, expand markets, create increased opportunities for trade, competition and specialization, which can lead to longer-term productivity gains.

Development of areas surrounding railway stations needs to be paid attention considering their impacts may cause uncontrolled development, as stated by Febri[5] that development of transportation, especially railway, causes the stopped by areas grow more rapidly as identified by the high land use and development of new economic centers which in turn will raise the number of population and resulting in the demand for land increases, it is therefore the area needs to be expanded. If this is not followed by spatial planning from the early stages, it may cause urban sprawl as consequences of uncontrolled expansion of areas surrounding the railway stations. It is therefore involvement is required for the areas surrounding railway stations to avoid problems associated with urban sprawl from arising.

Tony McCall [6] said that in 1950s economics basis used in indicated regional development, towards the end of the 20th century, regional development became far-more multidisciplinary in its approach, and in 21st , the focus of regional development is more on the spatial dynamics of regions. Furthermore Luthfi Muta'ali [7] stated that regional development potential is indicated by population, social – economic characteristics, land-use and infrastructure, and accessibility.

KRL Commuter Line Jakarta – Bogor began the operation in 2011.There are many stations between Jakarta – Bogor stop by the train, one of them is Cilebut Station. Since the existence of KRL Commuter Line Jakarta – Bogor, the area surrounding the station had become more densely populated. It is indicated by data of average population growth in Sukaraja District from 2011 to 2015 which is 4.6% per year. This high population growth has caused growth of residential areas, especially surrounding Cilebut station. This condition showed that the areas surrounding Cilebut station experienced rapid development.

This study is aimed to understand the development of areas surrounding the stations which is inter-urban transportation railway stopped by. Area surrounding Cilebut Station selected, located in Sukaraja District, Bogor Regency, West Java Province - Indonesia, which had rapid development since the station located in this district was stopped by KRL Commuter Line train Jakarta – Bogor as the inter-urban transportation railway. The results can be used for reference in the case of a region which facilitated station stopped by inter-urban transportation train, the region can be plan earlier to anticipate the bad impact of inter-urban transportation train to the region.

2. Data and methods

This study is limited to Sukaraja District – Bogor Regency as the area surrounding Cilebut Station. Data collection is through collection of primary data and secondary data. Primary data is collected through direct observation and documentation to land use activity in the surrounding Cilebut Station, it is used to obtain description of the existing condition of land use surrounding the station. Secondary data is obtained through institutional survey.

The analysis used for identifying the link between the railway and the regional development is analysis of correlation between the growth of passenger and population from year 2014 through 2017, as available data passengers, the formula in DewiSawitri and Sri Maryati [8] as follows:

$$r_{xy} = \frac{n\sum xy - \sum x\sum y}{\sqrt{n\sum x^2-(\sum x)^2}\sqrt{n\sum y^2-(\sum y)^2}} \tag{1}$$

Whereas if r value =$\pm$ 1 the correlation of the two variables is perfect, and ifr value = 0, there is correlation between the variables.

In identifying regional development, the analysis is done by land use change, the development and density of population, and the availability of economic facilities and the labor changed in each village in the Sukaraja District.

The analysis of land use change used before-after method by utilizing overlay analysis on the land use maps of 2011 and 2015. Variables used in the analysis are acreage and pattern of use variables. Whereas land use change analysis is started by preparing land use maps which covers preparation of area delineation, image cutting, image re-composition, geo-referencing on the images, digitizing on the images in such that it produces land use maps. Subsequently the produced land use maps will be overlaid to obtain data of acreage of change as results of overlaying 2011 map and 2015 map. In identifying population growth, calculation is done on the total population in 2011 through 2014, likewise of economic facilities and labor change. Data used from Potensi Desa 2011 [9] and Potensi Desa 2014 [10].

3. Results and discussion

3.1. Characteristic of Sukaraja District

Sukaraja District is located at 6°34'08.33 South latitude and 106°51'53.27 east longitude, at elevation of between 350-682 meters above sea level. Based on 2016 data, total acreage of Sukaraja District is 4200Ha. The biggest village is Nagrak with area of 800 Ha, the smallest village is East Cilebut with area of 100 Ha. Sukaraja District comprises 13 villages with administrative boundary as follows:

Northward	:	Cibinong District
Westward	:	Bogor City
Southward	:	Ciawi District
Eastward	:	Babakan Madang District

For further clarification see Figure 1.

IOP Conf. Series: Earth and Environmental Science **340** (2019) 012028 doi:10.1088/1755-1315/340/1/012028

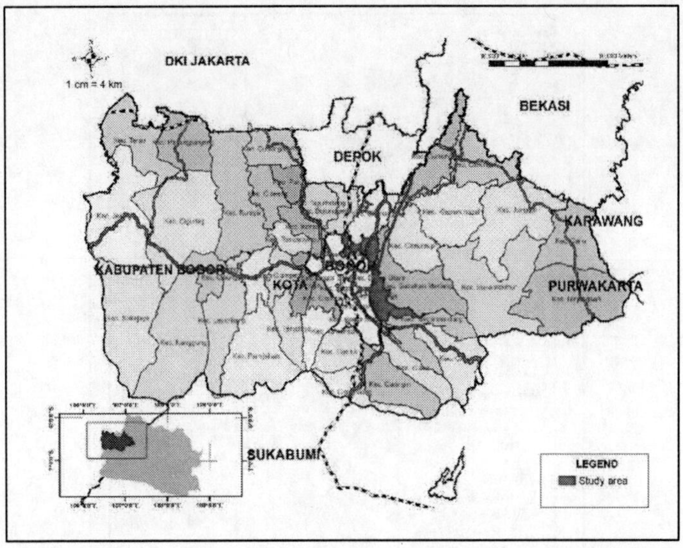

Figure 1. Administrative boundaries map of Sukaraja District

3.2.Correlation between inter-urban transportation railway and regional development

To understand the impacts of the inter-urban railway transportation to the regional development, firstly done is the correlation analysis between the two variables such as population and passengers which start and stop in Sukaraja District. The result strengthen the hypothesis regarding the existense of regional development in areas surronding Cilebut Railway Station as a result of inter-urban railway transportation. Table 1 show the available data used for correlation analysis.

Table 1. Total passengers and population in Sukaraja District from the year 2014 through 2017.

Year	Passengers	Population
2014	4,724,525	192,820
2015	5,384,099	197,168
2016	5,913,657	201,435
2017	6,466,896	205,599

Source: PT Kereta Commuter Indonesia 2018 and Sukaraja District in Figure year 2014 to 2017

Based on the result of correlation analysis above, the value of r is 0.99 which means the correlation of the growth of population, as one of regional development indicators, is positive with the growth of the number of inter-urban railway passenger. Therefore the impact of inter-urban railway transportation is significant to the regional development.

3.3. Analysis of land use change

In 2015, non built up area of Sukaraja District was 1,827.52 Ha (51%), whereas the built up area was 1,755.194 Ha (49%), mainly for urban residential with acreage of 1.453,7821 Ha or 41% of the total acreage of Sukaraja District. Majority of land use for urban residential is scattered surrounding Cilebut Station which is in the villages of West Cilebut, East Cilebut, Pasir Jambu, Mandala, and Cijunjung, whereas the areas farther away from the station which are the villages of Gunung Geulis, Cibanon, Nagrak and Sukatani are more dominated by non built up area, especially for padi fields, plantation and farming area. For further clarification regarding the land use in Sukaraja District see figure 2 and Table 1 below.

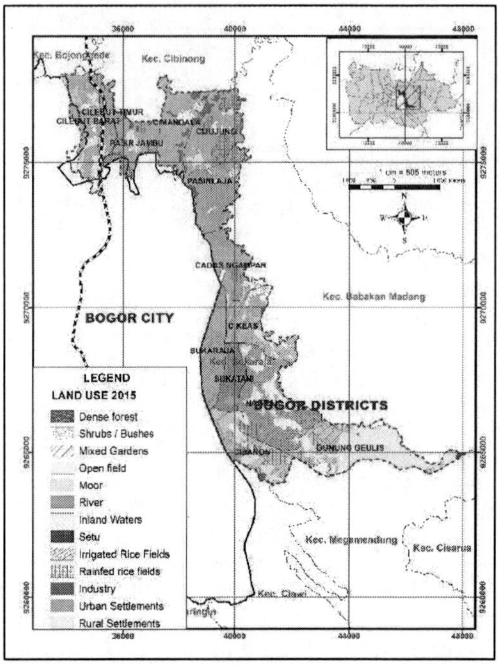

Figure 2. Sukaraja District land use map

Table 2. Percentage acreage of land use in Sukaraja District year 2015

Land Use	Percentage
Rural residential	7.09
Forest	0.11
Plantation	12.47
Mixed Plantation	14.88
River	0.68
Urban Residential	33.11
Rain dependent Padi Field	9.78
Industry	1.66
Land Water	0.01
Irrigated Padi Field	17.54
Bush	2.53
Lake	0.01
Open Area	0.14
Total	100

Source: GIS Analysis, 2015

Based on the results of land use analysis, significant change took place from 2010 through 2015, especially on land use of residential and industry. In 2010, residential area was 22.08% of the total area of the district, whereas in 2015 it was 40.20% of the total. Since the plantation area decreased from 45.37% in 2010 to 12.47% in 2015, it is logical to suspect that the area converted to residential area was from the area used to be plantation area.

3.3. Population analysis

One of the indicators that a region is developing is the increase of population. Villages in Sukaraja District experienced rapid population growth. Population of Sukaraja District in 2011 was 159,988 and in 2015 was 197,164, so annually average growth is 4,6%. Commensurate with this population growth, the population density in Sukaraja District increased from 3,809 people/sq.km to 4,694 people/sq.km.

To prove that this regional development was caused by the increase of population resulted by the existence stopped by Cilebut Station of inter urban-transportation railway KRL Commuter Line Jakarta – Bogor, it may be viewed from the population growth in the period of 2011 to 2015 in the villages located surrounding this railway station which are East Cilebut, West Cilebut, Pasir Jambu and Cimandala compared to population growth of villages farther away from station which are Ciujung, Pasir Laja, Cadas Ngampar, Cikeas, Sukaraja, Sukatani, Cibanon, Nagrak and Gunung Geulis. From the results of analysis found that villages surrounding the station tended to have high population density, whereas villages farther away from the station in the average have medium to low population density. For further clarification see Table 3 and Figure 3.

Results of this population analysis show that the existence of inter-urban transportation railway KRL Commuter Line Jakarta – Bogor influenced to the increase of population in Sukaraja District, especially in the area surrounding Cilebut Sation. This increase of population impacted regional development of areas surrounding the station.

Table 3. Population density of each Village in Sukaraja District, year 2011 and year 2015

Village	Population Density 2011 (people/sq.km)	Population Density 2015 (people/sq.km)
Cibanon	1703	1699
Gunung Geulis	1300	1473
Nagrak	1334	1570
Sukatani	2294	2473
Sukaraja	3665	4772
Cikeas	3469	3996
Cadas Ngampar	3113	3786
Pasirlaja	3883	4653
Cijunjung	5594	7526
Cimandala	7938	9120
Pasir Jambu	5185	6694
East Cilebut	18127	21551
West Cilebut	5706	7934
Sukaraja District	**3809**	**4694**

Source: Population analysis year 2011 and 2015

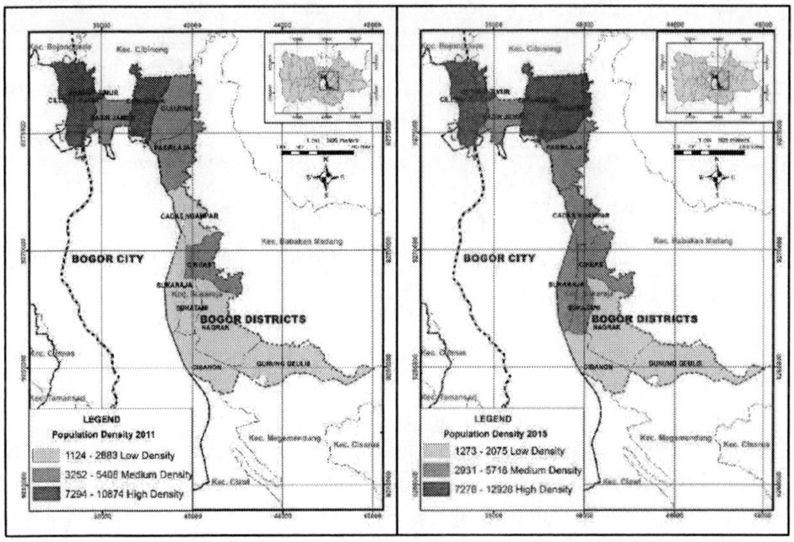

Figure 3. Maps of Population Density in Sukaraja District, 2011 and 2015

3.4. Economic Analysis

Economic analysis Sukaraja District will be approached by the changes of availability of facilities and the change of labor, show in Table 4 and Table 5.

Table 4. The changes of availability of economic facilities in Sukaraja District, 2011 – 2014

Villages	Industry	Grocery Store	Food stalls	Restaurant	Hotel	Mini Market	Bank
Cibanon	-1	29	-2	0	0	0	0
Gunung Geulis	0	0	13	1	0	1	0
Nagrak	5	1	-1	0	0	1	0
Sukatani	-6	-2	-3	0	0	0	0
Sukaraja	0	15	0	1	0	2	0
Cikeas	-3	34	-2	1	2	1	0
Cadas Ngampar	-1	45	2	0	1	0	0
Pasir laja	18	6	5	0	0	0	0
Cijunjung	6	300	11	1	0	1	1
Cimandala	2	8	-2	1	1	2	0
Pasir Jambu	7	137	0	0	0	4	0
East Cilebut	6	-8	-3	0	0	2	0
West Cilebut	1	185	-6	0	0	2	2
TOTAL	**34**	**750**	**12**	**5**	**4**	**16**	**3**

Source: Result Calculation from Potensi Desa year 2011 and year 2014

Table 5. The Changes of labor structure in Sukaraja District in the year 2011 - 2014

Villages	Agriculture	Industrial	Construction	Business and Hotel	Transportation	Services
Cibanon	-164	81	50	347	59	8
Gunung Geulis	313	46	541	250	-42	-396
Nagrak	-174	171	185	1836	597	1308
Sukatani	-32	19	117	184	1	-12
Sukaraja	-329	141	210	857	240	149
Cikeas	-263	80	235	953	228	438
Cadas Ngampar	-126	10	320	497	44	45
Pasir laja	60	83	249	1145	89	165
Cijunjung	37	454	301	895	213	397
Cimandala	19	501	409	908	166	346
Pasir Jambu	-210	523	149	1676	125	512
East Cilebut	-134	102	185	969	76	82
West Cilebut	109	140	218	833	105	418
TOTAL	**-894**	**2351**	**3169**	**11350**	**1901**	**3460**

Source: Result of Calculation from Potensi Desa year 2011 and year 2014

From Table 4, showed that the availability of facilities in some villages decreased, but the total availability of each facility in the whole district increased. From Table 5, showed that the agriculture labor in some villages decreased, but others are increased.

Based on the above data it is understood that during the period of 2011-2014 economic facilties increase across the board in Sukaraja District especially the grocery stores which increased by 750 units with highet increase in Cijujung village e.g by 300 units. And it strengthened by the changes of labor, from agriculture to labor that indicate urban area as industrial, construction, business and hotel, transportation and services. It shows that the existence of the inter-urban railway transportation has pushed up the economic of the region as indicated by the increase of economic facilties and changes of labor in the surrounding of stop-by railway station.

4. Conclusion

It can be conclude that there is correlation between the existence of inter-urban transportation railway with regional development. The changed of land use, population, labor and economic facility of rural to urban strengthen the prove impact of inter-urban transportation railway to regional development, especially Sukaraja District as area surrounding Cilebut Station.

5. References

[1] Republika online daily newspaper February 1st, 2018. Available from: https://www.republika.co.id/amp/p3gx3i428 (accessed 09 23, 2018).

[2] Stefancic, Gordana, Slavko Saric, and Robert Spudic. *Correlation between Land Use and Urban Transport: Case Study of Zagreb*. Traffic and Transportation, 2014: Vol. 1 26, No. 2, page 179-184.

[3] Mackenbach, Joreintje Dingena, Edward Randal, Pengjun Zhao, and Philippa Howden Chapman. *The Influence of Urban Land-Use and Public Transport Facilities on Active Commuting in Wellington, New Zealand: Active Transport Forecasting Using the WILUTE MODEL*.Sustainability, 2016: Vol. 8, page 242.

[4] Ministry of Transport New Zealand. *Contribution of transport to Economic Development: International literature review with New Zealand perspective*. November 2014 page 8.

[5] Fitrianto, Febri Nur. 2017. *Studi Dampak Pembangunan Stasiun Bandung Terhadap Daerah Sekitarnya*. Seminar Heritage IPLBI: ITB.

[6] Mc Call, Tony. 2010. *What do we mean by Regional Development?* Cradle Coast Campus University of Tasmania. Available from: http://www.utas.edu.au/_data/assets/

[7] Muta'ali, Lutfi. *Potensi Perkembangan Wilayah dan Kaitannya dengan Tata Ruang di Kawasan Lereng Merapi Propinsi Daerah Istimewa Yogyakarta.* Majalah Geografi Indonesia, 2005: Vol 19, no 1, page 63-87

[8] Dewi Sawitri and Sri Maryati. 2015. *Metode Analisis Perencanaan*. Penerbit Universitas Terbuka. Jakarta.

[9] Potensi Desa Badan Pusat Statistik, Jakarta. 2011 and 2014.

[10] Sukaraja District in Figure, Badan Pusat Statistik Kabupaten Bogor 2014, 2015, 2016, 2017.

Acknowledgments

Thanks to Kemenristekdikti which has provided funding this research for year 2018.

IOP Conf. Series: Earth and Environmental Science **340** (2019) 012029 doi:10.1088/1755-1315/340/1/012029

Simulation of Voluntary TDM scenarios on road service level in corridor of Nicolau dos Reis Lobato-Kolmera, Timor-Leste

K D M E Handayeni[1] and E F Gomes[1]

[1] Department of Urban and Regional Planning, Institut Teknologi Sepuluh Nopember, Surabaya, Indonesia

Email: erli.martha@gmail.com

Abstract. Nicolau dos Reis Corridor Lobato-Kolmera is one of the corridors in the center of Dili City with the dominance use of office, trade and services activities. This effects a high trip attraction, so the level of road service in this corridor starts to be unstable especially during peak hour. At present there is no policy to deal with congestion issues in terms of managing travel demand. Therefore, this study aims to develop Voluntary-based Travel Demand Management (TDM) scenarios that can improve road service in the Nicolau dos Reis Lobato-Kolmera Corridor. Scenario is developed through simulation of multiple linear regression models and calculation of degree of saturation (DS). The results show that the TDM scenario in the form of Vanpool is considered more effective in improving road level services compared to Carpool and Flextime scenarios. The scenario of Vanpool and Flextime can be developed in office activities, while in trade and service activities require a combination of Carpool and Flextime policies to significantly reduce DS values.

1. Introduction

The increase of vehicles number in urban areas is the dominant factor congestion problem at peak hour. Various efforts to solve congestion have been carried out, but most of the practices in cities in Indonesia put forward supply side rather than demand side [1]. This causes an imbalance between the provision of road networks and car-based travel demand. An approach that has often been applied in most cities in the world is the concept of Travel Demand Management (TDM). According to Ferguson [2], TDM is a concept which aims to increase the need for movement or the need for transfer in different space and time. TDM focus on people/goods mobility rather than vehicles movement. By TDM, transportation demand is managed through various strategies to avoid travel at the same time and / or occur at the location or place that is done simultaneously (Ferguson, 2000). In implementing TDM, a variety of specific strategies that have been developed in accordance with existing transportation in the city through four major strategies, namely time shift, route shift, modal shift, and location displacement [3].

Dili is the capital of the State of the Republic of Democracy of Timor-Leste (RDTL), which is currently the centre of activities for both government, offices and trade and services. Based on data from the Direcção Nacional Transportes e Terrestres (2015), in 2010 to 2013 the growth of motorized fleets increased rapidly where three quarters (3/4) of all vehicles were motorbikes which reached 31%. In addition, at the end of 2014 the number of private car vehicles reached 9,518 units and motorcycles reached 60,579 units. This is normal if there is congestion in the city of Dili, especially at peak hour.

This phenomenon often occurs in several roads in the Vera-Cruz sub-district (west of Dili), especially on the Nicolau dos Reis Lobato road and the Kolmera road section.

The problem of congestion can be reduced by providing a range of TDM strategies that can provide mobility options for travellers [3]. According to Ferguson (2000), the taxonomy of TDM consists of Voluntary, Market and Regulation. TDM Voluntary emphasizes changes in travel behaviour through various choices made voluntarily in the short term. Alternative modes and alternative time strategies are part of this TDM Voluntary [5]. In implementing the Voluntary TDM strategy, incentive efforts rather than disincentive efforts such as those proposed by Meyer (1999) regarding TDM strategies are grouped as carrot and stick strategies to influence travel behaviour. The same thing was stated by Taylor (2007) how the application of TDM Voluntary in Australia is effective in influencing changes in travel behaviour through an incentive mechanism (carrot approach) rather than a disincentive (stick approach) [6]. The results of studies in the US show that the effectiveness of Voluntary TDM implementation in the form of Vanpool and Carpool strategies can be seen from the reduction in the number and cost of travel [7]. One of the best practices from the TDM strategy that has been successfully implemented is in the city of Alameda, California, namely the strategy of transit (transit use and carpooling). Since 2006 in Alameda City, California has managed to offer an incentive approach for residents to change their daily travel patterns using bicycles, public transport and Carpools. The application has succeeded in changing travel behaviour using private vehicles reduced by 14%, public transport users have increased to 34% and 5% Carpooling. In addition, TDM has also been successfully implemented in other cities in the world, such as the cities of Cambridge, Singapore, Perth, Beijing, Dublin, Greater Vancouver Region, London, Trondheim, Portland and Auckland. The successful implementation of the TDM strategy can be seen from the decreasing level of travel length (VMT-Vehicle Miles Travelled), reduction of congestion (reduction in vehicle volume on roads), emission reduction, shifting mode choices, and so on [1].

The various effectiveness of TDM implementation is influenced by various factors, namely the level of public acceptance, coerciveness, the combination of strategies, the existence of political support and the level of publications / campaigns to the community [8][1]. In general, the TDM strategy without coercion is more easily accepted by the public rather than the TDM strategy by force [9]. TDM Voluntary is generally related to the TDM approach without coercion, meaning that travellers move voluntarily based on the choice of TDM strategies offered without limiting their freedom to use private vehicles. The TDM Voluntary strategy will be easier to implement in terms of public acceptance. Therefore, this study aims to identify various types of Voluntary TDM strategies that can be applied in the Nicolau dos Reis Lobato and Kolmera Corridors, and how this TDM Voluntary scenario affects the level of road services. The results of this study are expected to provide a new perspective on the handling of transportation issues in the territory of Timor-Leste using transportation management on demand side.

2. Research method

In this study using a type of quantitative descriptive research. Descriptive research aims to describe the nature of a situation that is temporarily running at the time the research was conducted and examine the causes of a particular symptom [10]. In collecting and expressing various problems and objectives to be achieved, this research uses a quantitative approach. According to Sugiyono (2009), quantitative research methods can be interpreted as research methods that are based on rationalistic philosophy, used to examine certain populations or samples [11]. Therefore, in this study using samples of 185 workers whose travel destination is corridor of Nicolau dos Reis Lobato and Kolmera. This sample size has met the requirements for data adequacy at an error rate of 10% on the basis of the absence of information on population in this corridor [12]. The sampling technique used is the purposive sampling where respondents was selected are those who have regular working movement in the study area.

There are three stages were carried out in this study. First, identification of respondent's preference to the kinds of Voluntary TDM strategy that is most appropriate. The second stage, the analytical method used in this study is multiple linear regression and calculation of road's Degree of Saturation. The multiple linear regression method is used to determine the relationship pattern between trip attraction of various types of land use and traffic volume in the study area. Furthermore, the method of calculating the degree of saturation is used to determine road's level of service (LOS) through a comparison of traffic volume with road capacity. The Voluntary TDM scenario simulation in this study was carried out by examining the Voluntary TDM options and their effect on trip attraction and traffic volume in the corridor of Nicolau dos Reis Lobato and Kolmera.

3. Result and discussion

3.1 Identification of worker's preference to Voluntary TDM strattegy

At first stage, this study identify the preference of travellers to Voluntary TDM options to be applied in the corridor of Nicolau dos Reis Lobato and Kolmera. Voluntary TDM options are Vanpool and Carpool as mode shift alternatives and option of Flextime, Staggered Shift dan Compressed Work Week as time shift alternatives. The options was cross tabulated with land-use and vehicle type. Table 1 shows the result of worker's preference tabulation where workers in office activity are more agreeing to the TDM option using Vanpool by 79% to go to work. Meanwhile, those who worked in the type of trade and service activity preferred with changes in mode using Carpool by 65%. In addition, to change the time or time-shift of both office workers chose Flextime option by 54% as well as those who work in trade and services activity chose Flextime option of 67% in order to reduce congestion. Most of travellers disagreed with option of Staggered-shift and Compressed work week to be applied in the corridor of Nicolau dos Reis Lobato and Kolmera.

Table 1. Cross tabulation between traveller's preference on Voluntary TDM Options with land-use and vehicle type

Voluntary TDM Option	Office Activity				Total of Respondents in Office Activity	Service and Trade Activity				Total of Respondents in Service and Commercial Activity
	Light Vehicle		Motorcycle			Light Vehicle		Motorcycle		
	Agree	Disagree	Agree	Disagree		Agree	Disagree	Agree	Disagree	
Mode Shift										
Vanpool	37%	8%	42%	13%	100%	5%	20%	40%	35%	100%
Carpool	14%	31%	26%	29%	100%	20%	5%	45%	31%	100%
Time-Shift										
Flextime	26%	19%	28%	27%	100%	18%	7%	49%	26%	100%
Staggered-shift	8%	37%	18%	37%	100%	4%	21%	32%	44%	100%
Compressed work week	8%	37%	17%	38%	100%	8%	16%	22%	53%	100%

Note:
- Selected TDM option
- Not selection TDM option

3.2 Simulation of Voluntary TDM's effect on level of road service

At this stage, multiple regression model was developed to catch the pattern of relationship between trip attraction and road level service. Trip attraction was calculated on each land-use type by traffic counting on weekday and weekend. Figure 1 shows the trips number of trade and service activity tends to stable every hour, but number trips of office activity are more bigger and tends to fluctuate.

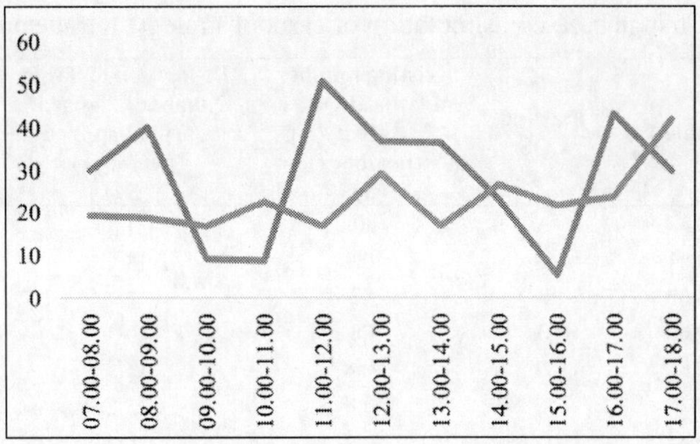

Figure 1. Comparation of Trip Attraction Number (pcu/hour) between Office (blue line) and Trade and Service Activity (orange line)

Furthermore, this study developed multiple regression model of trips number and road's degree of saturation. Saturation degree was calculated through traffic volume divided with road capacity. The highest traffic volume on Nicolau Labota-Kolmera road reached 1.745,8 pcu/hour with the capacity is 2.100,3 pcu/hour. The result shows that the saturation degree reached 0,83 or road's level of service reached D category which means unstable traffic flow. Using Pearson Correlation technique, this degree of road's saturation has significant relation with trips number of office activity, which the correlation value is 0.730 at p-value of 0.011. Furthermore, regression model of saturation degree and trips number as following equation:

$$Y = 0.589 + 0.004(X1) + 0.00042(X2) \tag{1}$$

where X1 is trips number of office activity; X2 is trips number of trade and service activity; Y is saturation degree (DS). This model has R-square of 53.4%. Based on the equation 1, simulation of TDM scenario aims to analysis of scenario effect on trips number change and also the change of road's Degree of Saturation (DS) and Level of Service (LOS). The following table describes simulation of Vanpool option at office activity and its effect on DS and LOS change. DS and LOS change were also simulated based on Carpool option at trade and service activity and on Flextime option at both type of land use.

Table 2. LOS Change based on simulation of vanpool strategy at office activity

Time	Existing DS	Existing LOS	Estimation of Trips Number Change if Vanpool applied (pcu/hour) (X1)	DS Change	LOS Change
07.00-08.00	0,57	A	16,8	0.67	B
08.00-09.00	0,81	D	13,3	0.66	B
09.00-10.00	0,63	B	3,1	0.61	B
10.00-11.00	0,69	B	3,5	0.61	B
11.00-12.00	0,81	D	21,7	0.69	B
12.0013.00	0,78	C	15,2	0.67	B
13.00-14.00	0,83	D	18,1	0.68	B
14.00-15.00	0,65	B	8,1	0.63	B
15.00-16.00	0,64	B	3,8	0.62	B
16.00-17.00	0,81	D	19,5	0.69	B
17.00-18.00	0,75	C	11,9	0.65	B

Table 3. LOS Change based on simulation of carpool strategy at trade and service activity

Time	Existing DS	Existing LOS	Existing number of trips at Office Activity (pcu/hour) (X1)	Estimation of Trips Number Change if Carpool applied (pcu/hour) (X2)	DS Change	LOS Change
07.00-08.00	0,57	A	30	21.5	0,73	C
08.00-09.00	0,81	D	40,2	15.8	0,77	C
09.00-10.00	0,63	B	9,2	12.7	0,63	B
10.00-11.00	0,69	B	8,8	19.3	0,64	B
11.00-12.00	0,81	D	50,8	13.5	0,82	D
12.00-13.00	0,78	C	36,8	25.2	0,76	C
13.00-14.00	0,83	D	36,4	13	0,76	C
14.00-15.00	0,65	B	23,2	22.9	0,70	C
15.00-16.00	0,64	B	5,4	18.8	0,62	B
16.00-17.00	0,81	D	43,2	18.3	0,79	C
17.00-18.00	0,75	C	29,8	27.2	0,73	C

Based on table 2 and 3, Vanpool option is more effective to change road's level of service rather than Carpool strategy. Implementation of Vanpool strategy at office activity is more possible to support the worker's travel pattern. In terms of providing transportation mode, government can supply the vehicles and require Vanpooling for workers in government sector. Different with Carpooling strategy, it is more difficult to be applied for workers in private domain. Flextime strategy which is applied at both type of land use can change LOS better rather than Carpool option as following table 4.

Table 4. LOS Change based on simulation of flextime strategy at both type of land use

Time	Existing DS	Existing LOS	Estimation of Trips Number at Office Activity if Flextime applied (X1)	Estimation of Trips Number at Trade and Servie Activity if Flextime applied (X2)	DS Change	LOS Change
07.00-08.00	0,57	A	35	24,52	0,74	C
08.00-09.00	0,81	D	35,2	13,68	0,73	C
09.00-10.00	0,63	B	9,2	16,8	0,63	B
10.00-11.00	0,69	B	24,16	26,68	0,80	D
11.00-12.00	0,81	D	35,44	12,72	0,78	C
12.00-13.00	0,78	C	38,38	34,6	0,76	C
13.00-14.00	0,83	D	34,82	11,4	0,73	B
14.00-15.00	0,65	B	23,2	26,6	0,69	B
15.00-16.00	0,64	B	18	25,94	0,67	B
16.00-17.00	0,81	D	30,6	19,66	0,71	C
17.00-18.00	0,75	C	29,8	42	0,75	C

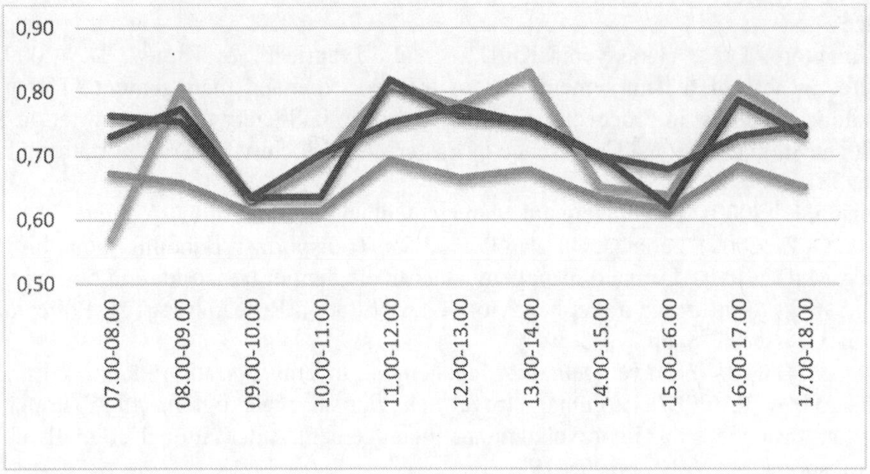

Figure 2. Comparation DS Change between Existing DS (orange line) and DS with Vanpool (blue line), Carpool (purple line), Flextime (red line)

Figure 2 describes that Vanpool option is most appropriate to change road's degree of saturation (DS) comparing with existing condition. Therefore, this option can change road's level of service (LOS) become better than initial condition, from D category change to B (stable flow) at peak hour. LOS change with Vanpool option is significant at any observed time of movement. Based on this simulation result, there are various strategies which appropriate to be implemented on each land use type as following:

1. In office activities, the scenario that is appropriate to be applied is Vanpool option. Workers can use a share vehicle that provide by government so that the use of private car can be reduced. In addition, the flexible time set in Flextime option encourage workers to shift depart and arrive time in corridor of Nicolau dos Reis Lobato and Kolmera. This flexible time keep considers the rule of working time which is eight hours per weekday.
2. In trade and service activities, the appropriate scenario to be applied is Carpool and Flextime. Because of DS and LOS change is low and only occur at certain time, it is necessary to combine those options in implementing Voluntary strategy. Depart and arrive time is arranged flexibly, but workers which the origin is close should travel using shared-car.

4. Conclusion

Voluntary TDM options which is preferred to be applied in corridor of Nicolau dos Reis Lobato and Kolmera are Vanpool, Carpool and Flextime. Time shift strategy with Staggered Shift and Compressed Work Week option is not suitable with the characteristic of worker's movement pattern. In office activity, only mode shift strategy is appropriate with applying Vanpool. Different with trade and service land use, both type of mode shift (Carpool) and time shift (Flextime) is suitable to be implemented. Work trip number in office activity are significant correlated with road's degree of saturation in corridor of Nicolau dos Reis Lobato and Kolmera. The effect of work trip total in office activity is greater than trade and service activity. Therefore, regression model of road's degree saturation (DS) formulates the coefficient of trip attraction number in office activity is greater than trade and service activity. This regression model is then used to simulate the change of road's DS and level of service (LOS). This study results that Vanpool scenario have significant effects on DS and LOS change rather than Carpool and Flextime scenario. At peak hour, Vanpool scenario can reduce the initial DS value by 0.83 to 0.68 or LOS of D to B (stable flow). Meanwhile, Flextime can only reduce the DS value to 0.75 and reach DS to 0.76 with Carpool. In the improvement of the road's level of service in corridor of Nicolau dos Reis Lobato and Kolmera, it is necessary to develop Vanpool and Flextime strategy in office activities. While in trade and service activities, a combination of Carpool and Flexible Time policy is needed to significantly reduce DS value.

5. Reference

[1] Kusumantoro, I. P., Handayeni, K. D. M. E., Syabri, I., & Kipuw, D. (2009). Level of effectiveness of the implementation of transport demand management (TDM) strategy in Indonesian cities. In Proceedings of the Eastern Asia Society for Transportation Studies Vol. 7 (The 8th International Conference of Eastern Asia Society for Transportation Studies, 2009) (p. 38).
[2] Ferguson, E. (2000). Travel Demand Management and Public Policy. Ashgate.
[3] Tamin, O. Z. (2000). Perencanaan dan Pemodelan Transportasi. Bandung: Penerbit ITB.
[4] Meyer, M. D. (1999). Demand management as an element of transportation policy: Using carrots and sticks to influence travel behavior. Transportation Research Part A: Policy and Practice. Vol 33, Issues 7-8, pp 575-599
[5] Meyer, M. D. (2016). Travel Demand Management. In Transportation Planning Handbook.
[6] Taylor, M. A. P. (2007). Voluntary travel behavior change programs in Australia: The carrot rather than the stick in travel demand management. International Journal of Sustainable Transportation. Vol 1, pp 173-192
[7] Wayne, B., & Laurel, R. (1997). Travel demand management in the USA: context, lessons learned and future directions. Energy Policy, 25(14–15), 1213–1215
[8] Gärling, T., Eek, D., Loukopoulos, P., Fujii, S., Johansson-Stenman, O., Kitamura, R. Pendyala R., Vilhelmson, B. (2002). A conceptual analysis of the impact of travel demand management on private car use. Transport Policy. Vol 9, pp 59-70
[9] Steg, L., Dreijerink, L., & Abrahamse, W. (2006). Why are energy policies acceptable and effective? Environment and Behavior, 38(1), 92–111. Available from: https://doi.org/https://doi.org/10.1177%2F0013916505278519
[10] Sugiyono. (2008). Statistika untuk Penelitian. Bandung: CV. Alfabeta.
[11] Sugiyono. (2009). Metode Penelitian Bisnis (Pendekatan Kuantitatif, Kualitatif, R&D). Bandung: Alfabeta.
[12] Lemeshow, S., Hosmer, D. W. J., Klar, J., & Lwanga, S. K. (1990). Adequacy of Sample Size in Health Studies. England: John Wiley&Sons Ltd.

IOP Conf. Series: Earth and Environmental Science **340** (2019) 012030 doi:10.1088/1755-1315/340/1/012030

Making the connection between accessibility and travel behaviour in university setting (Case study: Babarsari Area, Yogyakarta)

M K Devi[1], L M Fitria[1], M S Roychansyah[2] and Y Herwangi[2]

[1] Urban and Regional Planning Department, Sekolah Tinggi Teknologi Nasional Yogyakarta, 55281, Indonesia

[2] Department of Architecture and Planning, Universitas Gadjah Mada, Yogyakarta, 55281, Indonesia

Email: mutiasari@sttnas.ac.id

Abstract. Higher education in Indonesia, particularly Yogyakarta as City of Education, is continuously growing. Babarsari is one of higher education area with rapid growth and development in Yogyakarta. This area is surrounded by several universities which then form the agglomeration of higher education area. The influence of new activities as a growth pole like this educational activity has led to rapid change on land use and people's activity. This encourages the potential for high daily movement, especially the movement of students who are the dominant actors in this area. The majority of students are live near the university. However, this not lead to higher use of sustainable modes in this area. This paper tried to reveal the connection between accessibility and travel behaviour in university area. By using samples from three universities in this area (Atma Jaya University, STTNAS Yogyakarta, and UPN Veteran Yogyakarta), result shows that living in university surrounded area did not ensure students to have sustainable travel behaviour, particularly for their campus travel. The lack of non-motorized transport facilities is indicated as the main barrier for students to travel in more sustainable way.

1. Introduction

Accessibility has been a well-known concept in transportation planning field since 1950, and was defined as the ease of reaching desirable destination [1]. In other words, accessibility reduces the function of the size of the availability of activities in each zone and the cost of accessing these activities. It has been defined as the intensity of the possibility of interaction [2] and is determined by the patterns of its activities; quantity, quality, variety, and proximity; and the connectivity among them as provided by the transportation system. In the context of built environment, accessibility plays significant roles, specifically in shaping travel behaviours. Former studies have confirmed that accessibility to reach desirable destination strongly influences trip frequency, destination choice [3], mode choice and trip or tour complexity [4].

Drawing insights from that idea, university has a significant role in urban and transport development [5]. It comprises mixed land uses with vibrant built environment characteristics and offers more opportunities to participate in different types of activities with an accessible distance [6]. As a higher education, especially university, is continuously growing, changing and developing, this condition leads to higher traffic and denser population in university areas, where majority of the population is university students and commutes in surrounded area.

University students comprise a significant proportion of the travelling public [5] [7]. They have complex and distinct travel behaviours [7]. Therefore, it becomes vital to understand their travel behaviours particularly their accessibility to and from the campus. Former studies have found that, in term of spatial distance, access to universities matters since the spatial distance between residence of origin and the location where opportunities are located is regarded as an important factor in determining spatial accessibility [8]. However, since previous studies yield various findings and there is a little evidence whether the access to universities affects student travel behaviours, so this paper investigates how accessibility to the universities plays important roles in determining the patterns of student trips. This study expands former research by not only evaluating how the residential location affects travel behaviours, but also examining how the travel modes, such as bus, motorbike, cycling and walking, influence student desirable travel. This research also aims to assess how socio-economy moderates the relationship between accessibility to the universities and university student travel behaviours, specifically in developing countries, such as Indonesia.

This research was conducted in one of the most iconic cities in Indonesia, Yogyakarta. Yogyakarta has been known as the city of education. Currently, there is a growing number of educational institutes, especially higher education, and they attract thousands of students coming to this city. According to Yogyakarta Special Region's Spatial Planning Document [9], only several areas in this city are appointed to be higher education region, and one of them is Babarsari.

Babarsari is very distinct communities where students of different backgrounds, household incomes, lifestyle, attitudes mix together to study, live and recreate. This area is surrounded by several universities, such as Universitas Atma Jaya Yogyakarta, Universitas Pembangunan Nasional (UPN) Veteran, Sekolah Tinggi Teknologi Nasional (STTNAS) Yogyakarta, Universitas Proklamasi 45, and many more. The existence of these university groups forms an agglomeration of higher education areas. As a growth pole of new activities has led into rapid change on land use and people activities, it leads to a high potential for daily movement, especially student's movement as the dominant actors in this area. Therefore, this area can portray the university student travel behaviours and provide fundamental and valuable information regarding the relationship between campus environment and student travel demands.

The paper is organized as follows. The second section provides a literature review discussing the relationship between university and travel behaviour, the effect of residential location and travel mode towards student travel decisions, and the impacts of socio-economy in moderating the relationship between accessibility from and to campuses and university student travel behaviours. The third chapter describes methodology, including data collection and data analysis. The results of this study is presented in the fourth section followed by the potential discussion in the next chapter. Last section concludes the findings and proposes further research along with the limitations.

2. Literature Review

2.1. University and travel behavior

University population should be considered as a subpopulation with special travel behavioural characteristics in regional travel demand models. Cities which have higher population of university students will have a higher usage of public transportation, bicycle, sidewalk than cities that have lower number of university students [10]. A study which conducted at an urban university in Nigeria discovered that walking is the most common choice for students living on-campus and public bus is the most common choice for students living off-campus [11]. The proponents of this belief also argue that, in most developing countries, the actual condition is not common where most students reported commuting by motorized vehicles for their university trips [12] [13].

2.2. Residential location and travel mode

Residential location strongly affects people travel behaviour [14] [15] [16]. It is argued that most people who live close to their daily final destination will travel less by motorize vehicles [17]. In similar vein, people who stay near to the city centre where the main activities and attractions, tend to have simple pattern of movement, low use of motorized vehicles and high use of bicycle and walking [16]. University students come from different regions, thus the only option they have is to live in temporary residence. One of the main consideration in selecting the residential location is the distance to the campuses. This explains why there is a rapid growth of temporary residence, boarding house, apartments etc. because majority of university students are more likely to stay in university areas. Majority studies have reported that students who stay in the university area have a greater access to non-motorized transportation network than those who live far from the university [12] [18]. However, the lack of adequate infrastructure of non-motorized transports can be an indication that the access to more sustainable modes will be higher. This assertion will be addressed in the next chapter of this study.

2.3. Socio-economy and travel behavior

In addition to location elements, several theories also state that socioeconomic characteristics also influence travel characteristics [19] [20] [14] [21]. Daily community transportation activities depend not only on the relative distance of the location of the place of residence to the service facility. The destination location chosen to visit depends on the resources, needs, and interests of the traveler. The choice of modes of transportation is not only influenced by elements of the urban structure, but also greatly depends on the individual characteristics of each traveler, such as age, gender, economic conditions, family structure. In addition to the socio-economic factors mentioned earlier, one may have different perceptions of each mode of transportation and travel destination. This perception is generated from different levels of importance to several factors such as travel speed, comfort, flexibility of the mode of transportation or the district of the city [16].

3. Research methodology

3.1. Data collection

The purpose of this study was to focus on student travel behavior. Therefore, we conducted both primary and secondary approaches. First, in order to obtain students characteristics and their travel behaviors around Babarsari, we distributed questionnaires in three biggest universities, Universitas Atma Jaya Yogyakarta, Universitas Pembangunan Nasional "Veteran" Yogyakarta and STTNAS Yogyakarta. To represent student travel behaviour in Babarsari area, 300 questionnaires were collected from these three universities. Table 1 shows the sample size of the respondents in relative to their universities. The questionnaire has three sections. In the first section, questions related to socio-economy were asked. Next, the students had to answer their travel characteristics and followed by student residential location along with activity characteristics. Furthermore, we also observed pedestrian facilities in several spots around the universities to give the view regarding the quality of the facilities.

To support our primary data, we also derived relevant data regarding bus coverage service area together with its maps which shows public transport accessibility within the sub-district area of study from Department of Transportation Special Region of Yogyakarta and spatial planning documents along with its population distribution from Department of Land and Spatial Planning Special Region of Yogyakarta.

Table 1. Sample size of the respondents in relative to their university

	Students (%)
Universitas Atma Jaya Yogyakarta	33,5
Sekolah Tinggi Teknologi Nasional Yogyakarta	27,2
Pembangunan Nasional "Veteran" University of Yogyakarta	39,3

Source: Analysis, 2018

3.2. Variables and data analysis

Accessibility was measured by the distance between student residence and their universities, the number of walking and cycling facilities in university areas and the index of public transport accessibility in their campuses. The study used travel modes and travel frequency to describe student travel behaviours. Meanwhile, socio-economy characteristics could be defined from whether university students own motorized vehicles (car and motorcycle, respectively) and driving licensees. Since this study is descriptive, we examined these variables by looking at how strong and well they are related to each other.

4. Results and discussion

4.1. Built environment characteristics

Table 2. Summary of university characteristics

	Established	Population	Course Programs
Universitas Atma Jaya Yogyakarta	1965	approx. 11.405 students	12 undergraduate programs 6 postgraduate programs
Sekolah Tinggi Teknologi Nasional Yogyakarta	1986	approx. 4.840 students	6 undergraduate programs 2 diploma programs
Pembangunan Nasional "Veteran" University of Yogyakarta	1958	approx. 10.980 students	1 diploma program 20 undergraduate programs 8 postgraduate programs 1 doctoral program

Source: Kemenristekdikti, 2018

Table 2 summarizes the built environment characteristics in three universities. Univeritas Atma Jaya Yogyakarta and Universitas Pembangunan Nasional "Veteran" Yogykarta have approximately 11.000 students, while STTNAS only has 4.840 students. In term of course programs, Universitas Pembangunan Nasional "Veteran" Yogyakarta has the most study programs among them. Its students are enrolled in 1 diploma program, 20 undergraduate programs, 8 postgraduate program and 1 doctoral program.

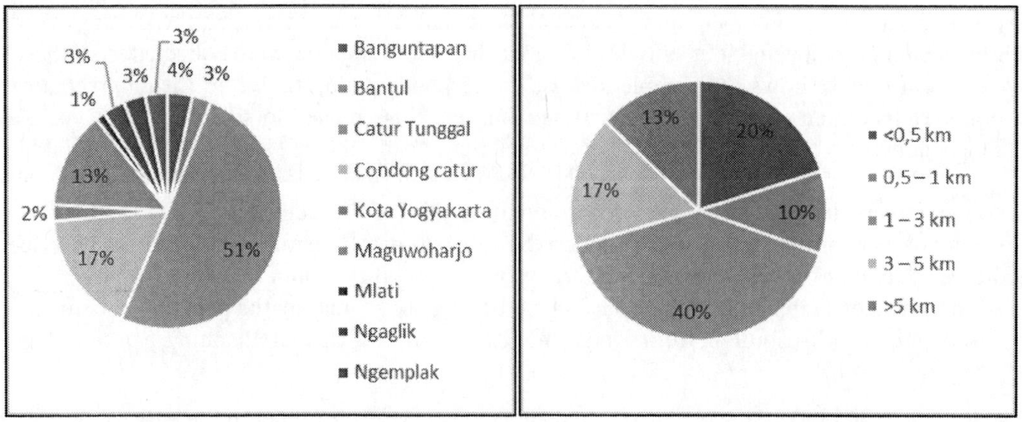

Figure 1. Distribution of students by residential location and its distance
Source: Analysis, 2018

Figure 1 demonstrates the distribution of students by residential location and its distance from their residence to their campuses. It can be seen that 51% of students are living in Catur Tunggal, followed by 17% of students are living in Condong Catur. In this case, Babarsari is located in Catur Tunggal, so

IOP Conf. Series: Earth and Environmental Science **340** (2019) 012030 doi:10.1088/1755-1315/340/1/012030

this result can represent the condition of student residential location in Yogyakarta. On the right side of the pie chart, it reveals that the distance from their residence to their universities, 40% of students are settling down between 1-3 km from campuses. Meanwhile, students who live less than 0,5 km and more than 3 km, 20% and 30% respectively. This finding support the argument for promoting the use of sustainable transportation, particularly walking and cycling, in university area as most of students live within walk and cycle distance to university.

The public transport service in Yogyakarta Special Province can be categorized into 4 type of services which are public transport (city bus and TransJogja), ojek, taxi, and non-motorized transportation (becak and andhong). TransJogja is expected to be the most reliable public transport services in Yogyakarta area which classified as Bus Rapid Transit (BRT) system. Accessibility towards public transportation service was measured through calculating the bus service coverage rate (BSCR) as the ratio between the length of TransJogja routes and total length of road within Babarsari area. According to the results of BSCR in all Yogyakarta urbanized area, we divided BSCR into three categories: low bus service coverage ratio (BSCR = < 0,1); middle bus service coverage ratio (BSCR = 0,1 – 0,3); high bus service coverage ratio (BSCR = > 0,3). As Babarsari area is located in Caturtunggal sub-district, the BSCR for this study use the value for this sub-district level. According to the value of BSCR, the public transport accessibility for the university area in this study is categorized as middle bus service coverage ratio with a value of 0,17.

Figure 2 describes the accessibility of TransJogja service in Caturtunggal sub-district area. Having middle bus service coverage ratio, this area has more than 30 shelters. However, even though this area was covered by good public transport service, but this type of mode still involves walking activity in which the pedestrian facilities are important for this context. Moreover, university students have high mobility throughout the day. The travel characteristics they might have were also a multiple trip which involve several transits before reaching their final destination location. Meanwhile, most of students live in university surrounded area. Thus, walking and cycling actually can be students travel mode choice to the university.

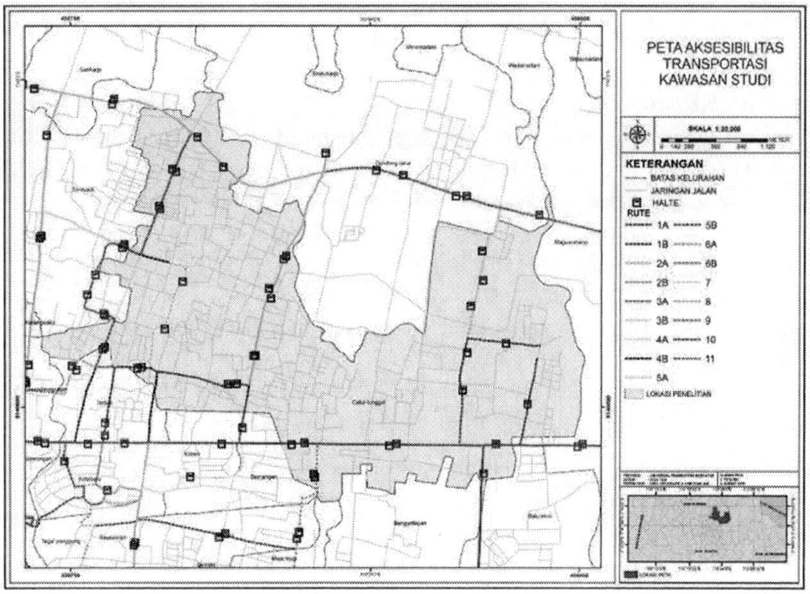

Figure 2. Public transport accessibility within the sub-district area of study
Source: Analysis, 2018

Meanwhile Figure 3 shows the condition of walking facilities in university surrounded area. Actually there are pedestrian infrastructure within area. However, it is founded that at some point, the pedestrian is not well connected in all of the area. Moreover, some spot was found to be used for other activities, making people cannot cross over in comfortable way. Vegetation was also not available all along the pedestrian, even it is considered as limit availability. The weather of Yogyakarta which is a tropical area with a high temperature particularly during the day, indeed will be the biggest barrier for student to walk around this area.

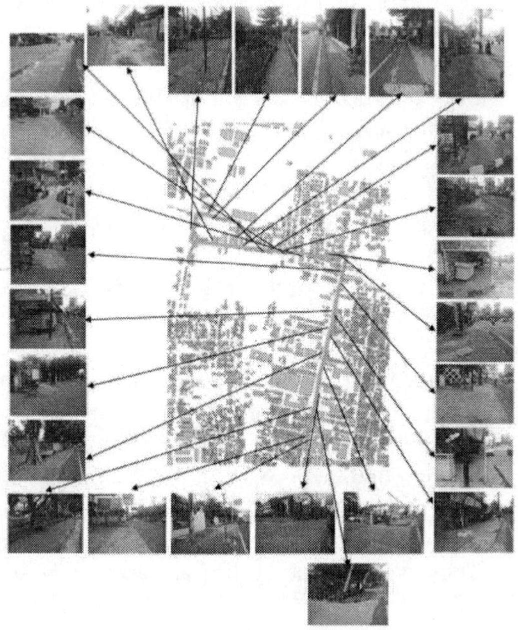

Figure 3. Pedestrian condition in Babarsari area
Source: Analysis, 2018

4.2. Socio-economy characteristics

Students who have higher pocket money and access to motorized vehicle are less likely to walk to university. It is obvious why greater vehicle availability as well as higher income would make walking less attractive relative to car travel. These two variables both individually and together may have strong enough influence on mode choice to overwhelm other factors favoring walk trips, such as a short distance to and from university. Students holding driving licenses are less likely to take public transport or even walking and cycling than those without driving licenses. Actually, students living far from university to walk or bike are prime candidates for public transport passenger if only their residential location have accessibility towards this type of mode. It is argued that students traveling through areas with sidewalks on main roads were also more likely to walk. Thus, this supports the arguments of improving non-motorized facilities to encourage walking and cycling.

Table 3. Student mode choice based on motorized vehicle and driving license ownership

Mode Choice	Motorized Vehicle Ownership				Driving License Ownership			
	No	Car	MC	Car and MC	No	SIM A	SIM C	Both
PT	-	1	1	-	-	-	2	-
Walking	27	-	7	-	18	-	16	-
Car	-	2	2	3	-	-	3	4
Bicycle	1	-	1	-	1	-	1	-
MC	11	2	160	6	12	4	112	51

Total	**39**	**5**	**171**	**9**	**31**	**4**	**134**	**55**

Source: Analysis, 2018

4.3. Travel characteristics

Travel behavior can be described through several approach. Travel mode, travel distance, travel time, and travel frequency are usually reported to be used in previous researches on travel behavior study. For this study concern, travel characteristic of students are described through student travel mode choice to university and travel frequency to university within a week. We tried to relate the travel characteristics with other variable, particularly regarding built environment and socio-economy variable.

Table 4. Distribution of student mode choice based on residential location distance to university

Residential Distance	MC	Car	Public Transport	Cycling	Walking
<0,5 km	27	0	0	0	18
0,5 – 1 km	19	1	0	1	3
1 – 3 km	73	3	1	1	12
3 – 5 km	32	3	1	0	1
>5 km	28	0	0	0	0
Total	179	7	2	2	34

Source: Analysis, 2018

Table 4 shows that motorcycle is used by students almost in all residential location distance groups. Even those who live within 0,5 km of the campus choose motorcycle for commuting to the campus. Actually, walking is also quite popular among those who live close to university. This can be explained by both proximity and the social economy characteristics which forced them to choose other than walking. Indeed, there are several factors which influence students travel behavior particularly their commuting choice to the university. It is not always about the distance they need to take. Other than this factor, several variables can be contributed to student decisions towards their travel mode. However, as the distance to the university increase, students are likely to use motorized vehicle for their university travel.

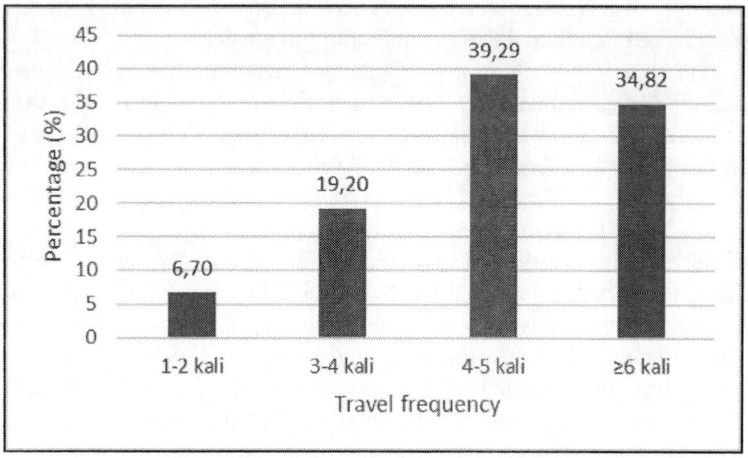

Figure 4. Statistical chart of student travel frequency in a week
Source: Analysis, 2018

Figure 4 shows the travel frequency distribution of students. The student travel frequency is categorized into 4 groups: 1 – 2 trips in a week; 3 – 4 trips in a week; 4 – 5 trips in a week; and ≥ 6 trips in a week. Most of students travel for 4 – 5 trips in a week which accounted for 39,3% of all students followed by 34,8% of students reported for travelling more than 6 times in a week to the university.

Table 5. Statistical information of student travel frequency regarding several categories

Group categories	N	1 – 2 trips/week	3 – 4 trips/week	4 – 5 trips/week	≥ 6 trips/week
By year of admission					
1st year stage	29	3	4	12	10
2nd year stage	62	2	9	24	27
3rd year stage	96	3	18	44	31
4th year stage or more	37	7	12	8	10
By motorized vehicle ownership					
Yes	185	11	39	74	61
No	39	4	4	14	17
By residential location distance					
<0,5 km	45	2	5	15	23
0,5 – 1 km	23	3	4	9	7
1 – 3 km	90	8	12	41	29
3 – 5 km	37	1	11	12	13
>5 km	29	1	11	11	6
By multiple trips need intensity					
Never	5	-	2	3	-
Sometimes	93	7	20	33	33
Often	65	6	11	30	18
Almost always	37	7	19	10	37
Always	24	3	3	17	24

Source: Analysis, 2018

We believe that at some point, the accessibility towards motorized vehicle can be contributed to people's mode choice. In this study, motorized vehicle ownership is used to represent student accessibility towards motorized vehicle. From Table 5, it is clear that more than 70% of students have own motorized vehicle. However, there is an interesting finding that not having motorized vehicle doesn't make the number of travel frequency low. There are 75% of students with no motorized vehicle access reported for having more than 4 trips to university in a week.

Regarding the involvement in multiple trips, students who sometimes do multiple trips on their daily travel tend to have high travel frequency in a week in which they travel to university even more than 6 times a week. The results even show that majority of students who almost always have multiple trips on their travel reported to travel for more than 6 times a week.

4.4. Accessibility and Travel Behavior

Discussing about accessibility towards travel behaviour is actually derived from the concept of integrated land use and transportation which has been studied by many researchers around the world. It is found that built environment shows a positive effect on travel behaviour. However, the findings in these researches are varies regarding how this built environment will determine people travel behaviour. Accessibility actually has been defined as the intensity of the possibility of interaction [2]. It is determined by the pattern of activities; their quantity, quality, variety, and proximity; and the connectivity between them as provided by the transportation system.

In order to measure level of accessibility in this study, the distance to reach destination location is considered. The main destination location of students is university. Therefore, we tried to relate how residential distance will determine on student travel decision. It is believed that living near the university will make students to choose more sustainable transport such as walking and cycling. Students will experience minimum effort to reach the university by walking and cycling since close distance are considered as high accessibility. Increasing of the distance number will result on the high proportion of motorized vehicle since students will find difficulties to travel to university by walking and cycling. However, the results in this study enrich the concept in which how this distance matter in university students travel decision, especially their travel mode choice. In fact, distance not always become the only factor to be considered by students in choosing their travel mode option. It can be seen through the findings that majority of student group who live close to the university chose motorized vehicle for their university travel. Students who travel by foot and cycle to the university don't own motorized vehicle. The accessibility towards motorized vehicle which in this study is measured through motorized vehicle ownership, have some roles in student travel decision.

University students have such complex travel behaviour. In daily basis, they tend to visit different location which making them to have travel chain. This condition forces students to be involved in such multiple trips. The need to do multiple trips in a day is reported to be one of the factor to be considered by students in choosing their travel mode. Students who typically have a high mobility in a day, will avoid a travel mode in which give them difficulties. Thus, they tend to choose motorized vehicle since this kind of mode have a high flexibility.

The results in this study which show high used of motorized vehicle even for those who live close to the university, support the arguments in which distance is not the only factor in determining students travel mode choice. As mentioned before, that accessibility comprises several dimensional concepts. The university area has good public transport accessibility which is measured through the BCSR result. There are also a high number of shelter which spread over the university area. This give an ease for student to reach the university, particularly for those who doesn't own motorized vehicle. However, the fact that majority of students live close to the university increase the concern that walking and cycling facilities within the area is important. An environment where there are a good walking and cycling facilities, will increase the willingness of its people to actively commute. Current condition of this facilities in the university area where several area is not covered by this facilities, could be the reason for the low used of active transport in this area.

5. Conclusion

Result of this study support the argument that at some point, residential location influences travel behavior, also when the socio-economy factors are taken into account. However, the choice of an individual is not only formed by one single factor, but it is usually formed through a combination of several factors. Among all of our respondents reported the high used of motorized vehicle for their university travel regardless their residential location distance. The variety of their socio-economy characteristics were assumed to be the reason for their travel mode decision. Moreover, the environment condition of university surrounded area, such as the availability of walking and cycling facilities, regional density, and public transport accessibility, are somewhat contributed to student travel decision, which in this context are the travel mode and travel frequency.

There are several limitations in data and methodical approach in this study. The first one, travel diary of student was collected for one day only in which there will be more fruitful findings if the data were collected on a different day within a week or even a year. Second, and most importantly, travel behavior data collected in this study did not report student origin and destination other than their residential location and university within a day in which understanding this information will elaborate how big the university area can generate movements and activities in surrounded area.

6. References

[1] Hansen, 1959 in Iacono, M., Krizek, K. J., & El-Geneidy, A. (2010). Measuring non-motorized accessibility; issues, alternatives, and execution. *Journal of Transport Geography*, 133-140.

[2] Hansen, 1959 in Handy, S. L. (1996). Understanding the link between urban form and nonwork travel behavior. *Journal of Planning Education and Research*, 183-198.

[3] Handy, S. L. (1996). Understanding the link between urban form and nonwork travel behavior. *Journal of Planning Education and Research*, 183-198.

[4] Hanson, S., & Schwab, M. (1987). Accessibility and intraurban travel. *Environment and Planning A*, 735-748.
[5] Zhan, G., Yan, X., Zhu, S., & Wang, Y. (2016). Using hierarchical tree-based regression model to examine university student travel frequency and mode choice patterns in China. *Transport Policy*, 55-65.
[6] Daisy, N. S., Hafezi, M. H., Liu, L., & Millward, H. (2018). Understanding and modeling the activity-travel behavior of university commuters at a large canadian university. Journal of Urban Planning Development
[7] Khattak et. al., 2011 in Nguyen-Phuoc, D. Q., Amoh-Gyimah, R., Tran, A. T., & Phan, C. T. (2018). Mode choice among university students to school in Danang, Vietnam. *Travel Behaviour and Society*, 1-10.
[8] Turk, U. (2017). *Socio-Economic Determinants of Student Mobility and Inequality of Access to Higher Education in Italy.* Verona: Department of Economics, Universita di Verona.
[9] Dinas Pertanahan dan Tata Ruang Daerah Istimewa Yogyakarta. 2017. Revisi Rencana Tata Ruang Wilayah
[10] Santos et. al., 2013 in Daisy, N. S., Hafezi, M. H., Liu, L., & Millward, H. (2018). Understanding and modeling the activity-travel behavior of university commuters at a large canadian university. Journal of Urban Planning Development
[11] Olawole and Olapoju, 2016 in Searcy, S. E., Findley, D. J., Huegy, J. B., Ingram, M., Mei, B., Bhadury, J., & Wang, C. (2018). Effect of residential proximity on university student trip frequency by mode. *Travel Behaviour and Society*, 115-121.
[12] Devi, M.K., Malkhamah, S., & Friman, M. 2015. Active commuting to the university.
[13] Devi, M. K. (2017, June). Potential to increase active commuting level in university area (Case study: Universitas Gadjah Mada). In *IOP Conference Series: Earth and Environmental Science* (**Vol. 70**, No. 1, p. 012022). IOP Publishing.
[14] Muliana, R., & Kustiwan, I. (2014). Perilaku Pemilihan Lokasi Tempat Tinggal dan Karakteristik Perjalanan Mahasiswa di Kota Bandung. *Jurnal Perencanaan Wilayah dan Kota 2*, 37-43.
[15] Ulfa, I. M., & Suwandono, D. (2014). Perilaku Perjalanan pada Akhir Pekan oleh Rumah Tangga yang Tinggal di Daerah Pinggiran Kota Semarang. *Jurnal Pembangunan Wilayah dan Kota*, 343-354.
[16] Naess, P. (2005). Residential location affects travel behavior-but how and why? The case of Copenhagen metropolitan area. *Progress in Planning*, 167-257.
[17] Tran, M. T., Zhang, J., Chikaraishi, M., & Fujiwara, A. (2016). A joint analysis of residential location work location, and commuting mode choices in Hanoi, Vietnam. *Journal of Transport Geography*, 181-193.
[18] Lundberg, B., & Weber, J. (2014). Non-motorized transport and university populations: an analysis of connectivity and network perceptions. *Journal of transport geography*, 165-178.
[19] Bamberg, S., Fujii, S., Friman, M., & Gärling, T. (2011). Behaviour theory and soft transport policy measures. Transport policy, **18(1)**, 228-235.
[20] Novitasary, S. (2015). Analisis Karakteristik Perjalanan Mahasiswa Perguruan Negeri berbasis Spasial di Kota Makassar.
[21] Whalen, K. E., Páez, A., & Carrasco, J. A. (2013). Mode choice of university students commuting to school and the role of active travel. Journal of Transport Geography, 31, 132-142.

Acknowledgement
This work was fully supported by the Minister of Research, Technology, and Higher Education of the Republic of Indonesia.

Placement of Ambarawa Railway Station in city spatial planning

N Solikhah[1]

[1]Department of Architecture and Planning, Universitas Tarumanagara, Campus 1, L Building, 7th Floor, Jl. Let Jen. S. Parman Number 1, West Jakarta 11440, INDONESIA

Email: nafiahs@ft.untar.ac.id

Abstract. The construction of the Ambarawa (Willem I) - Kedungjati railway line by the Nederlandsch Indische Spoorweg Maatschappij (NISM) is for military purposes and the transportation of the plantations of the Dutch colonial government so that it was significant to study. The successful placement of the train station is supported by the structural quality of the station building that was built as well as the city/ regional focal point. Then in the development of the regional spatial structure allows a shift in the role of the land route where the railway station building is located.This study aims to examine the placement of Railway Stations in the City Spatial Plan on the Ambarawa-Kedungjati Line because at present the Ambarawa train station still has a strong role because it is the starting point for the development of the city which produces a variety of morphological spatial expressions in the form of a compact form of square shape and Urban sprawl in the form of linear development. The results of the study can be taken into consideration in the planning of the area where the railway station building is placed in line with the development of the city and or district layout.

Keywords: *Railway station, Urban morphology, Urban sprawl, Ambarawa*

1. Introduction

The railroad network on Java is one of the most complete and densest railroad networks in Asia as one of the forms of the transportation equipment revolution in the early 19th century to complement the Grote Postweg (the Java-based highway from Anyer - Panarukan, currently more famous as the Pantura Line) built by Daendels in 1808-1810 [1]. The first railway line in Java, Semarang-Kedungjati, was inaugurated in 1871, then Batavia-Buitenzorg (Jakarta-Bogor) was opened in 1873, the Surabaya-Pasuruan route in 1878, the Buitenzorg-Bandung (Bogor-Bandung) route was completed in 1884, and then the Surabaya-Solo and Semarang lanes. In 1894, the Surabaya-Batavia railway line through Maos, Yogyakarta and Solo was successfully completed. In 1912 the second alternative route between Surabaya-Batavia, through Cirebon and Semarang was successfully completed. The Kedungjati-Willem I (Ambarawa) branch was inaugurated on 21 May 1873 following the inauguration of the Semarang-Surakarta-Yogyakarta line [1].

The construction of the railroad network in Java was inseparable from the Dutch strategic interests to exploit Java, so the placement of Railway Station buildings became a matter that required consideration [2]. This is reinforced by the opinion of Handinoto [3], that the railway station building is designed as a focal point so that it has an important role for the architecture of the city/ region.

The next step with the existence of the railway network is the placement of the train station in the city or district that is passed. The easiest tendency for railroad station placement is in the city center so that it is easily accessible by passengers. Placement of train stations in cities in Java in the past generally worked well, including train stations in Bandung, Tegal, Probolinggo, Pasuruan, Malang, Jombang. The successful placement of the train station is supported by the structural quality of the station building that was built as well as focal point of the city/ region [3].

As one of the initial routes built by Nederlandsch Indische Spoorweg Maatschappij (NISM), the Ambarawa-Kedungjati Line is significant to be studied. The construction of the Ambarawa Railway (Willem I) -Kedungjati railway is not separated from the military interests and transportation of Dutch colonial government plantations (Figure 1). Therefore, the placement of the railway station building on the Ambarawa (Willem I) - Kedungjati line is certainly on the basis of mature consideration.

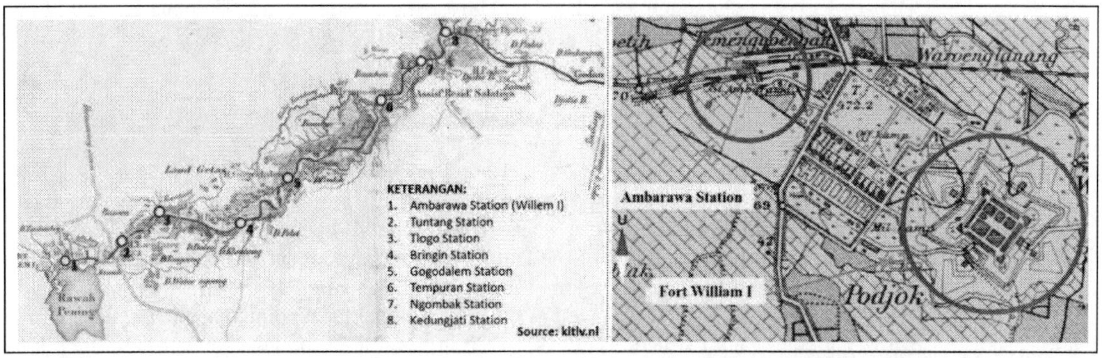

Figure 1. From the Ambarawa map in 1905, the placement of railway station buildings was close to military camps (military interests) and located on the plantation path (Source: kitlv.nl)

Along with the development of land transportation and the efficiency of the railway line and the collapse of the bridge that connects Ambarawa-Kedungjati, the Ambarawa (Willem I) - Kedungjati railway line was finally officially closed in 1976. Towards the end of the 20th century, road traffic on the Island Java is increasingly crowded, including the land crossing of Ambarawa-Kedungjati which is the main route connecting the City of Ambarawa-Grobogan Regency. The cities and regencies in the Ambarawa-Kedungjati railway line in general have developed rapidly, which has an impact on the placement of the train station which was originally planned properly in the city or district spatial layout. The development of the regional spatial structure allows a shift in the role of the land route that the railroad traverses. If initially the land route where the train station building is located is an important route, then in the development of the regional spatial structure allows a shift in the role of the land route where the railway station building is located [4]. With the plan of Directorate General of Railways of the Ministry of Transportation to reactivate Tuntang-Kedungjati railway line, an integrated system is needed.

Based on this background, this study aims to study the placement of Ambarawa Railway Station in City Spatial Planning. The results of the study of the placement of the Railway Station in the City Spatial Plan on the Ambarawa-Kedungjati Line can be taken into consideration in city and or district planning where the railway station building is placed in line with the development of the city and or district layout.

The approach method used in this study is a qualitative approach to find out the placement of Ambarawa Railway Station in the City Spatial Plan on the Ambarawa-Kedungjati Line because of its significant role in the history of Railroad transportation in Java. The study area boundaries is the Ambarawa Railway (Willem I) Station.

Primary and secondary data that has been evaluated using historical search methods (diachronic reading) and synchronic reading with the placement of railway station buildings as starting points will be analyzed evaluatively descriptively, namely problem solving procedures by describing the current state of the research object based on facts or what it is. Presentation of data will be supported with illustrations and graphic images to facilitate the delivery of information [5].

2. Architectural typology of Ambarawa Station

2.1. Spatial system

The main building space pattern of Ambarawa Station is arranged linearly (extends horizontally east-west) and symmetrical between the left wing and the right wing of the building. Linear pattern is one of the prototypes of train stations in Indonesia. Circulation flow for visitors to Ambarawa station is linear following the pattern of space. During the Dutch colonial period there was a difference in the flow of circulation between the indigenous population and the foreign population (European) which was seen in the difference between the waiting area and the toilet. The waiting area for the indigenous population is on the outside of the main building (Figure 2 point 2) while the toilet is on the outside of the main building (Figure 2 point 1). The waiting area for white people is on the inside of the main building which is equipped with bars and toilets in (Figure 2 points 3, 4, 5). [6]

Ambarawa Station applies Indische Architecture, marked by the presence of an open corridor (vestibule) that surrounds the main building. With the separation of building masses in accordance with the activity zone, the flow of passenger activity circulation is separated from the circulation flow of the main activity. The main building is oriented to the East-West axis horizontally. The opening in the main building of Ambarawa Station is oriented towards the platform and emplacement which is visible from the direction of the door opening, while the orientation to the front corridor is in the window area. When viewed from the space organization, Ambarawa Station is an island station type, where the main building of the station is flanked by an Emplacement/ railway line [6].

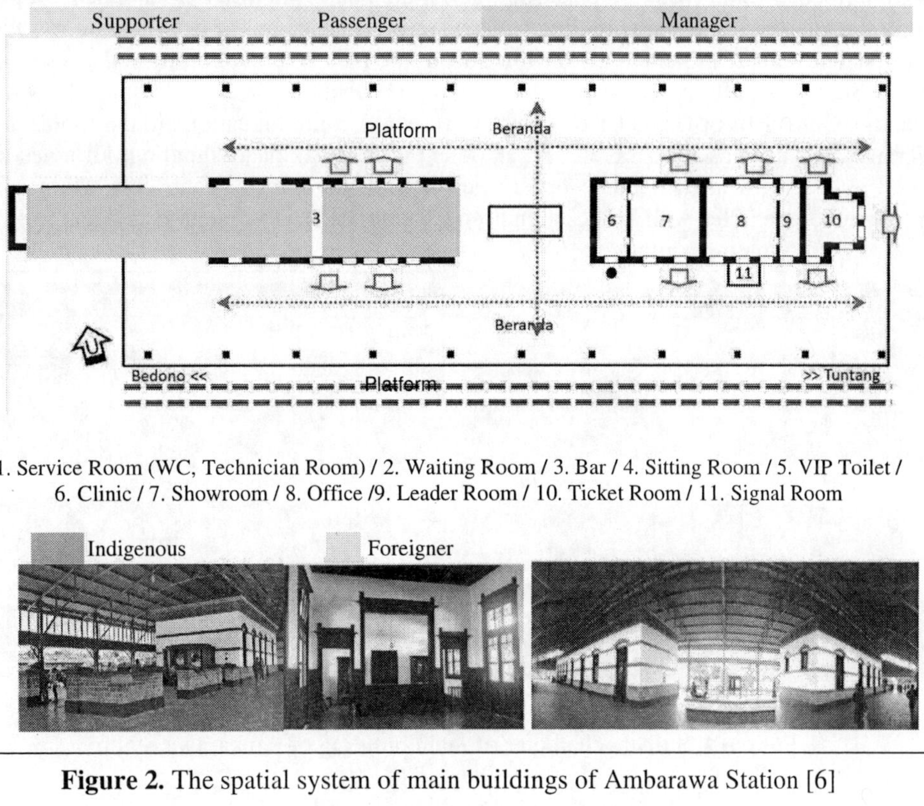

Figure 2. The spatial system of main buildings of Ambarawa Station [6]

IOP Conf. Series: Earth and Environmental Science **340** (2019) 012031 doi:10.1088/1755-1315/340/1/012031

2.2. Physical systems and figural quality

The physical form of the main building of the Ambarawa station is a combination of concrete structures with steel structures. The main building of the Ambarawa station is cube shaped using a concrete structure which is shaded by Overkapping in the form of a steel canopy with a wide span (about 22 m) of steel construction (Figure 3). Thickening of plastering on the main building as well as ornaments. Terracota couples as a waterproof coating as well as an aesthetic maker. The ceiling on the main building of the station using wood material that can reduce the heat that arises as well as a high ceiling allows hot air gathering place so that the room under it will become cooler, given the geographical climate of the tropics [6].

Figure 3. The figural quality of main buildings of Ambarawa Station [6]

2.3. Stylistic system

The stylistic system is analyzed based on indicators of roofs, columns, openings and decorative types. The main building of the Ambarawa station is a concrete roof which is shaded by Overkapping in the form of a steel canopy with a wide span (about 22 m) structured supported by a steel column (Figure 4). The shape and pattern of doors and windows has the same type, namely the Indis characteristic, using high and wide wood materials [8]. The most prominent ornaments in the main building of the Ambarawa station are terracotta frame with arc construction. At the end of the brick arch ended with molding of cement. In the interior of the main building get Art Deco influence on the details of doors, windows and floor motifs. The stilistic system is also formed from the means of supporting railway activities when Ambarawa Station is still active, namely: manual signal control made by Alkmaar (city name in the Netherlands), Genta PJL or Genta Cross Guard Officer is a communication aid that sends news related to the Train and Ancient Clock travel. The stations that use Alkmaar signaling, all notes use manual shifting levers that are near each money order (the notes are not operated centrally). Alkmaar's signaling system cannot be assembled with a block signal, so it cannot be used at a station adjacent to other stations that use electrical signaling equipment [6] [7].

Figure 4. Stilistic character of main building of Ambarawa Station

Source: [6]

3. Placement of Ambarawa Railway Station in city spatial planning

Ambarawa is one of the cities included in the first phase of the railway line construction by the Nederlandsch Indische Spoorweg Maatschappij (NISM). Ambarawa was originally a military city during the Dutch Colonial Government. For the operational benefit of the Dutch Colonial Government, King Willem I ordered the construction of a new railway station that would allow the government to transport its troops and the results of coffee plantations to Semarang. Ambarawa Railway Station was built adjacent to Benteng Willem I and was inaugurated on May 21, 1873 along with the opening of the Semarang-Kedungjati-Ambarawa line. With the deactivation of the Ambarawa-Kedungjati line in 1976, Ambarawa Station changed its function to the Ambarawa Railway Museum which was inaugurated on April 21, 1978. Ambarawa Station still serves the Ambarawa-Tuntang PP and Ambarawa-Bedono PP lines for tourism purposes.

The construction of the Ambarawa (Willem I) - Kedungjati railway line was inseparable from military interests and the transport of the products of the Dutch colonial administration. This can be seen from the selection of Ambarawa Station (Willem I) location adjacent to the Willem I fortress in the city of Ambarawa. When viewed from the station placement on city architecture, the morphology of the city of Ambarawa produces a variation of the city / region's morphological spatial expressions in the form of a compact square shape. A square city shows the opportunity to expand the city in a relatively balanced direction (Figure 5).

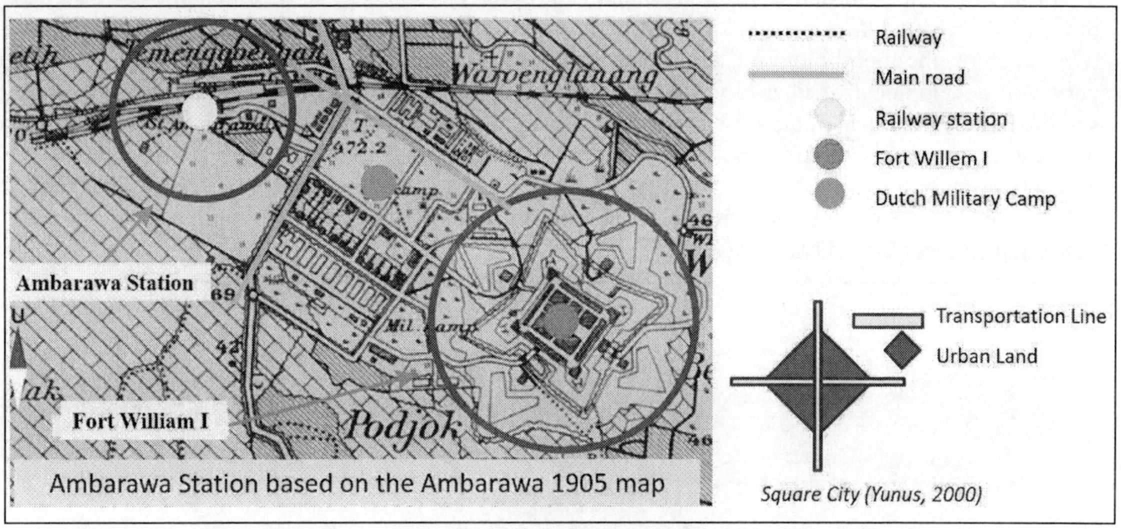

Figure 5. Position of Ambarawa Railway Station in city spatial planning
(Source: Author, 2018; [9])

Placement of the Ambarawa (Willem I) railway station is one example of the integration of station placement with the overall city layout. The Ambarawa train station is the main axis of the city which runs from North to South. The main building of the station is located at the intersection of the main road with the road leading to the fortress of Willem I so that the figural quality of the main building which is shaded by an overkapping roof forms a monumental impression of the station building as a focal point (Figure 6).

IOP Conf. Series: Earth and Environmental Science **340** (2019) 012031 doi:10.1088/1755-1315/340/1/012031

Figure 6. Figural quality of Ambarawa Railway Station and Fort Willem I.
Source: Author, 2018

In terms of city architecture, the placement of the Ambarawa train station is one of the determinants of the development of the city's morphology. Its location which is on the axis of the main road makes the surrounding area develop because it is now also the center of the city. Whereas the area around Fort Willem I is currently neglected because the fortress is not functioning. Part of the former military camp area was used as Ambawara's Jail Class IIA LP.

Furthermore, the morphology of the Ambarawa region developed, one of which was influenced by the development of transportation infrastructure. Judging from the development of transportation infrastructure, the morphology of the Ambarawa region in the period of lateral growth includes the period of development of inter-city transportation relations. Starting from the placement of the train station, based on the map of the Ambarawa region in 2011 it can be seen that the urban sprawl Ambarawa perforation process is in the form of linear development, which is a perforation that indicates the inequality of urban area spillages on all sides - the outer side of the main city area (Figure 7). Perforation is most quickly seen along the existing transportation routes, especially with the construction of toll roads that connect West Java to East Java.

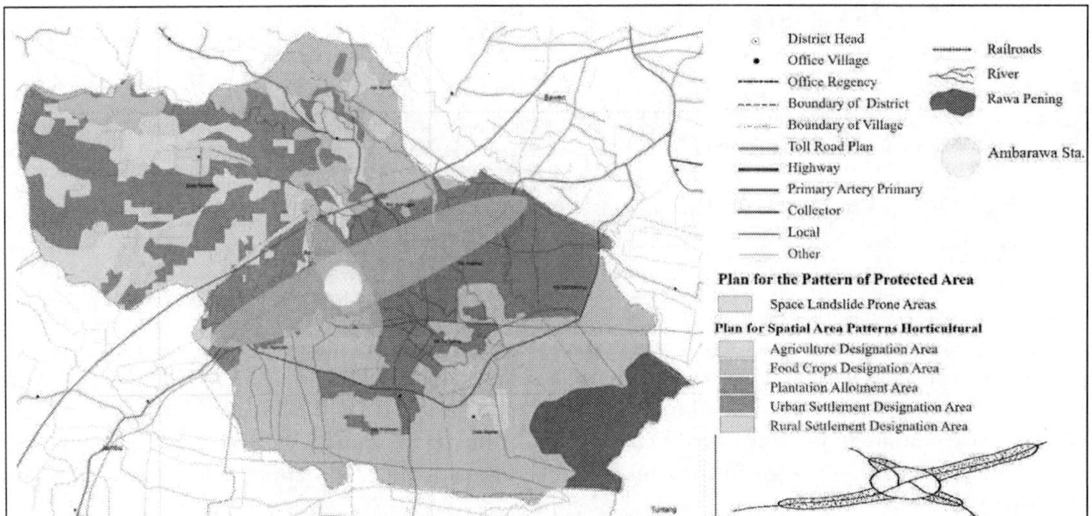

Figure 7. *Urban Sprawl* Ambarawa
Source: Author, 2018

4. Conclusion and recommendations

Placement of the ambarawa railway station building has an important role in the development of Ambarawa city. The placement of the Ambarawa train station building (Willem I) was initially not separated from the military interests and transportation of the plantations of the Dutch colonial government. In the development of the city of Ambarawa, the train station still has a strong role because it is the starting point for the development of the city which produces variations in the spatial morphological expressions of the region in the form of a square. Starting from the placement of the train station, it can be seen that the urban sprawl process of Ambarawa is in the form of linear development, which is a perforation that indicates the inequality of the urban area perforation on all the outer sides of the main area. city.

The results of the study of the placement of the Railway Station in the City Spatial Plan on the Ambarawa-Kedungjati Line can be taken into consideration in city and or district planning where the railway station building is placed in line with the development of the city and or district layout.

Based on the role of Ambarawa Railway Station building in the layout of the city of Ambarawa, it is necessary to manage an integrated historical system since the investigation phase (Heritage Management System), which includes: Significant Investigation Stage, Phase Assessing Significance, until Significance Management Stage, which is a step end to determine the direction of management policies and possible development. In addition, it is necessary to have an active role from the local government to provide learning to the public on the importance of the sustainability of buildings and historical areas.

5. References

[1] Lombard D 1996 *Nusa Jawa: Silang Budaya (Jilid 1)* (Jakarta: PT Gramedia Pustaka Utama)

[2] Dananjaya P. 2016 *Stasiun Beringin, Bukti Indonesia Kaya akan Hasil Bumi*. Retrieved from http://kebudayaan.kemdikbud.go.id/bpcbjateng/2016/09/22/stasiun-beringin-saksi-indonesia-kaya-akan-hasil-bumi/ Accessed on 7 March 2017.

[3] Handinoto 1999 Perletakan Stasiun Kereta Api Dalam Tata Ruang Kota-kota di Jawa (Khususnya Jawa Timur) Pada Masa Kolonial. *Jurnal Dimensi Teknik Arsitektur*. **27**, No. 2, DESEMBER 1999, 48 – 56

[4] Solikhah N., Kurnia A. S. 2018 *Kajian Perletakan Stasiun Kereta Api dalam Tata Ruang Kota di Jalur Ambarawa-Kedungjati*. Research Report. Not Published. Jakarta: Direktorat Penelitian dan Pengabdian kepada Masyarakat, Universitas Tarumanagara

[5] Moleong L. J. 2005 *Metodologi Penelitian Kualitatif* (Bandung: Remaja Rosdakarya)

[6] Solikhah N., Kurnia A. S. 2017 *Tipologi Arsitektural Stasiun Kereta Api Jalur Ambarawa-Kedungjati (Studi Kasus: Stasiun Ambarawa, Stasiun Tuntang, Stasiun Bringin, Stasiun Kedungjati)*. Research Report. Not Published. Jakarta: Direktorat Penelitian dan Pengabdian kepada Masyarakat, Universitas Tarumanagara

[7] Hartanti, Nurhikmah B. et.al 2010 *Stasiun Kereta Api di Pulau Jawa-Indonesia* (Jakarta: PT. Kereta Api (Persero) - Pusat Pelestarian Benda dan Bangunan afiliation with FTSP Universitas Trisakti)

[8] Junianto 2002 *Arsitektur Indis*. (Malang: Grup Konservasi Arsitektur Kota dan Lingkungan, Jurusan Arsitektur-Universitas Merdeka Malang)

[9] Yunus H. S. 2001 *Struktur Tata Ruang Kota* (Yogyakarta: Pustaka Pelajar)

IOP Conf. Series: Earth and Environmental Science **340** (2019) 012032 doi:10.1088/1755-1315/340/1/012032

The Jakarta Detailed Spatial Plan evaluation based on sustainable development principles

N C Drestalita[1] and R T Saputra[2]

[1] School of Architecture, Planning, and Policy Development, Institut Teknologi Bandung Multikampus, Bandung, 40132, Indonesia

[2] National Urban Housing and Development Corporation, Jakarta, 13340, Indonesia

Email: drestalita.naya@gmail.com

Abstract. Jakarta is one of the megacities in the world dealing with sustainability issues where urban planning plays a major role in achieving the city sustainability. This paper aims to analyze the extent to which sustainable development (SD) principles are embodied in the Jakarta Detailed Spatial Plan (RDTR). This aim is achieved by identifying SD principles through literature review based on the three pillars of SD: economy, environment, society. These principles are used to analyze the plan using Qualitative Content Analysis with scoring technique. The result shows that the SD indicators with the highest scores are sustainable transport system, green space and biodiversity, and waste management, while inclusive community, and safety and security have the lowest.

1. Introduction

In the last few decades, Sustainable Development (SD) issues have been one of the main concerns in planning practice. Technology development has formed a new modern and wealthy lifestyle in urban areas, which leads to rapid urbanization, environmental deterioration and social inequality [1]. Thus, the SD concept emerged with the intention to tackle the environmental, economy, and social challenges in the modern world. It is also related to the global warming and climate change phenomena.

As the capital city of Indonesia, Jakarta has transformed into one of the biggest megacities in the world with rapid economic growth, 26% in 2013-2017 [2]. Despite the major growth, the development has brought impacts such as economic and social disparities, inequalities, and environmental problems (slums, annual flooding, pollution, unemployment, and other challenges). The city, along with the rest of the country, needs development policies that covers these issues. Until the 1970s, Indonesian development policies focused on economic growth and economic development. Early 1980s was when the development started making environmental considerations, with the enactment of the first law of biodiversity [3][4]. Furthermore, after the United Nations set the 17 Sustainable Development Goals (SDGs), Indonesia has committed to contribute in achieving global sustainability using the SDGs as one of the development guidelines [5].

Planning plays a major role on achieving sustainable development, as developments in each governmental level refers to relevant plan documents. The Indonesian planning guideline implicitly embeds the sustainable development concept by setting the spatial plan analysis guideline based on the three pillars of sustainability, namely economy, environment, and society [6]. Furthermore, sustainability has been explicitly mentioned in most of the plan documents in different governmental levels (National, Regional, and Local), including Jakarta. The Jakarta Spatial Plan vision is to "*create a safe, pleasant, productive, and sustainable Capital City for Indonesia, align with world class cities which is occupied by prosperous citizens*" (translated from the Jakarta Local Regulation No. 1/2012 Article 3). This plan is further outlined in detail in the Jakarta Detailed Spatial Plan (RDTR), which explains the detailed plan for each municipality. This paper aims to explore the extent to which the Jakarta Detailed Spatial Plan covers the SD principles and indicators.

2. Sustainable Development

2.1. The concept

The exact definition of the SD concept is still on debates among scholars. Many argues that the definition "*the development that meets the needs of the present generation without compromising the ability of future generations to meet their own needs*" [7] even though being the most common definition, is not practical, as it is unclear for what the "needs" are [8][9].

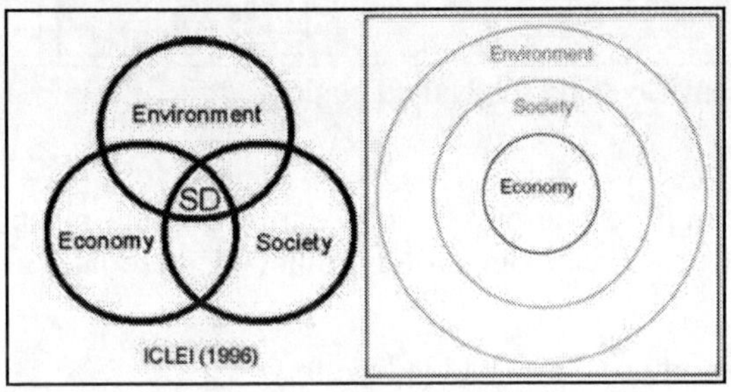

Figure 1. The overlapping circles of SD (left) [9] and the layered rings of SD [10].

Another common way to define the SD concept is through the three pillars of SD: economy, environment, and society. It is commonly illustrated as three circles that overlap in the middle, where the overlapping part of the three pillars is sustainable development. It shows that SD has a holistic goal of balancing those pillars in development [9]. However, the three circles concept is being criticized that the three pillars are seen as independent aspects. The three circles should be illustrated as layered rings, depicting that the pillars depend on each other and must not be treated separately (figure 1) [10].

The SD concept is also being illustrated using a triangle as a map [9]. Each corner of the triangle represents the sustainability pillar, where SD lies in the center of the triangle (figure 2). It can be used to map a policy regarding its emphasis on the three sustainability pillars. However, the positioning is subjective, as Connelly states

"*This clearly cannot be an exact and objective positioning, as no quantitative metric is possible—as with the other diagrams representing sustainable development, the field is a metaphor, an image used to illuminate an analysis*" (p. 268).

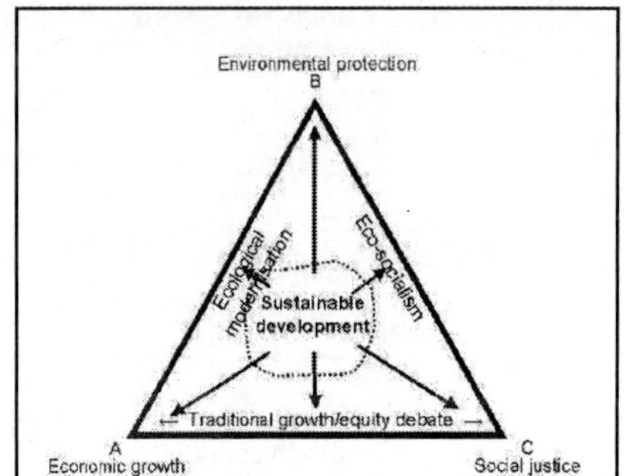

Figure 2. Mapping SD in the field [9]

2.2. United Nations' Sustainable Development Goals (SDGs)

The United Nation has set 17 Sustainable Development Goals (SDGs) to be achieved by 2030 by the UN countries. The countries should have their own national, regional, and local policies regarding SD, using the UN SDGs as guidelines. It is because each country is facing different economic, environmental, and social challenges. Generally, there are five main aspects to be satisfied by the SDGs for humanity, namely people, planet, prosperity, peace, and partnership, with the first and main goal is eradicating poverty [11].

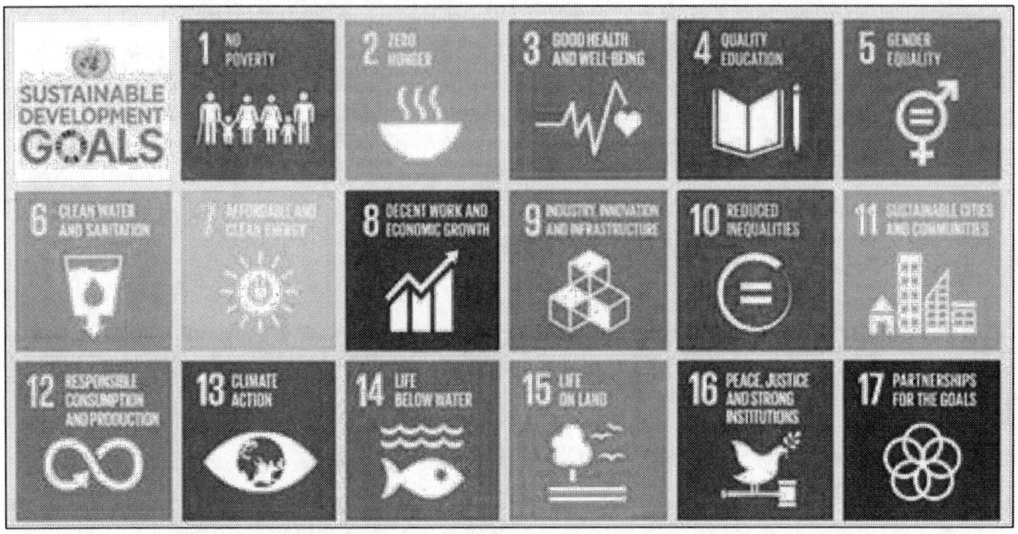

Figure 3. Sustainable Development Goals by the United Nations

2.3. Sustainable development principles and indicators

This section discusses how various researchers attempt to interpret the concept of SD into practice. Sustainability indicators are reviewed from previous research from various authors and are later used for analysis.

Table 1. Sustainable development principles and indicators

SD Principles	SD Indicators	Elements of the SD Indicators	Sources
Economy	Economic growth	- Identification and development of local economic potential - Easier access to economic activities	[11], [12], [13], [14], [15]
Environment	Sustainable transport system	- Development of public transport - Provision of supporting public transport facilities - Provision of pedestrian and non-motorized paths - Integration between public transport system and non-motorized paths with the local economic activities to reduce private transport dependency - High density and mixed use development to reduce mobility demand	[11], [12], [13], [14], [15], [16] [17]
	Green space and biodiversity	- Public green space provision and maintenance - Enhancement of local biodiversity (such as city forest, lake, river, etc)	[11], [13], [14], [15], [18], [19],
	Disaster management	- Identification of any potential disaster - Holistic disaster management strategies, including physical (infrastructure, etc) and non-physical (community, institutions, etc) aspects	[11] [20] [21] [22] [23]
	Water management	- Clean water quality control - Clean water distribution - Clean water consumption control	[11], [12], [13], [14], [15], [18], [19]
	Sustainable energy and energy	- Energy Conservation - Energy Efficiency	[11], [12], [13], [19], [24]

SD Principles	SD Indicators	Elements of the SD Indicators	Sources
	conservation	- Utilisation of Renewable, sustainable resource - Utilisation of other Low-GHG-Emitting Resources - Utilisation of Conventional Resources	
	Waste management	- 3R: reduce, reuse, recycle - Energy recovery - Waste treatment and disposal	[11], [13],[15], [17], [19], [25]
Society	Sustainable housing	- Housing provision based on the people's needs - Alternative home financing - Maximizing the land capacity - Affordable housing	[11], [12], [13], [18], [26]
	Inclusive community	- Assessing all spatial planning against diverse needs - Community participation - Considering the social impact of the development	[11], [27]
	Safety and security	- Addressing the issues of crime - Addressing the issues of vandalism - Healthcare - Smart traffic systems and routes - Smart safety systems for surveillance, search, detection and identification - Smart systems of crisis management to support decision making, early warning, monitoring and forecasting emergencies and environmental situation, - Operating units of police and Integrated Rescue System (IRS), - Safe internet connection and data protection, - Centres of data processing	[11], [12], [17], [28]

3. Research methods

The document used for analysis is the Jakarta Detailed Spatial Plan which consists of six municipalities (Central Jakarta, South Jakarta, East Jakarta, West Jakarta, North Jakarta, and Thousand Islands). A Qualitative Content Analysis (QCA) is used to help the researchers focus on the selected issues in the analyzed document. As researchers can easily get distracted in such a rich document, QCA helps create a structured analysis specifically to achieve the research aim and answer the research questions. Similar to the Quantitative Content Analysis, QCA also uses codes to analyze the data source. However, instead of measuring the frequency of each variable being mentioned in the document quantitatively, QCA aims to interpret the data and provide a deeper understanding [29][30].

One of the strengths of QCA is that it is flexible to use, depends on the research aim. This paper uses a scoring technique to measure the qualitative assessment quantitatively. The score 0 is given when the indicator is not mentioned in the document. The score 1 is given when it is mentioned, but without further explanation, or the explanation is not clear. Finally, the score 2 is given when the indicator is mentioned with a clear explanation which meets the guidelines in table 1.

Similar analysis method has been used by previous authors to evaluate plan documents [12] [18] [31]. This paper adopts Gough's scoring technique, where she also uses the scale of 0-2 to assess the plan documents in terms of sustainability and liveability, with two persons doing the analysis. Similarly, there are also two persons conducting the analysis in this paper, and the results are being compared and discussed for a reliability check. However, the results are still unavoidably subjective.

4. Analysis and discussion

The analysis is done for the Detailed Spatial Plan of six municipalities of Jakarta. There are three indicators with the score of 2, five with the score of 1, and two with the score of 0, as shown in table 2. Furthermore, the subsection below discusses the analysis for each indicator for every SD principle.

Table 2. Sustainable Development Indicators Scores

SD Principles	SD Indicators	Score
Economy	Economic development	1
Environment	Sustainable transport system	2
	Green space and biodiversity	2
	Disaster management	1
	Water management	1
	Sustainable energy and energy conservation	1
	Waste management	2
Society	Sustainable housing	1
	Inclusive community	0
	Safety and security	0

4.1. Economy

The plan for each district acknowledges its existing economic potential, as it is critical to recognize local potential and enhance it through innovative developments [14]. For example, Central Jakarta is generally designated as a commercial centre in Jakarta because it has several biggest commercial centres such as Tanah Abang and Roxy, which is being integrated with public transport systems to improve the accessibility. Thousand Islands, on the other hand, is designated as a tourism area which needs improvements in water-transport infrastructure. It shows that the plan has attempted to enhance local economic characteristics by improving public access to the economic activities.

The plan also acknowledges the informal commercial areas including its function and impacts, but there is no further explanation and direction on how to manage it. It has been recognized that the informal activities have brought negative impacts to the city, as quoted below.

> "*Informal commercials are taking up public spaces such as pedestrian ways, roadside, and even the road.*" (Translated from Menteng District RDTR, p.8)

The informal activities, despite having contributed to the city's economy and being needed by local citizens, also interfere the citizens' mobility by taking up roads and pedestrian ways, and the plan does not provide further action to address this problem. Therefore, the score 1 is given for this indicator.

4.2. Environment

4.2.1. Sustainable transport system

Public transport development is one of the main concerns in the transport section of the Jakarta Detailed Spatial Plan. The forms of public transport being developed and improved are road-based (busway) and rail-based (MRT, LRT, and trains). In certain spots, the Transit Oriented Development (TOD) concept is being implemented, such as in Harmoni, Gambir District. The development for each type of public transport is outlined in detail in terms of location and the supporting facilities criteria, as quoted below.

> "*The supporting facilities for public transport in the form of stop/shelter ... with the following criteria: 150-300 meter for busway stop; 400-500 meter for MRT shelter; 800 meter-1,5 km for LRT shelter.*" (Translated from Gambir District RDTR, p. 13)

The public transport is emphasized on the economic centres to reduce the use of private vehicles, as the plan states its target of public transport user. For example, Gambir District targets to have 60% of the road users use public transport by 2030. Pedestrian and non-motorized ways are also being developed and improved along arterial, collector, and local roads throughout all districts. For these reasons, the score of 2 is given for the sustainable transport system indicator.

4.2.2. Green space and biodiversity

Land use planning must not harm the nature, instead, it should enhance biodiversity and preserve the ecosystem. Preserving green space and biodiversity is essential for air pollution control, flood management, water quality, and providing homes to various specimens [18]. The Jakarta Detailed Spatial Plan has divided the public green space areas into two categories, namely conservated (includes riversides, rail borders, road medians) and non-conservated green space (includes city parks, cemetery, fields, etc). The conservated green spaces have a role of protection, such as protecting the river as the primary drainage system, while the non-conservated green spaces are used by citizens for daily life. Moreover, the plan also states the proportion of public and private green spaces for each district. It shows that green space is used to support other SD indicator which is disaster management (flood prevention). The score of 2 is given for this indicator.

> "*...increasing the blue open space by 5% and optimizing green open space as a water catchment area, and maintaining riversides and lakes as green open space for flood control*" (Translated from Johar Baru District RDTR, p. 27)

4.2.3. Disaster management

Jakarta is prone to flood that it becomes an annual disaster. The flood-prone areas in each district are identified, including the problem, cause, and mitigation strategies. Generally, the plan aims to drastically improve the drainage system and its capacity to accommodate overflows. This includes macro-drainage normalization, polders, pumps, infiltration wells, designation of green and blue open spaces as conservation areas, waterfront development, liquid waste management, and micro-drainage management. These measurements are classified as structural measures of flood management. The non-structural measurements such as flood forecasting and warning, public participation, and institutional arrangement [20], are however not featured in the plan.

> "*Drainage system in Kebayoran Baru District is planned to address the overflow and puddles which occur frequently with the following criteria ...*" (Translated from Kebayoran Baru District RDTR, p. 35)

One of the main aims of disaster management in sustainable development is to create a strong and sustainable community that is adaptable to climate change. That is why it is critical to involve the impacted people and related institutions in disaster management. Otherwise, the mitigation strategies would be unsustainable [20][21][22]. The plan identifies the evacuation sites in case a disaster occurs, however, there is no further explanation of how local citizens should participate. Local knowledge and familiarity of the disaster are critical, as people need to know what to do before, during, and after disaster strikes. Therefore, the score for this indicator is 1.

4.2.4. Water management

Water management means managing water as a resource from both supply and demand sides. The Jakarta Detailed Spatial Plan, however, emphasized the supply side instead of both. The citizens' water needs are outlined, followed by the aim of improving clean water distribution for all citizens. The plan does not provide actions to reduce water consumption and prevent over-consumption. The score of 1 is given for this indicator.

> "*Considering that almost all citizens have access to clean water, the clean water supply will be improved following the city's population growth.*" Johar Baru District, p. 14.

4.2.5. Sustainable energy and energy conservation
The type of energy being discussed in this section is limited to electricity as one of human's basic needs. It is also the only type of energy being mentioned in the Jakarta Detailed Spatial Plan. The plan examines today and future electricity demand based on population growth and mentions one alternative source of electricity aside from the State Electricity Company (PLN), which is solar panel, as quoted below. However, similar to water management indicator, the plan does not provide actions or strategies of demand control which is to reduce energy consumption. Accordingly, the score for this indicator is 1.

> "*Development of other energy in the form of solar cell utilization in Jatipulo sub-district, Kota Bambu Utara sub-district ...*" (Translated from Palmerah District RDTR, p. 35)

4.2.6. Waste management
The Jakarta Detailed Spatial Plan outlines the waste being produced in present day and the future, based on population growth. There are two types of waste identified in the plan, namely solid and liquid waste. The plan applies the 3R principles for treating solid waste in garbage collection points before moving them to a landfill or an integrated waste collection point. Every sub-district will have a 3R waste collection point with the minimum area is 500m^2.

Moreover, the plan assumes the amount of future liquid waste to be 70% of water consumption per capita. The plan specifies that off-site and on-site systems will be applied, with the detail explanation of maintenance and enhancement of the disposal network capacity in the on-site system. The score for this indicator is 2.

> "*Development of modular systems in multifunctional areas, environmental rejuvenation and new large-scale development.*" (Translated from Kebayoran Lama District RDTR, p. 36)

> "*....Revitalization of individual stool wastewater infrastructure with a multi-function septic*
>
> *tank treatment system for faeces and dirty water treatment (black and grey water)*" (Translated from Kebayoran Lama District RDTR, p. 36)

4.3. Society

4.3.1. Sustainable housing
Alongside the population growth, housing demand also arises. The land, however, does not. Sustainable housing solution is deemed necessary considering the limited space. The plan provides the present and future housing needs based on population growth and housing types (horizontal and vertical housing), specifying clearly the number of units that need to be built. The building intensity is being increased, mixed-use concept and zoning arrangement are also being applied (quoted below), which show that the plan has attempted to achieve sustainability by reducing mobility demand [17] [12]. However, the plan does not address the affordability of the property specifically. It only mentions the provision of a simple flat building for low-income communities in densely populated settlements, but not describe clearly the funding scheme for the project or the ownership unit. With the high price of housing property, not all citizens are able to afford a house, and the plan does not provide solutions for funding scheme to make the housing market affordable. Therefore, the score of 1 is given for this indicator.

> "*The alternative housing land provision that can be considered in planning the RDTR and the Jakarta 2030 PZ is to optimize the available land through its multi-functional use as a*

mixed designation. Similar to the criteria for increasing building intensity, the mix designation needs to consider the typology of the sub-district area and the locational advantage with advanced arrangements through zoning regulation techniques." (Translated from Pasar Minggu District RDTR, p.10)

4.3.2. Inclusive Community

Local communities and related stakeholders are involved in forming the plan, which is the basic procedure required by the government. The plan's substantial, however, does not describe the social impact of the development and community participation, therefore the score 0 is given. The point of view towards the community needs to be changed, as the plan sees the local community as an object only take benefit of the development. In reality, the local community needs to be developed as well. Public participation is critical to creating a strong and sustainable community, as outlined in the 11th SDGs. Furthermore the Community participation has so much advantage like improving democratic and service accountability, enhancing social cohesion, enhancing effectiveness to the regeneration process, making the policy more relevant to local communities, giving opportunity to the communities to develop the skills and network [32].

4.3.3. Safety and Security

Safety is one of the basic needs of Human being. The plan, however, mentions only building safety and not community safety, thus the score 0. With Jakarta being 4th out of 10 least secure city in the world based on infrastructure, digital, crime, road safety, to healthcare quality [33], as well as the terrorist attacks in the last decade, the government should consider safety and security indicator in plan documents. It is also critical to raise community awareness of safety and security issues.

5. Conclusion

Since sustainability issues emerge and are being recognized by the UN, many countries have tried to address the SD concept in their development policies. The term "sustainability" and "sustainable development" are being widely used, while the definition is still in debates that many scholars try to interpret the contested concept in their own understanding. "Sustainability" is mentioned in the Jakarta Spatial Plan vision, which is outlined further in the Jakarta Detailed Spatial Plan. However, the plan does not provide an explanation of what "sustainability" means and what it is to sustain.

Out of 10 SD indicators obtained from the literature review, three are explained in the Detailed Spatial Plan in detail, namely sustainable transport system, green space and biodiversity, and waste management. The other 5 are mentioned in the plan, however, either it is just briefly mentioned, or the explanation does not reflect sustainable development as outlined in table 1 (elements of the SD indicators). These indicators are economic development, disaster management, water management, sustainable energy and energy conservation, and sustainable housing. There are 2 indicators that are not mentioned in the plan, which are: inclusive community, and safety and security.

The scoring result above shows that the Jakarta Detailed Spatial Plan has not grasped the holistic understanding of the SD concept. Several indicators have indeed complemented each other (such as sustainable transport system and economic development), showing the 'overlapping circles' of the economic and environmental pillars. However, the social pillar is generally neglected. Figure 4 depicts the position of the Jakarta Detailed Spatial Plan if being illustrated with the sustainability triangle [9].

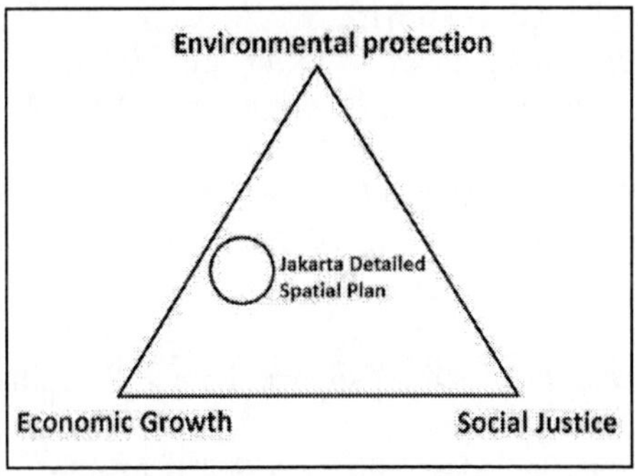

Figure 4. The Jakarta Detailed Spatial Plan in SD Triangle Map [9]

IOP Conf. Series: Earth and Environmental Science **340** (2019) 012032 doi:10.1088/1755-1315/340/1/012032

6. References

[1] Castells M 2000 *Urban sustainability in the information age* City. **4**(1) 118-122. doi: 10.1080/713656995

[2] Jakara Central Bureau of Statistics 2018 *Gross Regional Domestic Product of DKI Jakarta Province by Industry*. (Jakarta: Jakarta Central Bureau of Statistics 2018)

[3] Silalahi D 2001 Hukum Lingkungan dalam Sistam Penegakan Hukum Lingkungan di Indonesia ed.3 (Bandung, Indonesia: Alumni)

[4] Siahaan N H T 2004 *Hukum Lingkungan dan Ekologi Pembangunan* Ed.2. (Jakarta, Indonesia: Erlangga)

[5] President Joko Widodo. 2016. Pembangunan Berkelanjutan Sebagai Komitmen Global. Retrieved September 5, 2018 from http://presidenri.go.id/berita-aktual/pembangunan-berkelanjutan-sebagaikomitmen-global.html

[6] Peraturan Menteri Pekerjaan Umum 20/PRT/M/2007 tentang *Pedoman Teknis Analisis Aspek Fisik dan Lingkungan, Ekonomi, serta Sosial Budaya dalam Penyusunan Rencana Tata Ruang. (2007).* available online at: http://birohukum.pu.go.id/uploads/DPU/2007/PermenPU20-2007.pdf

[7] WCED 1987 Our Common Future (Oxford, England: Oxford University Press)

[8] Castro C J 2004 *Sustainable development, Mainstream and Critical Perspectives. Organisation & Environment*, **17**(2) 195–225. doi: 10.1177/1086026604264910.

[9] Connelly S 2007 *Mapping Sustainable Development as a Contested Concept Mapping Sustainable Development. Local Environment.* **12**(3) 259-78 doi: 10.1080/13549830601183289.

[10] Giddings B, Hopwood B & O'Brien G 2002 *Environment, Economy and Society: Fitting Them Together into Sustainable Development. Sustainable Development.* **10** 187–96. doi:10.1002/sd.199.

[11] The United Nations 2015 *Sustainable Development Goals* Retrieved September 6, 2018 from https://sustainabledevelopment.un.org/?menu=1300

[12] Gough M Z 2015 *Reconciling Livability and Sustainability: Conceptual and Practical Implications for Planning*. Journal of Planning Education and Research. **35**(2) 145-60. doi:10.1177/0739456X15570320.

[13] Lowe M, Whitzman C, Badland H, Davern M, Aye L, Hes D, Butterworth I, & Giles Corti B. 2015 *Planning Healthy, Liveable and Sustainable Cities: How Can Indicators Inform Policy? Urban Policy and Research*, 33(2), 131-44. doi: 10.1080/08111146.2014.1002606.

[14] Haughton G & Counsell D 2004 *Regions, Spatial Strategis and Sustainable Development. Oxon* (England: Routledge)

[15] Kenworthy J 2006 *The eco-city: ten key transport and planning dimensions for sustainable. Environment & Urbanization*. **18**(1) 67–85 doi: 10.1177/0956247806063947.

[16] Handayeni K D M E, Ariastita P G 2014 *Adopting TOD (Transit Oriented Development) for Sustainable Transportation in Surabaya City* Tata Loka. **16**(2) 108-15.

[17] Howley P, Scott M & Redmond D 2009 *Sustainability versus liveability : an investigation of neighbourhood satisfaction. Journal of Environmental Planning and Management.* **52**(6) 847-864. doi: 10.1080/09640560903083798

[18] Berke P R & Conroy M M 2000 *Are We Planning for Sustainable Development? An*

[19] Brugmann J 1997 *Is there a method in our measurement? The use of indicators in local sustainable development planning. Local Environment.* **2**(1) 59-72. doi: 10.1080/13549839708725512.

[20] Tingsanchali T 2012 *Urban Flood Disaster Management* Sciverce Science Direct Procedia Engineering. **32** 25 – 37 doi:10.1016/j.proeng.2012.01.1233.

[21] Cutter et al. (2008). *A place-based model for understanding community resilience to natural disasters*. Global Environmental Change. **18** 598–606. doi:10.1016/j.gloenvcha.2008.07.013.01

[22] Engle et al. (2010). *Unpacking governance: Building adaptive capacity to climate change of river basins in Brazil.* Global Environmental Change. **20** 4–13. doi:10.1016/j.gloenvcha.2009.07.001.

[23] Berman et al. (2012). *The role of institutions in the transformation of coping capacity to sustainable adaptive capacityThe role of institutions in the transformation of coping capacity to sustainable adaptive capacity*, Environmental Development. **2** 86–100. Doi: /10.1016/j.envdev.2012.03.017

[24] Arbon I M 2012 *The Energy Hierarchy Approach To Optimum Use Of Energy Infrastructure – Sharing Ideas From The UK And Other Parts Of Europe Proce of the International Symposium on Engineering Lessons Learned from the 2011 Great East Japan Earthquake*, March 1-4, 2012, Tokyo, Japan. 156-65

[25] Department for environment Food and Rural Affairs 2011 *Guidance on applying the Waste Hierarch.*

[26] Tan Teck-Hong 2011 *Journal of Strategic Innovation and Sustainability*. **7**(1) 62-71.

[27] Russel, Wynne. 2013. *Spatial Planning for Social Inclusion.* LGAT News March. 43-45

[28] Lacinák et al. Smart city, Safety and Security. *TRANSCOM 2017: International scientific conference on sustainable, modern and safe transport.* Procedia Engineering 192 (2017) 522–27. doi: 10.1016/j.proeng.2017.06.090.

[29] Mayring P 2000 *Qualitative Content Analysis. Forum: Qualitative Social Research* (2), Art. **20**. retrieved September 6, 2018 from http://www.qualitative research.net/index.php/fqs/article/view/1089/2385

[30] Schreier M 2012 *Qualitative Content Analysis in Practice* (England: London SAGE Publications)

[31] Berke P R 1994 *Evaluating Environmental Plan Quality: The Case of Planning for Sustainable Development in New Zealand* Journal of Environmental Planning and Management. **37**(2) 155-169 doi: 10.1080/09640569408711967.

[32] Burns et al. (2004) *Making community participation meaningful: A handbook for development and assessment* (Great Britain: Bristol).

[33] The Economist Intelligence Unit. Safe City Index 2017.

Travel behavior of urban kampong residents and formal housing residents for shopping activity: Case of Yogyakarta City

N L P H Dewi[1] and Y Herwangi[1]

[1] Urban and Regional Planning Program, Department of Architecture and Planning, Universitas Gadjah Mada, Yogyakarta, Indonesia

Email: lianahendi@gmail.com

Abstract. Shopping is one of the important activities that is carried out by the residents to meet their daily needs. It also has some relationship with the built environment condition around the resident's houses. If the residential area cannot provide the needs of the residents, then the residents will travel out of the residential area, especially using private vehicle. Yogyakarta is one of the city in Indonesia which has a type of settlement that can support sustainable development by increasing pedestrian trips. On the other hand, the number of private vehicles in Yogyakarta always increase year by year. Therefore, the aim of this study is to identify the travel behavior for shopping activity at two kinds of urban settlement which have some differences in their built environment condition. This study is conducted by quantitative method with data collection using questionnaires and it is assisted by field observation. The result of this analysis using Mann Withney U-test shows that there are some differences of shopping travel behavior between urban kampong residents and formal housing residents. However, the chi-square test shows some weak relationships between travel behavior and built environment. In the other words, built environment does not significantly affects the shopping travels in Yogyakarta city, as well as many developed theories.

Keywords: *shopping trip, travel behavior, transportation, urban settlement, built environment*

1. Introduction

Analysis of travel behavior is the first step in transportation modelling [1]. Spatial transportation modelling is closely related to the generation and attraction of trips that on the respective zones. Hence, travel behavior describes about human activities (working, shopping, school, etc) and how human use transportation such as destination, distance, travel time, and vehicles used [2]. Then, on travel behavior, the trip generation zone or the main origin zone is the settlement zone, which can generate 80-90% of trips in urban areas. Therefore, settlement as trip generation area will have a major influence on transportation in the city.

City is formed as a function of complex human activities. Activities within the city are the expression of the people's efforts to meet their needs. Therefore, the city cannot be separated from transportation system since transportation is a derivative need due to the need of daily needs fulfilment such as school, work, shopping, recreation and so on [3]. The relationship between travel behavior and daily need fulfillment can be shown by travel behavior that has been explained above.

Shopping is one of the important activities that is carried out by people almost every day. This is a way to meet their daily needs especially for food, drink, cloth, and so forth. Moreover, the major economic sector which dominates in a city is the trades and services sector (tertiary sector). Regarding to travel behavior, it will affect to the mobility within the city which is also dominated by shopping activities or shopping trips. But, especially in Indonesia, there has not been much researches on shopping travel behavior. It is suspected because shopping travel is not a routine trip like working or school [4].

Yogyakarta is one of the city in Indonesia which has a good economic growth, especially on the trades and services sector. The trades and services sector based on data on 2017 dominated the economic structure about 7,49 percent [5]. The growth rate for trades and services sector also increases in general. Thus, the effect of the increasing of the trades and services can affect to the enhancement of shopping trips in Yogyakarta. In addition, Yogyakarta City must be able to anticipate the impacts of economic growth to urban mobility so that the development in the city can run sustainably.

The condition of Yogyakarta City is well-known as a city of Indonesia, especially for higher education. It is not actually a big city, but it has a high spatial hierarchy and rapid growth that attracts researchers to conduct research in Yogyakarta City. The physical conditions of the settlements in the city of Yogyakarta is quite diverse and also have a form of traditional Indonesian settlements, namely "kampung kota" or urban kampong [6]. The location is around the city center with high density, diverse of land use, and interlinked area design. Those conditions should be able to generate more sustainable trips. On the other hand, some different conditions are found at formal housing that developed in Indonesia recently and tend to have homogeneous impression, the design is less connected and lack of accessibility due to its development leading to the suburbs.

As many developed theories state that the settlement form factors which are seen from built environmental factors such as density, land use diversity, accessibility, design and distance to transit facilities can influence travel behavior [7]. In more detail, dense settlements with mixed land uses will increase proximity so that the travel can be reached only on foot [8]. Hence, Yogyakarta City which has many settlements with mixed land use characteristics, namely the urban kampong, should be able to reduce travels by using private vehicles. However, in Yogyakarta City, there has been an increasing in the number of private vehicles every year, even in 2016-2017 the number of motorcycles increased around 211% [9]. Those conditions give rising to the initial suspicion that there are some differences on the influence of built environment factors on travel behavior in Yogyakarta City if it is compared to theories that have been developed.

2. Research design

2.1 Defining urban kampong and formal housing

Urban kampong is defined as a settlement that is built by community's initiation (self-help housing) and without being planned carefully through a planning process or can be called unplanned development. Urban kampong is usually used to explain dichotomies between villages and cities, namely cities that are interpreted as modernity and kampong which is defined as backwardness, inequality, and all things that are not good. On the other hand, urban kampong has diversity in land uses so that the local needs of the community can be fulfilled in their own environment. In addition, the suburban area has begun to be filled with a more modern form of housing, which is planned by the government or developer or in this study called formal housing (planned development). Formal housing is usually categorized as gated community which causes homogeneity and limited access to facilities in close proximity. Therefore, residents have to travel out of their settlement area to meet the local needs with longer distances.

2.2 Data collections

The data collection method is using questionnaires that randomly spread using simple random sampling to 200 respondents who live in urban kampong and formal housing in Yogyakarta City. The respondents are selected by giving number to each house in the case study's area, then the numbers are picked randomly. There are two locations of urban kampong and two locations of formal housing that are selected by some criterion such as location distribution, distance from core activity, public transportation affordability, building density, and street pattern. The selected area are Tukangan Kampong, Kricak Kidul Kampong, Puri Timoho Asri Residence and Green House Residence. This study is also complemented by field observation to assess the built environment condition. The criterion and the location of case study are shown below:

Table 1. Some criteria to choose the case study

Location	Distribution	Proximity to activity center	Public transport affordability	Building density	Street pattern
Kricak Kidul Kampong	North side	Far away from CBD and government center	Passed by public transport	Moderate	Organic
Tukangan Kampong	Middle side	Near the CBD	No public transport	High	Organic
Puri Timoho Asri Residence	North side	Near the government center	Passed by public transport	Moderate	Cul-de-sac
Green House Residence	South side	Far away from CBD and government center	No public transport	High	Grid

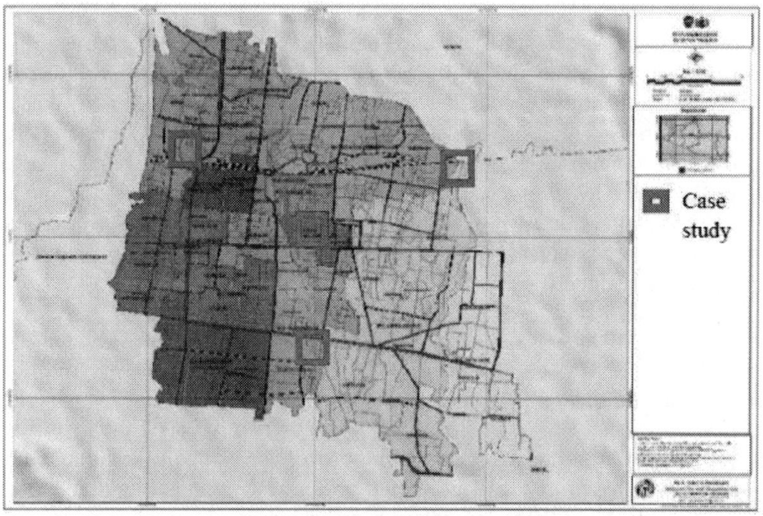

Figure 1. The location of case study

2.3 *Data analysis*

The result of the questionnaires are analyzed using Mann Withney U-test to assess if there are some differences or similarities of the mean between the frequency, distance and mode choices that done by urban kampong residents and formal housing residents on shopping purpose. Then, the relation between built environment (independent variable) and travel behavior (dependent variable) is analyzed using chi-square analysis with the formula as follows.

$$X^2 = \sum_{i=1}^{k} \frac{(Oi - Ei)^2}{Ei} \qquad (1)$$

X^2 : Chi square value Oi : Observed frequencies
k : The number of class Ei : Expected frequencies

This study also uses hypothesis, namely:

a. Null hypothesis: There are no differences between the dependent variable with independent variable of urban kampong residents and formal housing residents.

b. Alternative hypothesis: There are some differences between the dependent variable with independent variable of urban kampong residents and formal housing residents.

In addition, for the built environment measurement is using scoring method based on the theories and standards as follows. The unit of built environment measurement is by an area. In the other words, the score is given to each area (each case study) which is delineated using 100 meter radius.

a. Mix Land Uses analyzed using entropy with the formula:

$$Land\text{-}use\ mix = \frac{(-1)*[(\frac{b1}{a})ln(\frac{b1}{a})+(\frac{b2}{a})ln(\frac{b2}{a})+(\frac{b3}{a})ln(\frac{b3}{a})+\cdots]}{ln(n)} \quad (2)$$

Where is:
b = the area of each land use
a = the total area
n = the number of land uses

b. Building density analyzed using Building Coverage Ratio (BCR) with the formula:

$$BCR = \frac{Built\ area}{Total\ area}$$

c. Scoring analysis to know the rate of facilities affordability. The analysis based on Indonesian government standard named SNI 03-1733-2004.

3. Travel behavior for shopping purpose

3.1 Travel frequency

Frequencies are defined as the intensity of a travel during a week. Nonetheless, in this study the travel that is counted just the travel that always happened every week. The result from 200 questionnaires show that there are 162 shopping travels in urban kampong and 161 shopping travels in formal housing. It is happening because one family could go shopping in some difference places in a week, for instance to the public market, store, and *warung*. For shopping activity, the composition of travel frequencies are shown below:

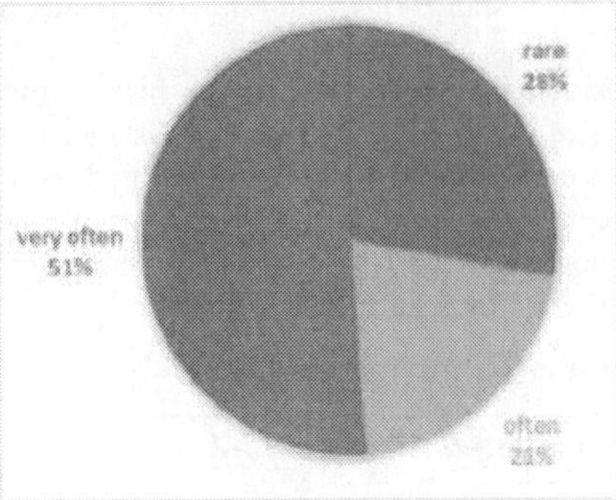

Figure 2. Composition of shopping frequencies of urban kampong residents

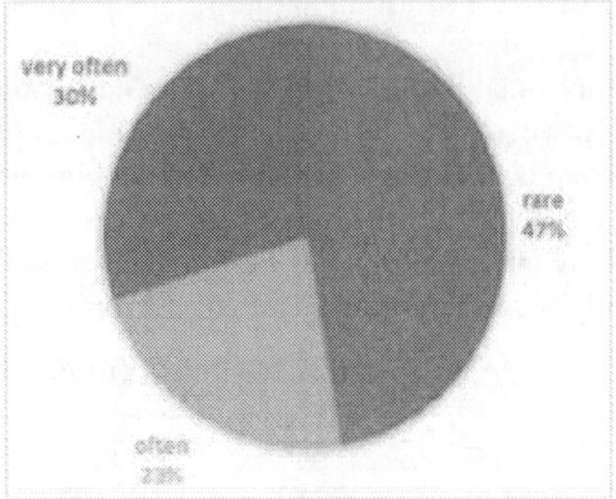

Figure 3. Composition of shopping frequencies of formal housing residents

From those figures, it can be concluded that urban kampong residents are more often travelling for shopping than the formal housing residents. About 57 percent of the shopping travels of urban kampong residents are carried out very often. However, only 30 percent of the shopping travels of formal housing residents are carried out very often, thus formal housing travels are more dominated by the rare frequencies. When the mean of frequency is compared using Mann Withney U-test, the result shows that there are some differences between the mean of shopping frequencies of urban kampong residents and formal housing residents as shown below:

Table 2. Mann Withney U-test result for travel frequency

α	p-value	Count-Z	Table-Z
0.1	0.000	4.736	1.67

The value of Count-Z (4.736) > Table-Z (1.67), thus the null hypothesis is rejected which mean there are some differences between the mean of shopping travel frequency of urban kampong residents and formal housing residents. This fact is happening allegedly due to shopping facilities that more variable in urban kampong rather that in formal housing. Moreover, it also allegedly due to the economic condition that affects to the buying capability namely the formal housing residents could buy more things at once rather than the urban kampong residents. Thus, the urban kampong will go shopping more often because they only can afford few things at once.

3.2 Travel distance
Travel distance is defined as the distance that is traveled by a person from their house to the location of destination. The distances are grouped into two categories such as near (less than 400 meter), moderate (400 meter-5 kilometer) and far (more than 5 kilometer). The analysis of shopping trip's distance categories is shown by the graphs below:

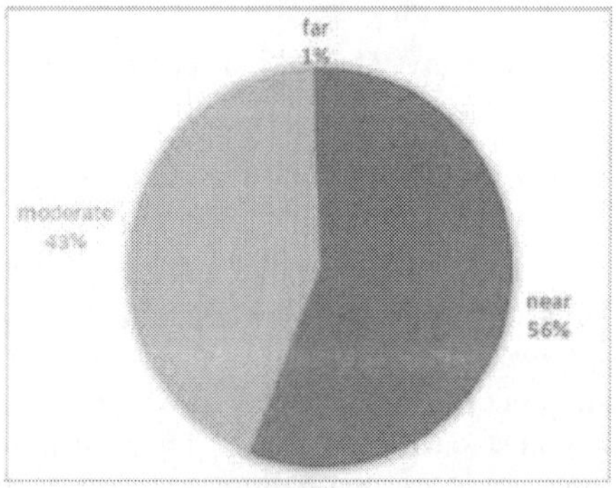

Figure 4. Composition of shopping distance of urban kampong residents

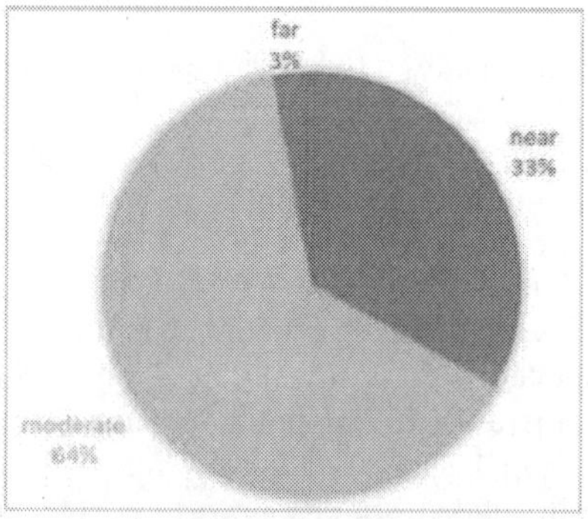

Figure 5. Composition of shopping distance of formal housing residents

The graphs above show that the urban kampong residents prefer to go shopping just around the walking distance. However, the formal housing residents prefer to go shopping on vehicle distance, although those are 33 percent travels that are traveled in walking distance. Although, the formal housing does not have some shopping facilities like *warung*, or store near it, the residents still can buy their daily needs in walking distance because there are some traders who move around the housing. Thus, the residents do not need to travel out from their housing area. This is one of the uniqueness that is found in Indonesia.

Table 3. Mann Withney U-test result for travel distance

α	p-value	Count-Z	Table-Z
0.1	0.000	1.825	1.67

The table above shows that the value of Count-Z (1.825) > Table-Z (1.67) which mean the null hypothesis is rejected. In the other words, there are some differences between the mean of shopping travel distance of urban kampong residents and formal housing residents. On the other hand, the difference of the travel distance is not to significant. It's only 433.5 meter.

4. Built environments that affect travel behavior

4.1 Built environment condition
Built environment conditions are analyzed using four variables which are selected from the previous theories. The selected variables that have been explained on part 3 namely: density, mixed land use, design, and destination accessibility. The measurement of those variables are using scoring method

based on the theories. Then, the result from those scoring method in each area would be compared each other to define the highest class until the lowest class. Scoring analysis results are shown below:

Table 4. Built Environment Condition

Variables	Kricak Kidul Kampong	Tukangan Kampong	Green House Residence	Puri Timoho Asri Residence
Density	high (92.84)	very high (202.4)	high (71.74)	low (52.35)
Mixed Land Uses	mixed (0.27)	low mixed (0.22)	moderate(0.23)	low mixed(0.20)
Design	good (350)	good (397)	moderate (314)	good (106)
Destination Accessibility	good (121.73)	good (121.33)	good (101.78)	moderate(79.66)

The result of built environment measurements can show that the urban kampong has better built environment condition rather than the formal housing, except the mixed land use in Tukangan Kampong which has a low score. This condition can be related with sustainable transportation with the indicators are compact, densify, transit, connect, mix, cycle, shift, and walk [10]. Based on the result of built environment analysis, it can be concluded that urban kampong area is more compact, dense, and connected. Hence, the condition of urban kampong area should promote sustainable transportation rather than the formal housing area.

4.2 The Relationship between built environment and travel frequency

The analysis of relation between built environment and travel frequency is using cross-tabulation analysis. The result of this analysis shows that all of the p-value in each variable is less than α (p-value α). It means that all of the variables affect travel frequency in shopping trips. On the other hand, the Somers' d coefficients (C) show very low until low connections. Some value of Somers' d coefficients are less than 0.199 so the connections are very low [11]. Moreover, some value of Somers' d coefficients are between 0.20-0.399 so the relationship are low[11]. The relationship table shown below:

Table 5. Chi-Square analysis for travel frequency

Variables	α	Travel frequency	
		p-value	**C**
Density		0.000	0.180
Mixed land uses	0.1	0.000	0.198
Design		0.000	0.220
Destination accessibility		0.000	0.263

The value Somers' d coefficients are showing positive value. It means that the relationship between built environment condition and travel frequency moves on the same side. If the value of built environment grows better, than the shopping travel will be carried out more often, and vice versa.

Between all of those built environment variables, destination accessibility and design are strongly affecting the travel frequency than others. It is supposed that the distance to the shopping facilities and the residential design which is comfortable to pedestrians could increase the shopping frequency. This condition could make the residents easily reach the shopping facilities so the trips would be carried out frequently.

4.3 The relation between built environment and travel distance

This analysis for the correlation between built environment and travel distance is using the same method with the previous analysis for the travel frequency. The result of this analysis also shows the same meaning. The p-value of all of the relation between built environment and travel distance are less than (p-value $< \alpha$) and some of the Somers' d coefficient are less than 0.199 and between 0.20-0.399. Those result show the conclusion that the built environment can affect the travel distance but not to significant as the table below:

Table 6. Chi-Square analysis for travel distance

Variables	α	Travel frequency	
		p-value	**C**
Density	0.1	0.001	-0.107
Mixed land uses		0.000	-0.096
Design		0.000	-0.159
Destination accessibility		0.000	0.263

The built environment variable which is strongest affecting the travel distance is the destination accessibility and the weakest is the mix land uses. It can be concluded that the distance to shopping facilities (accessibility) is more important than area of shopping facilities which is measured by entropy index. This study shows one of the uniqueness of shopping trips in Yogyakarta that the distance of shopping trips are not only determined by the location of built up facilities, but also the unbuilt facilities namely moving traders or "pedagang keliling". The moving trader comes to the residential area every morning to serve the residents with their daily needs such as vegetables, meats, fruits, and so forth.

However, the result of the Somers' d coefficients on travel distance analysis show a little difference. The value of Somers' d coefficients are negatively affecting travel distance. In the other words, if the value of built environment increase, than the travel distance will decrease but not to significant. This condition can shows that the built environments have not been affected Indonesian travel distance yet. But it can be a good beginning to promote sustainable transportation in Indonesia with some other ways.

5. Discussion

Based on the results of this study, the built environments are not significantly affecting the shopping trips as the previous theories have proved. For instance, the factor of land use diversity does not have a significant influence on shopping trips. This indicates that the diversity of land use which is assessed through entropy index is not the only indicator that determines the length of travel. Resident who lives in more homogeneous land use, namely in formal housing also can travel for shopping on short distance because there are traders who move closer to buyers. It shows that even though the entropy index is small, residents can still be served by commercial facilities properly. In the other words, this study want to give a new point of view namely to serve people with good access to facilities, not only thinking about the built facilities. It could be intervened by giving "moving facilities" to provide the local needs of people, for instance the moving trader or in Indonesia called "pedagang keliling".

6. Conclusion and recommendations

This study tries to compare the condition of shopping travel behavior of people who live in urban kampong as undeveloped area and people who live in formal housing as developed area. It is related with the built environment conditions of those settlement area. Urban kampong is more dense, mix and accessible than the formal housing which is more homogeny, less dense and lack of accessibility because it grows in suburbs. When the travel frequency and distance are compared, the result shows that urban kampong would generate more sustainable travels that the formal housing. It is proved by the mean of urban kampong's travel distance that is shorter than the formal housing because it will produce less emissions.

However, the correlation between built environment and travel behavior surprisingly shows some weak relations. The built environments are not significantly affecting travel behavior in Yogyakarta City as have been mentioned by the previous theories. It could be affected by the uniqueness of Yogyakarta's trading culture that moves closer to the buyer called "pedagang keliling". Moreover it is also suspected because the residents consider about the quality of the shopping facility so they may travel farther to get the facility they need and they want. Nonetheless, the weak relationships between travel behavior and built environment could be a good beginning for Indonesia especially for Yogyakarta City to pay more attentions to built environment conditions. The provision of shopping facilities is not only thinking about the number of the facilities but also the quality of those facilities.

7. References

[1] Dimotriou, H.T 1992 *Urban Transport Planning* (London: Rutledge)

[2] Stangeby I 2000 *Methodology of Travel Behavior Research* (Oslo: Institute of Transport Economics, University of Oslo)

[3] Tamin, Ofyar Z. 2000 *Perencanaan dan Permodelan Transportasi* (Bandung: Penerbit Institut Teknologi Bandung)

[4] Dewi H 2018 *Comparison of the Travel Behavior by Urban Kampong Residents and Formal Housing Residents in Yogyakarta City* (Yogyakarta: unpublished)

[5] BPS Yogyakarta 2017 *Produk Domestik Regional Bruto* (Yogyakarta: Badan Pusat Statistik)

[6] Heryati 2008 *Kampung Kota Sebagai Bagian Dari Permukiman Kota Studi Kasus : Tipologi Permukiman Rw 01 Rt 02 Kelurahan Limba B Dan RW 04 RT 04 Kel.Biawu Kecamatan Kota Selatan Kota Gorontalo* (Gorontalo: Universitas Negeri Gorontalo)

[7] Ewing, R. & Cervero, R 2010 Travel and the Built Environment vol 76 (London: *Journal of the American Planning Association*) p1-30

[8] Talen, E. 2009. *Urban Design Reclaimed,* Washington DC: American Planning Association.

[9] Wardhani M.2018.Wow, Jumlah Kendaraan di Kota Yogya Naik Hingga 200% [Online] available at : http://jogja.tribunnews.com/2018/02/26/wowjumlah-kendaraan-di-kota-yogya-naik-hingga-200 [Accesed 13 Juli 2018]

[10] ITDP 2017 *Principles for Transport in Urban Life: Better Together* (Jakarta: ITDP)

[11] Sugiyono 2012 *Metode Penelitian Kuantitatif Kualitatif dan R & D* (Bandung: Alfabeta)

Acknowledgements

The outcomes of this paper are results from a consultancy project supported by Urban and Regional Planning Program, Department of Architecture and Planning, Universitas Gadjah Mada. The project: Comparison of the Travel Behavior by Urban Kampong Residents and Formal Housing Residents in Yogyakarta City

IOP Conf. Series: Earth and Environmental Science **340** (2019) 012034 doi:10.1088/1755-1315/340/1/012034

Changes of landuse in the campus area and their implications toward traffic condition

O R Manullang[1], A R Rakhmatulloh[1], D A Sihaloho[1] and N M Samosir[1]

[1] Department of Urban and Regional Planning, Universitas Diponegoro, Indonesia

Email: oktomanullang73@gmail.com

Abstract. This study revisits how changes in the land use affect traffic condition around a rapidly growing university campus area. Changes in landuse, associated to the increase in activity development which affects the vehicle trip rate, have increased the traffic loading in the road network in the study area. At the same time, the growing number of road users in the road network results in increasing traffic conflicts negatively impacts the traffic condition. The phenomenon has been identified in and around Tembalang Area in Semarang City, with rapid changes in land use due to the increasing demand for supporting activities to provide the needs of students, especially in the main transport corridors. This study examined changes of landuse and their implication toward traffic loading as a first step to control the use of space in the campus area. Quantitative method was applied to analyze trip attraction in comparison to the volume of traffic. The results showed that eatery land use activities were the largest traffic load increase contributor (29-33%) burdening the road network in the study area. Other types of activity identified to negatively impact the service levels of the road network are supermarket, cafe, and laundry services.

Keywords: *landuse change, trip rate, traffic volume*

1. Introduction

Land use in Tembalang Subdistrict in Semarang City ("Tembalang" hereafter) are changing rapidly especially since Diponegoro University started moving its university activities to this locale. Together with three other higher education entities namely Semarang State Polytechnic (Polines), Semarang Health Polytechnic (Poltekkes), and Pandanaran University (Unpand), the development of activities have transformed landuse in the area, either in the form of changing building function or land utilization. These changes is driven by the increasing economic interaction, i.e. the growing demand for the fulfillment of basic needs from approximately 57.000 students. New activities have developed especially in the trade and service sectors.

Changes in landuse for trade/service activities happen mostly around the main access roads in Tembalang area, this has been observed to increase the traffic loading on the road network; the higher level of utilization of a land, the higher trip rate attracted [1]. The main access to the area, specifically Ngesrep Corridor as well as Sirojudin Corridor, has an assortment of trade/service activities, such as eatery, cafe, printing, general store, stationery store, attire store, laundry, motel, etc. It is expected that if changes in land use continues to occur, the trip rate will also increase significantly, potentially leading to congestion and reducing the corridors' level of service.

IOP Conf. Series: Earth and Environmental Science **340** (2019) 012034 doi:10.1088/1755-1315/340/1/012034

Differences in characteristics are expected to affect the level of contribution to the road loading in the two corridors in discussion. Ngesrep Corridor is the main access connecting the area to the arterial road as well as to Semarang City centers, it is understandable that this corridor develops earlier, and in the past 10 years (2008-2018), this corridor has experienced continuous changes and development of building functions. Meanwhile, changes of land use in Sirojudin Corridor are more related to activity developments along the corridor itself; this distinction of characteristic renders the two corridors very interesting as the study area.

This study compared the trip rate attracted by trade/service activities as well as calculate the additional traffic volume produced to describe the level of contribution of traffic impact on the respective corridor. The trip rate assessment of this study refers to the Institute of Transportation Engineers (ITE) methodology commonly used by researchers from various countries, both developing and developed country, as currently Indonesia have not developed its own manual. Although ITE have provided reference trip rates for various types of land use, they are not applicable as they do not reflect the lifestyle and travel pattern of people in Indonesia. To better suit the local context the trip rate is developed on a local scale [2]. The trip rate of the identified activities can be used to predict future traffic volumes, as part of traffic impact assessment (TIA) to determine the impact of a proposed land use development/change, especially in areas where higher education facilities are present. However, there is no agreement when the investigation would be fundamental and what approaches are regarded appropriate for the estimation of trip rates [3-5].

2. Trip rate of the corridors

The trip rate of the corridor was generated through a series of surveys, namely Identification of Landuse Change Survey and Trip Generation Survey. Types of trade/service activities sampled included eateries, attire stores, supermarkets, salons, cafes, cell stores, laundries, services, lens stores, shoes stores, printing places, motel, dorm and stationery stores. Each type of activity has a minimum sample size of five to strengthen the regression analysis [6]. Traffic counting was carried out during peak hours therefore the trip rate factor is used to generate trips during peak hours, not for the whole period of the way. The unit of factor was adjusted to the unit of predictor variable used.

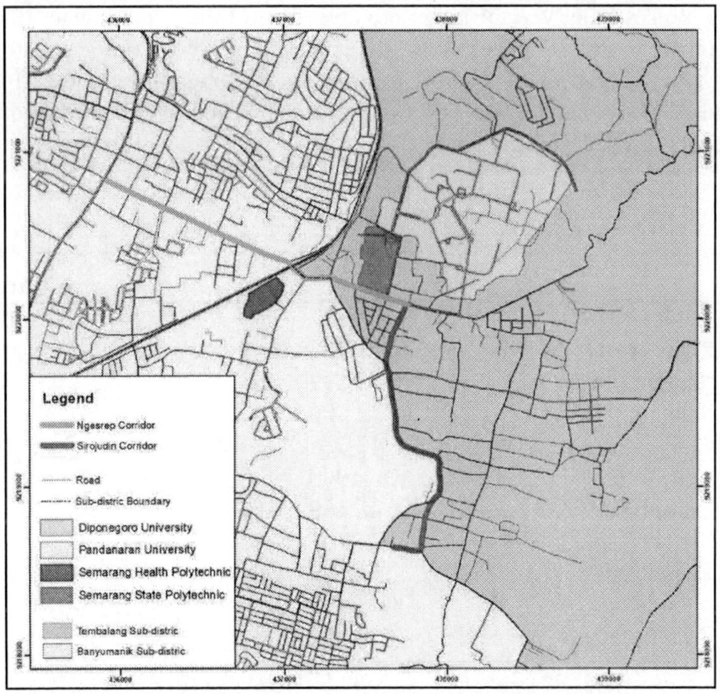

Figure 1. Study area *Ngesrep and Sirojudin Corridor*

IOP Conf. Series: Earth and Environmental Science **340** (2019) 012034 doi:10.1088/1755-1315/340/1/012034

2.1. Ngesrep corridor

In total, the Ngesrep corridor is 1.95 km long, consisting of two main road segments, namely Ngesrep Timur V Street (1.2 km long) and Prof. Soedarto SH street (0.75 Km long); it has dominantly 4/2 undivided road type with a width of ±16 meters. It has poor pedestrian path condition such as cracks, holes, and some part occupied by parked vehicles as well as street vendors. The land use along the corridor is dominated by trade and service activities, with a smaller proportion of it functioned for education, government offices, and house of worship. The trip rate modelling resulting from the analysis of the predictor factors on the Ngesrep Corridor can be observed below:

Table 1. Equation and trip rate factor of the activities on Ngesrep Corridor.

Num	Activity	Trip rate equation	Predictor Variable (x)	Trip rate Factor	Unit
1	Eatery	y = 0,182 x + 1,0959	Seats	0,208	vec/unit/peak hour
2	Attire store	y = 0,3368 x + 2,8076	Employees	1,417	vec/person/peak hour
3	**Supermarket**	y = 11,611 x + 10,944	Cashiers	**15,031**	vec/person/peak hour
4	Salon	y = 0,3904 x + 0,4605	Seats	0,500	vec/unit/peak hour
5	Café	y = 0,1327 x + 5,9813	Parking spaces	0,243	vec/m^2/peak hour
6	Cell store	y = 0,1563 x + 2,25	Parking spaces	0,357	vec/m^2/peak hour
7	Laundry	y = 2,125 x – 21,375	Operational hours	3,821	vec/hour/peak hour
8	Services	y = -1,0109 x + 15,728	Operational hours	0,357	vec/hour/peak hour
9	Lens store	y = 0,1319 x + 1,125	Parking spaces	0,444	vec/m^2/peak hour
10	Shoes store	y = 0,0695 x + 1,233	Floor areas	0,110	vec/m^2/peak hour
11	Printing place	y = 0,9083 x + 3,6667	Parking spaces	1,825	vec/m^2/peak hour
12	**Motel**	y = 0,021 x + 1,9824	Parking spaces	**0,045**	vec/m^2/peak hour

The independent variables forming the model vary for each type of activity. Variables are selected using an analogy by including variables that are considered to provide impact towards the number of trip generation and attraction. In this corridor, the presence of 75 eateries seems to have contributed to the fact that this type of activity as a whole (not per unit) attracts the largest trip during peak hours. However, individually, the largest trip rate for the one unit activity is produced by laundry activity, with supermarket activity following closely.

Table 2. Trip rate of the activities on Ngesrep Corridor.

Activity	Trip rate (y)	Sum	Trip rate of the corridor
Eatery	8,717	75	**653,8**
Attire store	3,683	22	81,0
Supermarket	48,099	7	336,7
Salon	2,100	11	23,1
Café	13,186	12	158,2
Cell store	4,000	12	48,0
Laundry	**48,150**	5	240,8
Services	4,102	10	41,0
Lens store	1,599	5	8,0
Shoes store	3,398	7	23,8
Printing place	7,299	9	65,7
Motel	3,696	9	33,3

IOP Conf. Series: Earth and Environmental Science **340** (2019) 012034 doi:10.1088/1755-1315/340/1/012034

2.2. Sirojudin corridor

The Sirojudin Corridor consists of KH Sirojudin street, Banjarsari Selatan street and Mulawarman Raya street with an overall length of approximately 1.89 km. This corridor has an average width of 8 meters and lacks pedestrian path. Ribbon pattern land use changes can be observed throughout the length of the corridor. There are many new developments and land use conversions on its first layer ribbon recently, especially on Mulawarman Raya street. The changes include land use functions such as eatery, attire store, laundry, motel, and stationery store. Several activities which were used as samples in the Ngesrep Corridor analysis were excluded from the analysis on this corridor due to the small number of occurrence. The independent variable of floor area is more suitable as a variable to predict the trip rate of each of the activity. The following is the trip rate equation, predictor variable, and its trip rate factor.

Table 3. Equation and trip rate factor of the activities on Sirojudin Corridor.

Num	Activity	Trip rate equation	Predictor variable (x)	Trip rate factor	Unit
1	**Dorm**	y = 1,035 + 0,008 x	Floor areas	**0,010**	vec/m2/peak hour
2	Laundry	y = 0.757 + 0.5 x	Floor areas	0,530	vec/m2/peak hour
3	Supermarket	y = 14,954 + 0,0363 x	Floor areas	0,156	vec/m2/peak hour
4	Attire store	y = 0,7532 + 0.0685 x	Floor areas	0,093	vec/m2/peak hour
5	Stationery store	y = -3,7977 + 2,2136 x	Employees	1,791	vec/person/peak hour
6	Eatery	y = 7,0842 + 0,1325 x	Parking spaces	0,340	vec/m2/peak hour
7	Café	y = 4,8153 + 0,543 x	Employees	0,900	vec/person/peak hour
8	**Printing place**	y = 1,684 + 2,467 x	Printers	**2,770**	vec/unit/peak hour

Printing place activity has the largest trip rate factor for each independent variable unit it has. Meanwhile, dormitory activity has the smallest trip rate factor. As a group of samples the largest trip rate in the corridor comes from the activity of the eatery, possibly due to the large number of sample. However, individually, the largest trip rate of one activity is owned by supermarket activity and the smallest by attire store activity.

Table 4. Trip rate of the activities on Sirojudin Corridor.

Activity	Trip rate (y)	Sum	Trip rate of the corridor
Dorm	5,802	12	78,1
Laundry	13,250	10	132,5
Supermarket	**19,500**	6	115,1
Attire store	**2,900**	6	17,0
Stationery store	16,119	4	60,9
Eatery	11,608	42	**475,9**
Café	12,240	6	73,4
Printing place	15,512	6	93,1

3. The comparison of trip rate toward traffic volume

Traffic volume was obtained from traffic counting on peak hour at 17:00 - 18:00. After the results of the traffic counting processed, it compared with the attracted trip from the activities on the corridor.

3.1. Ngesrep corridor

The activity of eatery was identified as the largest contributor producing the highest addition of road traffic load in the Ngesrep Corridor, with a trip rate that weighing at 29% of traffic. Characteristically, the activity of eating has a somewhat fixed time dimension, but with a flexible places dimension.

These characteristics are also observable for shopping at the supermarket activity; its trip rate contributes at 15% of traffic. Although the laundry trip rate reaches 11%, this activity is only done in a short time, so it is not significant to the road condition. Unlike cafe activity (7%), where visitors need an average of 2-3 hours to enjoy their time.

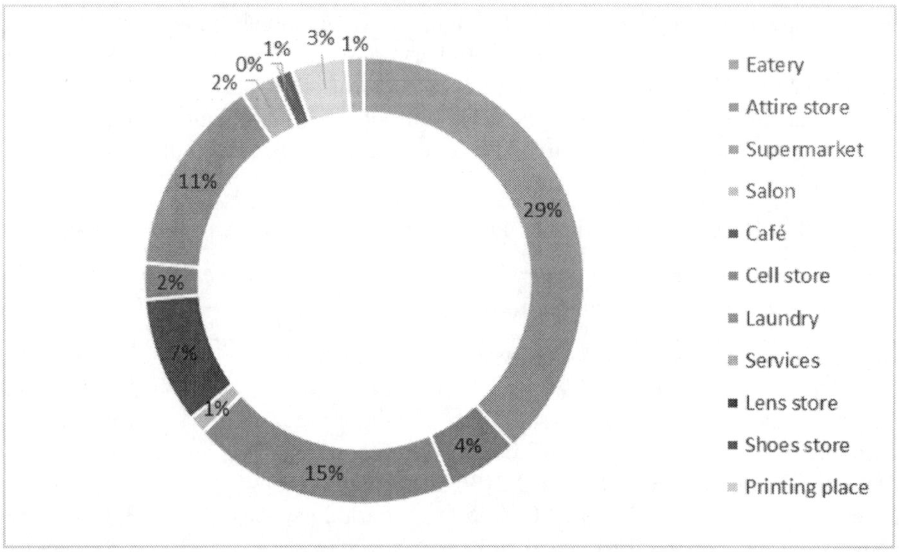

Figure 2. The contribution of trip rate of activities towards traffic on Ngesrep Corridor

The additional trips attracted by the land use activities are observable to have implications to the road network's level of service. The Ngesrep Corridor has a DS value of 0.81 with a level of road service at D level. The increasing number of conversions of buildings along the corridor into an eatery, supermarket, or cafe is expected to further affect the traffic volumes, predictably further reducing the level of service of the road network. It is also observable that lack of parking spaces also causes visitors to park on the road; despite the presence of parking attendants supposedly organizing the parking activities, it still impedes traffic flow as well as increases the potential for traffic conflicts along the corridor.

3.2. Sirojudin corridor

Eatery activities have been identified to contribute significantly to the increase of traffic load on the corridor during peak hour, reaching up to 32.7% of the total traffic. This is expected, because an eatery, or a place for eating, provides for a basic need which occurs at relatively the same period of time during the day for almost everyone. This activity has the characteristics of a fixed time and a flexible place so that accumulatively eatery will be visited two to three times a day.

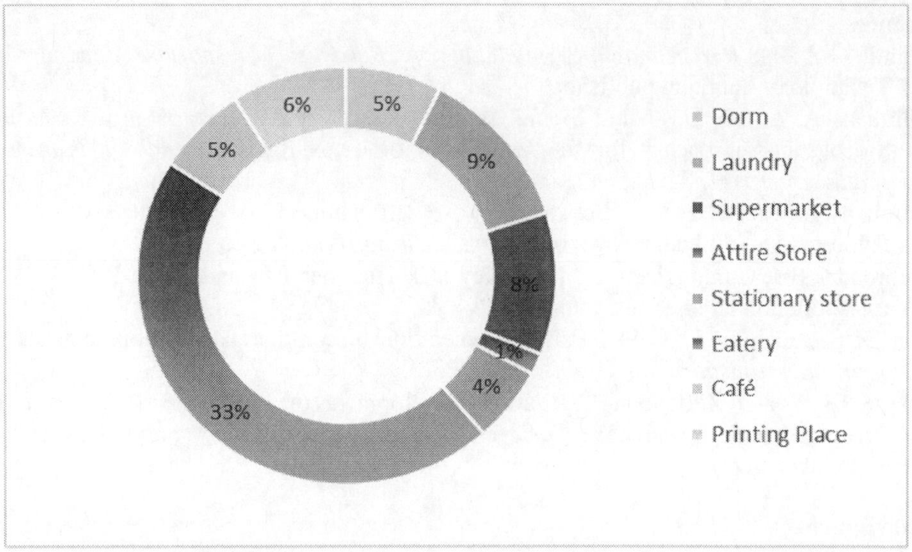

Figure 3. The contribution of trip rate of activities towards traffic on Sirojudin Corridor

On average, the trip rate of each trading/service activitiy on the corridor adds an increase of 9% in traffic volume during peak hour. At present, the Sirojudin corridor has a DS value of 0.676 with a level of road service at C level. If there is an increase in the number of eateries and supermarkets in the future, then it will increase the volume of traffic and affect the road service level.

The control of landuse utilization by the city government is also relatively low because 42% of trade/service activities have permission from the local government. Conversely, other business owners have not taken care of business licensing, but have engaged in activities or run their businesses. The landuse change from non-built to built up occurs a lot in Mulawarman Raya street. At present, the corridor still has potential to grow continually because there are still lots of non-built lands.

4. Conclusion

The college activities contribute to accelerating changes in landuse in order to provide students' needs. In the last 10 years (2008-2018), Landuse in Ngesrep Corridor encountered a shift in building functions. Meanwhile, the Sirojudin Corridor has encountered more land use changes from non-built to built up land.

Assessment of the trip rate explains that some activities can have a high trip rate value, but the population in the corridor influences the amount of contribution in road loading. For instance, the trip rate of eatery is not the highest, but due to the large population, this activity is the biggest contributor, specifically 29% in Corridor Ngesrep and 33% in Sirojudin Corridor.

The addition of activities that can affect the decline in the road service level is an activity that has a high trip rate. These types of activity are eateries, supermarkets, cafes, and laundry. Control of spatial use of these trade activities needs to be tightened, such as through the mechanism of granting business licenses. Although trading/service activity is a business interaction process, the number of supply should be monitored and restricted in order to reduce traffic impacts, such as congestion.

Trip rate equations can be used to predict the amount of permissible activity threshold. Even though the contributors who burdened the road did not only come from the first layer of the corridor, but almost 72-76% of the traffic volume during peak hours was influenced by activities located on the side of the corridor. This underlines that the importance of the threshold for knowing the capacity of the corridor in the development of new activity, especially for Higher Education Areas.

5. References

[1] Tamin O Z 2008 *Perencanaan, Permodelan, Dan Rekayasa Transportasi* (Bandung: Institute of Technology Bandung publisher)

[2] Minhans A, Zaki N H, Shahid S, and Abdelfatah A 2014 A Comparison of Deterministic and Stochastic Approaches for the Estimation of Trips Rates *Journal of Transport System Engineering 1* **1** 1-11.

[3] Minhans A, Zaki N H, and Belwal R 2013 Traffic Impact Assessment: A Case of Proposed Hypermarket in Skudai Town of Malaysia. *Jurnal Teknologi.*

[4] Schroeder B J, Cunningham C M, Findley D J, Hummer J E, and Foyle R S 2010 Manual of transportation engineering studies.

[5] Fan H S, and Lam S H 1997 Parking generation of commercial developments in Singapore. *Journal of transportation engineering*, 123(3), 238-242.

[6] Juette G W and Zeffanella L E 2007 Development of Philippine Trip Generation Rates *Presented at the Eastern Asia Society for Transportation Studies conferences*, Dalian, China, September 24-27

Acknowledgments

This research was financially supported by The Faculty of Engineering, Diponegoro University, Indonesia through Strategic Research Grant 2018.

The spatial configuration of crime in Surabaya

P R Satiawan[1], K P Tucunan[1] and R Y Azarine[1]

[1] Urban and Regional Planning Department, Institut Teknologi Sepuluh Nopember

Email: puturudy@yahoo.com

Abstract. As a second larger city in Indonesia, Surabaya is listed among highest crime rate city in this country. One of the prominent crime in Surabaya is a street crime. This is a type of crime that often occurs in in public spaces which can be experienced by anyone. One of the urban planner guideline in integrating spatial approach to prevent crime is defensible space theory, this research is deducting the theory as a variable as well as inducting variable that might vary between location to country. Using ethnographic approach, the interview was conducted with a crime offenders that later triangulate with the other conventional expert. This research found that there are two (2) types of space in each crime incident, which researcher defined as a action space and defense space. Action space consist of three types of space which are: the observation space, the execution space, and the escape space. Meanwhile, the defense space can be categorized as space that can be maintained positively and space that can be maintained neutral.

Keywords: *crime, spatial configuration, defensible space, action space*

1. Introduction

Surabaya is the second largest city in Indonesia after Jakarta. As a metropolitan city, Surabaya has a number of complete facilities and infrastructure such as ports, highways, highways, strategic infrastructure, where these advantages make Surabaya a trading city that plays a very important and strategic role in the national economic arena. With such potential, Surabaya is an attractive place for many people to live and work. Besides having a number of relative advantages, the large population in Surabaya also turned out to increase the chances of various crimes.

According to the Surabaya Police press release, in 2017 in the city of Surabaya there were indicated 15 spots that were prone to crime. To some extent, this is one proof that crime tends to occur repeatedly in the same place. On the other hand, city residents have their own perceptions of places that are prone to crime. Therefore it is necessary to do a study that raises the phenomenon of the relationship between the incidence of crime and the space for the occurrence of crime.

2. Literature review

2.1. Crime

Crime is all kinds of actions that are economically and psychologically harmful, which violate applicable laws and social and religious norms. It can be interpreted that, criminal acts are all actions in the form of words, actions, and behavior economically, and socially psychologically are very detrimental to society and make the safety of citizens threatened both already written in criminal law and not [1].

Sahetapy dan Reksodipuro (in [1]) said that crime is any act that is prohibited by public law to protect the public and given sanctions in the form of criminal acts by the state. These deeds are punished for violating the social norms of society, namely the existence of behavior that is considered inappropriate by its citizens. Meanwhile, Paul Tappan (in [2]) said that crime is an intentional act or omission in violation of criminal law (statutory or case law), committed without defense or justification, and sanctioned by the state as a felony or misdemeanor. Willlian Blackstone (in [2]) also define that a crime is an act committed or omitted in violation of a public law either forbidding or commanding it.

Forms of crime include [1]:

a. Theft
b. Immorality
c. Pickpocketing
d. Mugging
e. Stabbing with sharp weapons / fire
f. Persecution
g. Murder
h. Fraud
i. Corruption

2.2. Crime and the environmental design

Prevention of crime through the concept of Crime Prevention Through Environment Design (CPTED) covers the design, maintenance and use of the built environment to improve the quality of life and reduce the incidence and fear of crime. There are two important things in the CPTED concept [3], namely:

a. Activity Support
   Activity support involves passive and active efforts to promote the presence of responsible pedestrians in certain areas, thereby increasing the social value of the area
b. Management and Maintenance
   Landscape maintenance, lighting and other features are very important to ensure that the CPTED element meets its intended purpose

In the implementation stage of CPTED concept, the most important thing is that crimes against people and property tend not to occur if there are other people around. The presence of others will:

a. Prevent crime
b. Helps catch criminals
c. Calling for help from others
d. Report the crime and act as a witness

Meanwhile, there are 4 important aspects in the CPTED that are published by the Government of Singapore [4], namely:

a. Natural surveillance
b. Natural access control
c. Territorial reinforcement
d. Maintenance and management

Crowe (in [5]) highlight the assessment of CPTED with what he calls the "Three-D Approach". Essentially, the Three-D Approach is based on the notion that human space is designed to fulfill three functions:

a. All human space has some designated purpose
b. All human space has social, cultural, legal, of physical definitons that prescribe the desired and accepted behaviors
c. All human space is designed to support and control the desired behaviors

2.3. Crime in the context of space

The majority of the studies describe the causes of crime including non-physical constraints, such as economics, social, etc. However, the cause of the crime itself can come from factors that are generally recognized by the community. Brantingham & Brantingham (in [6]) requires that at least 4 things make a crime happen:

a. the existence of law
b. the presence of the attacker
c. the existence of the victim
d. the existence of the place to commit a crime

Paulsen & Robinson (in [6]) stated that crime must have a legal component to state that someone is defined as being against the law. Besides that the incidence of crime must also be accompanied by an attacker and the presence of victims, and the most important thing is the existence of an appropriate space for committing a crime.

The Spatial Theories of Crime based on Environmental Criminology states that there are two main components of crime, namely badness which includes motivation and opportunities, and aspects of crime location which include mobility, awareness space, right target and choice. The two aspects mentioned above have 4 dimensions namely:

a. Legal dimension
b. Victim dimension
c. Offender dimension
d. Spatial dimension

2.4. Defensible space

The concept of the defensible space area is a concept of space that is associated with resistance to crime. According to Tucunan [6] the concept of the defensible space area has 2 approaches, namely a mezo approach (related to the character of space and community) and a micro-approach (related to design - architectural aspects).

Tucunan [6] mentions that the concept of defensible space does not only explain architectural concepts, but defensible space is also a concept of how space can interact with its people to defend against acts of crime embodied in design. Newman [7] defined defensible space as a model for residential environments which inhibits crime by creating the physical expression of a social fabric that defends itself.

Defensible space is a theory that utilizes social control and good supervision by the local residents to maintain their living environment, and this is highly dependent on the design of the urban environment itself. A good urban environment will be able to generate informal social control which is at the core of the defensible space

Defensible space relies on self-help, not on government intervention, so it is not vulnerable to withdrawal of support from the government. This depends on the involvement of citizens to reduce crime and eliminate the presence of criminals. It has the ability to bring people from different economic and racial conditions together in a mutually beneficial entity. For low-income people, defensible space can benefit their lives and the opportunity to improve the world around them through their own actions and improve their quality of life

To create a defensible space, an environment must have a monitoring system. This system relies heavily on the design of circulation in public spaces and placement of family units around public spaces so that residents can look at the road or neighbors. In designing this space, it takes an understanding of territory, a real picture of the building, and clear restrictions on public space and private space.

The greater the number of people sharing communal space, the more difficult it is for people to identify ownership because the public feels they have the right to control or determine the activities that occur within them. It is easier for outsiders to gain access and linger in the area in a shared building 24-100 families, than in a building with 6 to 12 families.

2.5. Characteristics of space that can trigger crime

Elizabeth Wood (in [5]) said that there was a real difference in the level of crime between a quiet and poorly maintained environment with a crowded, well-maintained environment and supervision from its inhabitants. Oscar Newman states that attention to activities that occur on the road is termed 'the eyes on the street', which shows that occupants' attention to their environment is very effective in suppressing the potential for crime [7].

Crime is concentrated geographically and demographically [8]. Residents in urban areas are exposed to a higher risk of being victims of crime compared to residents in rural areas [8]. In this consideration, there is no doubt that the design of the built environment and spatial planning can influence behavior and opportunities for crime in society. According to Nes [9], human behavior that takes place in a space, and spatial planning provides various opportunities for interaction in society. Interaction between people can make a community safe or insecure which is a basic part of people's welfare [10].

The study by Oscar Newman [7] shows that tall buildings with lobbies, elevators, fire escape, tap and corridors isolated from public view, have high potential crime rates, rather than low buildings. The concept of territoriality can be used as a basis for creating informal social supervision spaces, so that people can avoid criminal acts by taking the necessary protective measures. Jacobs (in [10]) states that the more people can look towards the road, the less likely crime is on the road.

There are elements that influence the existence of crime [6], namely:

1. Lighting
   Sufficient lighting is very important so that people can see and be seen clearly. From a security standpoint, strategically placed lighting can have an effect in reducing the fear of criminal acts. Sufficient lighting available during the day can provide security and reduce the chance of crime compared to night. Because the lighting during the day makes people and goods clearly visible, and not too glare which can result in the invisible person being seen.
2. Natural Surveillance and Sight Lines
   Natural surveillance can be achieved through design and maintenance that allows people to carry out their normal activities while easily observing the space around them, as well as removing hiding places for people involved in criminal activities, while sight lines are defined as lines of vision. The inability to see what lies ahead along the route due to sharp angles, walls, fences, bushes or poles can be a serious obstacle to feeling safe. The high density of activities or buildings can reduce the level of sight lines, while with reduced sight lines, the level of natural surveillance will also decrease. Weak natural surveillance will create opportunities for crime.
3. Building Design
   Building design greatly affects the chances of crime. Characteristics of buildings or areas that do not allow owners or residents to monitor whoever enters will be vulnerable to crime. The number of spaces that are stuck or do not have alternative roads will cause a sense of being trapped or isolated. According

IOP Conf. Series: Earth and Environmental Science **340** (2019) 012035 doi:10.1088/1755-1315/340/1/012035

to Newman [7] tall buildings can cause crime because access to the building is too difficult to achieve if crime occurs. The smaller the public space in the area or building will lead to higher chances of crime.

4. Mixed Landuse
   The existence of mixed land use reduces the chances of crime. This is because mixed use will increase interaction within the community. In addition, activities in a land use that have a high chance of crime can be reduced to other activities with lower opportunities.
5. Landscaping
   The density of plants can affect the chances of crime. The more tight and high the plant will reduce visibility and this will reduce the ability to monitor the environment
6. Pathways/Accesibility
   Locations that have many choices of exits will have a greater chance of crime. This is due to the choice of ways to escape for the perpetrators, especially if the perpetrators recognize the area
7. Territorial Reinforcement
   Territorial reinforcement is a way to foster a sense of territory in an environment through urban design. What is considered important is the sense of ownership of the area, so that the criminals do not carry out their actions in the region. The area can be either a private area or a public area.

3. Methods

The method used in formulating a crime room configuration is based on the perceptions of criminals, victims, police, legal experts, and sociologists. Their perception is used to identify the configuration of "safe" space and configuration of "dangerous" spaces. This perception is adapted to the criminal mindset based on the experience of carrying out a number of criminal activities. Thus, the space configuration will be formulated based on several points of view.

The Content Analysis (CA) method is an in-depth analysis that is used to identify objectively, systematically and generally. Objectives means sticking to rules or procedures so that if done by other researchers will produce similar conclusions. Systematic means that the determination of content or categories is carried out in accordance with the rules that are applied consistently. General means that the invention must rely on relevant theoretical references. Information obtained from Content Analysis can be linked to other attributes of the document and has high theoretical relevance.

Table 1. The Variables Observed

Indicator	Variable
Place Isolated Impression Building Design	Isolated
Place	Crowd
Pathway/Accessibility Place Mobility	Accessibility
Territorial Reinforcement Space Awareness	Sense of Territory
Choise Safe Choice	Route Choosing
Social Control Activity Support Urban Area Land Use Mix	Interaction

Indicator	Variable
Land Use Mix Activity Support	Activity Balance
Managemen and Maintanance	Well-maintained
Managemen and Maintanance	Tidy
Society Existence	Public Space
Surveillance Building Design	Visibility
Natural Surveillance Landscaping	Building Density
Control Space Natural Surveillance Landscaping	Eyesight Angle
Territorial Reinforcement	Space Possessing
Lighting	Lighting

Sources: Analysis, 2018

4. Results

4.1. Spatial characteristics of crime

There are 15 crime-prone spots in Surabaya, those spots has various form of the space like corridor-shaped space and area-shaped space. The corridor-shaped space are located in Surabaya main roads such as in MERR (Ir. Soekarno St.), Dupak St., Mayjend HR. Muhammad St., Demak St., Kartini St., Diponegoro St., Tanjungsari St., Mayjend. Sungkono St., and Bubutan St. The area defined as area-shaped space are scattered randomly in several Surabaya sub district such as in Genteng, Tambaksari, Pakal, Benowo, Karangrejo, and Gubeng.

According to Newman [7], there are crime-related characteristic of space, namely:

a. No/less social control space;
b. Urban space;
c. Isolated space;
d. No/less controlling space;
e. Obscure or less lighting space;
f. Deserted space;
g. High/ low accessible space;
h. No/less surveillance;
i. Not/less maintained;
j. Mix land use;
k. High/low density buildings/vegetation,
l. Space that not public (private space).

Any space that match with one of those characteristics above mentioned mostly vulnerable to crime. However, the empirical evidence from 15 crime-prone spot's in Surabaya shown that not all of those characteristics above are affecting crime or are encouraging the criminals to commit the crime. In fact, most of all characteristics associated with crime obtained from the theory are not proven, and on the contrary, some are having opposite characteristic. All of the fifteen crime-prone spots in Surabaya are having their own route options and this is in accordance with the theory. Meanwhile, some characteristics such as isolated space, space that not public, and obscure or less lighting are totally in the opposite way with the empirical condition. Actually all of 15 crime-prone spots in Surabaya are not isolated space, it is also public space, and also has high level of brightness. The rest of the space characteristics identified above are not completely existed in 15 crime-prone spots in Surabaya.

On the opposite of the regular theory of defensible space area, the most common characteristic of crime-prone spots in Surabaya are crowded space and not deserted space and also has social control. It is also maintained space with a good surveillance and non mixed use of land-use. However some spatial crime characteristics are found proven in most crime-prone spots. Largely, some crime-prone spots in Surabaya have strategic accessibility, and in a high density buildings/vegetation environment.

Perceptions of the characteristics of space differ from one experts to another. This depends on their background and experience in building their own minds about space, especially the space for crime. Differences in background and experience can be specified in 5 categories. These categories are the criminals themselves, victims, police investigators, legal experts, and sociologists. The five categories represent their background and experience. Criminals have their own knowledge of location, who is the target, choice of routes, and so on. Such knowledge will be difficult to understand because their knowledge and thoughts about the crime space are built through a very long process of experience. The victims are community representations. From the character of the community's mind, we can find out whether or not there is a gap between villain's perception and public perception. The police investigator has his own point of view built on the investigation into the crime. Experts represent their respective scientific fields.

In certain cases, the theories believed by experts are not always true. During this time the police and experts tried to analyze where crime had occurred, so that further crimes were expected to be anticipated. But the construction of thinking is not the same as the construction of thinking of criminals so that anticipation does not always pay off.

4.1.1 The criminals

According to the perpetrators of the crime, there are a number of conditions considered before committing an evil act. The perpetrators of crime prefer to choose spaces that are not isolated, in the form of public spaces, and spaces with bright lights. This is contrary to the characteristics of space in the theory of evil. In the theory of crime, most cases of crime occur in isolated spaces, private spaces, and spaces that are not so clearly characterized. However, what is believed by criminals is different, according to them, isolated spaces, private spaces, quiet spaces, and unclear spaces tend to be avoided by society. Factually crimes occur in all types of spaces where there are potential victims of crime in the room, so that perpetrators of crimes will always follow the pattern of community activities in carrying out their evil actions. Crime incidents can be divided into three stages where these three stages occur sequentially.

The first stage is observation where the perpetrators of crime try to idealize space, potential victims, and the habits of prospective victims, the community, and every component of space, namely the design of buildings, CCTV, street furniture, even a spot on the road in the form of standing water. Criminals will go around two or three times to observe the conditions and circumstances around. This traveling activity is facilitated by certain road patterns. U-turn and ring road will make it easier for criminals to go around without stopping.

The second stage is the execution where at this stage of execution the main target of the perpetrators is the situation of crowds, open spaces, public spaces, and bright spaces. Private and quiet space tends to be avoided by society. The execution space is often different from the observation room, because in the process of execution, criminals use more instinct about space. After determining the target of the crime, the perpetrator will follow the target until the right time. The right time is defined as a calculation that takes into account the evil habits of the target, and the character of the space itself.

The last step is to flee after the perpetrator of the crime executes. The choice of route when escaping can differ from one execution to another. Criminals are faced with many route options to escape. In general, criminals prefer to choose wide streets or alleys because on narrow roads or alleys, usually social control is very high because of strong ties in the community in narrow spaces.

These three stages require and at the same time increase the knowledge of the perpetrators of the sense of territory (Figure 1.).

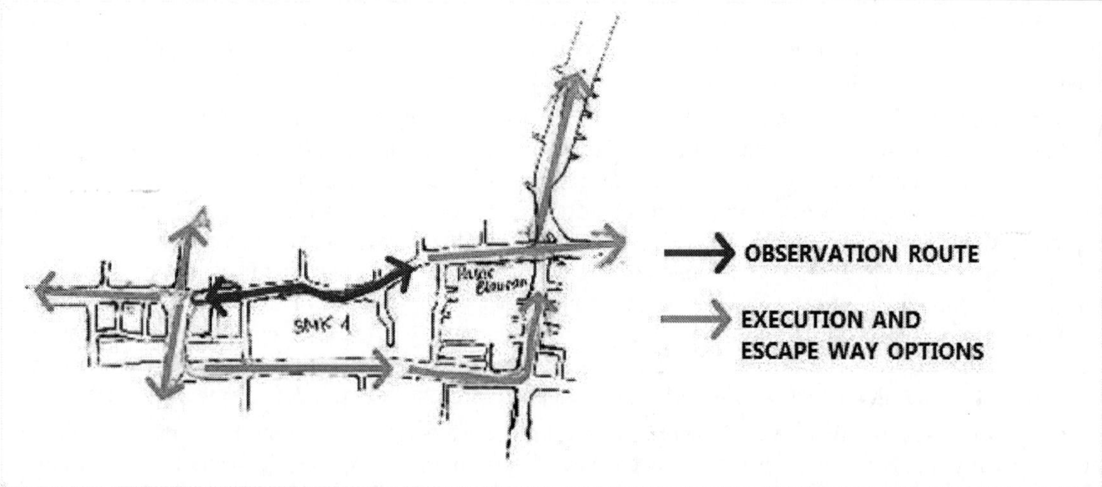

Figure 1. Example of Crime Territory in Surabaya Roads from The Criminal Point of View
Source: field survey, 2018

4.1.2 The victim/society

Public perception of space in general has similarities with the characteristics of space according to the theory of crime. Isolated space, accessible space, not tight social ties, smaller social controls, unclear spaces, less ownership of areas, and areas that are not well understood, can affect crime. According to public perceptions, the perpetrators of crimes generally come from low social strata, and this is synonymous with poorly maintained space. There is a perception in the perpetrators that victims generally live in relatively well-maintained spaces, because often victims are identified as humans from the upper economic class. Well-maintained spaces are generally characterized by low building density, where this causes social ties in the community to be often low. The low density of buildings will have implications for the low supervision of the surrounding environment by building owners.

4.1.3 The police investigator

From police records, most crimes occur in open spaces, public spaces, and during the day. Most of the perpetrators of crimes caught in these spaces were not due to high public awareness, but because of the fragility of the security system in the room. Inter-personal interactions in the public space have no effect to reduce or prevent crime. According to the investigator, if the criminals consider accessibility factors, the only accessibility that is considered when they escape is police accessibility.

Most crime incidents occur in residential areas and in road spaces. Crimes in road spaces often occur in commercial areas. Criminals will follow the target wherever they go and execute at the right time and in the right space.

Awareness in the community itself is one of the things that can reduce or prevent crime. The environment with caring people tends to protect their environment when there are strangers entering their neighborhood.

Awareness in the community will make every member of the community know each other, and quickly realize when there are strangers coming to their environment.

4.1.4 The law expert
What experts think and analyze is largely the same as the characteristics of space according to the theory of evil. Legal experts believe that crime takes place in isolated and lonely spaces, because this kind of space can reduce the possibility of finding evidence. Evidence can be in the form of people, recordings, or others. According to legal experts, accessibility is highly considered, where accessibility is not always related to road patterns, but also access. Limited areas or exclusive spaces and require special permits safer than inclusive spaces that can be accessed by anyone. A room that requires a special permit and has one gate system, it is much easier to supervise the surrounding environment. Fewer route options will be safer. Public space has the same opportunities and contributions in determining the success or failure of a crime event. Public space is an inclusive space, a space that can be accessed by everyone, has a high risk of crime because there is an opportunity to find evidence of crime.

Crime can also be caused by the community itself. When community ties are weak, people are indifferent, criminals will take advantage of their actions. Therefore the community must have concern for one another, feel ownership and responsibility to protect their territory, so that the possibility of crime in their environment can be reduced. The character of the location and design of the building also affects the ability of the community and also the apparatus to conduct supervision. In essence, crime is a battle between people (potential victims of crime) and perpetrators of crimes, especially related to the control and ownership of a region.

4.1.5 The sociology expert
From a sociological point of view, crime usually occurs in isolated spaces, because criminals feel that there are no rules they must obey in isolated spaces. Isolated is not only related to physical aspects or buildings, but also open space where it is very possible for someone to feel isolated in an open space. The more accessible a space, the more vulnerable the space is from crime activities.

Criminals have a number of options in carrying out their actions. Each option has its own advantages and risks. Route selection is included in the consideration of criminals. All ways to escape must be considered. The existence of route options is one of the factors causing crime. Social ties in society have a big impact to reduce the possibility of crime. An environment that is well understood by every member of the community is usually one of the characteristics of a community that has strong social ties. This social bond will increase their desire to own and protect their territory. Internal readiness is generally synonymous with a well-maintained environment so that a well-cared environment will be safer from crime. Basically, high-density settlements have stronger social bonds than low-density settlements. Low density buildings can increase their internal readiness through the provision of facilities such as CCTV. CCTV itself can be used as a preventive instrument that can symbolize the existence of a well-maintained space because it has good and adequate security facilities.

Criminals are often perceived as poor people who are identified as domiciled in slums, so that spaces that are identical to the domicile of rich people tend to be targets of crime. There is a stigma that states that criminals are always looking for what they do not have. There is a mindset in the community that if one room with high crime risks makes an effort to reduce risk, it will increase the risk of crime in the other space, because if one space increases internal preparedness, criminals will automatically try to find other potential spaces.

Criminals are easier to commit crimes in public spaces. Criminal instincts that will determine in public space are like what criminals will do. Some characteristics of the scheme of time and location that are targeted by criminals are the time of day, unclear space, crowded spaces by humans, empty space.

Both the community and criminals have their own perceptions in terms of territorial control, but the difference is that for criminals, the presence of the community is often used as a marker for mapping the territory of crime. These markers can only be seen and understood by fellow criminals. There is the possibility of overlap between space occupied by criminals and space controlled by the community. In this case, the level of overlapping is the beginning of the occurrence of crime.

4.2. Spatial configuration of crime

According to the assumption of criminals behavior and their criminal mind, crime pattern can be analysed which supported by several viewpoints, and spatial configuration of crime can be constructed in Figure 2. That spatial configuration consists of two kinds of space, namely defensible space, that was originally created by Newman [7], refer to the physical layout of communities to allow residents to control the areas around their homes, and action space, refer to a space that opposed to defensible space, which is favourable for criminals to commit their crime. The word "action" refers to the term developed by Wikström [11] on his research raised the concept of Situational Action Theory (SAT).

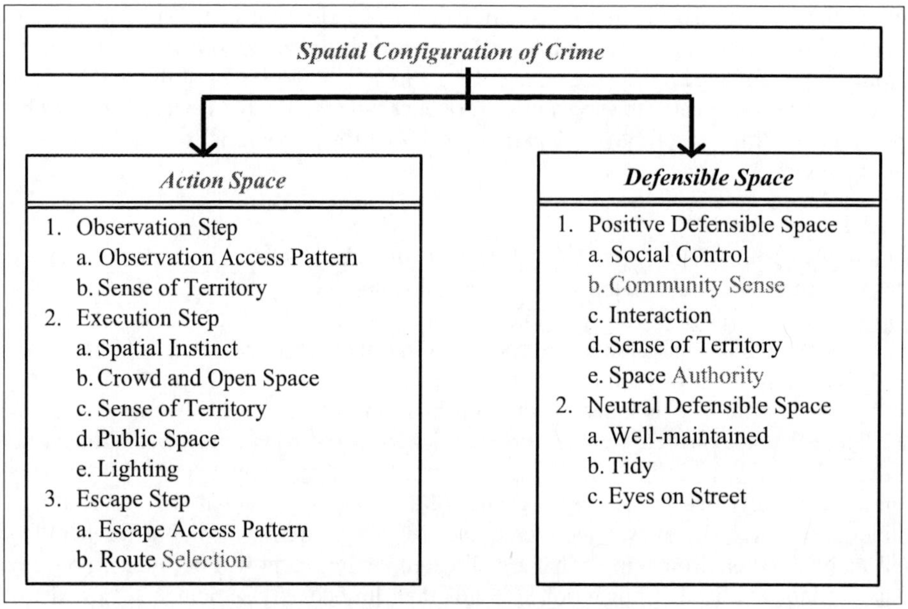

Figure 2. Crime Spatial Configuration
Source: analysis results, 2018

4.2.1 Action space

Action space is a conducive space for criminals to commit their crime. Action space boundaries exist only in criminals's state of mind. There are three steps in one crime incident, namely a) observation, b) execution, and c) escape. Observation, execution, and escape are three different activities which have their own specific characteristics. Observation is a process to figure out the condition at specific time concerning the opportunities, risks, space pattern, target selection, and observe target habit within a very short time. After selecting a target at observation step, criminals will follow their target until execution time. Afterwards, criminals get away through specific pattern of space in escape step.

1. Observation Step
   a. Observation Access Pattern

Basic Access	Observed Access Pattern
■ **U-turn/ring road patterns**	
● *Every road user can keep traveling without stopping on u-turn.*	● *Opportunity to criminals → they can figure out the condition at that time, find, and observe the target by committing a travel without stopping.*

The examples of space with u-turn/ring road and crime-prone spot in streets are like the spaces along the road of MERR, Demak. St., Diponegoro St., Genteng, and Gubeng.

b. Sense of Territory
In the observation step, sense of territory is a space defined by criminals to understand and memorize all the space characteristics, that includes street furniture, buildings, alleys, society habits, lighting, and rush hour habit It is useful for determining the timing, execution location, strategy, and motives in the execution step

2. Execution Step

a. Spatial Instinct
Criminals will execute and select their target according to their own instinct. These intuition/instinct is commonly obtained from their crime experience. The term intuition/instinct in this context means spatial considering-calculating knowledge that reside inside their own state of mind. Spatial elements in every criminals's intuition can be utilized on target selection as well as society activity type selection.

b. Crowd and Open Space

a1) Open Space: Criminals tend to commit their crime in a open space. It is a common situation that society tends to do their activity in an open space instead of in an isolated space. Furthermore, the tendency that many people prefer to be in open space will be affecting society's perception about feeling.

a2) Crowd: Since there are many people prefer to be in a crowd situation, the criminals used to commit their crime in crowd atmosphere as well. The criminals will always follow the society activity patterns. However, the execution process is much depends on their own criminals's intuition.

a3) The examples of crowded crime-prone space in streets are like spaces along the road of Dupak St., Kartini St., Tanjungsari St., Pakal, Benowo, Karangrejo. There are no crime-prone spaces in an isolated area.

c. Sense of Territory
Having the sense of territory in the observation step, which includes the memory, knowledge, and control of street furniture, buildings, alleys, society habits, lighting, and rush hour habit, the criminals will exploit them to commit the crime.

d. Public Space

Theory	Empirical
■ *The existence of public space reduces crime risk, because a lot of people can oversee*	■ *Criminals tend to commit crime in public space to keep up the pattern of society activities. Societies tend to do their activity in a public space for the reason that they will get secure feeling there.* ■ *Public space has easier access than private space*

Theory	Empirical
	■ *Mostly, all of the street crime incidents are taken place in public space.*

e. Lighting

Theory	Empirical
■ *Crime tends to occur in obscure space*	■ *Criminals choose bright space rather than obscure space, for the reason that society tend to do their activities in a bright space.* ■ *Mostly, all of the crime incidents are taken place in bright space.*

3. Escape Step
   a. Escape Access Pattern

Basic Access	Escape Access Pattern
■ *Criminals tend to pass public road or big alley*	■ *Public road and big alley identical with weak social bonding. Public road and big alley tend to have indifferent society.*

b. Route Selection

Theory	Empirical
■ *A space with more route options is more prone to crime incident, because criminals will have more strategies or plans in one series of crime.*	■ *Route selection is needed by every single criminal in syndicated crime platform, and most of street crime at this time is syndicated crime.* ■ *Route options can be a u-turn road or crossroads* ■ *All of crime incidents have their own route options pattern*

4.2.2 Defensible space

Defensible space is a space which can reduce crime risk by way of environmental setting and atmosphere. Defensible space is constructed under society's state of mind, so that crime prevention efforts start from the society itself.

1. Positive Defensible Space
   Positive defensible space is a space characterized where crime reducing and crime prevention can be achieved by increasing some certain factors. The factors are:
   a. Social Control
      Criminals tend to avoid space with strong social bonding within their society. Strong social bonding will increase the desire to possess and protect their surroundings.
   b. Community Sense
      Community sense is characterized by a society who uphold togetherness, without dissension. This condition brings positive impact to their internal readiness to prevent a crime.
   c. Interaction and communication
      The higher the interaction and communication within society, the more they will have mutual understanding among themselves. The society will quickly be aware of when there is stranger get into them.

d. Space authority
The authority of a territory is actually a situation which the competition between society and criminals to dominate the territory is existed and settled. If criminals dominate the space, those space will turn into action space. If a society dominate the space, the space will be secured from any crime.

2. Neutral Defensible Space
Neutral defense space is space characterized by a situation that increasing or decreasing of some certain factors is not affecting neither crime nor security. The factors are:
a. Well-maintained neighborhood
Well-maintained neighborhood is a situation rarely not considered by criminals, so that the changing of its pattern does not affect anything. Well-maintained neighborhood is an internal readiness to prevent a crime.
b. Tidy neighborhood
Neatness is an environmental condition rarely not considered by criminals as well, so that the increasing or decreasing of this facto does not affects anything. Neatness can be considered as a security mechanism in crime prevention.
c. Eyes on street
Regardless there are much people get together in a space, or in an open space, criminals will continue to do their crime. In the criminal's perspective, more people who most likely oversee does not affects the decision to do the crime.

5. Conclusion

Crime mostly happen on the same space. This prove that crime has its own spatial pattern as well as crime behavior has its own pattern. Spatial characteristics of crime mostly believed that a crime space consists of deserted space and obscure space. It can not be denied that deserted and obscure space give threatened feeling. That is why society tends to avoid deserted and obscure space to prevent them from crime incident.

Crime prevention is a useless thing if what we try to protect is not associated with something that criminals try to find. Therefore, it is important to consider criminals's viewpoint in this context. The way criminals see a space and a society, they have different mindset since they have various different experience in committing crime. Meanwhile, space characteristics theory which not based on criminals's mindset could not completely explore crime problem. The problem now is what kind of space that criminals most likely to do their crime, regardless their thinking is conflicting with the society's or not.

It is found that criminals have their own territory, they posses a specific set of territory, and they delineate their territory with an imaginary boundaries. Only the member of their community can recognize and identify where the boundary is. Their territory characteristics are commonly consist of bright space and crowded space. That is why the criminals territory is quite similar with the space where society feels secure.

Territorial controlling is a rivalry situation between society and criminals. The one who has stronger power and has better understanding, memorizing, defending, and protecting the territory will dominate the space. When criminals dominate the territory, the space will definitely turn into crime-prone zone, and vice versa, if society dominates the territory, the space will be a safe zone from crime.

6. References

[1] Pratiwi, N. N. (2014). Skripsi Kajian Sosiologis Bentuk-Bentuk Kriminalitas di Angkutan Umum. Universitas Lampung.
[2] Rusman, L. (2014). Skripsi Tinjauan Kriminologis terhadap Kejahatan Perjudian di Kota Bau-Bau. Universitas Hassanudin.

IOP Conf. Series: Earth and Environmental Science **340** (2019) 012035 doi:10.1088/1755-1315/340/1/012035

[3] Portland Government. (2015). Crime Prevention Through Environmental Design. Crime Prevention Program. City of Portland.
[4] Singapore Government. (2003). Crime Prevention Through Environmental Design Guidebook. National Crime Prevention Council of Singapore.
[5] Hill, Bryan and Paynich, Rebecca (2014). Fundamentals of Crime Mapping (second edition). John & Bartlett Learning, LLC an Ascend Learning Company
[6] Tucunan, K. (2011). Thesis Pola Ruang Kriminalitas Kota Studi Kasus Kota Surabaya. Universitas Gadjah Mada.
[7] Newman, Oscar (1996). Creating Defensible Space. United States Department of Housing and Urban Development. Office of Policy Development and Research.
[8] Higgins, N., & Millard, B. (2009). Geographic patterns of crime. In A. Walker, J. Flatley, C. Kershaw & D. Moon (Eds.), *Home Office Statistical Bulletin-Crime in England and Wales 2008/09 (Vol. 1(11/09))*. London: Home Office.
[9] Nes, A. V., Rueb, L. (2009). Proceedings of the 7th International Space Syntax Symposium. Eds.: Koch, D., Marcus, L., Steen, J. Trita-ARK. *Spatial behaviour in Dutch dwelling areas: How housing layouts affects the behaviour of its users*. Stockholm: Forskningspublikationer.
[10] Aldrin, A., Marzbali, M. H., Razak, N. A., & Tilaki, M. J. M. (2011). A Review of the Effetiveness of Crime Prevention by Design Approaches tobattled Sustainable Development. Journal of Sustainable Development, Volume 4, No 1
[11] Manzo, Gianluca, ed (2014). Analytical Sociology: Actions and Networks. First Edition. John Wiley & Sons, Ltd, 77-78

IOP Conf. Series: Earth and Environmental Science **340** (2019) 012036 doi:10.1088/1755-1315/340/1/012036

Analysis on calculation of Vehicle Operating Cost (VOC) at Gejayan intersection before and after fly over ring road operation in Yogyakarta

R E Wibisono[1], **M S D Cahyono**[2] **and A Muhtadi**[3]

[1] Program Study of Civil Engineering, Engineering Faculty, Narotama University
[2] Program Study of Civil Engineering, Engineering Faculty, Widya Kartika University
[3] Program Study of Civil Engineering, Engineering Faculty, Narotama University

Email: endro_wibisono89@yahoo.com

Abstract. Gejayan intersection is located in Sleman Regency of Yogyakarta. Fly Over will be constructed at the intersection in the ring road of Yogyakarta. The nearest intersection to the east is Gejayan Intersection with a distance of 1573.11 meters. Gejayan intersection performance in 2018 has reached saturation by having Level of Service E. The more severe conditions will occur in the next 10-20 years. If it is planned using Fly Over on the ring road in Yogyakarta connecting Gejayan intersection, the vehicles flow must have other alternative by road widening. The most important step is to predict the Vehicle Operating Costs (VOC). After the Fly Over operation, VOC can be calculated to determine the cost saving. The scenario to predict VOC was 80% of vehicle volume would pass the Fly Over, and the 20% would still use the existing road. The results showed that the cost saving of Vehicle Operating Costs (VOC) of the existing road compared to the fly over use in 2019 was IDR 37,941,324,952. In 2039 the saving would be IDR 240,232,694,488. If the road was only widened, the cost saving of VOC in 2019 would be IDR 9,084,204923 and in 2039 would be IDR 72,107,188,198.

Keywords: *Vehicle Operating Cost, Cost Saving, Fly Over, Intersection Performance.*

1. Introduction

Gejayan intersection is located in the series of northern ring road intersection in Sleman Regency of Yogyakarta. The most west are Jombor Intersection, Monjali Intersection, Kentungan Intersection, Gejayan/Condong Catur Intersection, and UPN Intersection. Gejayan intersection is located in commercial area and has 4 causeways, namely north (U), south (S), east (T), and west (B). The north – south causeway is Kaliurang Street and east – west causeway is north ring road. Here are 5 intersection networks in the north ring road.

IOP Conf. Series: Earth and Environmental Science **340** (2019) 012036 doi:10.1088/1755-1315/340/1/012036

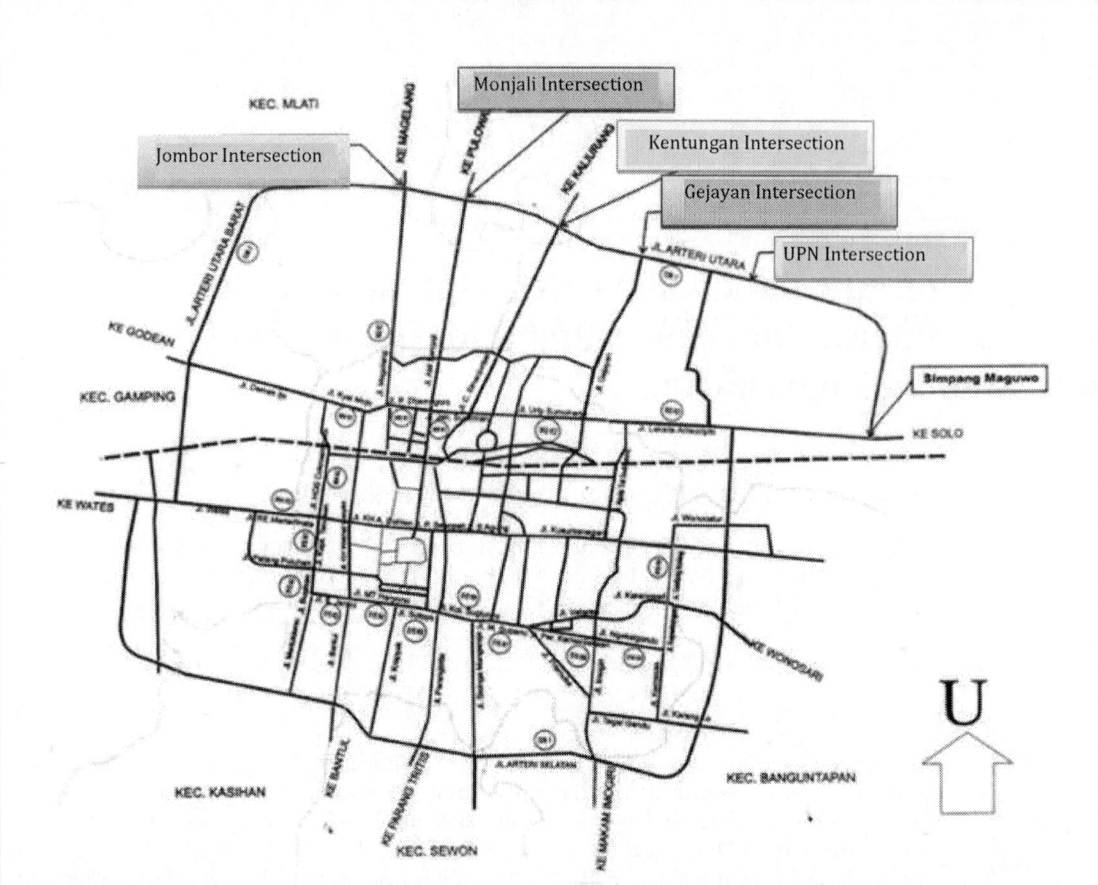

Figure 1. Situation map of 5 intersectionsat North Ring Road,Sleman Regency of Yogyakarta.

In 2018, Fly Over will be constructed at intersection in Yogyakarta ring road. In the area, there are 4 intersections. The middle one is Kentungan Intersection and the closest intersection to the east of Kentungan Intersection is Gejayan Intersection with a distance of 1573.11 meters. From signalized intersection performance of Gejayan Intersection in 2018, the signalized intersection of Gejayan Intersection saturation Yogyakarta already has value 1.33 in 2018 or in other word has LOS E (between 0.8 until 0.1). The prediction of signalized intersection performance in 2028 showed that DS value reached 2.76. It means that the intersection capacity is not capable to serve the number of vehicles passing through the intersection. Condition and situation in the intersection at peak hours will be more jammed, the speed will increase, and the travel timewill be longer.

The delay at the intersection for every vehicle in 2018 has a value of 223.3 second/smp. It indicates that every vehicle passing the unsignalized intersectionwill stop at the intersection for 223 seconds to allow vehicles from other direction to cross the intersection.

If an intersection has LOS Evalue, then the driverswill already feel uncomfortable when passing the intersection. Similar to the delay at Gejayan signalized intersection Yogyakarta, every vehicle has 50% chance or more to queue at the unsignalized intersection. The more severe conditions will occur in the next 10-20 years.

If it is planned using Fly Over or Underpass at the ring road in Yogyakarta connecting 4 intersections including Gejayan intersection, indeed we need to think the concept maturely to serve the vehicles flow from west to south and vice versa as well as east to south and vice versa. The vehicles flow must have other alternatives, namely by performing road widening according to the flow that will pass the road segments mentioned above. The most important step is to predict the Vehicle Operating Costs (VOC) and the profit after the Fly Over operation. After performing the Fly Over operation, the calculation of cost saving of VOC could be determined.

VOC was obtained from the calculation of every vehicle passing Gejayan intersection and predicted according to vehicle growth in Yogyakarta. From the result, it could be multiplied to all aspects of costs ranging from materials and mechanic wage. Finally, the load cost in certain speed in the desired year could be determined.

The scenario performed in predicting VOC was 80% of vehicle volume that would pass the Fly Over, and 20% of vehicle volume still used the existing road.

2. Literature review

The basic objective of planning is to estimate the amount and need for transportation in the future or in the year of the plan to be used for various types of investment. To better understand and find the best solution, a transportation system approach is needed. The transportation system in general (macro) can be broken into several smaller (micro) systems that each interact and influence. (Tamin, 1997).

Vehicle Operating Costs (VOC) is a value that states the amount of costs incurred for a vehicle operation. VOC consists of several components, namely:

2.1. Running cost

Fuel Cost
Oil / Lubricant Cost
Tire Usage Cost
Maintenance Cost (repair and maintenance, General Overhaul)
Over Head costs (Incidental expenses)

2.2. Fixed cost

Insurance
Capital Interest
Depreciation (Vehicle Depreciation)
Time Value

2.3 VOC calculation by using PCI model

Vehicle operating costs are costs used for vehicles from one place to another (transportation activities). The method used to calculate vehicle operating costs used today is the Pacific Consultants International Inc. method. Tokyo, Japan (PCI).

In this case the vehicle operating costs are the sum of the movement costs (operating costs) and fixed costs. The formulas used in calculating Vehicle Operational Costs based on the PCI Method are as follows:

- Equation for Fuel Consumption:

Sedan (PC)	: Y=0.03719S*S-4.19966S +175.9911
Mini/Medium Bus	: Y=0.06846S*S-8.02987S +340.6040
Large Bus	: Y=0.12922S*S-13.68742S+541.0279
Mini Truck	: Y=0.06427S*S -7.06130 S+318.3326
Large Truck	: Y=0.11462S*S-12.85594S+503.7179

Where:
Y = Fuel consumption (liter/1000 km)
S = Running Speed (Km/Hour)

- Equations for Engine Oil Consumption:

Sedan (PC)	: Y=0.00025S*S- 0.02664S+	1.441710
Mini/Medium Bus	: Y=0.00057S*S- 0.06130S+	3.317530
Large Bus	: Y=0.00030S*S- 0.12968S+	7.062390
Mini Truck	: Y=0.00048S*S- 0.05608S+	3.073830

Large Truck : Y=0.00100S*S- 0.11715S+ 6.409620

- Equations for Tire Usage:

Comparison of tire usage on toll road and ring road

$$Type : \frac{Cost\ on\ ring\ road}{Cost\ on\ toll\ road}$$

Passenger vehicle : 1.94
Bus : 1.10
Truck : 1.10
Equations for Tire Usage:
Sedan (PC) : Y=(0.0008848S–0.0045333)
Mini/Medium Bus : Y=(0.0012356S-0.0064667)
Large Bus : Y=(0.0012356S-0.0064667)
Mini Truck : Y=(0.0011553S-0.0005933)
Large Truck : Y=(0.0011553S-0.0005933)
Y'=Y*totaltire*tireprice/1000km

- Equation for Maintenance Cost:

Maintenance cost for spare parts
Comparison of spare part usage on toll road and ring road:

$$Type \quad \frac{Cost\ on\ ring\ road}{Cost\ on\ toll\ road}$$

Passenger vehicle : 1.73
Bus : 1.27
Truck : 1.26

- Maintenance costs for spare parts of vehicles passing toll roads:

Sedan (PC) : Y=(0.0000064S+ 0.0005567)
Mini/Medium Bus : Y=(0.0000320S+0.0020891)
Large Bus : Y=(0.0000320S+0.0020891)
Mini Truck : Y=(0.0000191S+0.0015400)
Large Truck : Y=(0.0000191S+0.0015400)
Where:
Y' =Y* vehicle value (/1000 km)
S = Running Speed (Km/Hour)

- Maintenance hours for workers

Sedan (PC) : Y=(0.00362S+0.36267)
Mini/Medium Bus : Y=(0.02311S+1.97733)
Large Bus : Y=(0.02311S+1.97733)
Mini Truck : Y=(0.01511S+1.21200)
Large Truck : Y=(0.01511S+1.21200)
Y'=Y* hourly mechanic fee (/1000 km)

- Equation for Vehicle Depreciation:

$$Sedan\ (PC) \quad : \quad Y = \frac{1}{2.5\ S + 125}$$

$$Bus \quad : \quad Y = \frac{1}{8.756\ S + 350}$$

Truk : $Y = \dfrac{1}{6.129\ S + 245}$

Where:
Y = Vehicle depreciation per 1000 km was multiplied with the vehicleprice
S = Running Speed (Km/Hour)

- Equation for Interest Rate:

Sedan (*PC*) : $Y = \dfrac{120}{500\ S}$

Bus : $Y = \dfrac{120}{2500\ S}$

Truk : $Y = \dfrac{120}{1750\ S}$

In which:
Y = Interest rate per 1000 km was multiplied with 0.5 from vehicle value. Interest rate
= 12%/year
S = Running Speed (Km/Hour)

3. Research method

Surveys were conducted for primary data in the form of counting the total vehicles at Gejayan intersection, Yogyakarta. Surveys were conducted during peak hours, morning and evening. These surveys took note on types and amount of vehicle that went in and out at each Gejayan intersection. In addition, the secondary data was also used to support the calculation of traffic performance with data in the number of vehicle in 2013 until 2016 in Yogyakarta. The number of vehicle calculation of each intersection compared to the growth rate of vehicle produced degree of saturation that could determine the *Level of Service* at each intersection. The maximum saturation degree then was predicted up to 10 years and the researchers predicted the Vehicle Operating Costs (VOC) profit after the Fly Over operation. By calculating VOC, it could be seen how much the cost saving of VOC after Fly Over operation compared to before operation. VOC was obtained from calculation if every vehicle passed the Gejayan intersection and predicted according to vehicle growth in Yogyakarta. Then the result could be multiplied to all aspects of costs ranging from materials and mechanic wage. Finally, the load cost in certain speed in the desired year could be found.

4. Result and Discussion

Several scenarios assumption performed for the calculation of Vehicle Operating Costs to determine the level of cost saving efficiency are:

- No Fly Over Construction
  Without fly over construction at Gejayan intersection of Yogyakarta, its performance is getting worse from the existing year (2018). The saturation degree will be higher, the queueis getting longer, and the longer of vehicles will stop at Gejayan intersection of Yogyakarta. This will increase the vehicle operating costs if passing Gejayan intersection of Yogyakarta with the current condition.
- Performing Road Widening
  It is performed on the West and East side of the ring road. The road widening is 3 meters on the left shoulder. The function is to add lane and capacity for vehicles. So, the vehicles can use the existing road if the fly over would be constructed later.
- Performing Fly Over Construction
  Fly Over at Gejayan intersection of Yogyakarta with distance of 0.8 km is assumed can be

operated in 2018. For this assumption, calculation simulation will be conducted where the entire flow of vehicles passing through the ring road from west to east will pass the planned fly over consisting of 51,081 passenger cars and 4,958 heavy vehicles (truck and bus) in 2015. While in 2018 (with the assumption that Fly Over has operated), the number of vehicles passing through the fly over are 12,810 passenger cars and 1,742 heavy vehicles (truck and bus).

Scenario performed in predicting the VOC was 80% of vehicle volume would pass Fly Over, and 20% of vehicle volume still used the existing road.
From the assumption results, the vehicle costs passing through Fly Over and those that still using the existing road can be determined. For more details, the table and graph of the total VOC/year, the cost saving of VOC/year due to the 3 meters road widening, and the cost saving of VOC/year if all vehicles passing the ring road use the planned fly over are presented. The following is presented in table 1.

Table 1. Cost saving of VOC of Gejayan Intersection of Yogyakarta for 30 Years (IDR)

Year	Existing VOC	VOC of Widening	VOC of Fly Over	Savingsof the Existing –Fly Over	Savings of the Existing – Widening
2018	74638656413	66062798235	38910424749	35728231664	8575858177
2019	78672095595	69587890672	40730770643	37941324952	9084204922
2020	85224899340	74957417722	42736380143	42488519197	10267481618
2021	89957361268	79072888340	44817236394	45140124874	10884472928
2022	97732325907	85389664304	47125155608	50607170298	12342661603
2023	103294923119	90201050069	49490825315	53804097804	13093873050
2024	112551199954	97646522332	52165875443	60385324511	14904677622
2025	119101883948	103279235147	54868442596	64233441351	15822648800
2026	130163769297	112073245080	58151644070	72012125226	18090524216
2027	137893656718	118676614364	61282615921	76611040796	19217042354
2028	151171210941	129086757604	64878655279	86292555662	22084453336
2029	160312758832	136839002262	68438840930	91873917901	23473756569
2030	176330949532	149192759858	72635169960	10369577957	27138189674
2031	187168689888	158307145900	76720402995	110448286893	28861543987
2032	206609625926	173007814946	81542242572	125067383353	33601810980
2033	219494530678	183740343880	86280394508	133214136169	35754186797
2034	243260435539	201288273429	91872606117	151387829421	41972162109
2035	258629579208	213947374535	97269256573	161360322635	44682204673
2036	287939797943	234968760059	103817908955	184121888987	52971037883
2037	306344007959	249927648135	110030554500	196313453458	56416359823
2038	342890154837	275214338232	117572455226	225317699611	67675816604
2039	365034171068	292926982870	124801476584	240232694483	72107188197
2040	411239332816	323492734124	133571282351	277668050464	87746598691
2041	438042444868	344514827565	141828103549	296214341319	93527617303
2042	497517096899	381676818293	152117860338	345399236561	115840278605
2043	530209104484	406693833615	161650117187	368558987296	123515270868
2044	608604178664	452194194141	173534315178	435069863485	156409984523
2045	648886232144	482059789615	184648586850	464237645293	166826442528
2046	752367066529	538252801649	198498138205	553868928324	214114264880
2047	805949792286	574042824216	211234572895	594715219391	231906968070
2048	953490298799	644194310489	227637197036	725853101762	309295988310

Volume of traffic using arterial roads. With the flyover, the level of network services will increase so that transportation costs decrease and demand will increase. If it is planned to use Simpang Tak Sebidang or flyover on the ringroad arterial road in Yogyakarta City that connects 4 Simpang including the Gejayan intersection. Another alternative is to widen the road according to the current that will pass through the roads mentioned above. BOK / th savings due to the widening of the 3 meter road in 2018-2048, and the BOK / th Savings which when all vehicles passing on the arterial road use / pass the Uneven Field which will be planned in 2018-2048. The following is presented in Figure 2.

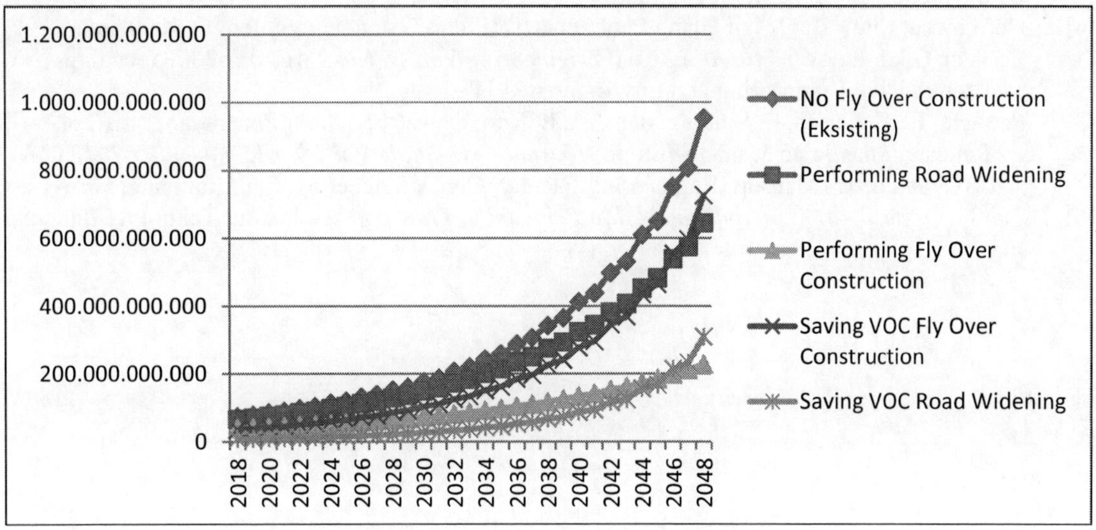

Figure 2. Relationship Diagram of the existing VOC, road widening VOC, Total Fly Over, and cost saving of VOC/year

5. Conclusion

- Total Vehicle Operating Costs (VOC) of the existing road in 2019 is IDR 78,672,095,595 and in 2039 is IDR 365,034,171,068
- Total Vehicle Operating Costs (VOC) of road widening in 2019 is IDR 66,062,798,236 and in 2039 is IDR 292,926,982,870
- Total Vehicle Operating Costs (VOC) of Fly Over in 2019 is IDR 38,910,424,749 and in 2039 is IDR 124,801,476,584
- The results show that the cost saving of Vehicle Operating Costs (VOC)of the existing road compared to the fly over existence in 2019, thecost saving is IDR 37,941,324,952 and in 2039 thecost saving is IDR 240,232,694,488
- If the road is only widened without fly over, then the cost saving of VOC in 2019 is IDR 9,084,204923 and in 2039 is IDR 72,107,188,198.

The above show that Vehicle Operating Costs (VOC) is higherwhile operating at the existing road than at fly over. So that the costsaving generated on the fly over operation is more efficient (large profit).

IOP Conf. Series: Earth and Environmental Science **340** (2019) 012036 doi:10.1088/1755-1315/340/1/012036

6. References

[1] Dewi, A.A.A., Suthanaya, P.A., & D.M.P. Wedagama 2013 Financial Feasibility Analysis of Benoa Bandara Nusa Dua Toll Road Development. *Journal Spektran Vol 1 No. 2 July 2013*

[2] Directorate General Bina Marga 1997 HCM (Highway Capacity Manual Project), Public Work Department

[3] Foundation for Research and Industrial Affiliation ITB, 1996, *Laporan Akhir Studi Perhitungan Biaya Operasi Kendaraan* (Bandung: PT. Jasa Marga, ITB)

[4] Hermawan, Rudy 2009 Kaji Ulang Penentuan Tarif dan Sistem Penggolongan Kendaraan JalanTol di Indonesia, *Journal Civil Engineering Vol. 16 No. 2 Agustus 2009* (Bandung: Faculty Civil Engineering and Environmental ITB)

[5] Petracia, P., F. Jansen, E. Lintong, dan A.L.E. Rumajar (2012), Studi Penentuan Tarif Tol Rencana Ruas Jalan Manado-Bitung, *Journal Civil Static Vol.1 No.1, November 2012 (27-33)Conclusion* (Manado: Engineering Faculty, Civil Engineering, Sam Ratulangi University)

[6] Tamin, Ofyar Z. 2000. *Perencanaan dan Pemodelan Transportasi*. Institut Teknologi Bandung.

Qualitative comparative assessment by fsQCA for Transit Oriented Development (TOD) area comparison

S Nurlaela[1], A Nadyla[1], A Y Zuhdi[1], K D M E Handayeni[1], I Fakhrianto[1], A R Nurkhariza[1] and L Yusuf[1]

[1] Department of Urban and Regional Planning, Institut Teknologi Sepuluh Nopember, Surabaya, Indonesia

Email: sitnurlael0@gmail.com

Abstract. The station development through transit oriented development principles in favor to transit patronage has emphasized the important of design, diversity and density. However, studies were lacking in identification of the level of combination of all these ingredients toward increasing ridership. Using the node-place balance perspective, the design, diversity and density are re-evaluated. Four transit terminals or stations are used as case study in the City of Surabaya. Each of the transit area is divided into several zones/subzones to evaluate the level of density, diversity and design into the value of node and place index. Each transit points would be regarded as having one among the five categories of node-place index. The level of the daily trip demand is identified as the target variable. Qualitative comparative analysis then is used to mapping the position of each station, suggested the level of combination of design, diversity and density in each node-place category, to explore which combinations that are found to be statistically significant to support the increases of trip demand. The research showed that employment density may has more impact than housing density, while node aspect maybe of more important compare to place aspect that explain high or low trip demand.

Keywords: *TOD, fsQCA, node-place index, 3D factor*

1. Introduction

The problem of congestion in Surabaya City has been among the worst in the world[1]. The private transport share was about 73.41% of total passenger transport[2]. Although the Suroboyo Bus is now operated in the City, but its network coverage is still limited. Often, public transport is unable to compete with cars due to: (1). Inability to compete in terms of speed or travel time; (2). Inability of public transport to connect the center of activities (residential, commercial, offices, etc) due to low accessibility.

Transit oriented development or TOD as one implementation of Land use and transport integration (LUTI) has successfully increased the public transport uses ([1], [2]). In Curitiba - Brazil, TOD has been successful in increasing the share of transit from 7% in 1970s to 75% in 2006 [3]. The strength of the TOD is in the implementation of the 3D development, such as: density - to increase the number of

[1] Castrol Magnatec, 2015
[2] Surabaya in Figure, 2016

IOP Conf. Series: Earth and Environmental Science **340** (2019) 012037 doi:10.1088/1755-1315/340/1/012037

residential and jobs in the transit area; the diversity - to mix the various land use types, and the various social economy of residents and buildings in the transit area; the design - to improve the pedestrian and bike services that are integrated with the transit accessibility. Density creates sufficient demand for mass transit, while design creates good accessibility for non-motorized transport (NMT) access and diversity creates activities opportunities.

Many scholars agreed on the idea that TOD encourages mixed use around the transit area [4], improved land potentials to develop activities opportunities [5], and can integrate the transit area with pedestrian, transport feeders, and mass public transport [3]. Nevertheless, these literatures perceived the 3D characteristics should fulfill certain parameters on density, diversity and design. For example [6] pointed out the minimum 10,000 employees or residents in a TOD station precinct. TOD design standard had been proposed by several research institutions, such as ITDP (the Institute for Transportation and Development Policy). In Indonesia, the regulation on TOD is documented in the Permen ATR No. 16/2017 about TOD classification. However, the implementation of 3D parameter standards may not be achieved due to constraints such as low density areas and lack of business attractiveness [7].

A node-place index has been an additional emerging concept in the literature that evaluate the level of integration between land use and transportation in TOD [8]. A balance node and place, according to [5], is a pre-condition to achieve the spatial development with a sustainable infrastructure provision. TOD with a balance node and place will improve the transportation system and land use integration in the area [5]. Using the node-place balance perspective, the design, diversity and density are re-evaluated in this paper.

Four transit terminals and stations in the City of Surabaya are used as a case study. Using a subzone analysis, each of the transit point is divided into several zones to evaluate the level of density, diversity and design into the value of node and place index. Each transit points would be regarded as having one among the five categories of node-place index. The level of the daily trip demand is identified as a target variable. Qualitative comparative analysis then is used to mapping the position of each station, suggested the level of combination of design, diversity and density in each node-place index category.

2. Literature Review

TOD defined as the development of transit area in within ½ mile from a transit point or station, with compact land uses, high density and intensity, mixed uses, designed to maximize pedestrian access to the transit. TOD was differentiated with TAD or Transit Adjacent Development, in which in TAD, the development of activity centers are in proximity or adjacent to the transit points (*transit-proximate development),* but these development in TOD are integrated or orientated to the transit points.

The 3D principles in TOD area are implemented in a way that density achieved, i.e., a sufficient density of buildings, jobs, and high access to jobs. Closer to stations, the density should be higher, means the building are higher. Diversity, a mixed use development determined by the proportion of land use of commercial or trade and service land uses. The mixed used is high enough in each subzone for direct mobility. Design, a provision of facilities to support the NMT for station access. In pedestrian area, the traffic speed is low, a connectivity between each subzone to the transit facilities and between activities centers to pedestrians are well maintained. An innovative parking should be implemented in the TOD area such as central parking and progressive payment system [9].

TOD planning principle consisted of the efficient location, rich mix of choices in a neighborhood unit, the value capture socially, economically and environmentally, the place making aspect and the node aspect [2]. TOD in a node and place perspective aims to integrate the role of transit as a transit point, a place for passenger boarding and alighting, and a place in that the transit area can become a center of activity that generating and attracting trips characterize with a sense of place attraction.

The node-place model for TOD was introduced by [9]. There is a relationship between the location as place or as the center of activities and as a node or the transit points for intermodal of public transport. The accessible node and accessible place attract people (trips). The accessible nodes are created by the high daily frequency of train services, the number of stations in within 45 minutes catchment, the number of destination served by the station, the number of connected access by buses, trams, and subway; the parking capacity, and the accessibility by walk or cycle. The accessible place is characterized by the degree of functional mix, the number of population and the number of employment of four sectors (retail, catering, education/health/culture, administration and services, industry and distribution).

The determinant factor for the node-place index, as summarized from ([10], [5], [11]) is described in the following table.

Table 1. Components of node and place model

The aspect	The variable
Node	The frequency of public transport leaving the station
	The type of station or terminal
	The length of travel
	The number of destination served by the transit service
	The level of pedestrian connectivity
Place	Land area ratio
	Floor space ratio
	Household density per hectare
	The number of employment or job per sector per hectare
	Degree of functional mix

In evaluating TOD according to the Node-Place model, [5] specified five types of possibilities in which the station may appear to have a balance node-place at the most ideal situation or unbalance node-place characteristic as the least to expect, and ranges of variation in between. Figure 1 illustrates these five possible situations.

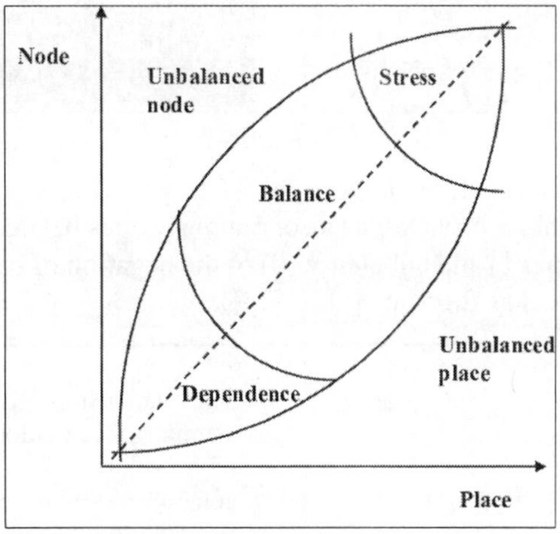

Figure 1. Node – Place Model
Source: [5]

3. Method

This research used QCA or Qualitative Comparative Analysis to analyze the relationship between the variable 3D TOD and the category in which the station would be included into the five categories of node-place indices. Unit analysis defined into micro scale i.e. the 800 meter radius of transit area are divided into several sub-zones, i.e. 5 zones for each station/terminal, resulted in 20 sub-zones altogether (see figure 2). The zone division was defined by main streets such as primary or secondary arterial and collectors, natural boundaries such as waterbody or rivers, and the distribution of land uses (See [12] for station precincts sub division).

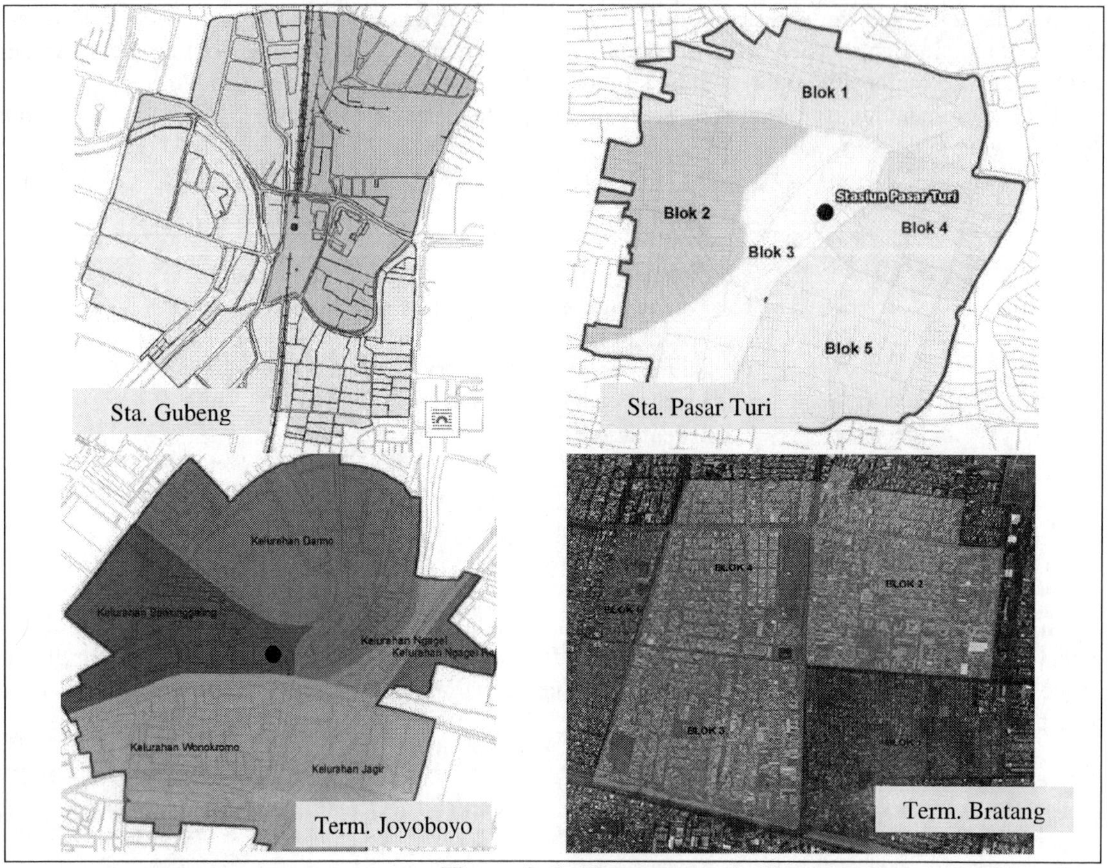

Figure 2. Study area

QCA used Boolean Algebra with the application of fsQCA (see www.compasss.org/) or fuzzy set QCA, in which the membership criteria fall as a continuous values between 0 to 1. The *y* category is the index balance (full membership 1) and unbalance (0) or the negation of balance. The process of analysis in QCA is explained in figure 3 as followed ([13], [14]).

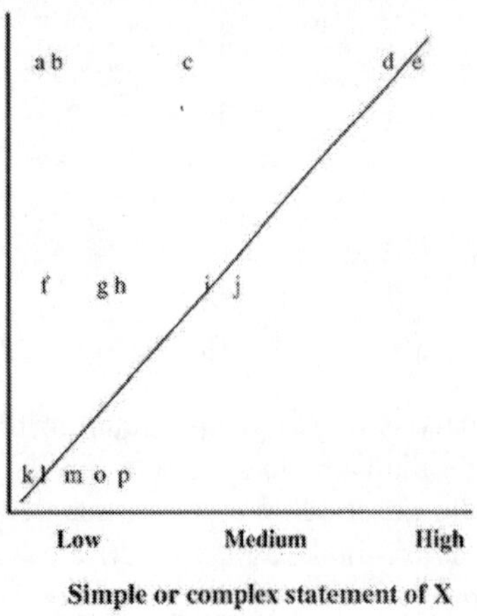

The relationship between X and Y in an asymmetric situation. Combination of high values of X is sufficient to generate a high value of Y; but combination of high values of X is not necessary for generating a high value of Y. Woodside, 2013 described the example from 15 cases as in the figure (case a to case p). Each of the case is mapped according to its X values to describe the instance of Y high (or low). (Woodside, 2013).

Figure 3. The relationship between X and Y according to asymmetric relationship (adapted from [14], p,3)

In this regards, there are two steps of analysis conducting in this research. First, to identify characteristic of 3D TOD, and secondly, to evaluate 3D TOD as a node place index and its relationship with trip demand.

In this paper, the *Y* values consisted of 5 possible situations [8]:

1. Unbalance node, i.e. a situation in which the node index (the transportation network system) is high but the land use or place is low.
2. Unbalance place, i.e. a situation in which the place index (the land use) is high but the transportation index or node is low.
3. Dependence, both the value of node and place are low. The transportation system is dependence on the external services, there is no competition for developing the vacant land.
4. Stress, is the opposite of dependence, the node and place index are both high.
5. Balanced, is a situation in that both node and place index can support the system, the environment, the need of the community without enforcement to extent the system. Node and place are supporting one another and integrated.

In [10], the range of value as criteria of categorization as above mentioned is explained. The values is presented in the table 2

Table 2. Node and Place Index Categorization

Class	Node and Place Index Range	Node Index	Place Index	Category
Very high	≥ 0,8	0,6 - 1	0,6 - 1	*Stress*
High	0,6 – 0,79	0 – 0,49	0 – 0,49	*Dependent*
Standard	0,5 – 0,59	0.5 – 0.59	0.5 – 0.59	Balance
Low	0,2 – 0,49	0,5 - 1	0 – 0,59	*Un-sustained node*
Very Low	≤ 0,19	0 – 0,59	0,5 - 1	*Un-sustained place*

Source: Adapted from [10].

The combination of *X*s configuration that may create a possible *Y* situation in this research is illustrated in the following table 3.

Table 3. The configuration of Y and X based on the criteria of 3D TOD

Possible situation Y	*Negation Y*	*X configuration for sufficiency condition*
Y1 = Stress	~ Y1 = Dependence ~ Y1 = Balance	dense*diversity*design ≤ y1
Y2 = Un-sustained node	~ Y2 = Un – sustained place ~ Y2 = Balance	dense*diversity~design≤ y2
Y3 = Un-sustained place	~ Y3 = Un – sustained node ~ Y3 = Balance	~dense~diversity*design≤ y3
Y 4 = Balance	~Y4 = All other situations	dense*diversity*design ≤ y4 ~dense~diversity~design ≤ y4

Source: Analysis

4. Result and discussion

The first finding is formed by analyzing the 3D components as part of node-place index. The method followed a framework offered in [15] (figure 4). The value of variables is measured based on these stages of calculation:

1. Measure the value of variable and compared it against the parameter.
2. Convert these values into the fuzzy index, ranged from 0 to 1.
3. Calculate the node place index from the combined value of variables according to the fuzzy index. The five node-place balance categories valued at 0.59 is the balance index.

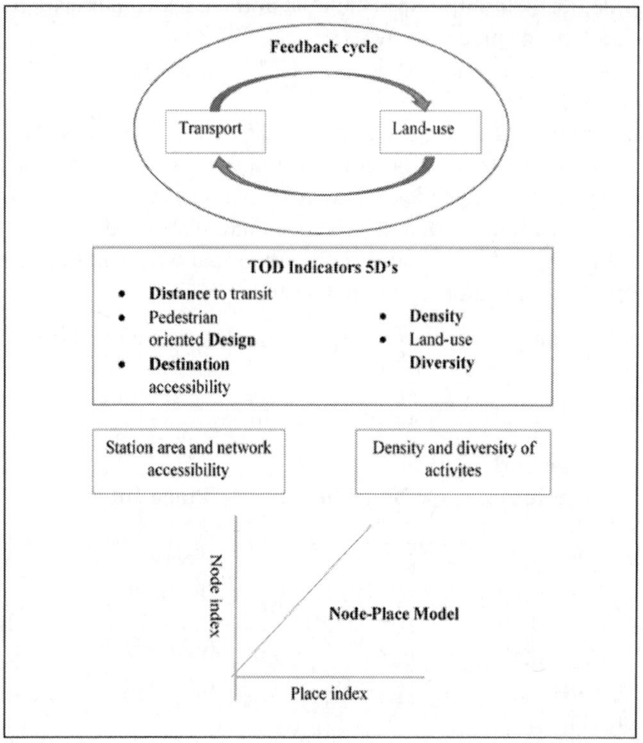

Figure 4. Node-Place Index based on the characteristics of 3D TOD (adapted from [15])

4.1 Node and place indices configuration in fsQCA

The node component according to 3D TOD consisted of design component, covered the variable of the availability of pedestrian (%) or *inde1*, pedestrian connectivity level or *inde2*, the width of pedestrian path or *inde3*, safety and convenience of sidewalk or *inde4*. The place component consisted of density component, covered the variable of the ratio of built area to the parcel area (KDB) or *inden1*, floor space ratio (KLB) or *inden2*, the residential density, i.e. the number of housing per hectare or *inden3*, employment density, i.e. the number of employment per hectare or *inden4*. The second component of place is diversity, measured by degree of functional mix or *indiv*. All of variable was calculated from each subzone and then aggregated for each station/terminal. In fsQCA, all node and place variables are configured to form the combination of variables according to the characteristics of case study and while maintaining the solution with consistency value recorded in fsQCA of at least 0.84. The results are showed in the following table (table 4 and 5) that calculated the calibrated values of node and place indices (symbolized with c) in Boolean algorithm.

Table 4. Configuration of variables for node aspect

Configuration of variables for node aspect	*Raw coverage*	*Unique coverage*	*Consistency*
~inde1_c*inde2_c*inde4_c	0.887	0.000	0.898
~inde1_c*inde3_c*inde4_c	0.896	0.009	0.858
inde2_c*inde3_c*inde4_c	0.969	0.08	0.922

Configuration of variables for node aspect	*Raw coverage*	*Unique coverage*	*Consistency*
Solution coverage	0.979		
Solution consistency	0.85		

Table 5. Configuration of variables for place aspect

Configuration of variables for place aspect	*Raw coverage*	*Unique coverage*	*Consistency*
inden1_c*inden4_c*indiv1_c	0.909	0.48	0.959
~inden2_c*~inden3_c*inden4_c*indiv1_c	0.430	0.002	0.921
Solution coverage	0.911		
Solution consistency	0.931		

According to table 4 and 5, the configuration of node variables with the highest consistency value consisted of configuration among *inde2, inde3,* and *inde4,* while for place variables consisted of *inde1, inden4,* and *indiv1.* From the model, high value of node index were composed by high value of pedestrian connectivity, the width of pedestrian path, and the convenience of sidewalk measured by lighting and shading. Meanwhile, high value of place index were composed by high value of building density (KDB), employment density, and mixed use.

4.2 Trip demand analysis in fsQCA

This section explained the relationship between the configured node and place index resulted from the first stage analysis and the trip demand for TOD area. Based on the results of analysis in fsQCA, four models were selected as follows:

1. **Model 1a** will include all variables. All configured variables of node and place (*inde2, inde3, inde4, inden1, inden4* and *indiv)* was plugged into the model. This model resulted in random distribution of stations, there was no clustered of subzones of station formed into a meaningful model in which the value of consistency is above 0.84. There was only one meaningful cluster of subzone, however, only covered one subzone, i.e. subzone 2 in Gubeng station.
   In developing the alternative model, the configured variables were divided into subset and superset of node and place variables to identify which configuration of variables resulted in the highest value of consistency and coverage index. The results are the subset node comprise of *inde2* and *inde3;* while the subset place comprise of *inden1, inden3, inden4* or configuration of *inden1, inden4* and *indiv.* Therefore, the alternate models consisted of as follows (model 1b, and model 2a and 2b) using these set of variables.
2. **Model 1b:** Model 1b consisted of configured variables *inde2, inde3, negation of inden3,* and *inden4.* This model has the consistency index 0.84 and the coverage value 0.56. The resulted is as follows: a cluster stations consisted of Joyoboyo terminal Subzone 3, Gubeng station Subzone 1, and Bratang terminal Subzone 2. (figure: 5). The model mathematically is written as follows.
   $$\boldsymbol{connectivity\,.\,pedestrian\ width\ {\sim}residential\ density\,.\,employment\ density\ \leq\ high\ trip\ demand}$$
   The dot ". "indicates the "AND" while the tilde "~" indicates the negation. The model said that high trip demand in the station area is determined by high connectivity of pedestrian, a wide pedestrian width, low residential density, and high employment density.

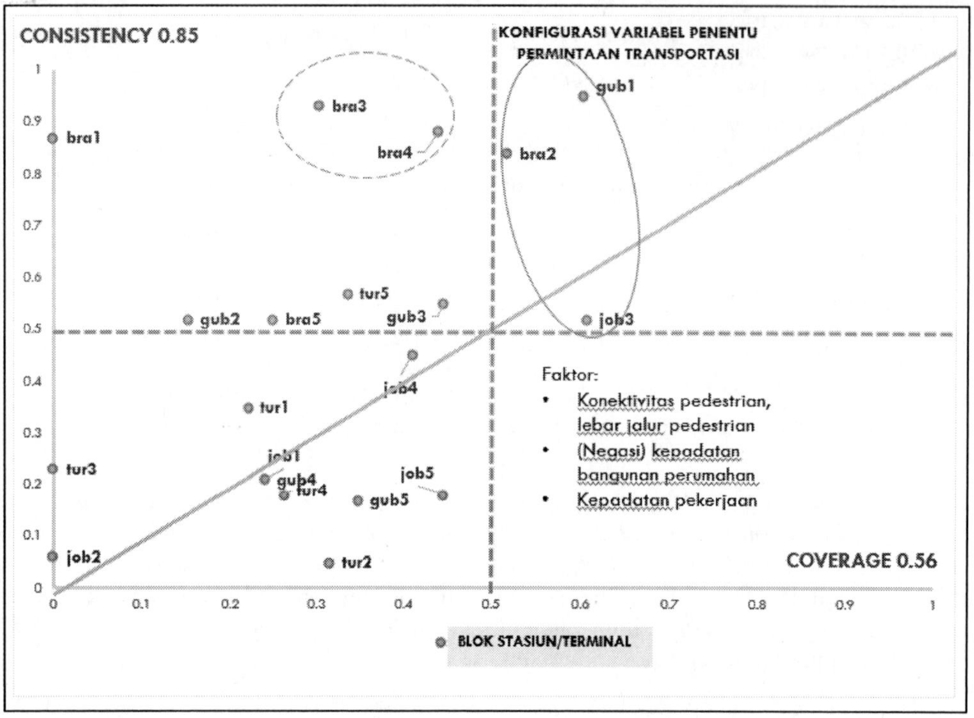

Figure 5. Model 2 Trip demand analysis based on the configured model: inde2, inde3, ~inden3, inden4 ≤ ttrip

Since "the coverage index in fsQCA assesses the degree to which a causal configuration accounts for instance of an outcome" ([13, p. 223]), the configuration resulted by the model 1b has higher proportion of cases that are consistent with that configuration; and model 1b is more informative than other models since the model has accounted for more cases: 56% out of cases with high outcome membership scores. However, only 3 subzones has appropriately indicated high trip demand, i.e. Stasiun Gubeng subzone 1, Terminal Bratang subzone 2 and Joyoboyo subzone 3. The model showed the combination of good pedestrian connectivity with appropriate pedestrian width and high employment density subzone supported for high trip demand. However, 7 subzones with high trip demand were not accounted for. Considering the density of housing and employment as two separate contested land use resulted in a configuration that has accounted for three subzones with high trip demand occurrence.

3. **Model 2a** tested the configuration for the low trip demand occurrence. The model consisted of configured variables *inde2, negation of inde3, inden1, negation of inden3 and inden4.* This model has the consistency index 0.878 and the coverage value 0.294. The resulted is as follows: a cluster consisted of Gubeng station Subzone 3. (figure: 6)
The model mathematically is written as follows.

$$\boldsymbol{connectivity \sim pedestrian\ width\ .\ KDB \sim housing\ density\ .\ employment\ density \leq \sim high\ trip\ demand}$$

The dot ". "indicates the "AND" while the tilde "~" indicates the negation. The model said that the low trip demand occurrence in the station area is determined by high connectivity of pedestrian with narrow pedestrian width, high KDB or building density, low residential density, and high employment density.
This model indicates more place variables than node variables, even though there is not much an issue of whether the pedestrian has appropriate width (as typically happening in the study area) as the connectivity is more important in determining low trip demand. Similarly to model

IOP Conf. Series: Earth and Environmental Science **340** (2019) 012037 doi:10.1088/1755-1315/340/1/012037

1b, the residential and employment density is contested, nevertheless, the building density is accounted for in the configuration, thus in supporting the presence of employment density, this variable is then interpreted as the non-residential building density. This model has much lower coverage value than model 1b. The model only resulted in one subzones Gubeng stasiun subzone 3 but erroneously with high trip demand. However, there are some subzones in clustered such as Pasar Turi subzone 2, Joyoboyo subzone 5, Gubeng subzone 4; all with low trip demand. It may be important to analyze alternative model that explained the low trip demand occurrence and identify a better fitted configuration. This will be discussed in model 2b.

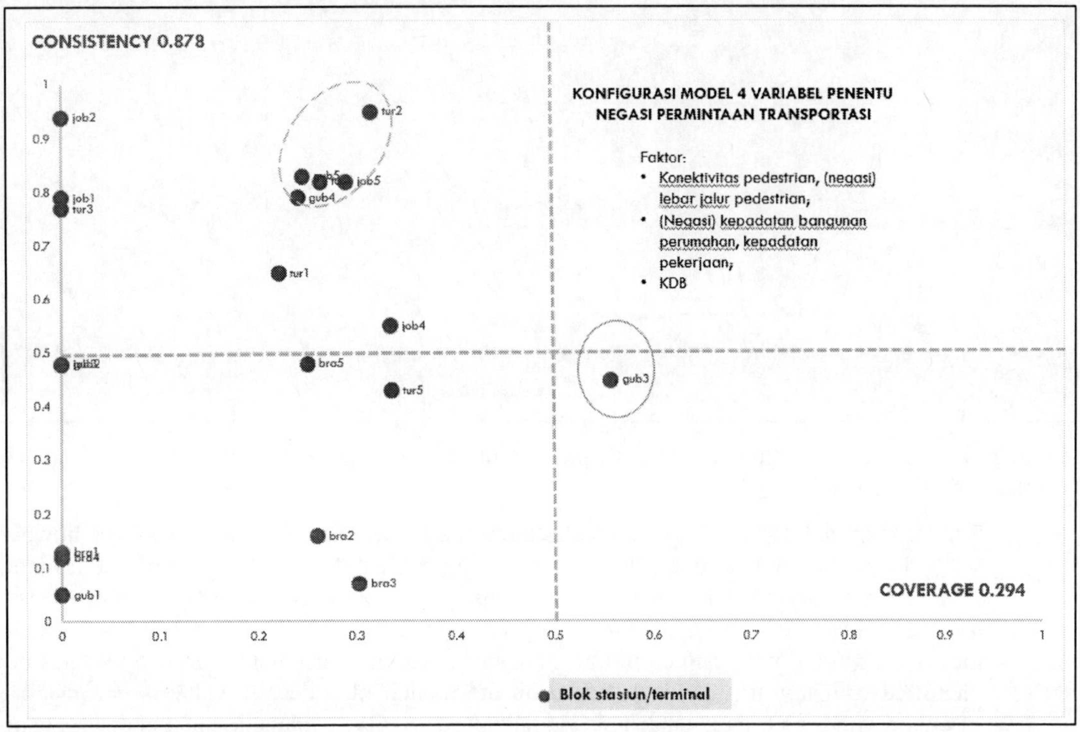

Figure 6. Model 3 Trip demand analysis based on the configured model: inde2, ~inde3, inden1, ~inden3, inden4 ≤ ~ttrip

4. **Model 2b** consisted of configured variables *negation of inde2, inde3, inden1, inden3 and inden4.* This model has the consistency index 0.89 and the coverage value 0.439. The resulted is as follows: a cluster consisted of Joyoboyo terminal Subzone 5. (figure: 7).
The model mathematically in figure 7 is written as follows.

$$\sim\boldsymbol{connectivity} * \boldsymbol{pedestrian\,width\,.\,KDB\ .\,housing\,density\,.\,employment\,density} \leq \sim\boldsymbol{high\,trip\,demand}$$

The dot ". "indicates the "AND" while the tilde "~" indicates the negation. The model said that the low trip demand occurrence in the station area is determined by low connectivity of pedestrian with good pedestrian width, high KDB or building density, high residential density, and high employment density.

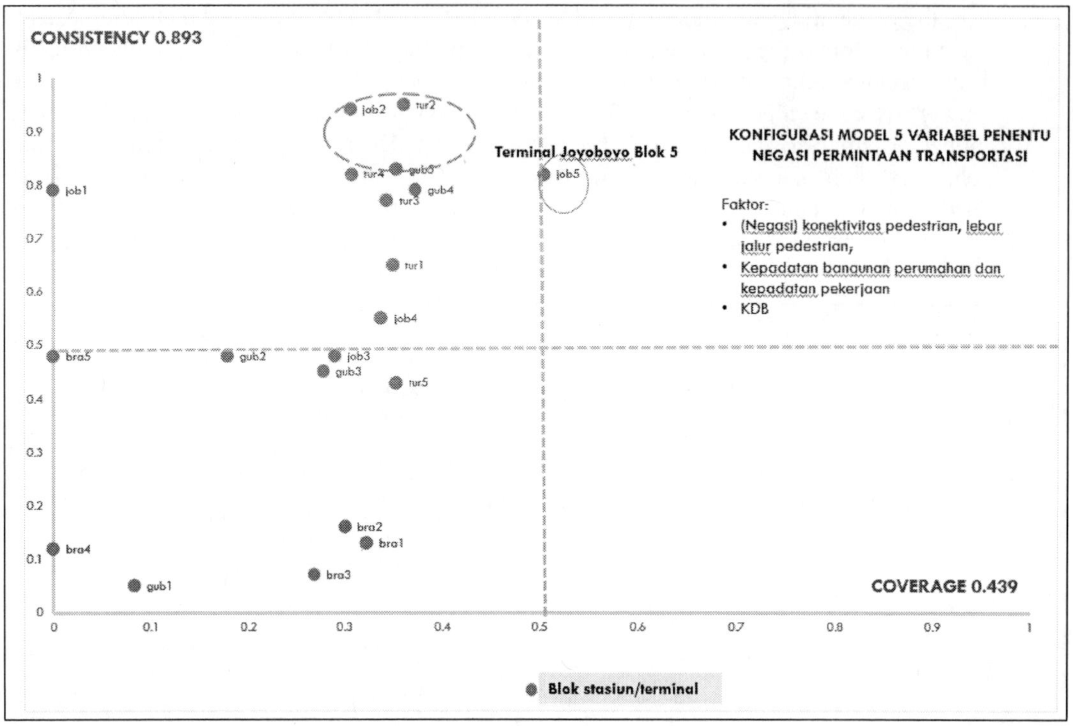

Figure 7. Model 4 Trip demand analysis based on the configured model: ~inde2, inde3, inden1, inden3, inden4 $\leq$~ttrip

Similar to model 2a, this model indicates more place variables than node variables, however, while the pedestrian has appropriate width, its connectivity level is low. In opposite to model 2a, all density variables (the residential and employment density and KDB or building density) has high values. This model has a modest coverage value (better than model 1a but lower than model 1b. Model only resulted in one subzones Joyoboyo terminal subzone 5 that has been identified with low trip demand. Based on the model, high density value is not necessary associate with high trip demand. This model gave a possible explanation that node aspect may be of more important than place aspect in determining the high or low value of trip demand. .

4.3 Discussion

The method of qualitative comparative analysis acted as an asymmetric modelling that may able to explain the prevalence of high and low trip demand occurrence according to the most fitted configuration set. In this paper, the high or low trip demand in the station area is tested with several set of configuration variables, accounted for the node and place characteristics. Model 1 and model 2 explained the determinant of trip demand. From model 1a, Gubeng station subzone 2 has been characterized with dominance floor space ratio and the width of the pedestrian. Both variables are high in within the subzone area. Model 1b, covered the area of Gubeng subzone 1, Bratang subzone 2, and Joyoboyo subzone 3. Among the all models developed in this paper, model 1b has the highest coverage value. The variable configured in this model consisted of connectivity and pedestrian path, also low housing density and high employment density. The interesting finding from model 1b, is that, the design variable, i.e. the connectivity of pedestrian and the width of pedestrian path, altogether support trip demand. However, the density variable, i.e. *housing density* maybe in contested to the number of employment density since low housing density combines with high employment density explain high trip demand. It is interesting to note that while employment and housing may represent different types of activities (residential vs non-residential), it may be true that the static frugal of space is not allowed the two competitive activities to use of the same limited space. This asymmetric notation may be a contradict to the common believe of mixed use principle of TOD, at least as a preliminary indication that in certain locations, the single most dominated land use may be of more important factor dictate demand for trip. The location mentioned in model 1b mainly the vocal point area of Surabaya City, such as area around the Surabaya Zoo in Joyoboyo, area around the Monumen Kapal Selam and Tunjungan Plaza in the catchment of Gubeng TOD area. These area have become the main generator of transportation that serve not only the neighborhood TOD trip demand, but also for the whole City and regionally.

On the other hand, model 2a and 2b explained factor influence the negation of trip demand or low trip demand. Model 2b has a better coverage and consistency than model 2a. Joyoboyo terminal subzone 5 has the lowest pedestrian connectivity compare to other subzones in all station/terminal. Both Joyoboyo terminal subzone 5 and Gubeng station subzone 3 have the lowest node index value compared to other subzones in its group. It seems the pedestrian facilities to access station/terminal precinct explained low trip demand and maybe of more important than that of the place aspect in explaining low trip demand.

Based on the model, density and design variable have important impact on transit demand in Surabaya City. Unlike the TOD in Western's cases, it is common that density and pedestrian friendly environment have significant effect on transit ridership in Asian Model of TOD e.g. Singapore and Seoul, Korea. The Singapore TOD experience shows that densely populated transit area and urban design applied to enhance a more transit supportive environment with walkability at micro-scale level have significant impact on increasing MRT ridership from 1990 to 2000 [16]. Density, pedestrian-friendly design and street network at rail station area are also evident as determinant factor of transit demand in Seoul City [17]. In this regard, some area of Gubeng, Joyoboyo and Bratang precinct provide density bonus, especially employment density, to attract more trip demand as emphasized in TOD planning. Pedestrian facility and access can be improved in terminal and station area in Surabaya to promote trip demand, as low quality of pedestrian facility may be regarded as the main factor causing low trip demand.

5. Conclusion

The node and place factor explaining TOD characteristics may be used as determinant of trip demand. The method of qualitative comparative analysis acted as an asymmetric modelling that may able to explain the prevalence of high and low occurrence according to the most fitted configuration set as explained in the subset of node and place variables. The result showed that employment density may has more impact than housing density, while node aspect maybe of more important aspect compare to place aspect that explain high or low trip demand. This findings is limited to the four stations/terminals being researched, as more case studies and more variances in the model may give better explanation about inconclusive findings about why some subzones precinct with high (or low) trip demand cannot be accounted for in the model. Nevertheless development strategy of high-employment density and design factors such as pedestrian connectivity and facility are prior factors to reduce automobile dependency and enhancing pedestrian accessibility to transit center. Improving the street network and a more transit supportive environment should be applied to promote transit users in Surabaya City.

6. References

[1] Cervero, R. et al. (2004). TCRP Report 102: Transit-Oriented Development in the United States: Experiences.Challenges. and Prospects. TRB. Washington.

[2] Dittmar, H., G. Ohland. (2004). The New Transit Town Best Practice in Transit-Oriented Development. Wasingthon, DC: Island Press

[3] Eltis. (2014). Transit Oriented Developmet in Curitiba, Brazil. Available from: http://www.eltis.org/discover/case-studies/transitoriented-development-curitiba-brazil

[4] Calthorpe Associates. (1992). City of San Diego Guidance System: Transit Oriented Development Guidelines. USA

[5] Reusser, D., P. Loukopoulos, M. Stauffacher, and R. Scholz. (2008). Classifying railway stations for sustainable transitions – balancing node and place functions. Journal of Transport Geography, 16(3):191–202. doi: 10.1016/j.jtrangeo.2007.05.004

[6] Newman, P. W. G., J. R. Kenworthy, et al. (1992). "The ecology of urban driving II--driving cycles across a city: Their validation and implications." Transportation Research Part A: Policy and Practice 26(3): 273-290

[7] Curtis, Carey & Mellor, Roger. (2011). Anticipating a New Railway: The Relevance of Transit-Oriented Development to Businesses in a Car-Oriented City. Urban Policy and Research. 29. 141165.10.1080/08111146.2010.540001.

[8] Bertolini, L. (1999). Spatial development patterns and public transport: an application of an analytical model in the Netherlands. Planning Practice and Research, 14(2): 199–210. doi: 10.1080/02697459915724

[9] Cervero, R. and K. Kockelman (1997). "Travel demand and the 3Ds: Density, diversity, and design." Transportation Research Part D: Transport and Environment 2(3): 199-219

[10] Moroj. 2005. Nodal Intensification Strategy: Evaluation of an Analytical Model in Metropolitan Cape Town. https://scholar.sun.ac.za/bitstream/handle/10019.13061/Moroj2.pdf?sequence=2

[11] Chorus, P. (2011). An Application of the Node Place Model to Explore the Spatial Development Dynamics of Station Areas in Tokyo. The Journal of Transport

[12] Curtis, C., and D. Olaru (2011). The relationship between accessibility and travel patterns of residents in three new station precincts. Paper Presented in Track 9 (Transport and Infrastructure) at the 3rd World Planning Schools Congress, Perth (WA), 4-8 July 2011

[13] Ferguson, G., C.M. Megehee and A.G. Woodside (2017). Culture, religiosity, and economic configural models explaining tipping-behavior prevalence across nations. Tourism Management, 62 (2017), 218 – 233, http://dx.doi.org/10.1016/j.tourman.2017.05.001

[14] Woodside, A.G (2013). Moving beyond multiple regression analysis to algorithms: Calling for adoption of a paradigm shift from symmetric to asymmetric thinking in data analysis and crafting theory. Article in Press. Journal of Business Research. Available from: http://dx.doi.org/10.1016/j.jbusres.2012.12.021

[15] Sanders, J. (2015). Linking Station Area Node and Place Functions to Traffic Flow. Master Thesis. Yokohama National University

[16] Lew, S.H and Pei-Ju, P., 2009. Transit-oriented development in a high-density city: Identifying its association with transit ridership in Seoul, Korea, Cities 28, pp 70-82

[17] Sung, H. and Oh, J.-T., 2011. An Asian Model of TOD: The Planning Integration in Singapore. Transit Oriented Development Making It Happen, Chapter 8. Edited by Curtis, Carey et al, 2011. (pp. 91-106). Ashgate Publishing Ltd, England.

IOP Conf. Series: Earth and Environmental Science **340** (2019) 012038 doi:10.1088/1755-1315/340/1/012038

Problem identification of micro hydro power plant program in East Java Province

S H Kusuma[1,2], B U Aulia[1,2], Sardjito[1,2] and N Farikha[1,2]

[1] Department of Urban and Regional Planning
[2] Institut Teknologi Sepuluh Nopember, Surabaya 60111, Indonesia

Email: surya.uya39@gmail.com

Abstract. The availability of electricity is an important necessity in life to support the economic and social activities of the community such as industry, agriculture, household and trade. Without sufficient access to electricity, the goals of community economic and social development will not be achieved. One of the East Java Provincial Government's efforts to meet its electricity needs is to build renewable energy sources such as micro-hydro power plants (or PLTMH in Indonesian) which are planned to be developed in 21 regencies or cities. However, until now the development of the PLTMH has not been implemented perfectly. Therefore, this research was conducted to determine the extent of the progress of the PLTMH development, factors that caused problems in the development of PLTMH, and recommendations for solving problems such as what could be formulated. The method used in this study is through qualitative descriptive analysis with indepth interview and using root cause analysis. Based on the results of this study it is known that the main thing that caused problems in the construction of the PLTMH was the existence of constraints in planning and financing. The recommendation for solving the problems obtained from this study is the need for technical planning and budgeting with financing schemes that can be sourced from the government or the private sector.

Keyword: *electricity, micro hydro power plant (PLTMH), root cause*

1. Introduction

Electrical energy is one of the main needs that can sustain economic and social activities of the community. The electricity needs of the people in Indonesia are increasing every year due to an increase in population and economic activities. As a state-owned enterprise that takes care of all aspects of electricity in Indonesia, the National Electricity Company, or PLN, has not been able to meet 100% of the electricity needs of the community. It is known that there are several regions that cannot be reached by the electricity supply from PLN due to the geographical conditions of the region. One solution in the supply of electrical energy, especially in remote areas, is to utilize the potential of natural resources as a renewable energy source which can be in the form of utilizing water flow as a micro-hydro power plant (or called PLTMH in Indonesian).

PLTMH itself is a system of small-scale power generation that uses hydropower as its driving force such as irrigation canals, rivers or natural waterfalls by utilizing water levels and discharge [1]. PLTMH is often referred to as a small-sized generator capable of producing electricity capacity of up to 100kW, besides that PLTMH is one of the renewable energy sources and is worthy of being called clean energy because it is environmentally friendly [2]. The construction of PLTMH continues to grow because it has a simple structure with high usage applications [3]. In East Java Province, there are plans for PLTH development in 21 districts / cities as stipulated in the 2011-2031 Regional Spatial Plan (RTRW) of East Java Province. However, until now the achievement of the PLTMH development has not been 100%, so there are still some regions, especially remote areas, which are not yet covered by electricity. This research was conducted to find out the extent of the achievement of MHP development in East Java Province, inventory problems in its development, and structure alternative solutions to these problems. With the knowledge of the problem of the development of the PLTMH, it is expected that the formulation of a handling strategy can be done more effectively and efficiently.

2. Research methods

The method used in this study is a qualitative method with an inductive approach. This method is used in order to produce a final formulation in a systematic, factual, and accurate manner regarding the characteristics of the object of research that will be used or better known as a decision-making approach from specifically to general [4]. The data analysis techniques used are qualitative descriptive analysis and root cause analysis.

Data collection itself is carried out by secondary surveys and structured interviews for approximately 2 (two) months in September-October 2017 to stakeholders in the energy network system in East Java Province. Data collection activities were carried out with the output of the achievement of the PLTMH development program in East Java Province as well as the problems faced in its development.

Meanwhile, in identifying the problems of PLTMH development in East Java Province, it was carried out using problem tree analysis. This analysis is used to determine the problems that occur in the realization of the program in a structured manner through a causal relationship [5]. It is expected that the use of this analysis technique can be known in detail the root causes of the problem of PLTMH development in East Java so that solutions that are right on target can be concluded.

3. Result

The plans for the development of PLTMH in East Java Province as stated in the spatial plans named Rencana Tata Ruang Wilayah Provinsi Jawa Timur 2011-2031 are spread across 21 regencies and cities including Nganjuk, Bojonegoro, Banyuwangi, Situbondo, Bondowoso, Jember, Lumajang, Malang, Probolinggo, Blitar, Tulungagung, Trenggalek, Pacitan, Madiun, Magetan, Pasuruan, Mojokerto, Ponorogo, Jombang, Gresik, and City of Batu.

Broadly speaking, based on the results of the in-depth interview, the researchers divided the achievement of the PLTMH development program in East Java Province over 3 categories that namely already implemented, process, and not yet implemented with the following explanation:

- A category has been implemented if there has been an implementation of PLTMH development in the district / city area.
- Process category if there are already detailed plans for the construction of PLTMH
- Categories have not been implemented if the planning and development stages have not been implemented

The PLTMH that has been built in East Java Province is only found in 9 regency/cities as seen in the data in Table 1.

IOP Conf. Series: Earth and Environmental Science **340** (2019) 012038 doi:10.1088/1755-1315/340/1/012038

Table 1. PLTMH data in East Java Province in 2017

Num	Regency/Cities	District	Sub District	Power (KW)	Number of families in the territory	Condition
1	Lumajang	Senduro	Burno	16	86	Not operating
		Tempursari	Kali Uling	25	80	Not operating
2	Pacitan	Nawangan	Tokawi	20	90	Not operating
3	Nganjuk	Sawahan	Bareng	12	56	Operate
4	Probolinggo	Krucil	Kalinan	40	350	Not operating
		Krucil	Watupanjang	34	58	Operate
		Tiris	Andung Biru	40	550	Operate
				2,5	20	Not operating
				13	135	Operate
5	Madiun	Kare	Kare	62	72	Operate
6	Trenggalek	Munjungan	Besuki	22	80	Not operating
7	Banyuwangi	Songgon	Sumber Arum	44	180	Operate
8	Situbondo	Sumbermalang	Sumberargo	15	170	Operate
9	Malang	Sampit	Sumber Suko	3	13	Operate

Sumber: Dinas Energi dan Sumber Daya Mineral Provinsi Jawa Timur, 2017

Based on the table above, it is known that from PLTMH that has been built in 9 Regency, there are still PLTMHs that are currently not operating, namely in Lumajang Regency, Pacitan Regency, Probolinggo Regency, and Trenggalek Regency. Apart from the PLTMH that has been built, currently the Blitar Regency Government is proposing the establishment of a PLTMH to the center due to limited district funds. For other districts / cities there has been no development of PLTMH as planned by East Java Province. An illustration of the achievement of the PLTMH development program in East Java Province can be seen in Figure 1.

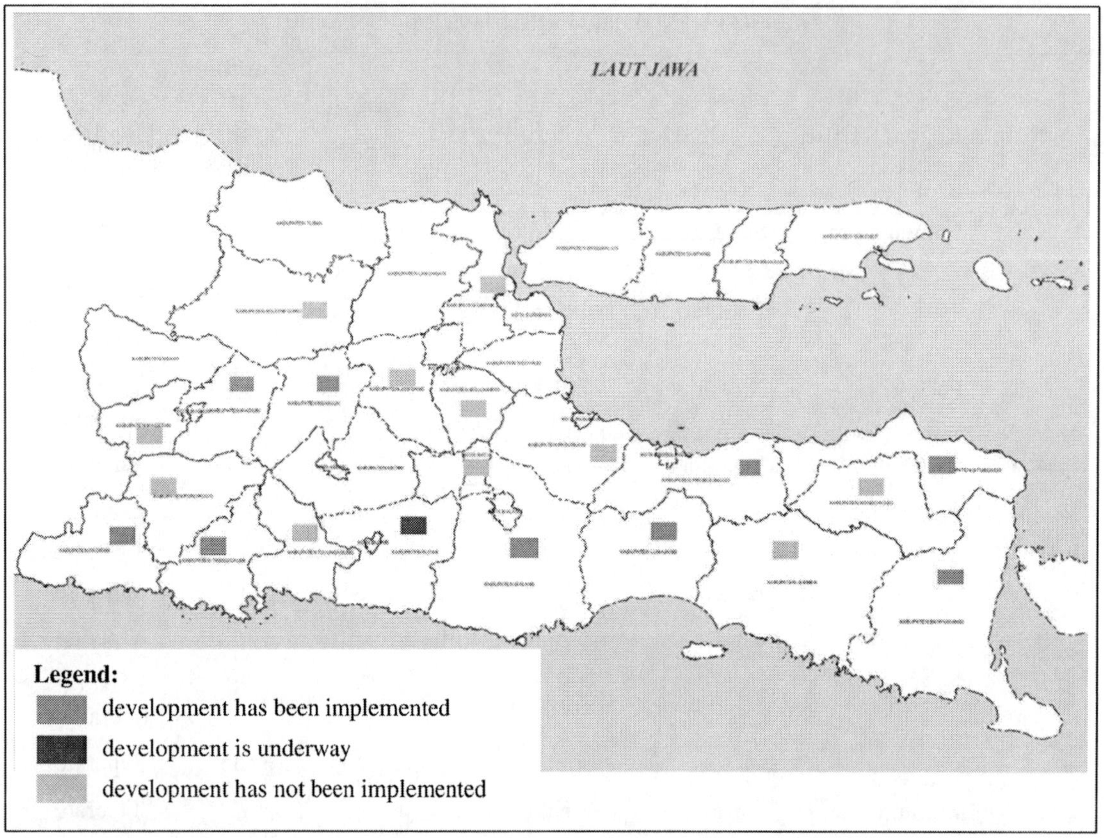

Figure 1. Achievement of the PLTMH Development Program in East Java Province

The next step that was carried out by the researcher was to identify the factors that caused the achievement of the PLTMH development program in East Java Province. Based on the in-depth interview, it is known that there are several obstacles that lead to unsustainable operation and the absence of PLMTH development which includes:

- The flow of electricity from PLN to remote villages in Lumajang Regency has been entered so that the PLTMH that has been built has not been reoperated. People prefer to use electricity from PLN because it is easier, no extra care is needed, and the power produced is stable.
- There is damage to turbines in PLTMH in Pacitan, Probolinggo and Trenggalek Regencies. As is known, PLTMH in this area was built from Provincial grants and private grants. This grant is only in the form of funds for the procurement of PLMTH and there are no funds for maintenance. So that local people are encouraged to take care of this PLMTH with limited knowledge and funds. This is what causes the PLTMH to be abandoned.
- Limited funding for PLTMH planning and construction in other locations.
- In some locations there are obstacles related to land acquisition due to the source of water flow for the PLTMH located in the forest area.
- There are still many PLTMH development plans that have not been accommodated in spatial planning books in each regency/city.

The problem factors presented above will be stated in a root cause analysis as follows:

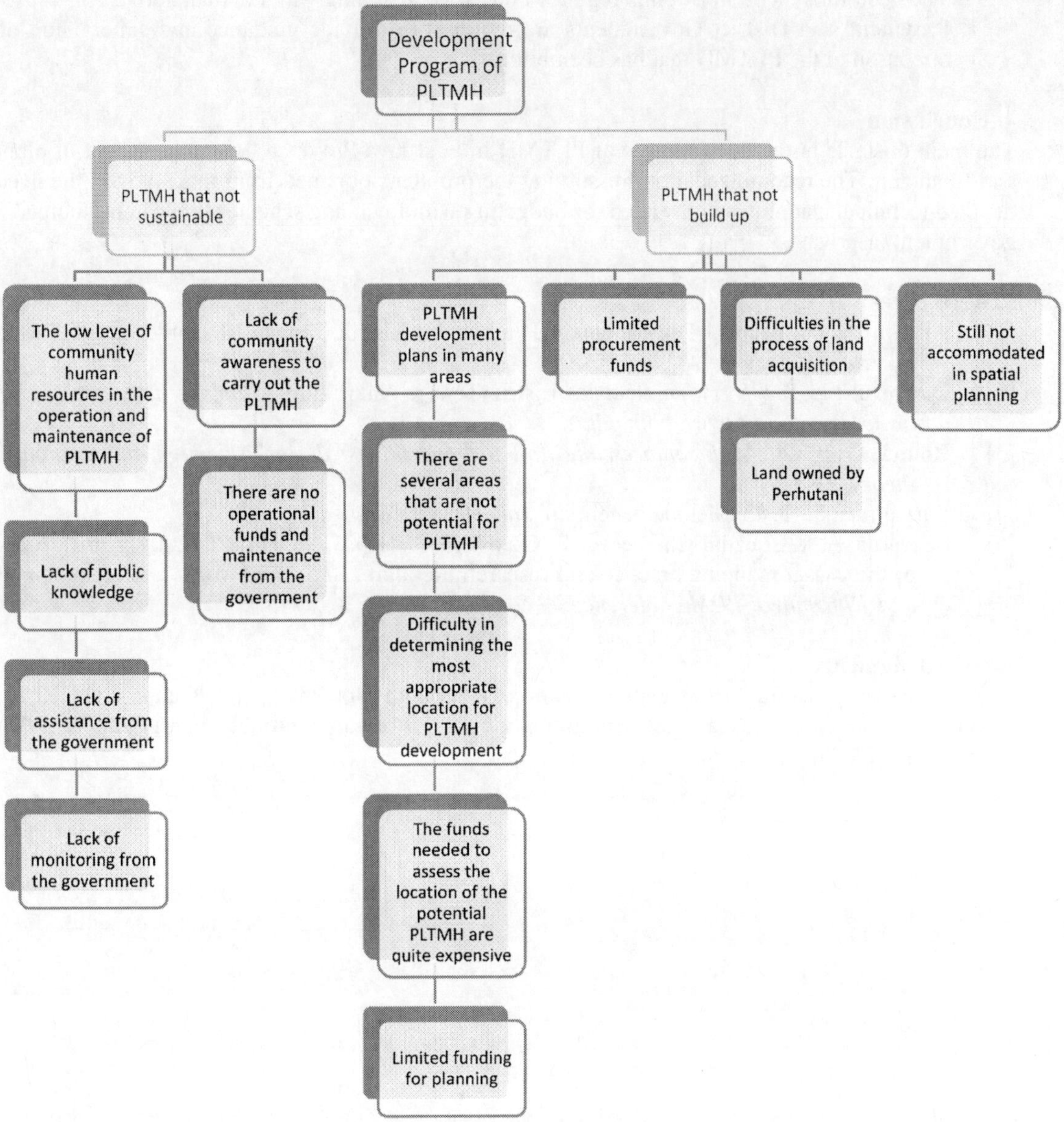

Figure 2. Root Cause Analysis of PLTMH Development Program in East Java Province

In order to resolving the problems of PLTMH development, researchers formulated recommendations as follows:

- The need for financing schemes originating from the Provincial Government, Regency / City Government, and funding assistance from private parties in funding PLTMH planning and development in East Java Province
- Funding in the operation and maintenance of PLTMH can be done by conducting electricity sale and purchase transactions from beneficiaries of PLTMH which in this case are the community. This is in accordance with what is stated in the Minister of Energy and Mineral Resources Regulation Number 39 of 2017.
- For PLTMH planning which is located in a forest area, it can use the scheme of forest use loan permit as stated in the Republic of Indonesia Presidential Regulation Number 4 of 2016

- The need for synchronization in regency / city level planning with the plan above
- Provincial and District Governments are required to provide guidance and supervision of the operation of the PLTMH that has been built.

4. Conclusion
The main obstacle in the development of PLTMH in East Java Province lies in the aspect of planning and financing. The recommendation for solving the problems obtained from this study is the need for detailed technical planning and the need for budgeting with financing schemes that can be sourced from government or private.

5. References
[1] Anonim. 2008. Manual Pembangunan Pembangkit Listrik Tenaga Mikrohidro. IBEKA-JICA. Jakarta.
[2] Damastuti, A. P. (1997). Pembangkit Listrik Tenaga Mikrohidro. *Sumber: http://www. elsppat. or. id/download/file/w8_ a6. pdf.*
[3] Round, G. F. (2004). *Incompressible flow turbomachines: design, selection, applications, and theory*. Elsevier
[4] Suryabrata, S., 1983. Metode penelitian. *Jakarta: CV Rajawali*.
[5] Scavarda, A. J., Bouzdin-Chameeva, T., Goldstein, S., Hays, J., & Hill, A. (2004, April). A review of the causal mapping practice and research literature. In *Second world conference on POM and 15th annual POM conference*.

Acknowledgments
The authors are grateful to the Institut Teknologi Sepuluh Nopember and Kementerian Riset dan Teknologi Republik Indonesia for providing funding for this research through *Hibah Pebelitian Pemula Dana Lokal ITS 2018.*

IOP Conf. Series: Earth and Environmental Science **340** (2019) 012039 doi:10.1088/1755-1315/340/1/012039

Importance performance analysis for smart city implementation in Surabaya, Indonesia

S Felasari[1] **and M S Roychansyah**[2]

[1] Department of Architecture, Faculty of Engineering, Universitas Atma Jaya Yogyakarta, Indonesia
[2] Department of Architecture and Planning, Faculty of Engineering, Universitas Gadjah Mada Yogyakarta, Indonesia

Email: s.felasari@uajy.ac.id ; saniroy@ugm.ac.id

Abstract. This paper presents a study to measure the implementation of smart city initiatives in Indonesia that has been applied by the city government in various city scales. Various benefits are believed to be gained by applying a smart city concept that emphasizes on the use of information and communication technology. However it depends on the readiness of the city government and its citizens in implementing the smart city concept. Importance Performance Analysis is used to measure the differences between public perception and the real conditions of the smart city implementation. Surabaya city, which was granted as one of the five winners of smart city awards in Indonesia for big city category, is used as a case study. The implementation of smart city's attributes is compared with the expectation of the citizens. The result of the research shows that the concept of smart city,which is used and implemented by the city government in providing services for its society is not always understood similarly to the societies' point of view. Involving the community in determining the priority of which attributes and dimensions of smartness in the provision of public services is very important to improve the level of implementation of smart city.

1. Introduction

The discourses of smart cities are getting more and more attention in the media, which show how in the future, the concept of 'smartness' can be applied in cities. According to Cohen [1], there are 3 generations of smart cities i.e smart city 1.0 which is technology driven, smart city 2.0 which is technology enabled and city led, and smart 3.0 which is citizen co-creation models. In the last several years the concept of smart city is no longer a discourse but has been applied by the city government including in Indonesia in various city scale either in big city, medium city or small town.

This paper discusses the implementation of the smart city initiatives in Indonesia particularly in a big city scale. Since 2015, the Indonesian Government in cooperation with a consortium has been trying to set out the smart city index for cities in Indonesia. In addition a roadmap for policy and strategy of urban development have been set up by The National Planning Bureau by introducing 'liveable', 'green', and 'smart' keywords [2] . It is generally believed that implementation of smart city will bring benefits for the urban residents in many ways.

Beside of that, good planning is a planning that is able to meet the needs of the urban residents. In this smart city planning, the demand of each urban resident need will be different among cities. It is based on the preferences of the people who live in it, so that the direct perception of the urban residents on their city conditions becomes an important process. A previous study has been done to measure the friendliness level for the medium city and small town in Indonesia like Yogyakarta and Magelang [3]. Nevertheless a study of the implementation of the smart city for the big city has not been done yet. Therefore Surabaya, one of the big cities in Indonesia is used as a study case. Surabaya is one of city in Indonesia, which is granted in a competition held by Kompas daily in collaboration with ITB and PGN as 'smart city' for the category of big cities with a population of more than 1 million. Apart from being a smart city, Surabaya also won an award as a smart city in the environmental category [4] .

2. Methods

2.1. Importance Performance Analysis

The Importance Performance Analysis (IPA) introduced by Martilla and James [5] was used to measure the differences between public perception related to the attributes of smart city dimensions (attribute's importance) and the real conditions of the smart city implementation in the field (attribute's performance). The results of IPA are expected to provide an overview of the gap between aspects of real community needs (importance) and services supplied from the city/government that can be assessed (performance) in the implementation of smart city. The real condition in the field shows how each attribute of this smart city dimension is perceived by its citizens as have already been applied by the city government in providing services for the community. The results are compared with the smart city dimension's attributes expected by citizens to be applied in the city in the future. The differences between performance and importance of the smart city's attributes reveal the implementation of the smart city concept. The attributes of smart city's dimensions are structured in a Likert Scale model's questionnaire (level 1-7).

2.2. Smart city dimensions and attributes

According to Cohen [6] there were 6 dimensions of smart city used to measure the smart city implementation i.e. (1) smart economy, (2) smart mobility, (3) smart environment, (4) smart people, (5) smart living, and (6) smart governance. The smart disaster management was added to fit to the context of Indonesia. Each smart city dimensions is divided into 6 attributes as below:

Table 1. Dimensions and attributes of smart city

	SMART ECONOMY (SE)			SMART LIVING (SL)	
attributes	SE-1	New startups	attributes	SL-1	Housing sufficiency
	SE-2	Internet-based full employment		SL-2	Gini Index
	SE-3	Innovative in business processing		SL-3	Investment in culture
	SE-4	ICT-based business license		SL-4	Smart crime prevention
	SE-5	Online transaction between institutions		SL-5	Accessible health and education services
	SE-6	Business information availability		SL-6	Life expectancy
	SMART MOBILITY (SM)			**SMART GOVERNANCE (SG)**	
attributes	SM-1	Clean-energy transport such as bicycle	attributes	SG-1	Online government services
	SM-2	Innovative and modern parking system		SG-2	Electronic benefit payments
	SM-3	Integrated multi-modal transit system		SG-3	WiFi coverage at public space
	SM-4	Real time public transport information services		SG-4	Sensor coverage
	SM-5	Smart cards for public transportation		SG-5	Integrated health and safety services
	SM-6	Online transportation apps		SG-6	Open data and apps

SMART ENVIRONMENT (SN)			SMART DISASTER MANAGEMENT (SDM)		
attributes	SN-1	Smart homes	attributes	SDM-1	Emergency warning system
	SN-2	Energy saving house		SDM-2	Responsive mitigation planning
	SN-3	Public building automation system		SDM-3	Sensor for monitoring disaster
	SN-4	Certified green buildings		SDM-4	Evacuation and rescue training
	SN-5	Solid waste and waste water treatment		SDM-5	Emergency information
	SN-6	Green open space		SDM-6	Disaster relief assistance
SMART PEOPLE (SP)					
attributes	SP-1	Accessible internet at home	attributes	SP-4	Voter participation in municipal election
	SP-2	Smart phone penetration		SP-5	Number of higher education degree
	SP-3	Civic engagement activities using ICT		SP-6	Creative industry jobs

Source: Adapted from [7]

It is important to know that choosing priorities, directions, policies, and utilizing information and communication technology with automatic computerization intelligently is the key strategy for the smart city implementation. In addition, more than that consideration, the citizen centric approach, which facilitates active participation of the urban residents in the city management, is also important way to create a successful implementation of urban strategy like smart city, bringing closer interaction between citizens needs and local government responses.

3. Smart city implementation in Surabaya

There were about 120 respondents participated in the survey. Various information related to gender, age, education level, occupation and the area where respondents live have been gathered through the questionnaire. Respondents consist of 45 % male and 55% female with level of education are 5 % secondary school, 78% high school and 17% higher education level.

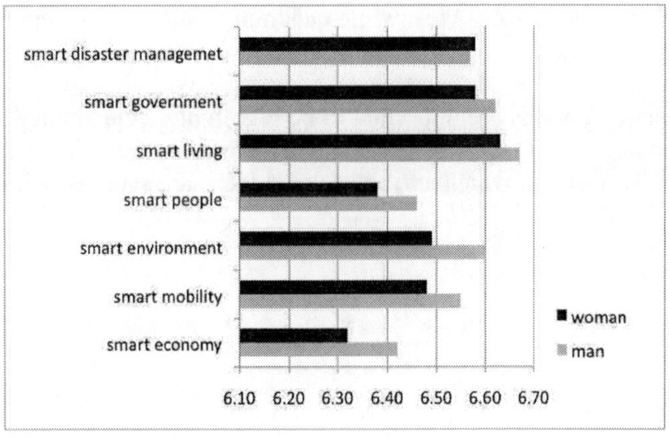

Figure 1. Expectation of smart dimensions according to gender

The importance of smart city attributes in Surabaya is perceived relatively the same between men and women. Nevertheless, female expect a little bit higher value for smart disaster management than male citizens, while other dimensions are in reversed (Figure 1)

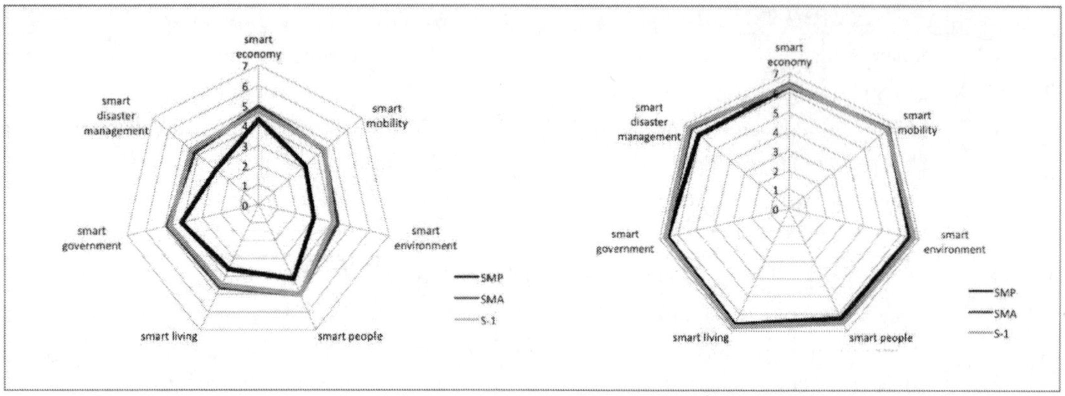

Figure 2. Performance (left) and importance (right) of attributes based on education level

Meanwhile education level influenced the perception of the smart city implementation in Surabaya. According to Surabaya citizens with primary education level the performance of the smart city's attributes is low especially in dimension of mobility, environment, and disaster management (Figure 2 -left). However, most of Surabaya citizens expect a high level implementation on all dimensions (Figure 2-right).

The result of the IPA of smart city implementation in Surabaya city can be seen from Figure 3. The figure represents the implementation of smart city initiatives in Surabaya city. The X- axis shows the performance of smart attributes implemented in the city while the Y- axis shows the importance of smart attributes perceived by the citizens. Each quadrant shows the gap between the performance and the importance of the attributes. Quadrant A shows high importance-low performance which can be interpreted that implementation is not as good as expected by the citizens. Quadrant B shows high importance - high performance. It means that the implementation is as good as expected by the citizens. Quadrant C shows low importance-low performance. It can be read that the attributes are as less important as city's performance. Meanwhile quadrant D shows low importance-high performance meaning that the city has a good implementation in some attributes although the citizens perceived it as not quite important.

In Figure 3, quadrant A and B are important to be highlighted. Quadrant A reveals obligation from the government to improve their services in some attributes that the citizens think they are important. Meanwhile Quadrant B shows in which attributes the citizens are satisfied with the services.

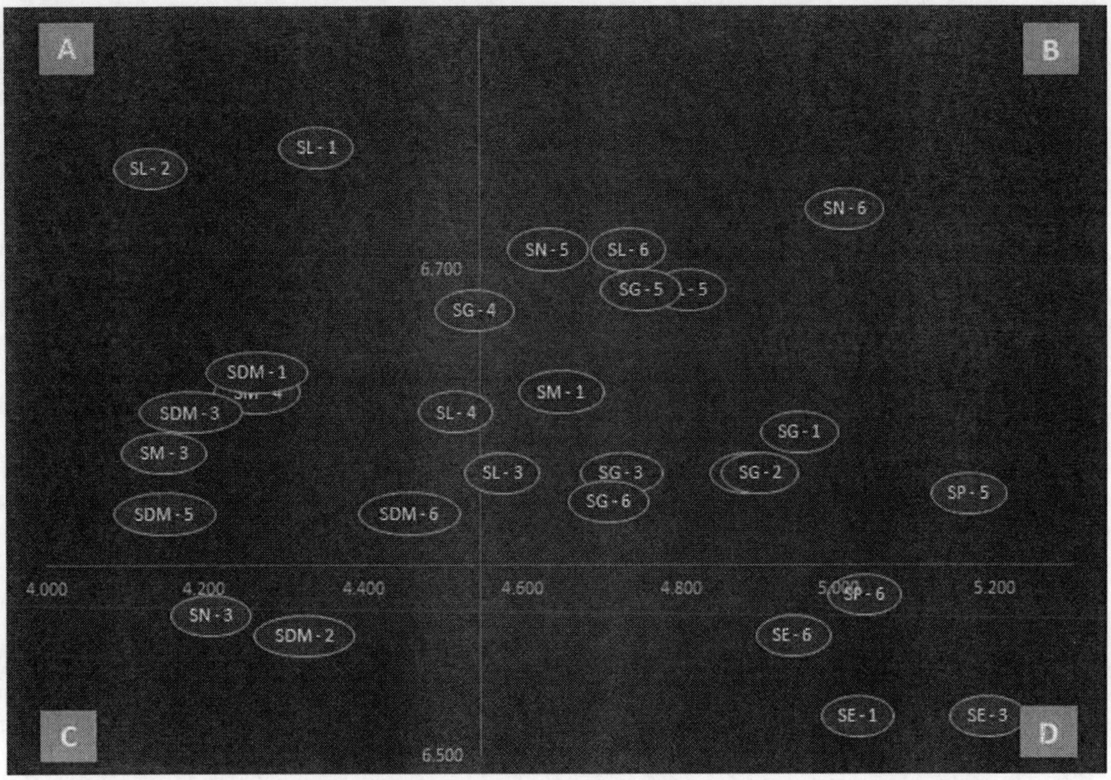

Figure 3. Importance Performance Analysis for Surabaya smart city's implementation

Surabaya residents consider their city government is fairly good at implementing the concept of smart governance by providing an online system and utilizing ICT for health and security services (SG-1, SG -2, SG -3, SG-5, and SG- 6). However, the community believes that Government needs to increase the use of sensors broadly to improve their services (SG-4).

In addition, the citizens are also pleased to the Government's effort in supporting smart living implementation by improving investment in culture (SL-3) and life expectancy (SL-6) and providing accessible health and education services (SL-5). However they have a highly expectation that the Government can provide sufficient housing (SL- 1), good gini index (SL -2) and smart crime prevention SL-4) as pointed out in quadrant A in (Figure 3)

Surabaya's achievements in winning smart cities in the environmental category seem to be proven by the satisfaction of its residents to the availability of open green spaces (SN -6), good handling of solid waste and waste water treatment (SN-5), and government support for the use of bicycles as a clean energy transportation (SM-1). Green buildings (SN-4) and energy saving house (SN-2) are two attributes that needs to be improved continuously.

Nonetheless, problems in the transportation sector such as the provision of parking and integrated multi-modal transit system (SM-2, SM-3), and the provision of modern and sophisticated public transportation are urged to be resolved (SM-4, SM-5) . Both Surabaya government and its citizens have the same opinion that internet-based employment, smart homes and building automation system, mitigation planning and evacuation training are attributes of smart city that get the last priority in their implementation.

4. Conclusion

The results show that about 64% attributes of smart cities were considered important by the residents of Surabaya to support community services. Among those attributes that are considered important, 48% of them have been implemented properly by the Government of Surabaya while 52% of others

IOP Conf. Series: Earth and Environmental Science **340** (2019) 012039 doi:10.1088/1755-1315/340/1/012039

need to be improved. Meanwhile 36% of other smart city's attributes are considered not too urgent to be implemented in the city of Surabaya. However, from the number of attributes that are not too important, 70% of them are applied very well by the government, for example in the economic dimension like the use of internet and ICT in economic and business activities.

This shows that the concept of smart city, which is used and implemented by the city government in providing services for its society like Surabaya, is not always understood similarly to the societies' point of view. Therefore involving the community in determining the priority of which attributes and dimensions of smartness in the provision of public services is very important to improve the level of implementation of smart city.

5. References

[1] Cohen B. The 3 Generations of Smart Cities: Inside the development of the technology driven city [Internet]. 2015. Available from: https://www.fastcompany.com/3047795/the-3-generations-of-smart-cities [cited 2018 Sep 27]

[2] Bappenas (National Planning Bureau). Kebijakan dan Strategi Pembangunan Perkotaan Nasional (KSPPN)(National Policy and Strategy for Urban Development) [Internet]. 2014. Available from: http://tataruangpertanahan.com/regulasi/pdf/kepmen/menppn/kepmenppn_65_2014.pdf

[3] Roychansyah MS, Felasari S. Measuring level of friendliness of smart city: a perceptual study. IOP Conf Ser: Earth Environ Sci. 2018;126(1):012172.

[4] Alexander HB. Surabaya, Yogyakarta, dan Magelang Kota Cerdas 2015 [Internet]. Surabaya, Yogyakarta, dan Magelang Kota Cerdas 2015. 2015 [cited 2018 Apr 4]. Available from: https://properti.kompas.com/read/2015/08/13/200801621/Surabaya.Yogyakarta.dan.Magelang.Kota.Cerdas.2015

[5] Martilla JA, James JC. Importance-Performance Analysis. Journal of Marketing. 1977;41(1):77–9.

[6] Cohen B, Cohen B. What Exactly Is A Smart City? [Internet]. Co.Design. 2012. Available from: https://www.fastcodesign.com/1680538/what-exactly-is-a-smart-city

[7] Anonym. Final Report of 'Jogja Istimewa' Smart City Study (Laporan Akhir Kajian Smart City Jogja Istimewa). Yogyakarta: Badan Perencanaan Pembangunan Daerah (Bappeda) Kota Yogyakarta and Pusat Studi Perencanaan Pembangunan Regional (PSPPR) UGM; 2016.

Analysis regarding modern market service level influence toward population movement patterns

T Poerwati[1], A H Imaduddina[1] and Y A L Putra[1]

[1] Urban and Regional Planning, Institut Teknologi Nasional (ITN) Malang, Jl. Bendungan Sigura-gura, no. 2, Malang

Email: tpurwati@rocketmail.com

Abstract. Modern market didn't differ too much from traditional ones, but this kind of market exists within buildings and self-serving in nature. In terms of modern market services like Alfamart and Indomart, it can be said that the service provided is relatively stellar for it's customers, whether it be the goods itself, the qualities of said goods, or the service system itself which includes home delivery. The purpose of this research is to find out the modern market service level influence toward population movement patterns. The research is done in Klojen District, Malang City. The analysis method used are qualitative descriptive method and Kendall's Tau correlation method. Based on analysis result it's concluded that minimarket service level didn't quite affect population movement patterns due to it's p-value which is 0,487 where the p-value is bigger than the significant value within this research which is 0,05. Population movement patterns were only influenced by type of goods needed (average of 61,7%), quantity of goods bought by the consumers, and price of goods (average of 29,3%).

1. Preliminary

Development of retail industry in Indonesia is rapidly increasing. The presence of modern retail enterpreneurs were giving it's own color for Indonesian retail industry development. Within relatively short time, few modern retail enterpreneurs with spectacular capitalistic capabilities made their mark in Indonesia. They manifests in form of minimarkets, supermarkets even hypermarkets which now scatters across Indonesian big cities. The change in shopping or spending pattern were due to the rapid development of modern market is rather facsinating to be observed, as it is estimated to last long into the future.

If seen more specifically about the development of modern market in an even smaller scope like within districts, especially in Malang city. Modern market development in Malang city gives significant enough contribution toward its regional economy. But this development isn't followed by fundamental change of services from stores, kiosk, or markets in the face of population shopping pattern changes, so that the development of modern markets in Malang city truly changes the population movement patterns in shopping, especially modern market developing around markets, shops, or kiosks.

Nowadays a lot of modern market, especially Alfamart and Indomart is developing in Malang city, Klojen district to be exact. The two modern market mentioned is rapidly developing due to Klojen

IOP Conf. Series: Earth and Environmental Science **340** (2019) 012040 doi:10.1088/1755-1315/340/1/012040

district population very high demands of everday needs. The intended goal of this research is to analyze and find out the influence of modern market service level toward population movement patterns.

Benefit of this research is to find out the influence of modern market service level toward population movement patterns.

2. Methodology

Methodical approach used within this research is deductive approach. Deductive approach means thinking method that demands factual apriori. Through this approach, it is expected that the result of this research is scientifically accountable and useful in development of research field. Data gathering method used is field surveys and questionnaire. Questionnaire is aimed at a number of corresponding group, which is: Klojen district populace which makes 75% of them and 25% which made up of enterpreneurs and or employees. As for the analysis method is done through qualitative and quantitative method. Qualitative analysis is done to spell out district characteristics and it's potentials.

Next, quantitative analysis is done with Relative Distribution Frequency method and Kendall's Tahu correlation method. Distribution frequency is used to classify the data collected based on determined research variables. For example, data classification of type of goods, facility distributions, accesibility, total population, etc. While Kendall's Tahu correlation method is used to find out the connection between the influence of modern market service level toward population movement patterns.

3. Analysis and interpretation

3.1. Analysis of modern market service level

Minimarket service level analysis is used to find out how big the service level of minimarket scattered on Klojen district toward fulfilling the populace life needs. This analysis is using viewing variables that could be used to find out the minimarket service level:

3.1.1. Type of goods

Goods sold in minimarkets (Alfamart/Indomart) were daily necessities, snacks and drinks, office supplies, cosmetics, and cigarettes. Type of goods often purchased by people were snacks and drinks were up to 27 people or average 36%, while those who shop for daily necessities were up to 23 people or average 30,7%.

Table 1. Goods Sold in Minimarkets (Alfamart/Indomart)

No	Type of goods	Total Respondent	Percentage (%)
1	Daily Necessities	23	30,7
2	Snack and Drinks	27	36,0
3	Office Supplies	5	6,7
4	Cosmetics	4	5,3
5	Cigarettes	16	21,3
	Total	**75**	**100,0**

Source : Populace Questionnaire Result

For snacks and drinks, people tend to prefer shopping in modern market like Alfamart/Indomart. This is due to the more variation presented by those stores. Snacks and drinks provided by shops and kiosks around it were lacking in comparison. People tend to shop on minimarkets compared to small kiosks, shops or the likes due to the diversity within the goods provided by said minimarkets, whether it be office supplies, cosmetics, snack and drinks, and daily necessities. (42% of respondents, more or less 56%).

Table 2. Reason Why Consumers Prefers Shopping in Minimarkets (Alfamart/Indomart)

No.	**Reasons**	**Total (Respondent)**	**Percentage (%)**
1	Goods Diversities	42	56,0
2	Prices	13	17,3
3	Quantity of Goods	2	2,7
4	Location	18	24,0
	Total	**75**	**100,0**

Source: Populace Questionnaire Result

3.1.2. Quantity of goods

In big grocery shopping in minimarkets, firstly consumers would look at availability of the goods they're looking for. If there's none available, then they would be looking toward shops, small kiosks, or markets in general that have it in bulks.

Table 3. Goods Availability in Minimarket (Alfamart/Indomart)

No.	**Availability**	**Total (Respondent)**	**Percentage (%)**
1	Large Amount	2	2,7
2	Available	18	24,0
3	Enough Goods	42	56,0
4	Not Available	13	17,3
	Total	75	100,0

Source: Populace Questionnaire Result

Goods available in modern market still can't quite adequately fulfill shopping needs of the population around it, especially if they would shop in bulk. This is due to the limited amount of goods that's being sold in modern markets. This is also due to drooping system from storage that would only provide limited amount of goods. This is also can be seen from total of goods normally sold in modern markets. The majority of existing modern market sold less than 5000 unit of goods each month (16 minimarkets or average of 64%).

Table 4. Amount of Goods Sold in Minimarkets (Alfamart/Indomart)

No.	**Goods Sold/Month (Pcs)**	**Total (Respondent)**	**Percentage (%)**
1	< 5.000	16	64,0
2	5.000 – 25.000	4	16,0
3	> 25.000	5	20,0
	Total	25	100,0

Source: Populace Questionnaire Result

3.1.3. Goods prices

With pricing system already regulated by company policies, the price of goods sold in modern market are relatively pricey, this is due to the goods prices is not in accordance to the general pricing, especially for specialized goods such as perfumes and underwears would be very expensive compared to shops that normally sell such items. This causes less interest of the populace to shop in modern markets, so they rather shop in shop or kiosks which would sell similar goods, but cheaper. Data from analysis table above proves that goods sold in such places were pricier than the general pricing or determined prices of shops and kiosks around it.

Table 5. Goods Prices in Minimarket (Alfamart/Indomart)

No.	Prices	Total (Respondent)	Percentage (%)
1	Very Cheap	0	0,0
2	Inexpensive	5	6,7
3	Pricy Enough	30	40,0
4	Very Expensive	40	53,3
	Total	**75**	**100,0**

Source: Populace Questionnaire Result

Table 6. Determined Prices in Minimarket (Alfamart/Indomart)

No.	Determined Prices of Goods	Total (Respondent)	Percentage (%)
1	Company Policy	25	100,
2	Market	0	0,0
3	Manufacturer	0	0,0
4	Distributor	0	0,0
	Total	**25**	**100,0**

Source: Populace Questionnaire Result

3.1.4. Demands

Goods with high number of demands were the kind of goods with plastic packages or bottles (19 respondents or average 76%). Goods with plastic or bottle packages like bottled beverages and snacks are often in demand due to it's practicality and convenience factor in carrying and throwing away it's packaging after consuming the content. But when shopping, consumers didn't shop for a whole lot, this is due to it being expensive and modern market didn't apply wholesale price onto it's products. Goods sold are counted with retail price when consumers buys the goods in bulks in modern market. But for certain goods there's wholesale prices applicable such as mineral water, instant noodles, or frying oil. But those are only applicable when there's promotion from the related modern market.

Table 7. Product Packaging in Minimarket (Alfamart/Indomart)

No.	Packaging	Total (Respondent)	Percentage (%)
1	Cardboards	3	12,0
2	Plastic-based or Bottle	19	76,0
3	Cans	3	12,0
	Total	25	100,0

Source: Populace Questionnaire Result

Table 8. Marketing Strategy in Minimarket (Alfamart/Indomart)

No.	Marketing Strategy	Total (Respondent)	Percentage (%)
1	Buy 1 Get 1 Promotion	5	20,0
2	Discounts	13	52,0
3	Weekend Special Discount	2	8,0
4	Customer Cards	2	8,0
5	Delivery Service	3	12,0
6	Cheap Bazaar	0	0,0
	Total	**25**	**100,0**

Source: Populace Questionnaire Result

3.2. Modern market service level influence toward populace movement patterns

This analysis is used to find out whether there's intervariable influence between minimarket service level variable and populace movement pattern variable, where each indicators from both variables would be associated one by one until it's found out whether there's influence or not, also to find out the magnitude of influence from each said variables' indicators.

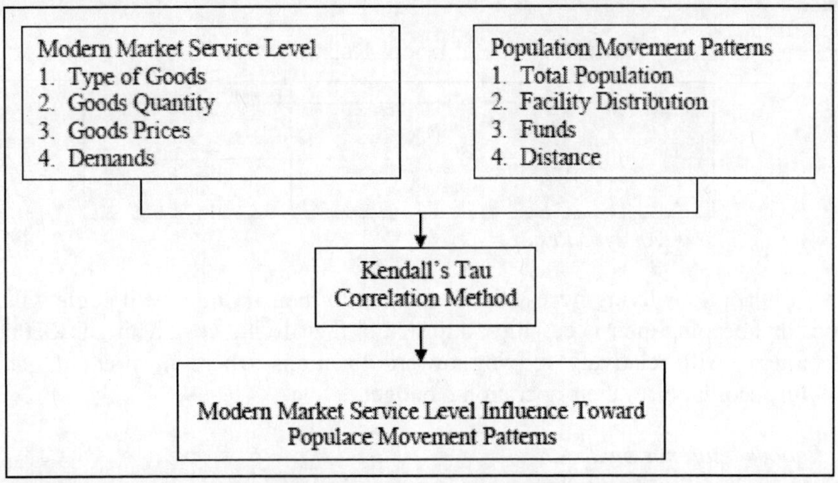

Diagram 1. Modern Market Service Level Influence Toward Populace Movement Patterns Analysis

3.2.1. Type of goods influence on population

According to Kendall's Tau analysis correlation analysis through SPSS, it is known that p-value significance score number is 0,695. This score is higher than the determined value score (0,05) in Kendall's Tahu test, so H0hypothesis is accepted and H1is rejected, which means there's No. relation between type of goods with total population. In conclusion there's no relation between type of goods sold in modern market with total population in Klojen district.

Table 9. Relation between Type of Goods with Total Population

			Jenis Barang	Jumlah Penduduk
Kendall's tau_b	Jenis Barang	Correlation Coefficient	1,000	-,037
		Sig. (2-tailed)	.	,695
		N	100	100
	Jumlah Penduduk	Correlation Coefficient	-,037	1,000
		Sig. (2-tailed)	,695	.
		N	100	100

Source: Analysis Result

3.2.2. Type of goods influence on facility distributions

According to Kendall's Tau correlation analysis through SPSS, it is known that p-value significance score is 0,839. This score is higher than the determined significancy score (0,05) in Kendall's Tau test, so H0hypothesis is accepted and H1is rejected, which means there's no relation between type of goods with facility distributions. In conclusion there's no relation between type of goods sold in modern market with facility distribution in Klojen district.

Table 10. Relation between Type of Goods with Facility Distribution

			Jenis Barang	Sebaran Fasilitas
Kendall's tau_b	Jenis Barang	Correlation Coefficient	1,000	,020
		Sig. (2-tailed)	.	,839
		N	100	100
	Sebaran Fasilitas	Correlation Coefficient	,020	1,000
		Sig. (2-tailed)	,839	.
		N	100	100

Source: Analysis Result

3.2.3. Type of goods influence on prices

According to Kendall's Tau correlation analysis through SPSS, it is known that p-value significance score is 0,000. This score is smaller than the determined significancy score (0,05) in Kendall's Tau test, so H1is accepted, which means there's relation between type of goods with facility distributions. In conclusion there's relation between type of goods sold in modern market with cost of living in Klojen district.

Table 11. Type of Goods Relation with Cost

			Jenis Barang	biaya
Kendall's tau_b	Jenis Barang	Correlation Coefficient	1,000	,410 **
		Sig. (2-tailed)	.	,000
		N	100	100
	biaya	Correlation Coefficient	,410 **	1,000
		Sig. (2-tailed)	,000	.
		N	100	100

Source: Analysis Result

People will fulfill their living necessities according to their living cost. People will look for the type of goods that become their necessities, adjusted to their living cost. This shows that population movement patterns will relatively moving toward locations where they could get their living necessities with accordance to their operational budget.

3.2.4. Type of goods influence on distance

According to Kendall's Tau correlation analysis through SPSS, it is known that p-value significance score is 0,415. This score is higher than the determined significancy score (0,05) in Kendall's Tau test, so H0hypothesis is accepted and H1is rejected, which means there's no relation between type of goods with distance. In conclusion there's no relation between type of goods that consumer needs with distance. People would move, looking for type of goods they need without looking about the distance travelled as long as they could get their hands on goods they're looking for, so distance travelled toward the location where they could get it wasn't a significant obstacle for people to shop for their life necessities.

Table 12. Relation Between Type of Goods with Distance

			Jenis Barang	Jarak
Kendall's tau_b	Jenis Barang	Correlation Coefficient	1,000	,077
		Sig. (2-tailed)	.	,415
		N	100	100
	Jarak	Correlation Coefficient	,077	1,000
		Sig. (2-tailed)	,415	.
		N	100	100

Source: Analysis Result

3.2.5. Amount of goods influence on population

According to Kendall's Tau correlation analysis through SPSS, it is known that p-value significance score is 0,561. This score is higher than the determined significancy score (0,05) in Kendall's Tau test, so H0hypothesis is accepted and H1is rejected, which means there's no relation between amount of goods with total population. In conclusion there's no relation between amount of goods sold in modern market with total population in Klojen district.

Table 13. Relation Between Amount of Goods with Total Population

			Jumah Barang	Jumlah Penduduk
Kendall's tau_b	Jumah Barang	Correlation Coefficient	1,000	,054
		Sig. (2-tailed)	.	,561
		N	100	100
	JumlahPenduduk	Correlation Coefficient	,054	1,000
		Sig. (2-tailed)	,561	.
		N	100	100

Source: Analysis Result

3.2.6. Amount of goods influence on facility distribution
According to Kendall's Tau correlation analysis through SPSS, it is known that p-value significance score is 0,005. This score is smaller than the determined significancy score (0,05) in Kendall's Tau test, so H1hypothesis is accepted, which means there's relation between amount of goods with facility distributions. In conclusion there's relation between amount of goods sold in modern market with facility distribution in Klojen district.

Table 14. Relation Between Amount of Goods with Facility Distribution

			Jumah Barang	Sebaran Fasilitas
Kendall's tau_b	Jumah Barang	Correlation Coefficient	1,000	-,272 **
		Sig. (2-tailed)	.	,005
		N	100	100
	Sebaran Fasilitas	Correlation Coefficient	-,272 **	1,000
		Sig. (2-tailed)	,005	.
		N	100	100

Source: Analysis Result

3.2.7. Amount of goods influence on costs
According to Kendall's Tau correlation analysis through SPSS, it is known that p-value significance score is 0,007. This score is smaller than the determined significancy score (0,05) in Kendall's Tau test, so H1hypothesis is accepted, which means there's relation between amount of goods with living costs. In conclusion there's a relation between amount of goods needed with cost of living. People in shopping for their living necessities has to be adjusted according to their operational cost, so their living cost isn't reaching a minus point even if the amount of goods they need to fulfill their necessities would increase. This is usually related to their priorities, people will make priorities in shopping, which item is more urgent than the others.

Table 15. Relation between Amount of Goods with Cost

			Jumah Barang	biaya
Kendall's tau_b	Jumah Barang	Correlation Coefficient	1,000	,262 **
		Sig. (2-tailed)	.	,007
		N	100	100
	biaya	Correlation Coefficient	,262 **	1,000
		Sig. (2-tailed)	,007	.
		N	100	100

Source: Analysis Result

3.2.8. Amount of goods influence on distance
According to Kendall's Tau correlation analysis through SPSS, it is known that p-value significance score is 0,577. This score is higher than the determined significancy score (0,05) in Kendall's Tau test, so H0hypothesis is accepted and H1 rejected, which means there's no relation between amount of goods with distance. In conclusion there's no relation between amount of goods sold in modern market with distance toward said market that has the goods.

Table 16. Relation Between Amount of Goods with Distance

			Jumah Barang	Jarak
Kendall's tau_b	Jumah Barang	Correlation Coefficient	1,000	-,053
		Sig. (2-tailed)	.	,577
		N	100	100
	Jarak	Correlation Coefficient	-,053	1,000
		Sig. (2-tailed)	,577	.
		N	100	100

Source: Analysis Result

People are more than willing to travel tremendous distances away if to get their hands on goods they needed in bulks. Distance wasn't that high of an obstacle for people in shopping for fulfilling their living necessities, due to this related to people's satisfaction level in fulfilling their living necessities. Even though the distance travelled is of significant range, as long as the amount of goods met the necessary total needed, it's not that high of an obstacle to tackle.

3.2.9. Price of goods influence on population
According to Kendall's Tau correlation analysis through SPSS, it is known that p-value significance score is 0,877. This score is higher than the determined significancy score (0,05) in Kendall's Tau test, so H0hypothesis is accepted and H1 rejected, which means there's no relation between goods prices with facility distribution. In conclusion there's no relation between price of goods offered in modern market with total population in Klojen district.

Table 17. Relation between goods prices with total population

			Harga Barang	Jumlah Penduduk
Kendall's tau_b	Harga Barang	Correlation Coefficient	1,000	-,014
		Sig. (2-tailed)	.	,877
		N	100	100
	Jumlah Penduduk	Correlation Coefficient	-,014	1,000
		Sig. (2-tailed)	,877	.
		N	100	100

Source: Analysis Result

3.2.10. Price of goods influence on facility distribution
According to Kendall's Tau correlation analysis through SPSS, it is known that p-value significance score is 0,201. This score is higher than the determined significancy score (0,05) in Kendall's Tau test, so H0hypothesis is accepted and H1 rejected, which means there's no relation between goods prices with facility distribution. In conclusion there's no relation between price of goods offered in modern market with facility distribution in Klojen District.

Table 18. Relation between goods prices with facility distribution

			Harga Barang	Sebaran Fasilitas
Kendall's tau_b	Harga Barang	Correlation Coefficient	1,000	,121
		Sig. (2-tailed)	.	,201
		N	100	100
	Sebaran Fasilitas	Correlation Coefficient	,121	1,000
		Sig. (2-tailed)	,201	.
		N	100	100

Source: Analysis Result

3.2.11. Price of Goods Influence on Costs
According to Kendall's Tau correlation analysis through SPSS, it is known that p-value significance score is 0,000. This score is smaller than the determined significancy score (0,05) in Kendall's Tau test, so H1hypothesis is accepted, which means there's a relation between price of goods with costs. In conclusion there's a relation between price of goods offered in modern market with living costs.

Tabel 19. Relation between goods price and cost

			Harga Barang	biaya
Kendall's tau_b	Harga Barang	Correlation Coefficient	1,000	-1,000 **
		Sig. (2-tailed)	.	,000
		N	100	100
	biaya	Correlation Coefficient	-1,000 **	1,000
		Sig. (2-tailed)	,000	.
		N	100	100

Source: Analysis Result

People's living cost is closely related with their income level. In attempt to fulfil their living necessities, populace would move to look for location that could give them reciprocal in form of income big enough to fulfill their various living needs. With the increasing price of goods needed to survive, people would work hard so that their income would cover their living cost.

3.2.12. Price of goods influence on distance

According to Kendall's Tau correlation analysis through SPSS, it is known that p-value significance score is 0,544. This score is higher than the determined significancy score (0,05) in Kendall's Tau test, so H0hypothesis is accepted and H1 rejected, which means there's no relation between goods prices with distance travelled. In conclusion there's no relation between price of goods offered in modern market with distance travelled to get said goods.

Table 20. Relation Between Price of Goods with Distance

			Harga Barang	Jarak
Kendall's tau_b	Harga Barang	Correlation Coefficient	1,000	,056
		Sig. (2-tailed)	.	,544
		N	100	100
	Jarak	Correlation Coefficient	,056	1,000
		Sig. (2-tailed)	,544	.
		N	100	100

Source: Analysis Result

3.2.13. Demands influence on population

According to Kendall's Tau correlation analysis through SPSS, it is known that p-value significance score is 0,878. This score is higher than the determined significancy score (0,05) in Kendall's Tau test, so H0hypothesis is accepted and H1 rejected, which means there's no relation between demands with facility distribution. In conclusion there's no relation between demands in modern market with facility distribution in Klojen District.

Table 21. Relation Between Demands and Population

			Jumlah Permintaan	Jumlah Penduduk
Kendall's tau_b	Jumlah Permintaan	Correlation Coefficient	1,000	-,014
		Sig. (2-tailed)	.	,878
		N	100	100
	Jumlah Penduduk	Correlation Coefficient	-,014	1,000
		Sig. (2-tailed)	,878	.
		N	100	100

Source: Analysis Result

3.2.14. Demands influence on facility distribution

According to Kendall's Tau correlation analysis through SPSS, it is known that p-value significance score is 0,847. This score is higher than the determined significancy score (0,05) in Kendall's Tau test, so H0hypothesis is accepted and H1 rejected, which means there's no relation between demands with facility distribution. In conclusion there's no relation between demands in modern market with facility distribution in Klojen District.

Table 22. Relation Between Demands and Facility Distribution

			Jumlah Permintaan	Sebaran Fasilitas
Kendall's tau_b	Jumlah Permintaan	Correlation Coefficient	1,000	-,019
		Sig. (2-tailed)	.	,847
		N	100	100
	Sebaran Fasilitas	Correlation Coefficient	-,019	1,000
		Sig. (2-tailed)	,847	.
		N	100	100

Source: Analysis Result

3.2.15. Demands influence on cost

According to Kendall's Tau correlation analysis through SPSS, it is known that p-value significance score is 0,417. This score is higher than the determined significancy score (0,05) in Kendall's Tau test, so H0hypothesis is accepted and H1 rejected, which means there's no relation between demands with cost. In conclusion there's no relation between demands in modern market with people's living cost in fulfilling their needs.

Table23. Relation between Demands and Costs

			Jumlah Permintaan	biaya
Kendall's tau_b	Jumlah Permintaan	Correlation Coefficient	1,000	-,079
		Sig. (2-tailed)	.	,417
		N	100	100
	biaya	Correlation Coefficient	-,079	1,000
		Sig. (2-tailed)	,417	.
		N	100	100

Source: Analysis Result

3.2.16. *Demands influence on distance*

According to Kendall's Tau correlation analysis through SPSS, it is known that p-value significance score is 0,936. This score is higher than the determined significancy score (0,05) in Kendall's Tau test, so H0hypothesis is accepted and H1 rejected, which means there's no relation between demands with distance. In conclusion there's no relation between demands in modern market with distance travelled by people to fulfill their living needs.

Table 24. Relation between Demands and Costs

			Jumlah Permintaan	Jarak
Kendall's tau_b	Jumlah Permintaan	Correlation Coefficient	1,000	,008
		Sig. (2-tailed)	.	,936
		N	100	100
	Jarak	Correlation Coefficient	,008	1,000
		Sig. (2-tailed)	,936	.
		N	100	100

Source: Analysis Result

3.3. *Early Conclusion*

Early conclusions to be pulled from minimarket service level (Alfamart/Indomart) analysis presented above is as follows:

Table 25. Minimarket Service Level Influence Toward Population Movement Patterns Analysis

No	*Influential Variables*	*p-value*	*Correlation Coefficient*	*Analysis*
1	Type of Goods with Population	0,695	0,037	No Influence
2	Type of Goods with Facility Distribution	0,839	0,020	No Influence
3	Type of Goods with Cost	0,000	0,410	Influenced
4	Type of Goods with Distance	0,415	0,077	No Influence
5	Amount of Goods with Population	0,561	0,054	No Influence
6	Amount of Goods with Facility Distribution	0,005	0,272	Influenced
7	Amount of Goods with Cost	0,007	0,262	Influenced
8	Amount of Goods with Distance	0,577	0,053	No Influence
9	Price of Goods with Population	0,877	0,014	No Influence
10	Price of Goods with Facility Distribution	0,201	0,121	No Influence
11	Price of Goods with Cost	0,000	1,000	Influenced
12	Price of Goods with Distance	0,544	0,560	No Influence
13	Demands with Population	0,878	0,014	No Influence
14	Demands with Facility Distribution	0,847	0,019	No Influence
15	Demands with Cost	0,417	0,079	No Influence
16	Demands with Distance	0,936	0,008	No Influence
	Average	0,487	0,188	No Influence

Source: Analysis Result

According to table above it is perceptible that :

1. If judged from it's 0,486 p-value, minimarket service level isn't affecting or there's no interplay relation with population movement pattern. Due to it's p-value score is bigger than significancy score within this research, which is 0,05.
2. Population movement patterns is only affected by type of goods needed, amount of goods bought by consumers, and goods prices. This is due to those indicators has significancy level of < 0,05.
   Which those influential indicators has moderate influence (*Correlation Coefficient* ; 0,486).
3. Modern market (Alfamart/Indomart) level of service in Klojen district "isn't satisfactory" or around 52,8%. This is due to few indicators not meeting consumer's expectations (goods availability,
   prices, marketing stategies) there's even indicators that's at odds with consumers (type of goods bought, determined goods prices, and sales climb.
4. Minimarket level of service that doesn't meet consumers expectation (average of 29,3%) is due to the determined pricing of goods sold is through company policy, so that the goods sold is pricier than general prices around said minimarket.
5. Level of service adjacent to consumers expectations (average of 61,7%) is due to the type of goods sold in minimarkets has a wide array of variations compared to goods sold in shops and kiosks etc,
   reason for shopping in minimarket (Alfamart/Indomart) by consumers, and goods sold according to companies monthly target.

4. Conclusion

Minimarket level of service is used to determine how big is the service level of minimarkets around Klojen district toward fulfilling the living necessities of the population around it. That aside, modern market level of service is also used to determine how far is the reach of it's service toward consumers. There's several indicators to determine minimarket level of service: type of goods, quantity of goods, pricing, and demands

1. If type of goods needed varies then living cost is increasingly high and on the other side if operational cost in household is big then the type of goods needed in modern market would increase. In other words, people would fulfill their living necessities in accordance to their living costs. People would also look for the type of goods that would be necessary for their living in accordance to their living costs. This shows that population movement patterns would be relative to the location where they could get their hands on all the necessary type of goods needed for their living necessities with taking their operational costs into account.
2. If a major amount of goods sold in modern market then the facility distribution isn't equally distributed and in reverse if the facility around the modern market is equally distributed then the amount of goods sold in modern market won't be so high.
3. People are shopping for their living necessities in accordance to their operational budget, so that their living cost didn't exceed their income even if the amount of goods needed for fulfilling their living needs would increase. This is usually related to their priorities in fulfilling said needs, people would make priorities in spending, which goods is truly needed and which isn't quite so needed for the moment.
4. Population's living cost is tightly related to their income level. In fulfilling their living needs, people would move to look for a location that could give them income that could be used for shopping for their living needs. With the increasing price of goods needed, then people would look for work so their income could cover their living cost so the necessary goods they needed could be bought.
5. If seen from it's p-value 0,487, then minimarket level of service didn't influence or there's no interrelation with population movement patterns. Due to p-value is higher than significancy score within this research which is 0,05.

6. Population movemen pattern is only affected by the type of goods needed, amount of goods bought by consumers, and price. This is due to those indicators have significancy score of < 0,05 where the influencing indicators has medium effects (*Correlation Coefficient* ; 0,486).
7. Modern market (Alfamart/Indomart) level of service in Klojen district "isn't satisfactory" or around 52,8%. This is due to few indicators not meeting consumer's expectations (goods availability, prices, marketing stategies) there's even indicators that's at odds with consumers (type of goods bought, determined goods prices, and sales climb).
8. Minimarket level of service that doesn't meet consumers expectation (average of 29,3%) is due to the determined pricing of goods sold is through company policy, so that the goods sold is pricier than general prices around said minimarket.
9. Level of service adjacent to consumers expectations (average of 61,7%) is due to the type of goods sold in minimarkets has a wide array of variations compared to goods sold in shops and kiosks etc, reason for shopping in minimarket (Alfamart/Indomart) by consumers, and goods sold according to companies monthly target.

5. Bibliography

[1] Setiawan, Jeri, dkk. 2012. "Pengaruh Keberadaan Minimarket Terhadap Pendapatan Pedagang Kelontong Dikelurahan Klender Kecamatan Duren Sawit Jakarta Timur". Jakarta. SPATIAL Wahana Komunikasi dan Informasi Geografi. Vol. 10. No.1(pg. 1-7).

[2] Wijayanti, Pardiana dan Wiratno. 2011. "Analisis Pengaruh Perubahan Keuntungan Usaha Warung tradisional Dengan Munculnya Minimarket (Studi Kasus Di Kecamatan Pedurungan Kota Semarang)".Undip (pg. 71-85).

[3] Widya, Utami Christina, 2006, Manajemen Ritel, Strategi dan Implementasi Ritel Modern, Jakarta, Salemba Empat

[4] Widya, Utami Christina. 2008. Manajemen Barang Dagang dalam Bisnis Ritel. Malang: Bayu media Publishing.

[5] Tjiptono, Fandy. 2008. Strategi Pemasaran, Edisi 3. Yogyakarta: ANDI.

[6] Sarwono, Jonathan. 2006. Metode Penelitian Kuantitatif dan Kualitatif. Yogyakarta. Graha Ilmu.

[7] Tarigan, Robinson. 2005. Perencanaan Pembangunan Wilayah. Jakarta. Penerbit Bumi Aksara

[8] Republik Indonesia. No. 107/MPP/Kep/2/1998 Keputusan Menteri Perindustrian dan Perdagangan R.I. Ketentuan dan Tata Cara Pemberian Izin Usaha Pasar Modern. Jakarta. Menperindag R.I.

[9] Republik Indonesia. No. 112 Tahun 2007. Peraturan Presiden Penataan Dan Pembinaan Pasar Tradisional, Pusat Perbelanjaan Dan Toko Modern. Jakarta. Sekretaris Negara.

IOP Conf. Series: Earth and Environmental Science **340** (2019) 012041 doi:10.1088/1755-1315/340/1/012041

Analysis of spatial parameter formulation in the effect of Green House Gas (GHG) emissions on food security in Surabaya

U F Kurniawati[1,2] **and H Idajati**[1,2]

[1] Department Urban and Regional Planning
[2] Institut Teknologi Sepuluh Nopember, Surabaya 60111, Indonesia

Email: ummi_fk@urplan.its.ac.id; fa.ummi99@gmail.com

Abstract. Surabaya is an urban area that has a population growth rate increasing every year, which is 0.52% in 2010-2013 and increased to 0.55% in 2013-2014. Increasing population growth has an impact on increasing urban food availability. Identification of food availability in Surabaya consists of agriculture and livestock sectors, namely rice production (X1), corn production (X2), cassava production (X3), beef cattle production (X4), dairy cow production (X5), buffalo production (X6), goat production (X7), horse production (X8), and poultry production (X9). Data reduction in sub-districts that have no contribution to the agricultural and livestock sector emissions results in 15 sub-district input data that have a contribution. The spatial regression process. Significance test on the independent variable shows global variables (G), namely rice production (X1), corn production (X2), cassava production (X3), dairy cow production (X5), buffalo production (X6), goat production (X7), and poultry production (X9). While the independent variables that show local variables (L) are beef cattle production (X4) and horse production (X8).

1. Introduction

Global climate change is caused by increased emissions of Green House Gases (GHG) due to various activities that encourage an increase in the earth's temperature. Increasing the temperature of the earth's surface is generated by the presence of sunlight radiation into the Earth's atmosphere, then part of this light changes into heat energy in the form of infrared rays absorbed by the air and the surface of the earth [1]. To resolve this issue, at the 2002 Earth Summit in Rio, was born climate change convention to stabilize the concentration of greenhouse gases in the atmosphere at a level does not endanger the climate system [2].

Climate is the main element of the metabolic system and plant physiology, so that global climate change will adversely affect the sustainability of agricultural development [3]. The agricultural sector, especially the food crops sector, is the sector that has the most impact on climate change. There are three main factors related to global climate change that impact on the agricultural sector are : changes in rainfall patterns, increased extreme climate events (floods and droughts), and increased air temperature [4]. Climate change has an impact on the food system in a number of ways including direct impacts on crop production to changes in markets, prices of food products, and supply chain infrastructure [5]. The largest distribution of GHGs in Indonesia are carbon dioxide (CO_2), methane

(CH_4) and dinitrogenoxide (N_2O) [6]. In addition to CO_2, the second largest greenhouse gas that contributes to global warming in Indonesia is CH_4, the majority of which come from the agricultural sector, including livestock activities. The disruption of the food system due to the increase in urban temperature as a result of the increase in GHG emissions has a negative impact on the food security of the population. During the period 1981-1990, every district in Indonesia each year experienced a decline in rice production of 100,000 tons and in the period 1992-2000, the amount of this decline increased to 300,000 tons [7].

Surabaya is an urban area that has a population growth rate increasing every year, which is 0.52% in 2010-2013 and increased to 0.55% in 2013-2014 [8]. Increasing population growth has an impact on increasing urban food availability. Food availability in Surabaya consists of agriculture and livestock sectors. Agriculture sector consist of rice, corn, cassava and sweet potatoes. Whereas livestock sector consist of meat, eggs, milk and fish. Food supply activities that have a contribution to GHG emissions are activities arising from the agricultural sector including livestock. In this study will be studied further about how the effects of GHG emissions from the agricultural sector and livestock on food availability in Surabaya by analyze spatial parameter formulation from the GHG emissions.

2. Identification of food availability in Surabaya

Identification of food availability in Surabaya City through descriptive statistics from the data of the Surabaya Food and Agriculture Security Office. Food availability in Surabaya is divided into 2 sources, namely vegetable food production and animal food production. Land limitations in the city of Surabaya and the tendency of the characteristics of urban areas narrow the production of plant and animal foods in the districts of Surabaya.

The amount of production of the availability of vegetable and animal food in the city of Surabaya in 2016 was shown by the production of vegetable food, consist of: rice production (X1); corn production (X2); and cassava production (X3). Whereas animal food production consists of: beef cattle production (X4); dairy cattle production (X5); buffalo production (X6); goat production (X7); horse production (X8); and poultry production (X9). From the overall data in each district, the availability of vegetable food is dominant in the type of rice production. While the availability of animal food is dominant in poultry production. Food availability in Surabaya is concentrated on the West and South Surabaya regions, such as the Lakarsantri District and Karang Pilang District.

3. Calculate production of Green House Gas Emmisions (agricultural and livestock sector) in Surabaya

GHG emission production in Surabaya is calculated from 2 sectors, namely the agricultural sector and the livestock sector. Emissions from the agricultural sector are calculated from the total area of agriculture and emissions from consumption of urea fertilizer. This emission is obtained by multiplying the paddy field with the emission factor of rice field agricultural activity, which is 1.3 tons CH_4 / Ha of paddy field (Calculator GHG v1.0, 2012). For fertilizer consumption, to get the total emissions multiplied by 0.20 tons of CO_2/ton of fertilizer consumption. The calculation of total emissions from agricultural activities in the each district as follows:

Table 1. GHG emission production from agricultural sector

Districts	Agriculture Area (Ha)	Emission Factor (CH_4 / Ha)	Urea Consumption (Kg)	Emission Factor (CO_2 / ton)	Emission Production (Kg/year)
Asemrowo	0	0	0	0	0
Benowo	217	282,1	42.120	8,424	290.524,0
Bubutan	0	0	0	0	0
Bulak	109	141,7	51.041	10,2082	151.908,2
Dukuh Pakis	0	0	0	0	0
Gayungan	9	11,7	1.800	0,36	12.060,0

Districts	Agriculture Area (Ha)	Emission Factor (CH_4 / Ha)	Urea Consumption (Kg)	Emission Factor (CO_2 / ton)	Emission Production (Kg/year)
Genteng	0	0	0	0	0
Gubeng	0	0	0	0	0
Gunung Anyar	5	6,5	4.950	0,99	7.490,0
Jambangan	4	5,2	0	0	5.200,0
Karang Pilang	85	110,5	21.883,5	4,3767	114.876,7
Kenjeran	0	0	0	0	0
Krembangan	0	0	0	0	0
Lakarsantri	817	1.062,1	388.671	77,7342	1.139.834,2
Mulyorejo	20	26	4.950	0,99	26.990,0
Pabean Cantikan	0	0	0	0	0
Pakal	352	457,6	172.360	34,472	492.072,0
Rungkut	7	9,1	5200	1,04	10.140,0
Sambikerep	361	469,3	158.644	31,7288	501.028,8
Sawahan	0	0	0	0	0
Semampir	0	0	0	0	0
Simokerto	0	0	0	0	0
Sukolilo	48	62,4	7.350	1,47	63.870,0
Sukomanunggal	0	0	0	0	0
Tambaksari	0	0	0	0	0
Tandes	5	6,5	3.200	0,64	7.140,0
Tegalsari	0	0	0	0	0
Tenggilis Mejoyo	0	0	0	0	0
Wiyung	76	98,8	17.100	3,42	102.220,0
Wonocolo	8	10,4	1.800	0,36	10.760,0
Wonokromo	0	0	0	0	0

. Source: Analysis, 2018

Districts that do not have agricultural land, have 0 Kg emission. Whereas the districts with the most GHG emission production from the agricultural sector are Lakarsantri District with 1,139,834.2 Kg of emissions / year. In addition, the production of GHG emissions from the livestock sector is obtained by multiplying the total livestock by livestock emission factors from the GHG Calculator. Each cow is multiplied by the emission factor of 1 Kg / head, each dairy cow is multiplied by an emission factor of 31 Kg / head, each buffalo is multiplied by an emission factor of 2 Kg / head, each goat is multiplied by an emission factor of 0.2 Kg / tail, each horse is multiplied by an emission factor of 2.19 Kg / head, and each poultry is multiplied by the emission factor of 0.03 Kg / head. Following are the results of the calculation of emissions from the livestock sector in the city of Surabaya.

Table 2. GHG emission production from livestoct sector

Districts	Beef Cattle Emissions (Kg)	Dairy Cow Emissions (Kg)	Buffalo Emissions (Kg)	Goat Emissions (Kg)	Horse Emissions (Kg)	Poultry Emissions (Kg)	Emissions Total (Kg/year)
Asemrowo	0	0	0	0	0	9,69	9,69
Benowo	0	0	0	0	0	7,32	7,32
Bubutan	0	0	0	0	0	10,68	10,68
Bulak	0	0	0	0	0	18	18

Districts	Beef Cattle Emissions (Kg)	Dairy Cow Emissions (Kg)	Buffalo Emissions (Kg)	Goat Emissions (Kg)	Horse Emissions (Kg)	Poultry Emissions (Kg)	Emissions Total (Kg/year)
Dukuh Pakis	0	0	0	2,4	0	12,06	14,46
Gayungan	0	0	0	0	0	8,43	8,43
Genteng	0	0	0	0	0	5,1	5,1
Gubeng	0	2976	0	0	0	1,44	2.977,44
Gunung Anyar	0	0	0	9	0	8,82	17,82
Jambangan	0	0	0	4	2,19	18,66	24,85
Karang Pilang	64	0	16	43,6	6,57	14,61	144,78
Kenjeran	0	0	0	0	0	4,98	4,98
Krembangan	0	0	0	0	0	16,26	16,26
Lakarsantri	25	682	44	52	0	58,05	861,05
Mulyorejo	0	0	0	5,8	0	5,67	11,47
Pabean Cantikan	0	0	0	0	0	2,34	2,34
Pakal	21	775	0	10	0	3,33	809,33
Rungkut	0	0	0	0	0	41,07	41,07
Sambikerep	16	0	0	9,6	0	2,34	27,94
Sawahan	0	0	0	7,8	0	11,58	19,38
Semampir	0	0	0	0	0	8,28	8,28
Simokerto	0	0	0	0	0	3,3	3,3
Sukolilo	0	0	0	5,4	0	33	38,4
Sukomanunggal	0	0	0	0	0	36,12	36,12
Tambaksari	0	0	0	0	0	27,48	27,48
Tandes	6	62	0	8,8	0	10,08	86,88
Tegalsari	0	0	0	0	0	12,09	12,09
Tenggilis Mejoyo	0	0	0	0	0	17,4	17,4
Wiyung	8	248	0	2	0	3,96	261,96
Wonocolo	0	6.975	0	0	0	15,87	6.990,87
Wonokromo	0	0	0	0	0	9,33	9,33

Source: Analysis, 2018

Production of livestock GHG emissions in the Surabaya is shown by the large number of livestock in each district. The most dominant livestock contribution to emissions in each -district is the type of poultry and the least is the horse. District with the most GHG emission production from the livestock sector are Wonocolo District with 6,990.87 Kg of emissions.

From the two sectors that show agricultural and livestock activities, the two emissions are summed up to get the total production of agricultural sector GHG emissions in the Surabaya.

Table 3. GHG Emissions from livestock and agricultural sectors

Districts	Agricultural Emission (Kg)	Livestock Emission (Kg)	GHG Emission Total (Kg)
Asemrowo	0	9,69	9,69
Benowo	290.524,0	7,32	290.531,32
Bubutan	0	10,68	10,68
Bulak	151.908,2	18	151.926,20
Dukuh Pakis	0	14,46	14,46

IOP Conf. Series: Earth and Environmental Science **340** (2019) 012041 doi:10.1088/1755-1315/340/1/012041

Districts	Agricultural Emission (Kg)	Livestock Emission (Kg)	GHG Emission Total (Kg)
Gayungan	12.060,0	8,43	12.068,43
Genteng	0	5,1	5,10
Gubeng	0	2.977,44	2.977,44
Gunung Anyar	7.490,0	17,82	7.507,82
Jambangan	5.200,0	24,85	5.224,85
Karang Pilang	114.876,7	144,78	115.021,48
Kenjeran	0	4,98	4,98
Krembangan	0	16,26	16,26
Lakarsantri	1.139.834,2	861,05	1.140.695,25
Mulyorejo	26.990,0	11,47	27.001,47
Pabean Cantikan	0	2,34	2,34
Pakal	492.072,0	809,33	492.881,33
Rungkut	10.140,0	41,07	10.181,07
Sambikerep	501.028,8	27,94	501.056,74
Sawahan	0	19,38	19,38
Semampir	0	8,28	8,28
Simokerto	0	3,3	3,30
Sukolilo	63.870,0	38,4	63.908,40
Sukomanunggal	0	36,12	36,12
Tambaksari	0	27,48	27,48
Tandes	7.140,0	86,88	7.226,88
Tegalsari	0	12,09	12,09
Tenggilis Mejoyo	0	17,4	17,40
Wiyung	102.220,0	261,96	102.481,96
Wonocolo	10.760,0	6.990,87	17.750,87
Wonokromo	0	9,33	9,33

Source: Analysis, 2018

4. Spatial parameter formulation in the effect of Green House Gas emmision on food security in Surabaya

Analysis of spatial parameter formulation in the influence of GHG emission production on the food availability of Surabaya City was carried out by regression analysis. Independent variables used are food availability in each district from the agricultural and livestock sectors. Independent variables consist of rice production (X1), corn production (X2), and cassava production (X3). While animal food production consists of beef cattle production (X4), dairy cattle production (X5), buffalo production (X6), goat production (X7), horse production (X8), and poultry production (X9). While the dependent variable used is the total emissions from the agricultural and livestock sectors. Determination of spatial parameters for GHG variables is carried out on district data which has the contribution of the agriculture and livestock sectors. After getting the data that has been reduced, a regression analysis is performed to formulate variable parameters.

4.1. Data input cluster

The following data is the input of the cluster analysis process with 10 clusters. Then, it is obtained the districts that has a contribution to GHG through the production of the agricultural and livestock sectors.

Table 4. Dependent dan Independent Variables

Districts	X1	X2	X3	X4	X5	X6	X7	X8	X9	Y
Asemrowo	-	-	-	-	-	-	-	-	323	9,69
Benowo	1041,16	87,91	-	-	-	-	-	-	244	290531,3
Bubutan	-	-	-	-	-	-	-	-	356	10,68
Bulak	583,73	-	-	-	-	-	-	-	600	151926,2
Dukuh Pakis	-	-	-	-	-	-	12	-	402	14,46
Gayungan	45,84	-	-	-	-	-	-	-	281	12068,43
Genteng	-	-	-	-	-	-	-	-	170	5,1
Gubeng	-	-	-	-	96	-	-	-	48	2977,44
Gunung Anyar	24,61	-	-	-	-	-	45	-	294	7507,82
Jambangan	20,92	-	-	-	-	-	20	1	622	5224,85
Karang Pilang	424,57	14,65	-	64	-	8	218	3	487	115021,5
Kenjeran	-	-	-	-	-	-	-	-	166	4,98
Krembangan	-	-	-	-	-	-	-	-	542	16,26
Lakarsantri	4145,7	58,61	-	25	22	22	260	-	1935	1140695
Mulyorejo	90,16	-	-	-	-	-	29	-	189	27001,47
Pabean Cantikan	-	-	-	-	-	-	-	-	78	2,34
Pakal	1773,44	107,45	-	21	25	-	50	-	111	492881,3
Rungkut	34,26		-	-	-	-	-	-	1369	10181,07
Sambikerep	1719,39	234,43	-	16	-	-	48	-	78	501056,7
Sawahan	-	-	-	-	-	-	39	-	386	19,38
Semampir	-	-	-	-	-	-	-	-	276	8,28
Simokerto	-	-	-	-	-	-	-	-	110	3,3
Sukolilo	225,51	-	-	-	-	-	27	-	1100	63908,4
Sukomanunggal	-	-	-	-	-	-	-	-	1204	36,12
Tambaksari	-	-	-	-	-	-	-	-	916	27,48
Tandes	25,68	-	-	6	2	-	44	-	336	7226,88
Tegalsari	-	-	-	-	-	-	-	-	403	12,09
Tenggilis Mejoyo	-	-	-	-	-	-	-	-	580	17,4
Wiyung	388,25	-	8,69	8	8	-	10	-	132	102482
Wonocolo	25,36	-	-	-	225	-	-	-	529	17750,87
Wonokromo	-	-	-	-	-	-	-	-	311	9,33

Source: Analysis, 2018

From the data above, it shows that there are several districts that do not have a basis in both aspects of agriculture and livestock emissions, so they need to be reduced. The reduction process is based on cluster analysis by forming the tendency of 10 clusters.

Table 5. District Reduction Result

Districts	Clustering	Result
Asemrowo	1	Reduce
Benowo	2	No
Bubutan	1	Reduce
Bulak	3	No
Dukuh Pakis	1	Reduce

IOP Conf. Series: Earth and Environmental Science **340** (2019) 012041 doi:10.1088/1755-1315/340/1/012041

Districts	Clustering	Result
Gayungan	1	Reduce
Genteng	1	Reduce
Gubeng	1	Reduce
Gunung Anyar	1	Reduce
Jambangan	4	No
Karang Pilang	5	No
Kenjeran	1	Reduce
Krembangan	4	No
Lakarsantri	6	No
Mulyorejo	1	Reduce
Pabean Cantikan	1	Reduce
Pakal	7	No
Rungkut	8	No
Sambikerep	7	No
Sawahan	1	Reduce
Semampir	1	Reduce
Simokerto	1	Reduce
Sukolilo	9	No
Sukomanunggal	8	No
Tambaksari	9	No
Tandes	1	Reduce
Tegalsari	1	Reduce
Tenggilis Mejoyo	4	No
Wiyung	10	No
Wonocolo	4	No
Wonokromo	1	Reduce

Source: Analysis, 2018

There are 15 districts data that are not reduced or have emission contributions in the agriculture and livestock sectors, namely Benowo District, Bulak District, Jambangan Subdistrict, Karang Pilang District, Krembangan District, Lakarsantri District, Pakal District, Rungkut District, Sambikerep District, Sukolilo District, Sukomanunggal District , Tambaksari District, Tenggilis Mejoyo District, Wiyung District, and Wonocolo District.

4.2. Regression spatial analysis

Spatial regression analysis with GWR4 software to obtain the formulation of spatial parameters from variables that have an effect on GHG in Surabaya City. The regression model used is Gaussian. For the first step, all independent variables fall into the Local (L) category. Kernel type used is Adaptive bi-square. Identification of variable X included in global (G) and local (L) parameters through t test. With a value of df = n - x = 15 - 9 = 6 and a probability of 15%. Identification provisions with t test: (1) There is no Global influence / variable (G): - t table < t value <t table; (2) There is influence / Local (L) variable: -t value <-t table or t value> t table.

Table 6. Spatial regression on independent variables with t-test

Variables	t value	t tabel (15%)	Global/Local
X1 : Rice production	0,624059	1,65017	Variabel Global (G)

Variables	t value	t tabel (15%)	Global/Local
X2 : Corn production	-0,558852		Variabel Global (G)
X3 : Cassava oroduction	0,026812		Variabel Global (G)
X4 :Beef cattle production	-2,812646		Variabel Local (L)
X5 : Dairy cow production	-0,457341		Variabel Global (G)
X6 : Buffalo production	-0,488448		Variabel Global (G)
X7 : Goat production	1,118236		Variabel Global (G)
X8 : Horse production	1,839031		Variabel Local (L)
X9 : Poultry production	-0,779965		Variabel Global (G)

Spatial regression process with Gaussian regression model and Adaptive bi-square Kernel type produces significant values. Significance test on the independent variable shows global variables (G), namely rice production (X1), corn production (X2), cassava production (X3), dairy cow production (X5), buffalo production (X6), goat production (X7), and poultry production (X9). While the independent variables that show local variables (L) are beef cattle production (X4) and horse production (X8).

5. Conclusion

The process of formulating the spatial parameters of the variables that have an effect on GHG in the Surabaya by reducing the data in districts that have no contribution to the agricultural and livestock emissions of the. There are 15 district input data that have contributed to the emissions of the agriculture and livestock sectors. The spatial regression process with GWR4 software uses a Gaussian regression model. The selected kernel type is Adaptive bi-square. Significance test on the independent variable shows global variables (G), namely rice production (X1), corn production (X2), cassava production (X3), dairy cow production (X5), buffalo production (X6), goat production (X7), and poultry production (X9). While the independent variables that show local variables (L) are beef cattle production (X4) and horse production (X8).

6. References

[1] IPCC 2001 *Climate Change 2001 : The Scientific Basis* Contribution of Working Group I to the Third Assessment Report of the Intergovernmental Panel on Climate Change [Houghton, J.T., Y. Ding, D.J. Griggs, M. Noguer, P.J. van der Linden, X. Dai, K. Maskell, and C.A. Johnson (eds.)] (Cambridge, United Kingdom and New York : Cambridge University Press) 881pp

[2] Kementerian Lingkungan Hidup 2012 *Status Lingkungan Hidup Indonesia*. (Jakarta: Kementerian Lingkungan Hidup RI)

[3] Las I 2007 *Strategi dan Inovasi Teknologi Pertanian Menghadapi Perubahan Iklim Global* (Badan Litbang Pertanian) presentation

[4] Salinger M J 2005 *Climate Variability and Change; Past, Present, and Future Over View* (Climate Change 70 : 9-29)

[5] Gregory, P J, Ingram J S I, and Brklacich M 2005 *Climate Change and Food Security* (Philosophical Transactions of The Royal Society) 360 pp 2139 – 2148

[6] PEACE 2017 *Indonesia and climate change : current status and policies* (PT. Pelangi Energi Abadi Citra Enviro)

[7] United Nation Development Programme (UNDP) 2007 *Human Development Report 2007/2008* (New York : United Nations Development Programme)

[8] BPS 2018 *Surabaya Dalam Angka* (Surabaya : Badan Pusat Statistik)

IOP Conf. Series: Earth and Environmental Science **340** (2019) 012042 doi:10.1088/1755-1315/340/1/012042

Mapping of triple helix cooperation to develop local economic activation in Kampong Dolly, Surabaya

V K Siswanto[1], Sardjito[1] and F E Nugroho[1]

[1] Department of Urban and Regional Planning, Institut Teknologi Sepuluh Nopember, Surabaya, Indonesia

Email: velykazu@gmail.com

Abstract. Kampong Dolly was popular back in the days when this district in Surabaya well-known as biggest prostitution area in South East Asia. It was not an easy job to drastically change from prostitute area into tourism destination. This collaboration is very important to do so that the potential of Kampong Dolly can be used as effectively and efficiently as possible in order to improve the welfare of the community. Triple Helix collaboration designed as a capital for developing a Local Economic Activation to support the branding of Dolly Tourism Village, improve community welfare and enhance the competitiveness of the creative industries in Surabaya City to national and international levels. The government uses SMEs scheme to employ ex-prostitute workers, therefore, it need strong commitment and hard work from many stakeholders. SMEs in Kampong Dolly has collaborated with the government, private sector and academics. However, the role of the government and academics still very weak in developing resources and technology. On the other hand, as for the collaboration of the private sector, low investment in terms of technology, which was developed to improve the advantages possessed by SMEs in Dolly Village.

1. Introduction

Kampong Dolly was popular back in the days when this district in Surabaya well-known as biggest prostitution area in South East Asia. Mayor of Surabaya, Tri Rismaharini has changed the negative branding of Dolly Village with made it into tourist destination lately. A tourist village with a variety choices of attraction ranging from mural art, traditional attraction, shopping area, and culinary. New branding program expected to improve and activate the local economic in Kampong Dolly.

It was not an easy job to drastically change from prostitute area into tourism destination. The government uses SMEs scheme to employ ex-prostitute workers, therefore, it need strong commitment and hard work from many stakeholders. On the other hand, the government use that scheme because Indonesia needs to improve the competitiveness of local products in order to compete in the global market for the sake of improving people's welfare. One of them is the need to increase the development of creative industries which encourage people to work independently [1]. Hasibuan argues that economic growth will move quickly if it is driven by creativity and innovation [2].

The potential of the creative economy highly developed with the micro-small and medium enterprises (MSMEs) based. MSMEs are able to absorb a large amount of workforce and provide an opportunity for businesses to develop thus compete with companies that more likely to use large capital [3]. MSMEs people are proven to be able to survive and become the economic driving force, especially after the economic crisis in 1998 [3].

However, the five years process of development in Kampong Dolly did not significantly improve the inhabitant welfare there. Former research at the same theme in Dolly stated that each SMEs have its potentials and problems. In this study, three active MSMEs based on creative economy identified in previous study in Kampong Dolly with each potential characteristics [1]. That three enterprises that also the object of this study are Batik, Shoes, and 'Puja' Printing.

Even though the SMEs have formed around five years ago, the majority have problems from the production process to marketing. One of the factors that caused this problem was the lack of cooperation between the government, academics and the private sector in developing this creative economic potential [4]. Cooperation between these stakeholders is usually called Triple Helix. Triple Helix is a component of government, academia and the private sector that forms collaborative relationships to be able to increase knowledge and innovation in order to solve some of the main problems that exist [5] [6] [7]. The lack of Triple Helix cooperation can be seen from the lack of use of innovative technology from academics for the production process, marketing that does not involve the private sector and government policy, especially in terms of managing less-than-optimal permits [4].

This collaboration is very important to do so that the potential of Kampong Dolly can be used as effectively and efficiently as possible in order to improve the welfare of the community. With this research, the Triple Helix collaboration can be designed as a capital for developing a Local Economic Activation to support the branding of Dolly Tourism Village, improve community welfare and enhance the competitiveness of the creative industries in Surabaya City to national and international levels.

The aim of this study are mapping and scoring triple helix cooperation in term of activating the local economic trough SMEs. Furthermore, to develop and increase community welfare in kampong neighborhood.

2. Methods

This research used qualitative approach for determine of sampling, data record, and for the analysis. Firstly, limitation needed to define theories and empirical facts relating to the development of the local economy, especially in Surabaya as metropolitan city. The empirical facts will relate about creative industries in common, SMEs, and triple helix method. Furthermore, the theory will formulate to conceptualization the production of research variables with descriptive methods.

Population of this study are the stakeholders related to creative industries development in Kampong Dolly. Kampong Dolly was taken into consideration by the object of the study because this area faced too many problems in terms of the efforts and commitment to develop the local economy. The main problem as stated from the former study was the lack of cooperation among stakeholders. This study will try overcome these problems with developing a Triple Helix Cooperation Scheme for Local Economic Activation in Kampong Dolly, Surabaya City.

Samples in this research were taken using non probability sampling methods through snowballing sampling. Snowballing is sampling technique whom the informant has been contacted first and met with researchers then they will recommend another person which researcher don't know. It is important process to use their social networks to refer researchers to other people who have the potential to participate or contribute, therefore, the information learnt or provided to researchers will widely varied [8]. To determine the respondents who will be involved in in-depth interviews were first taken from representatives of the government, the private sector / entrepreneurs, and the community that met the specified criteria.

Two main related process conducted to reach the goals of this research. Firstly, mapping analysis of the triple helix collaboration that has been formed in the development of Local Economic Activation in Kampong Dolly. That objective reach by casual loop diagram analysis with data from observation, structured interviews, and secondary data which own by local government agencies.

Secondly, assess triple helix cooperation level from in-depth interviews that held on each stakeholders. The analysis will conduct with linkert analysis, therefore, the result will be the cooperation level that has been established in Kampong Dolly.

3. Results and discussion

3.1 Local economic and creative industries

Local economic development aims to build the economic capacity of an area where communities, NGOs and sector partners work to create better employment and economic conditions. The potential of Dolly's ex-localization area is the presence of SMEs in Surabaya that are capable of supporting the economic activities in the small level community [1]. On the other hand, problem related to cooperation with government, private sector, and academics leads to lack of coordination between the stakeholders and the unavailability of investment coming from NGOs or private sectors [4]. Notwithstanding Surabaya City Government commitment are relatively high, their role in generating economic activity in Dolly with providing solutions and strategies are minor. The statement is the result of the community which lack of workers who capable in production or marketing, low level of capital, and weak assistant from academics [1].

The existence of these potential and problem related to stakeholder coordination, indicates that every MSME in Dolly's localization has a desire to improve the business to grow better. Therefore, in terms of specify the role of each stakeholder (government, private sector and academics) this study held triple helix mapping. That idea comes from creative industries that drastically rose in recent years.

The creative industry is an industry that originates from the utilization of creativity, skills and individual talents to create prosperity and employment through the creation and utilization of the creative power and data of the individual [9]. The definition of Creative Industry according to the government is an industry that relies on human creativity, utilizing the talents and skills possessed so as to improve living standards through the creation and exploitation of [10]. Business sectors including the creative industry are advertising, architecture, art, craft, fashion, video, film, photography, game, entertainment, publishing, software, broadcasting, and R&D.

3.2 Triple Helix Method

The theory of Triple Helix was originally from Etzkowitz & Leydersdorff as an innovation-based policy development method. This theory which reveals the importance of creating three-pole synergies, namely academics, business and government - in Indonesia is known as the ABG-concept. From his theory, the aim of ABG is science-based sustainable economic development. From this synergy, it is expected that scientific circulation will lead to innovation. Innovation that has economic potential, or capitalization of knowledge (knowledge capital)

Then according to Wishnu in the 2013 APEC CEO Summit Indonesia [11], the correlation between the role of universities, businesses and the government in increasing economic growth has the relationship of the three integral actors in the context of public-private partnership in a Triple Helix concept, known as ABG or Academic, Business & Government. In the concepts of Academic, Business & Government, industry acts as a production house, while the government is a source of contractual relations that ensures stable interaction and exchange, and the university as a source of new knowledge and technology. Synergy of these three sectors is a generative principle in building a knowledge-based economy, which enables closer economic integration.

Triple Helix as the main actor must always move to make circulation to form knowledge spaces, knowledge spaces where all three actors have equal understanding & knowledge, which will direct these three actors to form consensus space, an agreement space where the three actors begin to make agreements and commitments for something that will eventually lead to the formation of innovation spaces, innovation spaces that can be packaged into economically valuable creative products. This circulation always tries to create newness (innovation) and innovation often changes existing structures (making it unstable). Renowned economic scientist Joseph Schumpeter (Schumpeter, 1934) mentions this converting factor as Creative Destruction which means that the emergence of new innovations in the industry will displace old industries that are not creative and replace them with creative ones.

Based on the blue print of the Indonesian creative economy in 2025 issued by the Ministry of Trade of the Republic of Indonesia states that the main actors who can move the creative industry are scholars, business and government or also called triple helix [12]. The roles of each actor include:

a) The Role of Scholars: acts as an agent who disseminates and implements science, art and technology, as well as agents who form constructive values for the development of creative industries in society.
b) Role of Business: it acts as a business actor, investor and creator of new technologies and is also a consumer of creative industries. In an effort to develop creative industries business people play a role in creating products and creative as well as forming community and creative entrepreneurs.
c) Government's Role:
   i. Catalysts, facilitators and advocates that provide stimulation, challenges, encouragement for business ideas to move to a higher level of competence.
   ii. Regulators that produce policies relating to people, industry, institutions, intermediation, resources and technology.

4. Discussion

Based on the data recapitulation of the Putat Jaya Village Office, Surabaya, Dolly Village has 42 types of MSMEs based on the creative industry. These MSMEs are included in the culinary, fashion and printing sub-sectors. But the SMEs that are the objects in this study are three, namely Batik, Shoes, and printing. the collaboration and role of each stakeholder in triple helix will be discussed in detail in each SMEs for the Local Economic Activation in Kampung Dolly.

4.1 SMEs for the Local Economic Activation in Kampung Dolly

a. SMEs Batik Tulis Jarak Arum

Dolly batik has three pattern, namely 'arum batik', 'canting suryo', and 'alpujabar' which produces fabric and clothes that are ready to use. Right now, the production activities are only based on orders. Raw materials and tools for batik production include the tools and fabrics are from Pabean and candles to make the pattern (batik candles) from Madura. The following is a triple helix mapping for UKM Tulis Batik Arum. The following is the role of each stakeholder in every aspect.

Table 1. Triple Helix Mapping for Each Stakeholders Role in Batik SMEs

Aspek	Government	Academics	Private Sector
People	Training in batik production and diversification of batik products	Providing typical batik motifs and design training by ITS.	• Batik production training conducted by GMH • This Batik UKM has diversified its products such as batik cloth, batik necklaces, batik scarves, wall hangings and batik clothes
Industry	• Provision of batik houses for batik production sites. • Granting permission for SIUP. • Surabaya City Government provides a marketing place at DS Point. • The government also helped marketing this product several times through fashion fairs and	-	• Has collaborated with a mall in Surabaya to help with marketing by taking part in existing fashion exhibitions • In collaboration with e-commerce BukaLapak for batik marketing

Aspek	Government	Academics	Private Sector
	SME center exhibitions in the city of Surabaya.		
Technology	Giving technological assistance	There is a research on waste treatment results from the production.	There has been technological assistance provided by the private sector
Resources	-	-	- Purchasing fabric and canting from Pabean, Surabaya. - Another material from Madura
Institution	The government has formed Surabaya City Cooperative and MSME Office	Several institutions focus on Kampung Dolly's research development	There has been an association of SMEs in Dolly to further increase their potential together
Financial Intermediary	The Surabaya City Government has provided financial assistance to these SMEs -	Academics provide training on financial management for SMEs	- Providing financial assistance carried out by the NGO Gerakan Melukis Harapan - Infak management institution has conducted business and administration training for SMEs in all SMEs in Dolly

Source: analysis, 2018.

Based on the explanation of each stakeholders above, it can be clearly seen that the development of Batik is stronger in the business itself but weak in marketing role. This can be seen from the lack of interference from the government and academics in the development of this SME.

b. Shoes SME

SME Shoes is an SME engaged in fashion, especially producing leather shoes, flip flops, and hotel sandals since 2014. The sandals supplied to the Grand Surabaya Hotel, Bekizar Hotel, Night and Day Hotel, life hotel and Patata. Furthermore, the market for leather shoes from SMEs has reached outside the islands, such as Kalimantan and Jayapura. The following is a triple helix mapping for Mampu Jaya Shoes SME.

Table 2. Mapping Triple Helix UKM Sepatu Mampu Jaya

Aspects	Government	Academics	Private Sector
People	- Product diversification training has been carried out such as making hotel slippers and leather shoes.	-	- This SME has collaborated with the Indonesian Footwear Development Center in the production of leather shoes.
Industry	- The government also helps market the products at the UMKM centers in the city of Surabaya	-	- This SME has collaborated with several hotels in Surabaya in terms of marketing
Technology	- Granting permission for SIUP.	-	-
Resources	- The government has provided assistance with production equipment in the form of screen printing tools and tools for making shoes	-	- For raw materials, leather comes from sacred hanging - For soles from Seruni, Sidoarjo - For sponges that are used as raw material for hotel slippers, they come from Wedoro
Institution	- The government has established the Surabaya City Cooperative	- Some institutions focus on Kampung Dolly's	- There has been a community of SMEs in Dolly to further increase

Aspects	Government	Academics	Private Sector
Financial Intermediary	and MSME Office - The government has provided financial assistance to these SMEs given.	research development. - Academics provide training on financial management for SMEs	their potential together - Infak management institution has conducted business and administration training for SMEs in all SMEs in Dolly

Source: analysis, 2018.

Based on the explanation above, it can be seen that the development of Batik Jarak Arum UKM is stronger for the business itself. This can be seen from the lack of interference from the government and academics in the development of this UKM.

c. Printing SME

Printing SME not only produces unique Dolly shirts from but also helps UKM Shoes in Dolly in the process of producing hotel slippers. The workforce involved in this UKM came from youth clubs in the village. This UKM has also collaborated with several agencies, including Collaboration with ART Generation in Balai RW. The following is a triple helix mapping for Screen Printing UKM.

Tabel 3. Mapping Triple Helix UKM Sablon

Aspek	Government	Intelectual	Business
People	-	-	- Only make plain shirt printing and there is no diversification
Industry	- The Surabaya City Government provides a marketing place in DS Point and SME centers - fashion exhibition and SME center exhibition in the city of Surabaya	-	- Marketing is only by giving business cards and from mouth to mouth
Technology	-	-	-
Resources	-	-	- Supplier of fabric comes from Kapasan
Institution	- The government has established the Surabaya City Cooperative and MSME Office	- Some institutions focus on Kampung Dolly's research development	- There has been a community of SMEs in Dolly to further increase their potential together
Financial Intermediary	-	- Academics provide training on financial management for SMEs	- Capital only comes from self-help Chairperson RW

Source: analysis, 2018.

4.2 Triple Helix cooperation level assessment

The following is an assessment of the level of triple helix collaboration that has been formed in the development of the Local Economic Activation in Kampong Dolly.

a. Government Role

To assess the role of the government in the effort to develop the local economy in Dolly Village, this is done using a radar diagram which will be explained as follows. This assessment was carried out by summarizing the relationship between the government and 7 SMEs in Dolly Village.

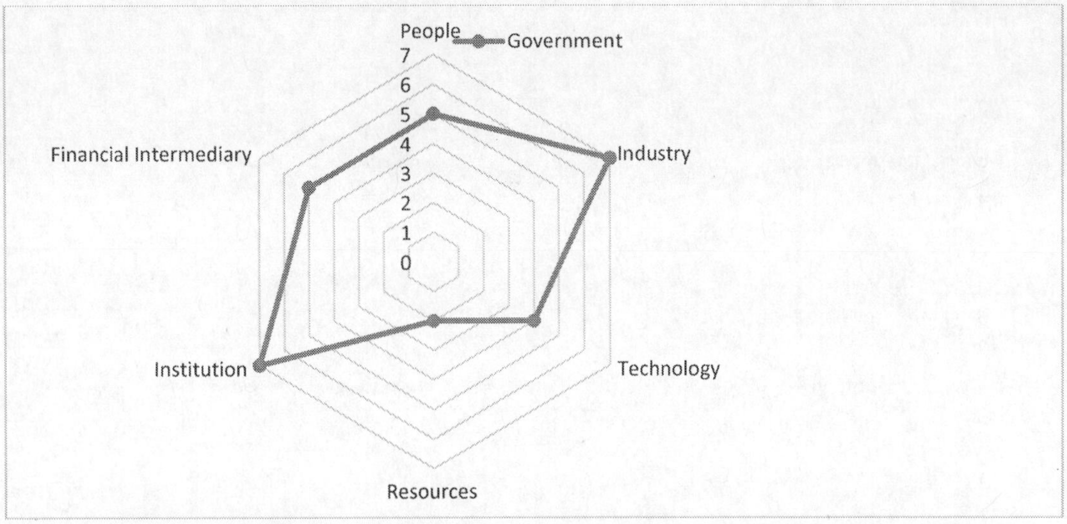

Figure 1. The role of government in the development of local economic activation

Based on the graph above it can be seen that the government role were still weak in helping the development of the local economy in Kampong Dolly in the aspect of "Resources". This happened because the majority of SME business owners in Kampong Dolly searched for their own raw materials. At present the Surabaya City Government has provided thematic village policies and this greatly facilitates SME entrepreneurs to obtain their raw materials.

Government assistance is mostly in the aspects of industry and institutions. This happened because the government already had a Cooperative and UMKM Service that overshadowed all UKM activities in Kampong Dolly. Both in terms of community empowerment, giving capital, arranging permits to marketing products. The Surabaya City Government has prepared a marketing location for all products produced in Kampong Dolly both in the DS Point building which is the marketing center for typical Kampong Dolly products as well as all SME centers spread throughout the city of Surabaya.

b. Academics Role

To assess the intellectual role in the effort to develop the local economy in Dolly Village, it is done by using a radar diagram which will be explained as follows. This assessment was carried out by summarizing the intellectual connection with 3 SMEs in Kampong Dolly.

IOP Conf. Series: Earth and Environmental Science **340** (2019) 012042 doi:10.1088/1755-1315/340/1/012042

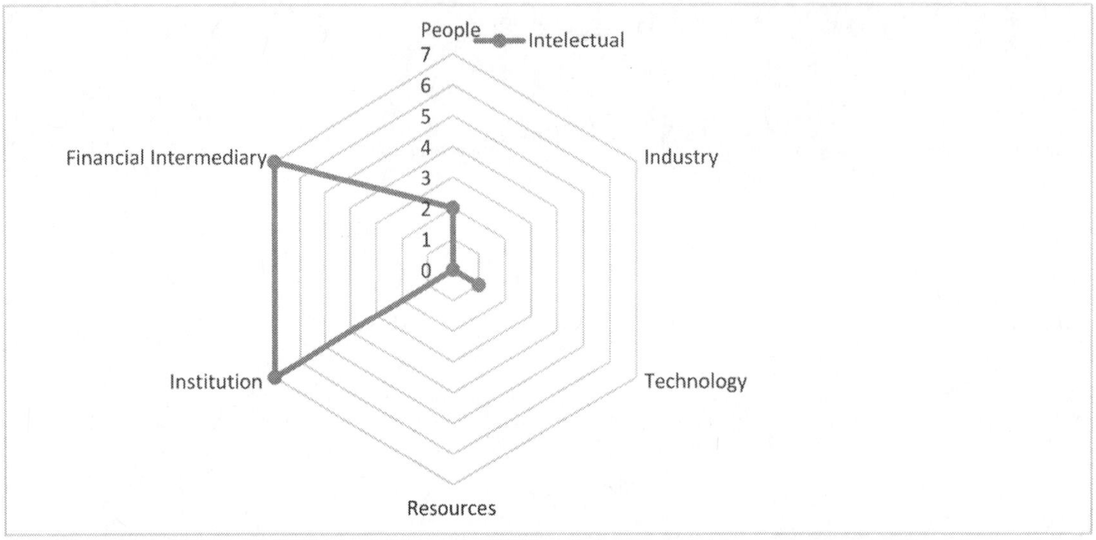

Figure 2. Academics role in developing local economic activation in Kampung Dolly

Based on the graph above it can be seen that for academics the highest is the provision of financial and institution assistance. This happened because many universities had provided assistance in the form of training on financial management for SMEs and had directed the research core to the efforts of developing MSMEs in Kampong Dolly as held by the Ten November Institute of Technology of the PDPM Study Center. As for the lowest role of academics is the provision of resources. This happens because SMEs are looking for their own resources. Higher education has not yet sought the substitution of raw materials that are easier to find, environmentally friendly and can further increase selling prices so that they can improve the welfare of Dolly's people.

c. Private Sector Role

To assess the role of business in the effort to develop the local economy in Kampong Dolly, a radar diagram will be explained as follows. This assessment was carried out by summarizing the relationship of business with 7 SMEs in Kampong Dolly.

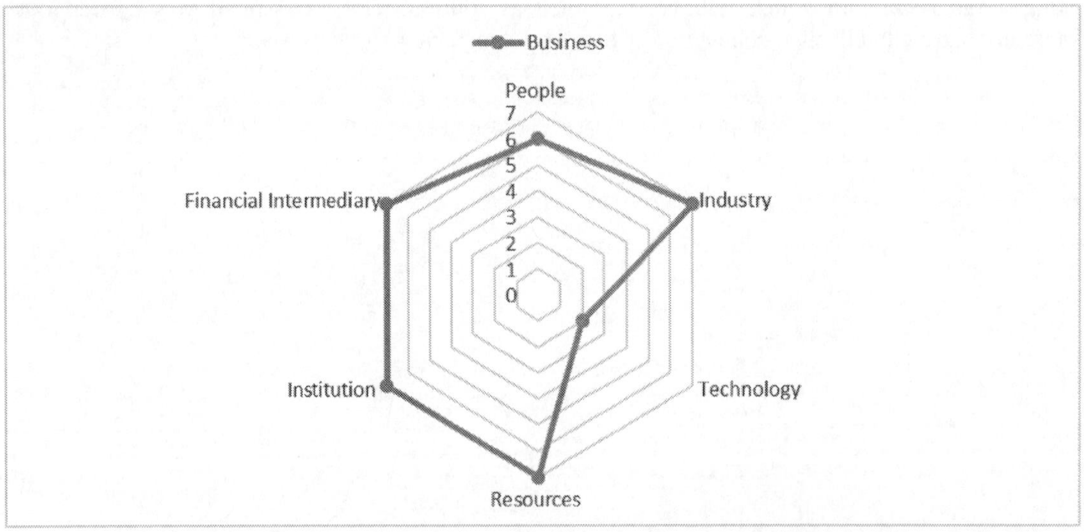

Figure 3. The Role of Business in Developing Local Economic Activation at Kampung Dolly

Based on the graph above, it can be seen that the role of business is very large for the development of SMEs in Dolly's village. This happened because many Dolly villagers took their own initiative both in terms of production, distribution and marketing. The other business or private sectors that have helped to develop the local economy in Kampong Dolly are the Gerakan Melukis Harapan (GMH). GMH is a community formed because of the awareness of the young generation of Surabaya, at the discretion of Surabaya Mayor Tri Rismaharini about closing Dolly's localization. The community that was founded on September 10, 2014 has moved to educate, develop the economy, and provide enthusiasm for the people of former Dolly localization. Through this GMH, Dolly Village UKM began to move well with training in SME processing, providing capital to assist in marketing. Not only GMH, some SMEs have also collaborated with private sector, hotels and malls in Surabaya to help the supply their products. In addition, the Infak Management Institute, which is one of the NGOs in Surabaya, has conducted business and administrative training for SMEs in all SMEs at Dolly.

5. Conclusion

The conclusion obtained from this research is that the development of SMEs in Kampong Dolly has collaborated with the government, private sector and academics. However, the role of the government and academics still very weak in developing resources and technology. On the other hand, as for the collaboration of the private sector, low investment in terms of technology, which was developed to improve the advantages possessed by SMEs in Dolly Village.

6. References

[1] B. U. Aulia and V. K. Siswanto, "Determination of micro and medium enterprises development," *CITIES2017*, no. Earth and Enviromental Science, 2018.
[2] Z. Hasibuan, "Pengembangan Ekonomi Kreatif Berbasis Teknologi Informasi dan Multimedia: Peluang dan Tantangan.," *Seminar Nasional di STMIK AMIKOM*, 2016.
[3] R. Sudayanto and R. Rina, "Strategi Pemberdayaan UMKM Menghadapi Pasar Bebas ASEAN.," 2014.
[4] V. K. Siswanto, "Pemetaan Dan Penyusunan Strategi Pengembangan Industri Kreatif Pada Usaha Mikro Kecil Menengah (UMKM) Di Kampung Dolly Kota Surabaya," *ITS*, 2017.
[5] J. S. C. B. L. S. R. A. Bergek, "Analysing the dynamics and functionality of sectoral innovation systems – a manual. In: 10 Year Anniversary DRUID Summer Conference, Copenhagen. June 27–29.," 2005.
[6] Carlsson and R. Stankiewicz, "On the nature, function, and composition of technological systems.," *Journal of Evolutionary Economics*, vol. 1, p. 93–118, 1991.
[7] J. S. H. M. R. A. Carlsson, " Innovation systems: analytical and methodological issues. „" *Research Policy*, vol. 31, p. 233–245, 2002.
[8] Bungin, Penelitian Kualitatif, Jakarta: Kencana Perdana Media Group, 2007.
[9] M. Widiasari, "Identifikasi Industri Kreatif Pada Subsektor Kerajinan Blangkon Guna Menghadapi Asean Economic Community (AEC) Studi Kasus: Kecamatan Serengan, Surakarta.," *Universitas Muhammadiyah Surakarta*, 2015.
[10] M. Syarif, " Analisis Perkembangan Dan Peran Industri Kreatif Untuk Menghadapi Tantangan ME 2015.," *Seminar Nasional Inovasi dan Tren (SNIT),* 2015.
[11] APEC, "APEC CEO Summit 2013 Indonesia. Kuliah Umum ABAC di UNAIR.," 2013.
[12] D. P. R. Indonesia, "Pengembangan Ekonomi Kreatif Indonesia 2025. Jakarta," 2008.

IOP Conf. Series: Earth and Environmental Science **340** (2019) 012043 doi:10.1088/1755-1315/340/1/012043

Empirical analysis of the relationship between ring road development and land use transformation in Salatiga City

V C Nasukha[1] and Y Herwangi[1]

[1] Urban and Regional Planning Program, Department of Architecture and Planning, Universitas Gadjah Mada, Yogyakarta

Email: virgari14@gmail.com; yherwangi@ugm.ac.id

Abstract. As many as 99% of national logistics transportation using land routes. To accommodate the need for logistics, the development of ring road is an efficient solution, especially to prevent the distortion of activities within the city. Likewise with Salatiga City, the construction of Salatiga's South Ring Road aims to reduce congestion on the main road in the city center. Of course, the existence of the ring road is affecting land use change in Salatiga City, especially in areas passed by the ring road. Therefore, the research question is how the relationship between the ring road and land use transformation in Salatiga City. While the purpose of this research is to identify the form of relationship between the ring road and land use transformation in Salatiga City.

Keywords: *ring road, land use transformation, relationship*

1. Introduction

99% of logistics transportation in Indonesia urban uses road network [1]. In addition, national economic mobility also depends on the urban road network. The urban road network not only accommodate logistics route, but also accommodate city activities. Consequently, that along with the development of the city, urban activities and logistic transport activities cause new problems in the city, congestion. Therefore, to accommodate these needs, ring roads is an alternative way for logistics routes in urban areas [2]. The ring road, which is generally built on the outskirts of the city, certainly causes the surrounding land use to develop, based on theory by Babcock (1960) [3]. The Sxial Babcock Theory states that the transport axis is the main factor that influences mobility, especially the relationship between Central Bussiness District (CBD) and its outer area. Zones along this transportation axis will growth faster than other zones.

As one of the cities passed by the main logistics route of Java Island, Salatiga City also uses ring roads as an alternative circulation of the city, which was built on the southern lane of the city. The southern ring road began operating in 2012, which aims to break congestion on the Salatiga City arterial road. In addition, arterial roads in the city of Salatiga also do not meet the capacity as the main logistics line, due to the increasing activity of the city community. Over time, transportation activities on the southern ring road are increasing. This causes increased activity around the ring road. The activities of the ring road and land use transformation are certainly interrelated.

Relationship between the axis of transportation and changes in land use can be seen in the phenomena around us, that land use changes depending on the activity of the road that passes through it. Likewise vice versa, road activities will adjust the land use around it [4]. But, the road does not fully affect the land use change, there are external factors that influence, such as policy, topography, and constellation with other regions [5]. Therefore, this paper will discuss the relationship of ring road and the transformation of surrounding land use.

2. Method

2.1. Study area

This research was conducted in Salatiga City, which was located in Central Java Province. Salatiga is a small city with a population of 186,420 people. Morphologically, it is located in the foot basin area of Mount Merbabu, and is located between small mountains, such as Gajah Mungkur, Telomoyo, and Payung Rong [6]. Salatiga City is passed by transportation routes that have quite high activity, because the main road in the City of Salatiga connects the National Activity Center (PKN) and the Regional Activity Center (PKW). The density of the main road activity is the background of the Central Java Provincial Government to build the South Ring Road (JLS) Salatiga in the interests of logistics transportation. The road network system in the City of Salatiga is also passed by the Semarang-Solo Toll Road, especially in Bugel Village, Kauman Kidul Village, and Tingkir Tengah Village.

Salatiga's South Ring Road (JLS) has a length of + 11.3 kilometers with a road width of 21 meters, and passes through 3 sub-districts with 7 villages. The JLS development aims to anticipate traffic congestion in Salatiga City, optimize transportation networks in Salatiga City, and spur economic growth in Salatiga City and its surroundings. Other than that, it's aim to increase the level of service for road users passing through Salatiga City, developing the area, and adding to the national road network system. In addition, JLS is also expected to change the pattern of traffic movements in Salatiga City. In addition, as explained in the Salatiga City Regional Spatial Plan (RTRW) 2010-2030, the development of the ring road network is one of the strategies for developing a transportation network aimed at facilitating the inter-center movement of activities.

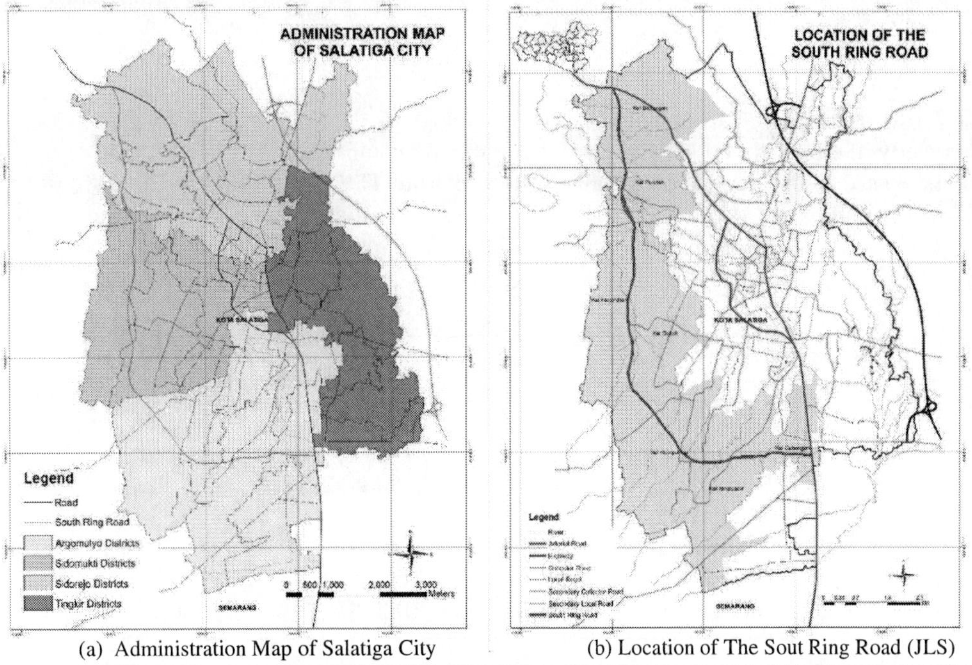

(a) Administration Map of Salatiga City (b) Location of The Sout Ring Road (JLS)

Figure 1. Illustration of the study area

IOP Conf. Series: Earth and Environmental Science **340** (2019) 012043 doi:10.1088/1755-1315/340/1/012043

2.2. *Delineation area*
In identifying the relationship between the ring road and land use transformation, the observation unit used is the Zone, so the research area is delineated into eight zones. This delineation is decided based on spatial characteristics, distance from the ring road, and administrative boundaries. These eight zones are then given the same treatment in accordance with the predetermined relationship parameters. These zones include:

Table 1. Delineation Area

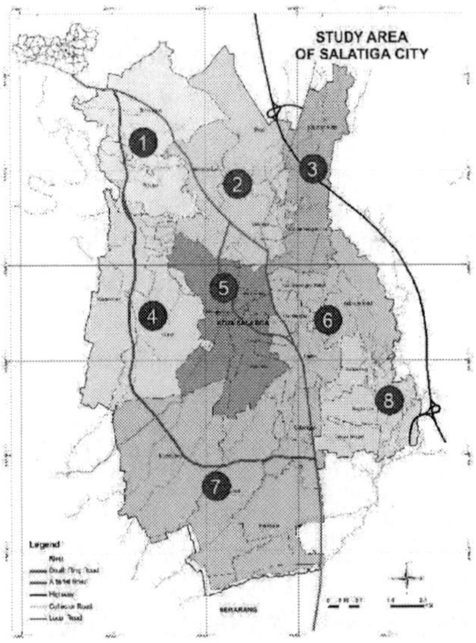

Figure 2. Illustration of the Zone

Zone	Area	Distance from Ring road	Administration Boundaries
	(ha)	(m)	(Kelurahan)
1	633.35	500	Pulutan, Blotongan
2	756.28	3000	Bugel, Salatiga, Sidorejo Lor
3	406.07	5000	Kutowinangun Lor, Kauman Kidul
4	771.82	500	Dukuh, Kecandran
5	547.76	3000	Tegalrejo, Mangunsari, Kalicacing
6	623.83	5000	Gendongan, Sidorejo Kidul, Kutowinangun Kidul, Ledok
7	1396.43	500	Kumpulrejo, Cebongan, Randuacir, Noborejo
8	412.15	3000	Tingkir Tengah, Tingkir Lor, Kalibening

To verify the relationship, we used scoring method. The parameters used are:

2.3. *Trip attraction*
In this study, the trip attraction calculated using the indicator of generation and area of land use. The generation used is the generation standard adopted from ITE 9th Edition in 2012, illustrated in the following formula:

$$\textit{Trip Attraction} = \textit{building floor area} \times \textit{trip rate} \quad (1)$$

Table 2. Trip Generation Rate

Land Use	ITE 9th Edition Rate	
Industrial	1000 SF	0.73
Residential	Dwelling Units	1
Tourism	Acres	4.3
Conservation Area	Acres	0.84
Public Facility	1000 SF	7.3
Office	1000 SF	1.21
Service	1000 SF	26.15

2.4. *Accessibility index*

Accessibility analysis aims to identify the level of community's affordability to facilities in a zone. In this study, researchers used a simple calculation adopted from the Minister of Settlement and Regional Infrastructure's Regulation Number 534 / KPTS / M / 2001. The calculation uses the parameters of road length and area.

2.5. *Population density*

Population density is used to identify activities in a zone. The more densely populated, the more generated generation. In this study, population density was calculated using the formula:

$$\textit{Population Density:} \quad \frac{\text{population}}{\textit{area}} \tag{2}$$

2.6. *Distance from ring road*

The distance from the south ring road is used as a relationship parameter, which is delineated using a buffer of 500 m, 3000 m, and 5000 m from the ring road. The buffer was calculate using administration boundaries and transport activity from each zone.

From these parameters, researcher determines the weight of each. Weighting the parameters uses the Ranking Method, and the determination of the value of each parameter uses the Guttman Scale. After that, the researcher calculates the multiplication score between weight and value. In determining the weight, the researcher uses Number Ranking method, with the formula:

$$W = \frac{(n - r + 1)}{\sum(n - r + 1)} \tag{3}$$

Where*:*
w *:* weight
n *:* number *of parameters*
r : ranking position of a parameter

Table 3. Weight and Value of the Parameters

Parameter	Weight	Range	Class	Value
Population density (population/ha)	0.1	0-2.33	Low	1
		2.34-4.67	Medium	2
		4.68-7	High	3
Distance from ring road (meter)	0.4	0-500	Near	3
		501-3000	Medium	2
		3001-5000	Far	1
Trip attraction (trip/hour/ha)	0.2	0-509	Low	1
		510-1018	Medium	2
		1019-1526	High	3
Accessibility index (meter/ha)	0.3	0-0.05	Low	1
		0.06-0.1	Medium	2
		0.11-0.15	High	3

IOP Conf. Series: Earth and Environmental Science **340** (2019) 012043 doi:10.1088/1755-1315/340/1/012043

From these parameters, researcher divides three typology of relationship, which were low relationship which had a total range of scores from 0 to 1.7, medium relationship which had a total score of between 1.8-2.2, and high relationship which had a total score of 2.3-2.6.

3. Relationship between South Ring Road Development and Land Use Transformation

3.1. Land use transformation in Salatiga City

Land use transformation in Salatiga City is relatively fast, indicated by significant land use increase figures. The most significant transformation is in the use of conservation areas, housing and industrial area. The conservation area has a growth index of 12.6 from an area of 39.14 hectares to 532.48 hectares. Beside, residential use has an area of 1,740.24 hectares in 2011 and increased to 2,338.14 hectares in 2016. In addition, the industrial has an increase index of 2.06 which has an increase from 67.6 hectares to 207.18 hectares.

Furthermore, land use which has the lowest transformation index is public facilities which only increased by 0.1 hectares and the military area which only increased by 3.39 hectares. Whereas there is land use that has decreased, namely office land use from 27.29 hectares to 21.33 hectares, forest land use from 183.55 hectares to 167.92 hectares, and agricultural land use from 3,234.46 hectares to 1,384.36 acres. For more details, see the following maps and graph below:

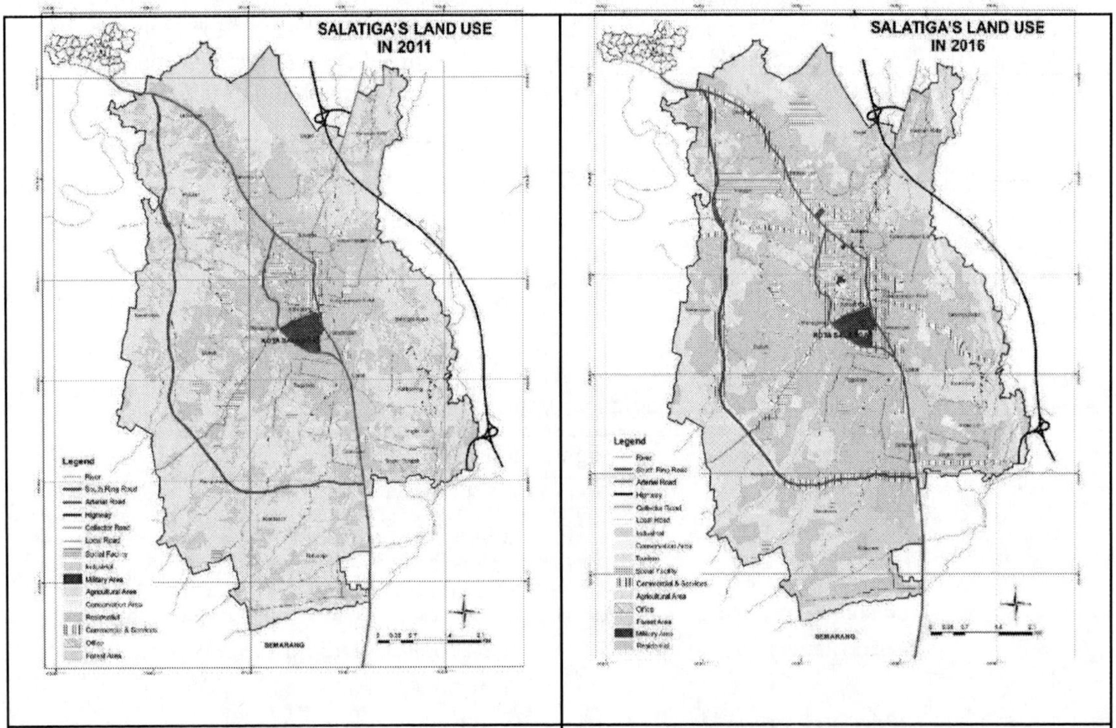

(a) Salatiga's land use in 2011 | (b) Salatiga's land use in 2016

Figure 3 Land Use Transformation in Salatiga City (map version)

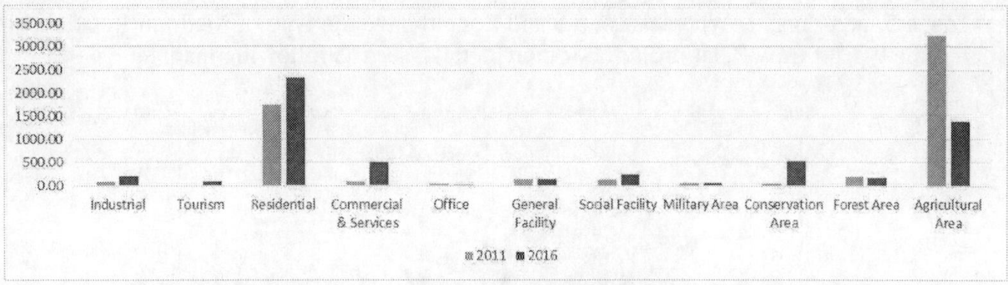

Figure 4. Land Use Transformation in Salatiga City (graph version)

3.2. Scoring the relationship

Based on the formula specified in the method section, the results of the calculation are as follows.

Table 4. Results of The Calculation from Each Zone Based on The Formula

Zone	Population density	Distance from RR	Trip attraction	Accessibility index
1	2.292	500	1221.743	0.081037449
2	5.189	3000	1526.497	0.151728954
3	4.2245	5000	259.8225	0.113590286
4	2.375	500	968.08	0.068615178
5	6.92	3000	893.4204	0.140372613
6	5.853	5000	987.6508	0.121898882
7	1.9025	500	1132.242	0.058785453
8	2.772	3000	96.64958	0.086481666

Identification of relationship using the scoring method described in the research method section. Parameters used in determining the score for each zone are: population density with a weight of 0.1, distance from ring road with a weight of 0.4, accessibility index with a weight of 0.2, and trip attraction with a weight of 0.3. The results are as follows.

Table 5. Relationship Scoring

Weight	0.1		0.4		0.2		0.3		Total Score	Relationship status
Zone	Population density	Value	Distance from RR	Value	Trip attraction	Value	Accessibility index	Value		
1	Low	1	Near	3	High	3	Medium	2	2.5	High
2	High	3	Medium	2	High	3	High	3	2.6	High
3	Medium	2	Far	1	Low	1	High	3	1.7	Low
4	Low	1	Near	3	Medium	2	Medium	2	2.3	High
5	High	3	Medium	2	Medium	2	High	3	2.4	High
6	High	3	Far	1	Medium	2	High	3	2	Medium
7	Low	1	Near	3	High	3	Medium	2	2.5	High
8	Medium	2	Medium	2	Low	1	Medium	2	1.8	Medium

From the table above, three types of relationship were identified, namely High Relationship, Medium Relationship and Low Relationship. Zones that have high relationship are Zone 1, Zone 2,

Zone 4, Zone 5, and Zone 7. Whereas Zone 6 and Zone 8 have the type of Medium Relationship, and Zone 3 has the type of Low Relationship. For more details, see the following map:

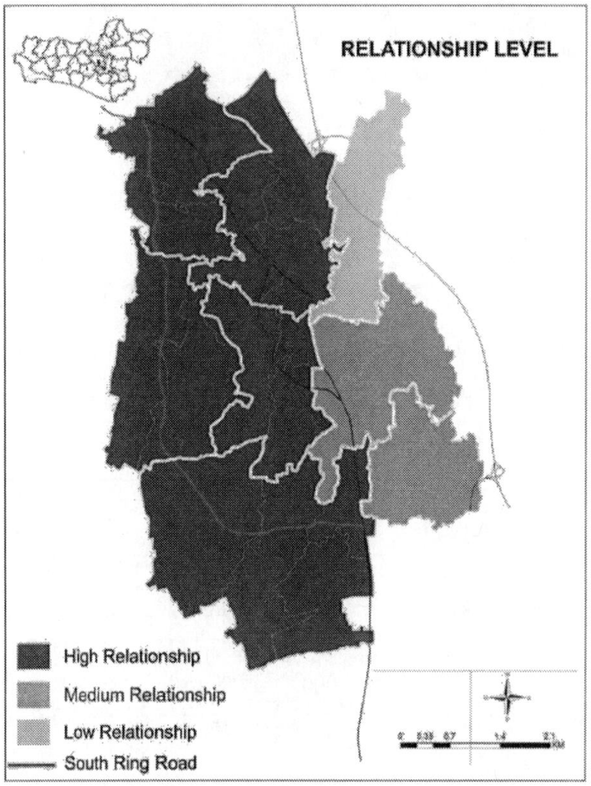

Figure 5. Map of the relationship level

4. Discussion

Land use transformation in Salatiga City is dominated by the interaction of land use in residential, commercial & services, social facilities, and agriculture. This land use transformation has led to an increase in attraction in certain zones, due to the growth of new activity center that have considerable attraction, such as education and commercial & services. In addition to trip attraction, the transformation of land use also affects the level of accessibility of a zone because of the increasing activity center, the road infrastructure also increases, and accessibility also increases. This increase in attractiveness and accessibility affects the relationship between transportation and land use axis.

Therefore, there are also some findings that affirm Babcock's Axial Theory, but there are also some that are not suitable with the theory. As mentioned in the previous section, the transportation axis will affect the changes in land use around it and increase the level of accessibility in the area. This is evident in the case of this study, that the South Ring Road (JLS) was able to attract a fairly high trip, clarified by the addition of social facilities (education) and commercial & services that began to popping up after the ring road operated.

Based on the discussion in the above points, there are also some findings that affirm Babcock's Axial Theory, but there are also some that are not in harmony with the theory. As mentioned in the previous section, the transportation axis will affect the changes in land use around it and increase the level of accessibility in the area. This is evident in the case of this study, that the South Ring Road (JLS) was able to attract a fairly high trip, clarified by the addition of social facilities (education) and trade & services that began to mushroom after JLS operated.

On the other hand, the results of the study show that not all zones that are close to JLS have the highest relationship between the other zones. In addition, the zone close to JLS has a moderate level of accessibility. This contrasts with Babcock's statement and affirms Apituley's statement that there are many factors related to road and land use.

5. Conclusion and recommendations

Based on the results of the discussion, not all zones that are close to JLS have the highest relationship compared to the other zones. It is precisely the zone that is at the center of city activity and is 500-3000 meters from JLS which has the highest connection. This can be indicated by a higher score of population density, trip attraction, and accessibility compared to other zones. In contrast to Babcock's statement in which states that the closer to the axis of transportation, the higher the relationship. In addition, the zone located on the JLS axis has a high attraction, because it is in entrance and exit of the JLS.

The zone near JLS actually has a low population density, but the pull score is high. This affirms Babcock's statement that activity will be increasingly crowded into the transport axis. In this case commercial & service activities, not housing activities. It can be seen in the spatial transformation around JLS that is significant in the use of commercial & services.

On the other hand, the results of the study show that not all zones that are close to ring road have the highest relationship between the other zones. In addition, the zone close to ring road has a medium level of accessibility. Therefore, there needs to be an evaluation of the South Ring Road (JLS), because it can relate to controlling the use of land use around the road, which should not have high activity that disturbs the activities of JLS.

6. References

[1] Munawar 2005 *Dasar-dasar Teknik Transportasi* Beta Offset Yogyakarta

[2] Colonna, Berloco, and Circella 2012 The Interaction between Land Use and Transport Planning: A Methodological Issue *Procedia: Social and Behavioral Sciences*: 84-95

[3] Babcock's Axial Theory in Yunus, Hadi Sabari 2005 *Klasifikasi Kota* Yogyakarta: Pustaka Pelajar

[4] Budi, Irawan Setia 2007 *Pengaruh Penggunaan Lahan Terhadap Bangkitan Dan Tarikan Pergerakan Di Sepanjang Jalan Gadjah Mada Kota Batam* Universitas DIponegoro

[5] Apituley Maxmillian 2003 *Keterkaitan Antara Perkembangan Kawasan Terbangun Dan Guna Lahan Dengan Perkembangan Jaringan Jalan di Kota Merauke Tahun 1990-2000* Universitas Gadjah mada

The analysis of supply chain risk logistics in implementation of West Sumatera - Riau toll road development

Yosritzal[1,3], B M Adji[1,3] and F Rizola[2,3]

[1] Lecturer at Civil Engineering Department, Engineering Faculty
[2] Master Student at Civil Engineering Department, Engineering Faculty
[3] Universitas Andalas, Indonesia.

Email: yosritzal@eng.unand.ac.id

Abstract. The aim of this research is to develop a risk mitigation strategy in supply chain process of West Sumatra-Riau toll road pavement material. Data for this study was obtained through brainstorming techniques, questionnaires, field surveys and literature review. The data was analyzed using House of Risk (HOR) method. There were 18 items of risk event's and 21 items of risk agent's identified from the HOR phase 1, whereas risk agent handling priority data and risk mitigation actions are 19 items by using HOR phase 2. The results show that there are 2 (two) risk agents ranked 1 and 2 for the most critical, namely A1 (poor supplier performance) with ARP of 860 and A2 (Low of Contractor Capability) with ARP of 507, which can cause ± 80% of risk events where the remaining 20% is caused by other agents. There are 2 (two) mitigation actions were obtained from HOR phase 2, namely PA1 (appropriate supplier selection method) and PA4 (appropriate contractor selection method) which can prevent the emergence of ± 80% of these risk agents and ± 20% is overcome by 17 other mitigation actions.

Keywords: *risk mitigation, supply chain, toll road pavement material, House of Risk, West Sumatera-Riau Toll Road.*

1. Introduction

The physical improvement of road infrastructure in term of quantity and quality at some points is critical, to elude transportation problems such as traffic disruption. In West Sumatera, one of the infrastructure improvement plans which will be constructed soon is the West Sumatera - Riau Toll Road. This will be the first Toll Road in West Sumatera. As the first toll road construction in West Sumatera, a mitigation to the potential of disruption to the construction process due to incompetent personnel of contractor, consultant and owner, and/or the failure in supply chain of material for the pavement is needed. The competency of the personnel of contractor, consultant and owner has been evaluated in previous studies [1][2][3]. Therefore, a study to predict potential problem with material supply should be taken before the construction is started.

The aim of this study is to develop a risk mitigation analysis which can emerge in the supply chain of pavement material in the construction of the West Sumatra-Riau toll road that will soon be started. It is expected that a mitigation of the emergence of risks of the pavement material could be initiated. This strategy can be made by first inverting the risk event and risk agent data which can appear in the supply chain of pavement material, which is then be processed by utilizing the help of the House of Risk (HOR) method. The HOR method will identify potential risk agents based on the value of ARP (Aggregate Risk Potential) and the priority of implementing mitigation actions based on ETD (Effectiveness to Difficulty) value.

It is expected that, the potential and critical risk agents toward the supply chain process of pavement material for the construction of the West Sumatra Toll Road - Riau Km 0 to 4.2 in the first phase; Padang - Sicincin could be identified and prevented.

2. Literature review

2.1. The analysis and risk management of construction project

Risk analysis is a process of identifying and assessing subjective and objective risks, while risk management is a formula consisting of responses and actions taken to reduce and manage the risks that have been analyzed [4].

2.2. Risk definition

Based on risk definitions by [5], [6] and [7], Risk can be defined as a condition where there are possibilities and opportunities for occurrence of deviations which are hazardous, damage, loss, accident, or any other unexpected consequences and are uncertain about the results to be achieved and also can cause unpleasant (harmful and harmful) consequences of an ongoing process or upcoming events.

2.3. The supply chain of construction project

According to [8], supply chain is a system which includes actors, suppliers, makers, transportation, distributors, vendors, and guarantors created to convert basic materials into a product and supply these products to users based on the requested value. In the context of construction, supply chain can be defined as a process of a set of activities which change natural materials to become the final products (e.g. roads, buildings, planning services), to be used by service users by ignoring the existing organizational boundaries.

2.4. House of Risk (HOR) method

The HOR method is a method which is often used in supply chain management. Previous researchers used the HOR method in their research, including [9] and [10].

The HOR (House of Risk) method is divided into 2 (two) phases, namely phase 1 and phase 2. Phase 1 is used to calculate the value of the ARP (Aggregate Risk Potential) of each risk agent and rank the risk agents to determine priority handling. Table 1 shows matrix model for HOR Phase 1.

Table 1. HOR Phase1 Matrix Model (Source: [9])

Supply Chain Side	**Risk (Ei)**	**Risk Agent (Aj)**							**Risk Impact (Si)**
		A1	**A2**	**A3**	**A4**	**A5**	**A6**	**A7**	
Supply	E1	R11	R12	R13	.	.	.	.	S1
	E2	R21	R22	.	.	.	.	.	S2
Control	E3	R31	.	.	.	.	.	.	S3
	E4	R41	.	.	.	.	.	.	S4
Process	E5	.	.	.	.	.	.	.	S5
	E6	.	.	.	.	.	.	.	S6
Demand	E7	.	.	.	.	.	.	.	S7
	E8	.	.	.	.	.	.	.	S8
Risk Agent Emergence		O1	O2	O3	O4	O5	O6	O7	
Aggregate Risk Potential		ARP1	ARP2	ARP3	ARP4	ARP5	ARP6	ARP7	
Risk Agent Priority Rank									

Similar to [10], in this research, identification of risk events and risks agent were conducted on the four side of supply chain: supply, control, process and demand.

HOR phase 2 method is used to calculate the ETD value (Effectiveness to Difficulty) of each action that has been inverted to mitigate potential risk agents and rank the mitigation actions. Table 2 shows Matrix Model for HOR Phase 2.

Table 2. HOR Phase 2 Matrix Model (Source: [9])

Risk Agent to be handled	Prevention Action (PAk)					Aggregate Risk Potentials (ARP)
	PA1	PA2	PA3	PA4	PA5	
A1	E11					ARP1
A2						ARP2
A3						ARP3
A4						ARP4
Total Effectiveness of action	TE1	TE2	TE3	TE4	TE5	
Degree of difficulty performing action	D1	D2	D3	D4	D5	
Effectiveness to difficulty ratio	ETD1	ETD2	ETD3	ETD4	ETD5	
Rank of priority	R1	R2	R3	R4	R5	

3. Research methodology

3.1. Research design

The object of this study was the planning of West Sumatera - Riau toll road construction project. House of Risk (HOR) method was used in this study. The variable used in this study was identified based on literature review and existing supporting data.

3.2. Data sources type

3.2.1. Primary data

Data for this study was collected through interviews (interviews), brainstorming techniques, field investigations, and questionnaires. The data is the opinion of the experienced and competent (expert) parties toward possible major risk events and risk agents that could influence the construction of the West Sumatera – Riau toll road.

Respondents were selected from the different sides which was assumed to have understanding and/ or experience in toll road construction or rigid pavement road project elsewhere such as: (the owner in this case a key person in Ministry of PUPR); Academicians; Contractors; Suppliers. Each party was represented by 3 people.

The questionnaire used in this research consists of 3 (three) main parts. The first part contained of the identity consisting of name, position, agency, address, length of work. Part two, was the procedure for filling out the questionnaire, and Part three, was the contents of the questionnaire; the level of probability of the occurrence of a risk agent, the level of correlation between risk and risk agent and the severity of each risk.

3.2.2. Secondary data

Secondary data such as: data of risk event and risk agents and Detailed Engineering Design (DED) were obtained from the literatures and project owner documentations.

3.3. The data analysis of questionnaire

First step in data analysis is to determine the level of severity risk events and risk agent's occurrence. After that, the severity risk events and level of risk agents' occurrence data are evaluated in terms of their agreement to each other and those which are non-sync data will be eliminated. A correlation scoring was then conducted to obtain the correlation level between both data.

3.4. Risk evaluation

3.4.1. HOR phase 1

HOR phase 1 consists of some processes. The first step of this phase was Risk Identification. In risk identification, the risks which can occur on each side of the supply chain, namely supply, control, process and demand were identified. The impact (severity) of the risk and the likelihood of the risks agent occurrence were assessed in a 5-point scale as shown in Table 3 and Table 4.

Table 3. Risk Event Severity Value

Scale	Information	Impact
1	Very small impact	0% ≤ - < 5%
2	Small impact	5% ≤ - < 15%
3	Average impact	15% ≤ - < 45%
4	Big impact	45% ≤ - < 80%
5	Very big impact	≥ 80%

Table 4. Occurrence Risk Agent Value

Scale	Information	Emergence
1	Very rare	0% ≤ - < 20%
2	Rare	20% ≤ - < 40%
3	Sometimes	40% ≤ - < 60%
4	Often	60% ≤ - < 80%
5	Very often	80% ≤ - ≤ 100%

A correlation matrix between each risk agent with each risk occurrence were scored as 0 if there is no correlation and 1, 3, and 9 indicate a low, moderate, and high correlation respectively. The next step is to calculate the Aggregate Risk Potential of Agent (ARP) and rank risk agents based on ARP values and choose several risk agents with the highest ARP to be used in HOR Phase 2.

3.4.2. HOR phase 2

In HOR Phase 2, the risk agents with high ARP values were selected using the Pareto Diagram. Further, identify possible actions to prevent the emergence of those and then, determine the correlation between each preventive action and each risk agent (E_{bc}) with a score 0 if there is no correlation and score 1, 3, and 9 for a low correlation, medium, and high correlation respectively.

For the next step, calculate the Total Effectiveness (TE_k) of each action using the eq. 1.

$$TE_k = \sum ARP_b * E_{bc} \quad (1)$$

Where:
TE_k = Total Effectiveness
ARP_b = Aggregate Risk Potential
E_{bc} = The Correlation between Preventive Action with each Risk Agent.

The difficulty level in carrying out each mitigation action (Difficulty - Dk) was then be predicted using scale 1 or 2. The score = 1 if there are additional cost of mitigation actions and score = 2 if no additional cost needed.

Total Effectiveness Ratio (TEk) to the Difficulty Level (Difficulty - Dk) was calculated using eq. 2.

$$ETD_k = TE_k/D_k \quad (2)$$

Finally, determine the Priority Rating of each mitigation action based on the ETD value. Figure 3 shows all of the steps mentioned in the section 3.

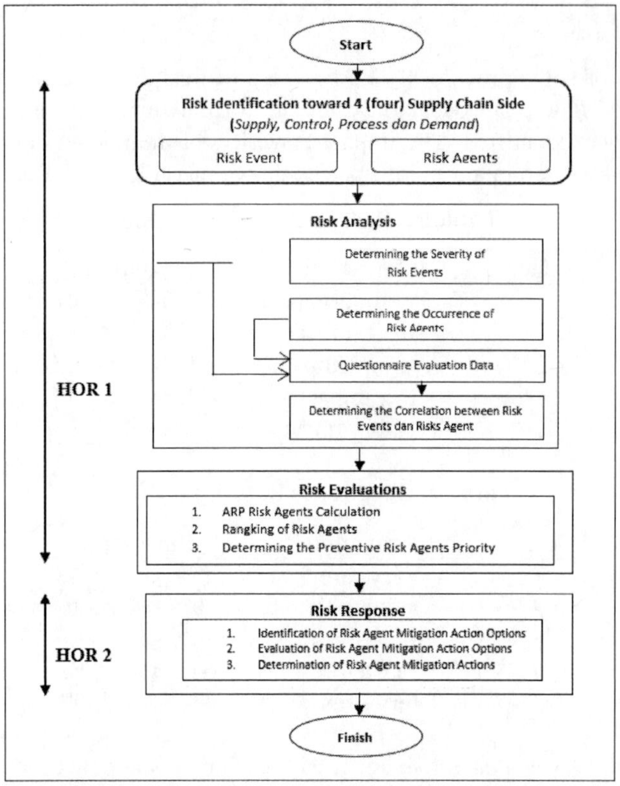

Figure 1. Research Methodology Flowchart

4. Results and discussions

4.1. Risk identification toward 4 (four) supply chain side

Based on literature review, 18 of risk events and 21 of risk agents were identified in the supply chain process of the pavement process for the Toll road planning in West Sumatra – Riau as shown in Table 5 and Table 6.

Table 5. Identified Risk Events

No.	*Risk Events*	*Severity*
E1	Material quality is not in accordance with the order specifications and need to be resubmitted.	4.07
E2	Delayed material due to the production constraints at the factory.	3.80
E3	Damaged Material during the supply.	3.87
E4	Delivery delay of material due to the material limitations stock from supplier.	3.80
E5	Material delivery was canceled because the supplier does not have the transportation facility.	3.73
E6	The waiting duration for the material to meet the required quantity number due to the limitation of the available transportation equipment.	3.80
E7	The supplier does not commit to the agreement in the contract.	4.27
E8	The volume of material coming does not meet the specification quantity.	4.13
E9	Sudden changes to the project activity schedule.	3.80
E10	Changes in logistics demand due to the miscalculation of the works schedules of the contractors.	3.67

No.	*Risk Events*	*Severity*
E11	Lack of managerial skill which led to a delay in finishing a task which cause the material supply being delayed.	3.60
E12	Material is resent as the previous delivery was failed due to inaccurate instructions received.	3.67
E13	The contractor was negligence in doing the payment to the sub-contractor.	3.33
E14	Delayed material due to accidents when shipping.	3.33
E15	The Difficulties in getting material.	3.80
E16	Additional material ordering due to the changes in technical specifications.	3.40
E17	Delays in material shipping due to the traffic congestion.	3.87
E18	No constant price for material.	3.40

Table 6. Identified Risk Agents

No.	*Risk Agents*	*Occurrence*
A1	Supplier bad performance.	3.13
A2	Low contractor capacity.	3.13
A3	Suppliers is not consistent.	3.00
A4	The slow response of suppliers.	3.13
A5	Shipping is not on time.	3.00
A6	The appropriate and accurate information from suppliers.	3.33
A7	The amount of project material needs.	3.40
A8	The scarcity of material.	3.07
A9	Sudden request.	3.20
A10	The appropriate and accurate information from consultants.	3.40
A11	Count on one supplier.	2.93
A12	Project team competency.	3.33
A13	The appropriate and accurate information from owner.	3.27
A14	Logistics prices reference is not accurate.	2.53
A15	A long technical evaluation and needs an adjustment of the budget.	2.93
A16	Impact toward local communities.	3.20
A17	Safety Factor process.	3.53
A18	Natural disaster factor.	2.93
A19	Damage to supply transportation.	2.80
A20	Delay related to the changes in specification	2.67
A21	Rupiah exchange rate instability against dollar	2.40

4.2. Reliability test

Reliability test was conducted to test the data consistency. Risk events reliability (r_{count}) was 0.737 with number of data (N) is 18. Compared with r_{table} of 0.468, the r_{count} was higher than r_{table}, so that the data is said to be consistent. Similarly, risk agents reliability test shows that r_{count} is 0.697 which was higher than r_{table} of 0.433 so that the agent risk data is also consistent.

4.3. Data processing with HOR phase 1 method

Before doing the tabulation through HOR matrix phase 1, the correlation level between risk events against risk agents was decided first with the standard provisions of values 0, 1, 3, 9, where 0 indicates no correlation (no correlation) and values 1, 3, and 9 show low (low), moderate and high (high) correlations. The result is shown in Figure 2.

Risk Event (Ei)	Risk Agent (Aj)																					Sev.
	A1	A2	A3	A4	A5	A6	A7	A8	A9	A10	A11	A12	A13	A14	A15	A16	A17	A18	A19	A20	A21	
E1	9	1	3	0	3	0	1	0	3	3	3	0	0	0	0	0	0	3	3	0	0	4.1
E2	9	0	3	3	0	0	0	3	0	3	0	0	0	0	3	1	1	0	3	0	1	3.8
E3	9	0	0	1	3	0	1	0	0	0	0	1	0	0	0	0	1	9	9	0	0	3.9
E4	3	0	3	1	0	0	1	1	0	3	0	0	0	0	0	1	0	3	1	3	1	3.8
E5	9	1	3	3	0	0	0	0	0	3	0	3	0	0	0	0	0	1	3	1	0	3.7
E6	3	1	1	3	1	0	9	0	3	3	0	1	0	0	0	0	1	1	1	0	0	3.8
E7	9	0	9	3	0	0	0	1	3	1	1	3	0	0	0	1	3	3	3	1	3	4.3
E8	9	1	9	1	0	1	3	1	1	3	3	3	1	0	0	0	1	1	0	3	0	4.1
E9	1	0	1	1	0	0	1	1	3	1	1	1	0	1	0	9	3	3	1	3	0	3.8
E10	3	9	3	3	0	0	3	1	1	1	0	3	3	1	1	0	1	1	1	1	0	3.7
E11	0	9	0	0	0	0	1	0	0	0	3	3	0	0	0	0	0	0	0	0	0	3.6
E12	0	9	0	0	0	3	0	0	0	0	1	9	3	0	0	0	0	0	0	0	0	3.7
E13	0	9	1	0	0	0	3	0	0	0	0	3	1	0	3	0	0	1	0	0	0	3.3
E14	1	0	1	1	0	0	0	0	0	0	0	0	0	0	0	0	1	3	1	0	0	3.3
E15	3	9	1	3	0	0	3	9	3	3	1	3	3	1	1	0	0	3	0	1	3	3.8
E16	0	0	0	0	0	0	0	0	0	0	0	0	0	0	0	3	1	3	0	3	0	3.4
E17	1	3	3	0	0	0	1	0	0	0	0	3	0	0	9	0	1	1	1	1	3	3.9
E18	1	1	3	3	0	0	0	3	3	0	0	3	0	3	3	0	3	9	0	0	9	3.4
Occ.	3.13	3.13	3.00	3.13	3.00	3.33	3.40	3.07	3.20	3.40	2.93	3.33	3.27	2.53	2.93	3.20	3.53	2.93	2.80	2.67	2.40	
ARP	860	607	518	308	83	51	347	232	248	317	150	487	134	54	217	180	228	492	294	173	178	
Rank	1	2	3	8	19	21	6	11	10	7	17	5	18	20	13	14	12	4	9	16	15	

Figure 2. Correlation level between risk events and risk agents

Paretto diagram was developed to sort the score of ARP and the percentage cumulative of ARP. The result is shown in Figure as following.

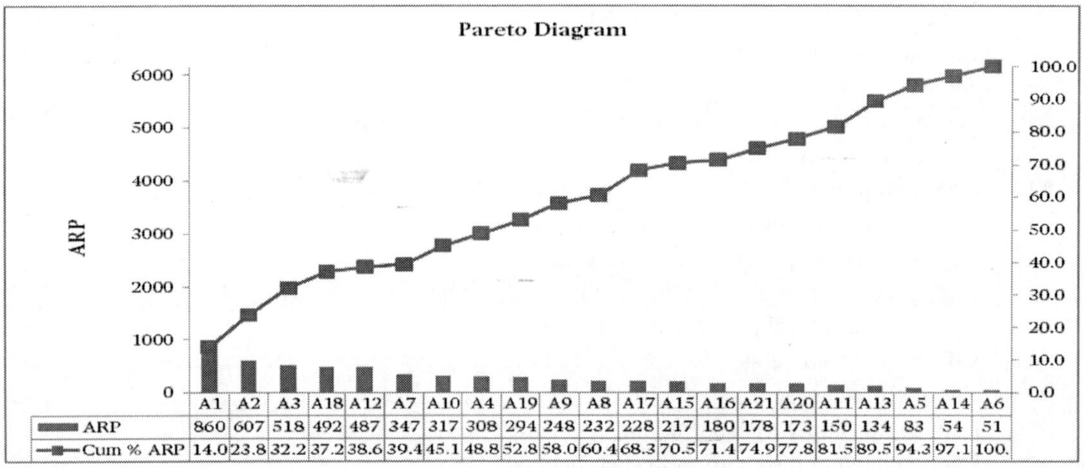

Figure 3. Paretto Diagram HOR Phase 1

Based on Figure 3, with the Paretto 80/20 diagram principle (means that 80% of risk events are caused by 20% of risk agents), it was found that 80% of these risk events are caused by 2 risk agents, namely A1 (poor supplier performance) and A2 (Low contractor capacity).

4.4. *Identification of mitigation actions*

Identification of mitigation actions was carried out to mitigate the 15 risk agents, which constitute 75% of the total ARP accumulation. Table 8 shows all the 15 (fifteen) items according to rank are A1, A2, A3, A18, A12, A7, A10, A4, A19, A9, A8, A17, A15, A16 and A21 respectively.

Table 7. Risk Agents Mitigation Actions Identified

Code	Mitigation Strategy
PA1	Proper supplier selection method.

Code	Mitigation Strategy
PA2	Alternative suppliers preparation.
PA3	Observe and monitor tightly toward the supplier performance.
PA4	Proper contractor selection method.
PA5	Provide the management project assistance.
PA6	Bad weather predictions, see the BMKG Weather prediction time scale starting from the daily prediction to 6 months prediction.
PA7	Prepare for disruption or natural disaster handling facilities.
PA8	Adding the resources and schedules when natural conditions are good.
PA9	Conducting the strict evaluation during procurement process.
PA10	Creating a schedule list according to the suppliers' capacity.
PA11	Strengthen the management tools.
PA12	Conducting a strict schedule recording.
PA13	Early coordination with related parties.
PA14	Creating a better schedule process.
PA15	Good information and communication system.
PA16	Adjusting the level of joblessness based on the predicted fluctuations of rupiah exchange rate toward dollar.
PA17	Adjusting the team strength with compiling projects.
PA18	Preparing SOP with a time limits.
PA19	Anticipating the changes before starting the project.

4.5. The data processing by HOR phase 2 method

HOR Phase 2 processing matrix was produced based on the score of correlation between the risk agent and mitigation action. The result is shown in Figure 4.

	Mitigation Action																			
Risk Agent	**PA1**	**PA2**	**PA3**	**PA4**	**PA5**	**PA6**	**PA7**	**PA8**	**PA9**	**PA10**	**PA11**	**PA12**	**PA13**	**PA14**	**PA15**	**PA16**	**PA17**	**PA18**	**PA19**	**ARP**
A1	9	3	0	3	0	0	0	0	0	0	0	0	0	0	3	0	0	0	0	860
A2	0	0	0	9	3	0	0	0	9	0	0	0	0	0	3	0	0	0	0	607
A3	9	9	9	3	0	0	0	0	3	0	0	0	0	0	1	0	0	0	0	518
A18	1	0	1	1	0	9	3	9	1	0	0	0	0	9	1	0	0	0	0	492
A12	0	0	0	0	0	0	0	0	9	0	0	0	0	0	1	0	9	0	0	487
A7	9	9	3	1	0	1	0	0	0	9	0	0	0	9	1	0	0	0	0	347
A10	0	0	0	0	0	0	0	0	1	0	0	0	0	0	9	0	0	0	0	317
A4	3	0	9	0	0	0	0	0	0	0	0	9	0	0	1	0	0	0	0	308
A19	3	0	0	0	0	0	0	0	0	0	9	0	0	0	0	0	0	0	3	294
A9	0	0	0	0	0	0	0	0	0	0	0	0	0	9	9	0	0	0	3	248
A8	3	1	0	3	0	0	0	0	0	0	0	0	0	3	3	0	0	0	0	232
A17	9	0	1	3	0	1	0	0	0	0	0	0	9	3	1	0	0	0	0	228
A15	0	0	0	0	0	0	0	0	0	0	0	0	0	1	0	3	0	9	9	217
A16	3	0	9	9	3	0	0	0	3	0	0	0	0	3	0	0	0	0	0	180
A21	1	0	0	3	0	0	0	0	0	0	0	0	0	0	3	9	0	0	9	178
Total Effectifness (TEK)	21,289	10,597	10,815	13,970	2,361	5,003	1,476	4,428	12,749	3,123	2,646	2,772	2,052	11,920	13,096	2,253	4,383	1,953	5,181	
Degree of Difficulty (Dk)	1	2	1	1	2	1	2	1	1	1	2	1	2	1	1	1	1	1	1	
Effectifness to Difficulty (ETD)	21,289	5,299	10,815	13,970	1,181	5,003	738	4,428	12,749	3,123	1,323	2,772	1,026	11,920	13,096	2,253	4,383	1,953	5,181	
Rank of Priority	***1***	***7***	***6***	***2***	***17***	***9***	***19***	***10***	***4***	***12***	***16***	***13***	***18***	***5***	***3***	***14***	***11***	***15***	***8***	

Figure 4. Correlation level between risk agent and mitigation action

Based on the HOR matrix phase 2 in Figure 4, a Pareto diagram was produced and shown in following Figure 5.

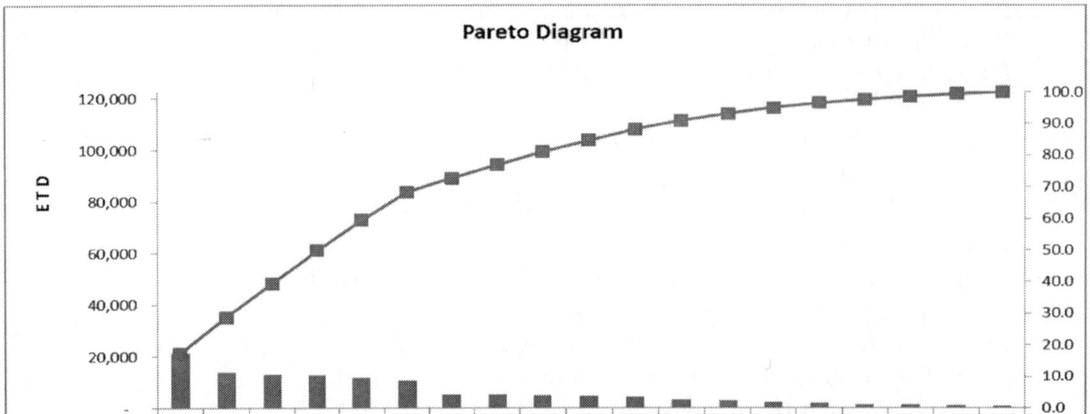

	PA1	PA4	PA15	PA9	PA14	PA3	PA2	PA19	PA6	PA8	PA17	PA10	PA12	PA16	PA18	PA11	PA5	PA13	PA7
ETD	21,2	13,9	13,0	12,7	11,9	10,8	5,29	5,18	5,00	4,42	4,38	3,12	2,77	2,25	1,95	1,32	1,18	1,02	738
Cum % ETD	17.4	28.8	39.5	49.9	59.6	68.4	72.8	77.0	81.1	84.7	88.3	90.8	93.1	94.9	96.5	97.6	98.6	99.4	100.0

Figure 5. Paretto Diagram HOR Phase 2

Based on Paretto diagram in Figure 5, the 80/20 principle suggests that 80% of risk agents can be overcame by 20% accumulation of mitigation strategies which represented by ETD, namely PA1 (the right method of supplier selection) and PA4 (the right method of selecting contractors). These strategies are inline with the result of HOR phase 1 in Figure 2, where 80% of risk events are caused by 2 risk agents namely: Poor supplier performance (A1) and Low contractor capacity (A2).

However, as the studied project will be the first toll road project in the region, it is better to anticipate as much as possible of the risks. The authors suggested to take some more actions such as: Good information and communication system (PA15), Conducting the strict evaluation during procurement process (PA9), Creating a better schedule process (PA14), Observe and monitor tightly toward the supplier performance (PA3), Alternative suppliers' preparation (PA2), and Anticipating the changes before starting the project (PA19). All of these mitigation actions will cover 75% of the total accumulated ETD value and overcome more than 80% of the risk events.

5. Conclusions and recommendations

An analysis of supply chain risk of pavement material for West Sumatera – Riau toll road project has been conducted using House of Risk (HOR) Method. It was found that the critical risk agent on supply chain materials process of West Sumatera – Riau toll road project were supplier bad performance and low contractor capacity. In order to overcome the risk, some mitigation actions were proposed such as to conduct a proper supplier selection method and proper contractor selection method. Both action could overcome 80% of risk agents that appears on supply chain materials process of West Sumatera – Riau toll road project. However, as the project will be the first toll road project in the region, it suggested to do 75% of the possible mitigation actions such as applying a good information and communication system, conducting the strict evaluation during procurement process, creating a better schedule process, observing and tightly monitored the supplier performance, preparing alternative suppliers, and anticipating the changes before starting the project. All of these actions could overcome more than 80% of risk events.

6. References

[1] Yosritzal, Purnawan, Putri, E. E., and Ratu, E. K., 2018, Priority setting for competency development training topics for road construction site managers to reduce the risk of

construction failure. MATEC Web of Conference 229, 01003 (2018), https://doi.org/10.1051/matecconf/201822901003.

[2] Yosritzal, Purnawan and Asmery, V., 2017, Competency of West Sumatera Provincial Road Construction Project Supervising Consultants in Owner Perception. International Journal of Civil Engineering and Technology (IJCIET) Vol. 8, Issue 10, October 2017, pp. 743-752. ISSN Print: 0976-6308 and ISSN Online: 0976-6316.

[3] Juwita, W., Yosritzal and Rinaldi, D., 2016, The Leadership Assessment Study on the Level of Road Technical Personnel Competence at Dinas Prasarana Jalan Tata Ruang dan Pemukiman (Road Infrastructure, Spatial and Settlement Agency) Province of West Sumatera Using Requirement-Satisfaction Analysis. Proceeding of Sustainable Civil Engineering Structures and Construction Materials; Special Topics in Public Works Research and Technology, September 5, 2016 in Bali.

[4] The Association for Project Management (APM), 2000, *Project Risk Analysis And Management*, APM, Buckingham, 12p, ISBN: 0953159000

[5] Vaughan, E. J., T. M. Vaughan, 2007, *Fundamentals of Risk and Insurance, 10th edition*, John Wiley & Sons, Inc, United States, 745p, ISBN-13 978-0-470-08753-4

[6] Manuj, I., J. T. Mentzer, 2008, Global Supply Chain Risk Management, *Jurnal Logistik Bisnis* University of North Texas Denton, Vol. 29 No. 1, p133-155

[7] KBBI, 2018, *Pengertian Risiko* (Definition of Risk), (https://kbbi.web.id/risiko), accessed on 2nd February 2018

[8] Badan Pembinaan Konstruksi Kementerian Pekerjaan Umum, 2013, *Kajian Rantai Pasok Materi dan Peralatan Konstruksi Dalam Mendukung Investasi di Bidang Konstruksi Berkelanjutan (Construction Material and Equipment Supply Chain Study to Support Investment in Sustainable Construction)*, Pusbin SDI, Bandung, 142p

[9] Pujawan, I.Ny., L.H. Geraldin, 2009, House of The Risk: A Model for Proactive Supply Chain Risk Management, *Jurnal Management Proses Bisnis* Teknik Industri ITS Surabaya, Vol. 15 No. 6, p953-967

[10] Hatmoko, J. U. D, F. Kistiani, 2017, *Model Simulasi Risiko Rantai Pasok Material Proyek Konstruksi Gedung* (Simulation risk model of Supply Chain of Material for Building Construction Project), *Jurnal Teknik Sipil* Undip Semarang, Vol 23 No. 1, p1-13

IOP Conf. Series: Earth and Environmental Science **340** (2019) 012045 doi:10.1088/1755-1315/340/1/012045

Bureaucratic reform of tourism sector public services in Tana Toraja Regency

Yusriadi[1], U Farida[2], S Z Bin-Tahir[3] and Misnawati[4]

[1]Sekolah Tinggi Ilmu Hukum Pengayoman, Bone, Jl. Yos Sudarso, Bone, 92713, Indonesia
[2]Sekolah Tinggi Ilmu Ekonomi Amkop, Makassar, 90111, Indonesia
[3]Universitas Iqra Buru, Jl. Prof. Dr. Abdurrahman Baslamah, Maluku, 97571, Indonesia
[4]PPKH, Bone, Jl. Andalas, Bone, 92700, Indonesia

Email: yusriadi.yusriadi@uqconnect.edu.au

Abstract. Bureaucratic reform's goal regarding the tourism is to increase tourist attractiveness by analyzing cultural tourism development efforts from public services. The type of research used is phenomenology. Determination of informants was determined by the snowball method, with data collection techniques carried out by observation, semi-structured interviews, and documentation. This study used 9 informants, 3 domestic tourists, 2 foreign tourists and 4 people from the local government. The data analysis technique used follows an interactive model which includes collection, data condensation, data presentation, and verification/drawing conclusions. The results showed that good service by the regional government of Tana Toraja regency would be directly proportional to the development of tourism objects through the application of information technology as a marketing medium, namely by using online media to be a guide to the progress of the tourism sector. Service quality affects the attractiveness of tourist objects, organizations that manage tourist attractions, and tourists themselves. The success of the bureaucratic reform carried out by the local government of Tana Toraja Regency in the tourism sector regarding public services has a positive impact in increasing the number of tourists.

Keywords: *Bureaucratic Reform, Public Service, Tourism, Tana Toraja*

1. Introduction

Bureaucratic reform is a systematic process, integrated, and comprehensive efforts aimed at realizing good governance [1]. Administrative reform as the artificial measurement of administrative transformation against resistance [2]. The Bureaucratic Reform of the Ministry of Tourism began in 2010, as an implementation form of Presidential Regulation mandate Number 81 year 2010 which concerning the Grand Design of Bureaucratic Reform 2010-2025. The Government of South Sulawesi positively responded the implementation of bureaucratic reform in tourism sector.

One of the most popular tourist areas in South Sulawesi frequented by domestic and foreign tourists is Tana Toraja district with an area of 3,203 km² and a population of 221,081 people (2010). The Tana Toraja tribe community still adheres to a unique belief and lifestyle that is very similar to the culture of Nias, which makes Tana Toraja become one of the world cultural heritage sites registered at UNESCO.

Some of the famous tourist destinations in Tana Toraja are *Tongkonan* houses where *Puang Sangalla* (King of Sangalla') dwells as the resting place and is also the palace where the government of the *Sangalla* kingdom was managed at that time, *Tongkonan Buntu Kalando* titled "*tandoan tananan langi 'loud Kaero tongkonan layuk*". Currently, *Tongkonan Buntu Kalando* is used as a museum to store prehistoric objects and relics of the *Sangalla* kingdom. Graves of unborn teeth (aged 6 months down) placed in a perforated living tree. *Tongkonan Pallawa* is one of the very interesting *tongkonan* or traditional houses and is located between bamboo trees on the top of a hill. The *tongkonan* is decorated with some buffalo horns that are plugged in the front of the traditional house, located about 12 km to the north of *Rantepao*; this place is often referred to as the house of spirits. At the *Lemo*'s cemetery, it can be seen the corpses that are stored in the open air, in the midst of steep rocks. This burial complex is a combination of death, art, and ritual. At certain times clothes from corpses will be replaced by *Ma 'Nene* ceremony. Located on the Burake Hill, Tana Toraja has built a statue of Jesus which is claimed to be the highest in the world. That is the location of the statue is at an altitude of 1100 meters above sea level or the location of the statue is the highest in the world even though the size of the statue itself is not the highest in the world.

General objectives in the field of tourism are the achievement of tourist satisfaction, providing services to tourists as consumers, regulating how the quality of services provided by local governments, besides the success rate of a tourist place also depends on tourist satisfaction that has been achieved during a visit to Tana Toraja. To accomplish this satisfaction, the quality of service plays an essential role in its management towards good governance. The concept of services provided to tourists is emphasized on infrastructure which is an integral part of achieving the expected quality of service so that every available facility must be attractive so that tourists do not feel bored when visiting Tana Toraja. In referring to the good quality of service, it indeed demands the regional government be able to manage it in a planned manner, especially in the existing tourism infrastructure.

The Culture and Tourism Department (Disbudpar) of Tana Toraja district acknowledged that infrastructure and supporting facilities still constrain tourism development. Indeed, the various national and international tourism events have been held to increase tourist visits. One of the constraints of the Tana Toraja tourism sector today is the lack of lodging. Regional tourism development is inseparable from the availability of adequate accommodation facilities. Based on data from the Department of Culture and Tourism in Tana Toraja, the number of hotels was three units, and the guesthouse/inn is ten units. The number of hotel rooms and inns is 243 rooms with some 430 beds. The number of rooms and beds available in various hotels and inns in the Tana Toraja region is not comparable to tourist visits during the peak tourist season, i.e., June to October and December and when tourism events are held [3].

In addition to lack of lodging, Tana Toraja tourism sector also still faces transportation problems. Although Tana Toraja already has a pioneer airport, only certain types of aircraft can land. For land transportation, tourists need seven to eight hours from Makassar to Tana Toraja. As a result, tourists spend a lot of time on the trip while the time for visiting tourists to Tana Toraja is limited. At present, the government is building a new international airport in the Mengkendek sub-district. The existence of the new airport will support the improvement of the Tana Toraja tourism sector because it can land wide-body aircraft. Although the number of buses serving the Makassar-Tana Toraja route and vice versa quite a lot. However, the distance traveled by land still takes seven to eight hours [3]. Based on the above facts, this phenomenon is unusual to develop, wherein the public service in the tourism sector gives a signal of obstacles to infrastructure and transportation. This phenomenon shows evidence that from various types of public services, tourism services are still interesting to study.

2. Research Method

This research was carried out in the area of Tana Toraja Regency, Government of South Sulawesi Province. This object was selected because of Tana Toraja district as a district with famous tourism object in South Sulawesi Province certainly has a very high volume of tourist activity; Tana Toraja has more heterogeneity, especially cultural heterogeneity so that relationships are created which can

IOP Conf. Series: Earth and Environmental Science **340** (2019) 012045 doi:10.1088/1755-1315/340/1/012045

neutralize the influence of primordial's which might affect the analysis. The type of research used is phenomenology which is expressing someone's experience [4]. Determination of informants was determined by the snowball method where technique of informants determining was smaller then enlarges, with data collection techniques carried out by observation, semi-structured interviews, and documentation. This study used 9 informants, 3 domestic tourists, 2 foreign tourists and 4 people from the local government. The data analysis technique used follows an interactive model which includes collection, data condensation, data presentation, and verification/drawing conclusions.

3. Results

Tourism products compared to other products have different characteristics, tourism products as an object of tourism marketing have main elements that comprise of tourist destinations interest, facilities and the easy access in reaching these destinations [5]. To determine the quality of public services in the tourism sector in Tana Toraja Regency, researchers used three dimensions of the quality of public services namely tangible, reliability, responsiveness, assurance and empathy [6], this study only used tangible, reliability and assurance dimensions:

a. Tangible

Tangible dimensions related to equipment, personal and communication media in service. If tourists feel this dimension as users of the service is good, then tourists will judge well and feel satisfied with the services provided by the local government. Local governments as public service providers must provide comfort for visiting service users, meaning that the government in providing services to the tourism sector must try to make tourists feel comfortable and at home during their tour in Tana Toraja. Tourists in tourism activities also need facilities supported the trip, to fullfil the travel needs government need to provide various facilities from departing tourists' residence, staying in tourism destinations and returning back to their place. From the results of interviews with the informant (SS) revealed that:

> "...Place of creation in Tana Toraja, I feel quite at home and want to longer there, the city is very beautiful, as well as the comfort given to me as a tourist, I feel protected as a domestic tourist, everywhere can be found at affordable costs (Interview, February 9, 2018)"

The local government of Tana Toraja district has provided convenience for service users who want to tour the place. One example of the comfort offered for tourists who visit is, for tourists who wish to stay, they only need to show their Identity Card (KTP) for domestic tourists and for foreign tourists to show their passports. This is consistent with the results of interviews with informants (IK) saying that:

> "I stayed at this place, the conditions that were asked for enough ID cards and because I brought my wife and asked for a marriage certificate, I think the rule made it very easy for me as a tourist and not complicated, and also good for those who bring the couple to ask for proof of marriage certificate. I think this is to avoid free sex for tourists (Interview, February 17, 2018)"

Related to work discipline is needed by every employee in carrying out services. The results of the interview with the informant (SY) said that:

> "I always recreation every year to Tana Toraja, where the government really appreciates the tourists who visit, I feel the comfort when visiting Kete Kesu village, the manager kindly approached me and guided me as a tour guide and explained the history in the village (Interview, February 17, 2018)"

Regarding assistive devices, it is very supportive of the service process so that services can be done quickly. An aid in the service process is needed for the smooth operation of the facility. The tools used by the local government in introducing existing tourism objects in Tana Toraja Regency have been based on e-government, to facilitate the traveler who wants to go to Tana Toraja, the government makes an official website for tourism, everything about Tana Toraja tourism can be found on the www.site.visittoraja.com. Now a traveler can find various information about tourism in Tana Toraja.

Starting with tourist attractions, various cultural arts, traditions, and activities that can be done while in Tana Toraja [7].

The obstacles encountered from the results of research conducted in Tana Toraja Regency, especially for foreign tourists are transportation problems. For foreign tourists who will visit Tana Toraja can only be done by land transportation, because in Tana Toraja district there is no airport yet. So that for domestic tourists who visit must go through Makassar City and the journey taken is around 7 hours 47 minutes. Interview results with foreign tourist informants (LW) revealed that:

> "I went to Tana Toraja through the city of Makassar. First, I was a foreign tourist from France, taking a plane from Jakarta to Makassar City took 2 hours, then after that, I took another bus to Tana Toraja which took about 8 hours, so it was quite tiring that must be taken (Interview, February 23, 2018)"

The use of transportation aids in the service process is significant because with the help of tools will facilitate service employees and make it easier for tourists to get services. In a tourist trip, there are also important factors that influencing tourist satisfaction, that is an available access to reach the destinations that sometimes ignored by tourists in planning a trip, so it generally will affect the travel budget.

b. Reliability

The reliability dimension is the ability of the service unit to provide promised services immediately and satisfy consumers. Tourist destinations can raise feelings towards needs related to hospitality through someone or something. The Tana Toraja Regency Government already has clear standards related to improving service quality for tourists; this is evident from the Tana Toraja Regency tourism development strategy stated in the Tana Toraja District Regional Development Master Plan (RIPPDA) in 2011-2016.

In the case of clear standards related to interviews with local government officials as informants (JM) said that:

> "The direction of tourism development policy in Tana Toraja Regency has been well planned regarding its service standards going forward, on what should be done by the local government, this can be seen from the stipulation of the Regional Tourism Development Master Plan, the government fully supports Tana Toraja as a tourist center in South Sulawesi (Interview, February 9, 2018)"

Having a public service standard is important for the guidance of local governments in serving tourists in the service process because by referring to RIPPDA, the service process can run well to achieve service goals, especially in Tana Toraja District.

c. Assurance

The assurance dimension includes knowledge, capability, politeness, and trustworthiness that employees have, free from danger, risk and doubt. The Tana Toraja Regency Government, through the Regional Revenue Agency and the Tourism Office, implemented the recreation levies on Buntu Burake, Makale Tourism Objects, starting June 1, 2017. The amount of fees for school children and students is Rp. 2,000 per person. Researchers, Tourism Work, and Local Government Guests are set at Rp. 5,000 per person. For domestic tourists, adults, and public is Rp. 10,000 per person. While foreign tourists applied higher rates, namely Rp. 20,000 per person, for parking fees for motorcycle vehicles Rp. 1000 per unit. For private car and minibus is Rp. 2000 per unit, and for bus or particular vehicle Rp. 5000 per unit [8]. With the guarantee of costs from service personnel, transparency as a form of good governance works well. Local governments provide timely guarantees in service; this is in accordance with what was disclosed by informants (KM) saying that:

> "I think the committee at each tourist location in Toraja Regency is very timely, around 7:00 a.m. I walked to Keta Kesu Village, where the serving staffs was ready, even though it was very

early, but the employees already standby at the ticket purchase point to enter the existing tourist attraction (Interview, February 19, 2018)"

However, different things expressed by informants (EM) said that:

"Last year I visited Tana Toraja Regency, on December 31, to celebrate the New Year at the place, but when I arrived in Tana Toraja, it turned out that the accommodation was full, so I was forced to not get lodging (Interview, February 23, 2018).

The results of the two informants are very different, but this can be understood because one of them explained that they did not get lodging, but that was because the tourists were more crowded on that day because of the approaching New Year celebrations. Favorite tourist destinations are still constrained by transportation infrastructure. The government just realized that the completeness of the facility greatly supported the increase in the number of visits. One of them is tourism in Tana Toraja, South Sulawesi.

The budget for tourism development comes from the central and regional governments. Central government expended for airport expansion and runway. Whereas the local government is focused on community preparation such as making home stays, which are now available for around 1,022 rooms. The budget for tourism development comes from the central and regional governments. The central government for airport expansion and runway, while the regional government is focused on community preparation such as making home stays which are now available around 1,022 rooms (Jun, 2017).

4. Discussion

Indonesia's tourism sector has a long history; Bali, Jakarta, Medan, Yogyakarta, Surabaya and Makassar are becoming known by foreign tourists. Besides natural tourism, Indonesia has a variety of cultures and historical heritage. The weakness of Indonesia's tourism industry mainly lies in infrastructure availability, the safety or comfort image, marketing system and promotion. The effect of infrastructure is too vital to build tourism sector connectivity. During this time, the infrastructure problem is a major weakness. Likewise, tourism services must improved continously through bureaucratic reform.

Service is an economic activity that creates added value and provides benefits to consumers at a specific time and place. From this research, the quality of public services is measured in three dimensions, tangibility, reliability and assurance. From the tangible dimension is the ability of a service provider to show its existence in society. The comfort and ability of physical facilities/infrastructure and the circumstances of the surrounding environment are concrete evidence of services provided by service providers which include physical and technological facilities. Tangible dimensions are determined by indicators, namely the convenience of the place of service, the ease of the service process, the discipline of employees in performing services, the ease of access to customers. The public sector of the tourism sector in Tana Toraja Regency has applied tangible dimensions and indicators. The application of e-government will encourage cultural change, which also means a change in the value system [9]

The evaluation of public quality service that has been running in line with tourist expectations in this dimension, but in its implementation, there are still indicators that have not been run to facilitate tourists such as transportation problems. The Office of Culture and Tourism recognizes that Tana Toraja Regency's tourism development is still constrained by infrastructure and supporting facilities, even though various national and international tourism events are held to increase tourist visits to Tana Toraja [10]. In this case, the government must negotiate and collaborate with various tourist interests. So that public services have the quality that is expected by tourists. Efforts made by the regional government in providing good quality services related to the transportation problems were responded to positively, the construction of the airport proved this. The government is building a new international airport in Mengkendek Sub district, the existence of the new airport will support the improvement of the Toraja tourism sector because it can land wide-body aircraft [10]. The positive thing that is done from the tangible dimension is the use of e-government in introducing Tana Toraja as the best tourism area in South Sulawesi by providing www.visittoraja.com now a traveler can find various information about tourism in Tana Toraja. Making a city smarter is an obligation and cannot be opposed to its existence [11].

The second dimension is reliability is the ability of service providers to provide services as promised in a timely and reliable manner. In this study, the aspects of reliability are determined by standard indicators that are clear or not, the public service of the tourism sector of Tana Toraja Regency has implemented the dimension of reliability, this is seen by the existence of the Regional Tourism Development Plan (RIPPDA) of Tana Toraja Regency in 2011-2016 as a tourism development standard. The implementation of public services must have service standards and be published as a guarantee of certainty for service recipients. Certainty must be created so that the community and bureaucratic apparatus can interact well [12]. Public service standards as guidelines in public services by government agencies and can be used as indicators of assessment of the quality of services that have been provided. With the existence of standards in public service activities, it is expected that the community can get services that are in accordance with the needs and processes that are satisfying and do not complicate the community as service users [13].

The third dimension is assurance a guarantee and certainty that is knowledge, politeness, and ability of service employees to foster a sense of trust for service users to service providers. The public service sector of the Tana Toraja tourism sector has implemented the assurance dimension along with its indicators. Assessment of public quality that has been running in accordance with the expectations of the community in this dimension, among others, employees provide timely guarantees in service, employees offer guaranteed costs in services; there will still be indicators that have not been implemented to facilitate tourists such as limited lodging problems, if entering certain months. Many tourists do not get hotel/lodging rooms, so they stay in the neighboring district, North Toraja [10]. Public servants in the community will be able to walk as expected if the supporting factors are sufficient and can be used effectively and efficiently. In certain months such as June to September to December, the entire inn is not able to accommodate guests, because at that time many nomads from Toraja returned to the village to hold a *Rambu Solo* ceremony and *Rambu Tuka* death party and *Rampanan Kapa* or the traditional Tana Toraja wedding party [14] The lodging problem is understandable considering that these particular months are crowded months of tourist visitors and also because of a large number of native residents going home, but in the usual months the lodging in Tana Toraja can accommodate both domestic and foreign tourists. Quality is a dynamic condition that deals with products, services, people, processes, and environments that meet or exceed expectations [15].

To fix the weaknesses, government must seriously carry out bureaucratic reform. During this time, the bureaucratic culture in this country has become an obstacle to develop in various sectors, bureaucracy problem, corruption and infrastructure, are a major obstacle to national development. So far, Indonesian bureaucratic culture has become weak point of Indonesian tourism. The bureaucratic apparatus seems rigid and not dynamic in addressing world development.

The government must embrace business world, starting from business travel agents, hotels, airlines, and business actors related. The government must be able to prove that tourism management in Indonesia is not inferior to other countries.

5. Conclusion

A good tourism sector service by the Regional Government of Tana Toraja Regency will be directly proportional to the development of tourism objects in Tana Toraja Regency. Toraja tourist destinations in South Sulawesi have progressed viewed quantitatively and in particular from the data on the number of tourists collected by the Tana Toraja Regency Tourism Office in recent years. In 2013 19,324 tourists were visiting Tana Toraja, the previous year 13,532 and the last year 9,005. In 2010 and 2009 the number of tourists visiting Tana Toraja was 5,634 and 5,607 [16]. The success of the bureaucratic reform carried out by the local government of Tana Toraja Regency in the tourism sector regarding public services has a positive impact in increasing the number of tourists. Efforts to reform

IOP Conf. Series: Earth and Environmental Science **340** (2019) 012045 doi:10.1088/1755-1315/340/1/012045

the government bureaucracy in Indonesia will never be successful if it is done in a linear mindset that does not address the root problem [17]. The success of bureaucratic reform can only be felt if the public service is considered good by the community [18].

6. References

[1] Sedarmayanti, *Sumber Daya Manusia dan Produktivitas Kerja*. Bandung: CV Mandar Maju, 2009.

[2] G. E. Caiden, *Administrative Reform Comes of Age*. W. de Gruyter, 1991.

[3] J. Lembang, "Pariwisata Tana Toraja Masih Terkendala Infrastruktur," Makassar, 2013.

[4] A. Edgar and P. Sedgwick, *Key Concept in Cultural Theory*. London and New York: Routledge, 1999.

[5] A. Y. Oka, *Perencanaan Strategi Pemesaran Daerah Tujuan Wisata*. Jakarta: PT. Pradaya Pramita, 2002.

[6] et all Zeithaml, "Measuring the quality of relationship in customer service: An empirical study," *Eur. J. Mark.*, 1996.

[7] F. Shafa, "Kini Toraja Punya Situs Pariwisata Resmi," *Detik Travel*, Jakarta, 2016.

[8] Arthur, "Mulai 1 Juni 2017, Masuk Objek Wisata Buntu Burake Harus Bayar Retribusi," *Kareba Toraja*, Tana Toraja, 2017.

[9] Yusriadi, "Reformasi Birokrasi Indonesia: Peluang dan Hambatan," *J. Adm. Publik (Public Adm. Journal)*, vol. **8**, no. **2**, pp. 178–185, 2018.

[10] Jun, "Pengembangan Wisata Terkendala Infrastruktur," *Indo Pos*, 2017.

[11] Yusriadi, A. Sahid, I. Amirullah, A. Azis, and A. A. Rachman, "E-Government-based Bureaucratic Reform in Public Service," in *2nd International Conference of Communication Science Research (ICCSR 2018)*, 2018, pp. 68–70.

[12] Yusriadi, "Tantangan Reformasi Birokrasi Berbasis E-Procurement di Indonesia," *Publikauma J. Adm. Publik Univ. Medan Area*, vol. **6**, no. **2**, 2018.

[13] J. Ridwan and A. Sodik Sudrajat, *Hukum Administrasi Negara dan Kebijakan Pelayanan Publik*. Bandung: Nuansa, 2009.

[14] I. M. Ashdiana, "Lovely December Diharapkan Picu Kunjungan Wisatawan," *Kompas*, 2013.

[15] F. Tjiptono, *Manajemen Pemasaran*. Yogyakarta: Andi Offset, 2000.

[16] Editor, "Wisatawan Toraja Meningkat Tajam," *Fresh News Story on Travel & Tourism of Indonesia*, 2014.

[17] Yusriadi, "Bureaucratic Reform to the improvement of public services Challenges for Indonesia," *Publikauma J. Adm. Publik Univ. Medan Area*, vol. **6**, 2018.

[18] H. MENPAN RB, "Pelayanan Publik, Indikator Keberhasilan Reformasi Birokrasi," *Kementerian Pendayagunaan Aparatur Negara dan Reformasi Birokrasi*, 2014. [Online]. Available: https://www.menpan.go.id/site/berita-terkini/pelayanan-publik-indikator-keberhasilan-reformasi-birokrasi.

AUTHOR INDEX